Lecture Notes in Mechanical Engineering

Lecture Notes in Mechanical Engineering (LNME) publishes the latest developments in Mechanical Engineering—quickly, informally and with high quality. Original research or contributions reported in proceedings and post-proceedings represents the core of LNME. Volumes published in LNME embrace all aspects, subfields and new challenges of mechanical engineering.

To submit a proposal or request further information, please contact the Springer Editor of your location:

Europe, USA, Africa: Leontina Di Cecco at Leontina.dicecco@springer.com
China: Ella Zhang at ella.zhang@springer.com
India: Priya Vyas at priya.vyas@springer.com
Rest of Asia, Australia, New Zealand: Swati Meherishi at swati.meherishi@springer.com

Topics in the series include:

- Engineering Design
- Machinery and Machine Elements
- Mechanical Structures and Stress Analysis
- Automotive Engineering
- Engine Technology
- Aerospace Technology and Astronautics
- Nanotechnology and Microengineering
- Control, Robotics, Mechatronics
- MEMS
- Theoretical and Applied Mechanics
- Dynamical Systems, Control
- Fluid Mechanics
- Engineering Thermodynamics, Heat and Mass Transfer
- Manufacturing Engineering and Smart Manufacturing
- Precision Engineering, Instrumentation, Measurement
- Materials Engineering
- Tribology and Surface Technology

Indexed by SCOPUS, EI Compendex, and INSPEC.

All books published in the series are evaluated by Web of Science for the Conference Proceedings Citation Index (CPCI).

To submit a proposal for a monograph, please check our Springer Tracts in Mechanical Engineering at https://link.springer.com/bookseries/11693.

Mohammad Altamush Siddiqui · Nadeem Hasan ·
Andallib Tariq
Editors

Advances in Heat Transfer and Fluid Dynamics

Select Proceedings of AHTFD 2022

Springer

Editors
Mohammad Altamush Siddiqui
Zakir Husain College of Engineering and Technology
Aligarh Muslim University
Aligarh, Uttar Pradesh, India

Nadeem Hasan
Zakir Husain College of Engineering and Technology
Aligarh Muslim University
Aligarh, Uttar Pradesh, India

Andallib Tariq
Department of Mechanical Engineering
Indian Institute of Technology Roorkee
Roorkee, Uttar Pradesh, India

ISSN 2195-4356 ISSN 2195-4364 (electronic)
Lecture Notes in Mechanical Engineering
ISBN 978-981-99-7212-8 ISBN 978-981-99-7213-5 (eBook)
https://doi.org/10.1007/978-981-99-7213-5

This Springer imprint is published by the registered company Springer Nature Singapore Pte Ltd.
The registered company address is: 152 Beach Road, #21-01/04 Gateway East, Singapore 189721, Singapore

Paper in this product is recyclable.

Contents

About the Editors

Dr. Mohammad Altamush Siddiqui is currently Professor in Department of Mechanical Engineering, Zakir Hussain College of Engineering and Technology, Aligarh Muslim University (AMU), India, and also serving as Dean Faculty of Engineering and Technology, AMU. He obtained his B.Sc. in mechanical engineering from AMU in 1982 and M.Tech. from Indian Institute of Technology (IIT) Kanpur in 1985. He further obtained his Ph.D. from AMU in 1992. His research interests include convective heat transfer, energy conversion systems, refrigeration and air-conditioning, etc. He has developed heat transfer and solar energy laboratory in the initial year of services at AMU. He has published around 43 papers in various international journals of repute and 2 chapters.

Dr. Nadeem Hasan is currently Professor in Department of Mechanical Engineering, Zakir Hussain College of Engineering and Technology, Aligarh Muslim University (AMU), India. He obtained his B.Sc. and M.Sc. in mechanical engineering from AMU in 1991 and 1996, respectively, and Ph.D. from Indian Institute of Technology (IIT) Delhi in 2006. His research interests include computing of incompressible fluid flow and heat transfer, proper orthogonal decomposition, modeling and computation of low Mach number flows subjected to large temperature differences using non-Oberbeck-Boussinesq models, modeling, and computation of high-speed compressible flows. He has developed novel particle velocity upwind family of schemes for computing of compressible flows right from very low Mach (incompressible limit) to very high supersonic/hypersonic continuum flow regimes. He has also developed novel far-field boundary conditions for incompressible flows around immersed bodies. He has published around 33 papers in various international journals of repute and 4 chapters.

Dr. Andallib Tariq is currently Professor and Head, Department of Mechanical and Industrial Engineering, Indian Institute of Technology (IIT) Roorkee, India. He obtained his BE in mechanical engineering from Bangalore University and obtained M.Tech. and Ph.D. degrees from IIT Kanpur. His research interests include experimental fluid mechanics, turbulence, heat transfer enhancement, thermal contact conductance, inverse problems, gas turbine blade cooling, nuclear safety research, and optical diagnostic tools like LCT, PIV, and IR thermography. He has established and implemented a critical heat transfer enhancement technology based on Liquid Crystal Thermography (LCT) and was the foremost to initiate the implementation of LCT for gas turbine heat transfer research in India. He has published around 39 papers in various international journals of repute along with 2 chapters.

Flow Phenomenon and Computational Fluid Dynamics

Control of Suddenly Expanded Flow with Cavity at Sonic Mach Number

Muhammad Ikhwan Fiqri, Khizar Ahmed Pathan, and Sher Afghan Khan

Abstract This study focuses on the control of the pressure in the base corner as well as flow quality in the duct during a numerical simulation of sonic airflow from a convergent nozzle exhausted into the circular duct with annular rectangular cavities. Three variables were found to significantly affect the base pressure: expansion level, area ratio, and duct size. For a lower area ratio, the passive control with cavities increases the pressure in the wake corner. Wake pressure also depends critically on the geometry and dimension of the cavity. In this investigation, we took into account the flow and geometrical factors of nozzle pressure ratio (NPR), cavity aspect ratio (ASR), and cavity location (CL). At sonic Mach numbers with the sudden enhancement of the flow area, the computational fluid dynamics method employs flow from a convergent nozzle. The Mach number of the study was M = 1.0 and an exit area ratio of 2.25 was found between the duct and the nozzle. The duct length considered was L = (1D, 2D, 3D, 4D, and 6D), and NPRs considered (1.5, 2, 3, 4, and 5) were used for simulations. The ASR 1 cavities are annular rectangles, 3 mm wide and 3 mm high. The outcomes of the base pressure are compared for ASR 2 with dimensions of 3 mm in width and 6 mm in height. In this study, a turbulence model based on the K-standard wall function was used to simulate and examine the nozzle. The results demonstrate that pressure in wake was considerably inspired by NPR and the cavity positioning from the base.

M. I. Fiqri · S. A. Khan
Mechanical and Aerospace Engineering Department, Faculty of Engineering, International Islamic University, 53100 Malaysia, Kuala Lumpur, Malaysia
e-mail: sakhan@iium.edu.my

K. A. Pathan (✉)
Department of Mechanical Engineering, CSMSS Chh. Shahu College of Engineering, Aurangabad 431011, India
e-mail: kapathan@csmssengg.org

M. A. Siddiqui et al. (eds.), *Advances in Heat Transfer and Fluid Dynamics*, Lecture Notes in Mechanical Engineering, https://doi.org/10.1007/978-981-99-7213-5_1

1 Introduction

In recent times, sudden flow expansion has been a fundamental problem in various applications. These types of challenges are prevalent in both the automotive and aerospace industries. The goal of regulating these flows is to enhance engineering productivity, technical simplicity, economic viability, and norms compliance. The bluff body flow that develops when the viscous and inviscid regions come into proximity to one another is another crucial feature. Separate shear layers emerge from the tail of the body, further defining the shape. These shear layers are intrinsically unstable, so when they contact, potent vortices are formed. The low-pressure zone produced by the vortices around the body's base enhances its resistance to forward motion. A variety of flow control methods have been employed to reduce the inherent resistance. These two approaches are often referred to as "active control" and "passive control," respectively. Passive control, an energy-free strategy for controlling suddenly expanded flow, is the primary focus of this paper.

Additionally, they are referred to as the cavity at the base. This study explores the role of a cavity as a control mechanism for wake pressure at Mach unity. In the investigation, we analyze how cavities influence base pressure at $M = 1$ while using a passive flow regulator.

Fluid detachment, reattachment, and recirculation characterize the suddenly expanded flow. Such a fluid flow is segmented by the shear layer into a recirculation region and a free stream region. The reattachment line is defined as the point where the dividing streamline re-attaches with the duct. The basement was fashioned into a storage tank with room for two full supplies.

The purpose of this research is to demonstrate why it is crucial to manage flow fields experiencing rapid growth. One of the responsibilities of the controller is to adjust the mixing and flow rates. However, active and passive control is used in all flow management strategies. Passive flow control manages energy directly from the flow, while dynamic control requires an exterior energy source to control the flow. Research is also conducted to improve performance, save energy, and reduce the base drag. Seventy percent (70%) of the resistance force is due to the sub-atmospheric pressure at the base corner, so decreasing it will increase the range of missiles, rockets, and fighter planes while also saving on energy and fossil fuels (diesel).

This sudden expansion flow field has been used in numerous experimental examinations focusing on base pressure. Meanwhile, research into the cavity's impact on sonic-Mach-number flow through a suddenly widened nozzle is essential. Thus, this research aims to explore the use of passive methods, specifically a cavity, for regulating the base pressure of suddenly expanding flows.

2 Modeling and Meshing

This sudden expansion flow field has been used in many experimental investigations focusing on the ability of pressure regulation in the corner as well as flow quality. Meanwhile, research into the cavity's impact on sonic-Mach-number flow through a suddenly expanded nozzle is essential. Thus, this research aims to explore the use of a passive mechanism, specifically a cavity, for regulating the base pressure of suddenly expanding flows.

2.1 *Mesh Independence Test*

The verification of mesh-independent status employs identical geometry. Three element sizes are simulated i.e. (small, medium, and large). In Table 1, we can see the element size, as well as the total number of nodes and elements.

The right element size is computed by verifying the mesh's independence. More iterations and computational time are needed to achieve a higher NPR. More accurate results can be obtained with less computational effort if the appropriate element size is used. Figure 1 shows the structure of the structured mesh density for the element size given in Table 1.

The density of a mesh is proportional to the size of its elements; a finer mesh requires smaller element sizes and vice versa. Table 3 shows that the number of

Table 1 Various element sizes and their associated attributes are based on the similar geometry model

Element size (mm)	Nodes	Element
1.0	1970	1757
0.5	7095	6671
0.3	19,048	18,340

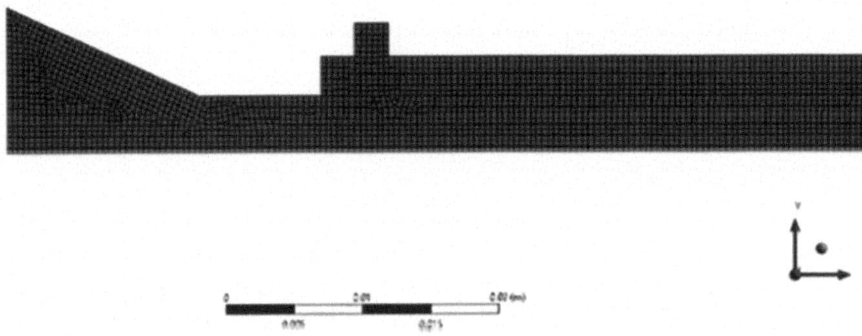

Fig. 1 The structured meshed model for mech element size 0.3 mm

elements increases dramatically for increasingly trace elements. Elements with identical NPR and geometry should have a variation in element size of less than 20 percent to determine the appropriate element size for meshing. The recapitulations are increased to 20 k to guarantee the solutions converge with high correctness, although in most cases, just 10 k would be enough. Table 1 suggests a 0.3 mm element size, leading to the selection of roughly 19,048 nodes and 18,340 elements for further investigation. This is because there is less than a 20% difference in base pressure ratio between 0.5 mm and 0.3 mm elements. The computational time and number of iterations both increase dramatically if the element size is reduced to 1 mm. Therefore, the best element size is 0.3 mm.

The purpose of this geometry validation with previous experimental work [1] is to compare the results of the simulation concerning the conjectural flow variables for the prototypical with cavity AR1 and 0.5. The parameter of interest is the pressure at the bottom wall. Validation geometry is based on the diagram in Fig. 2. The length of the duct varies from 1 to 6 for L/D values of 1, 2, 3, 4, and 6.

2.2 *CFD Analysis*

Since the nozzle profile is symmetric about the axis, the 2-D plane was also made axisymmetric. The density-based solver was chosen for this simulation since there is a change in the density. Since heat transfer was involved in the flow, the model is opted to use the energy equation to describe it. In the materials, the ideal gas was treated like any other fluid. Assuming that fluid viscosity is independent of temperature, Sutherland set the viscosity constant. Density, specific heat, and thermal conductivity were the only other properties that didn't change. According to a study [2–16], the 2-D axisymmetric problem is solved to perform the analysis when the geometries are symmetric about the axis. Many authors did simulations to examine nozzle flow, with results that show " "excellent agreement" with experimental data [17–35]. During geometry generation, we gave each segment of the axisymmetric nozzle that abruptly grew into a circular duct with an annular rectangular cavity a name and assigned the zone type in Fluent software for analysis.

3 Results and Discussion

A convergent nozzle is considered connected to a suddenly expanded circular duct with control validated from previous experimental work. For geometry validation, the L/D values used are 1, 2, 3, 4, and 6. Furthermore, the current work is simulated using the same model but with varied dimensions based on Mach number 1 and varying NPR. The NPR values chosen are 1.5, 2, 3, 4, 5, while the L/D selected as 1, 2, 3, 4, and 6 with 0.5D and 1D locations at two different cavity aspect ratios (AR) 3:3 and 3:6. At the end of this section, the variations in the base pressure is discussed.

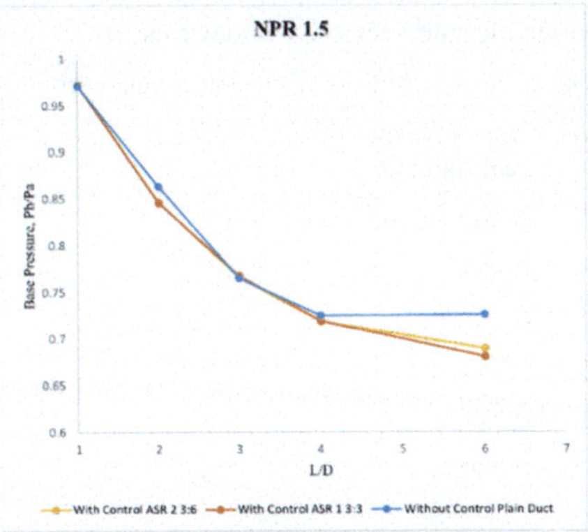

(a) NPR = 1.5

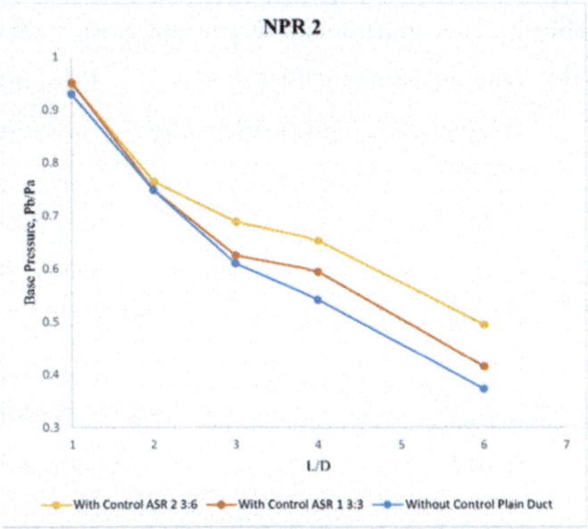

(b) NPR = 2

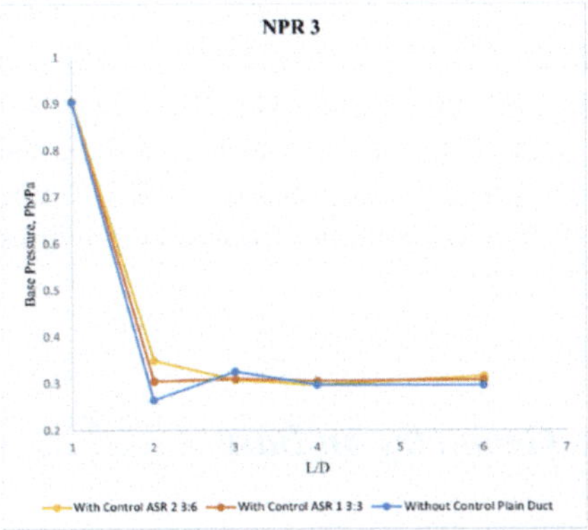

(c) NPR = 3

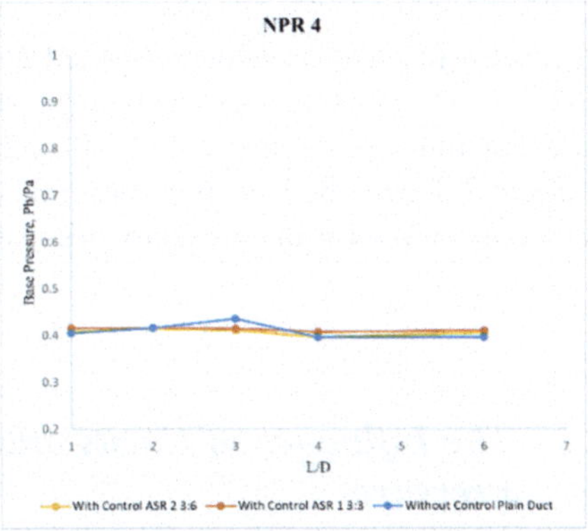

(d) NPR = 4

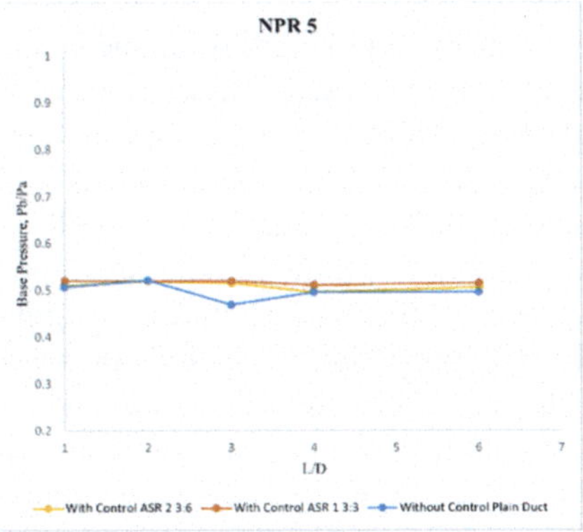

(e) NPR = 5

Fig. 2 The base pressure ratio variation with L/D for plain duct and duct with control

Table 2 The relationship of present work with previous literature review without control

L/D	Rathakrishnan (2001)	Present work		Percentage error (%)
	Base pressure ratio (literature paper)	Pressure (Pa)	Base pressure ratio (validation)	
1	0.990	−5066.25	0.952	4.2
2	0.891	−15,198.75	0.853	4.8
3	0.839	−20,771.62	0.795	5.5
4	0.813	−24,925.85	0.754	7.8
5	0.792	−26,040.52	0.743	6.6
6	0.789	−26,445.82	0.739	6.8

3.1 Validation of Results

According to [1], tests were performed at various rib aspect ratios at NPR of 2.458 and L/D = 1 to 6. Thus, for the validation parts, the parts considered were the convergent nozzle with the plain duct with NPR of 2.458, and L/D varied from 1, 2, 3, 4, 5, and 6. Table 2 shows the base pressure ratio for the present and previous work [1]. The percentage error for the present work was less than ten percent through the previous studies.

3.2 The Influence of Cavity and Its Geometry on Base Pressure

3.2.1 At Constant Nozzle Pressure Ratio

The wake pressure alteration with the length-to-diameter ratio at a fixed NPR, as depicted in Fig. 2. Circular ducts with cavity aspect ratios of 1 and 0.5 are compared with a convergent nozzle flow which exhausted into the circular pipe. The NPRs chosen for this investigation are 1.5 when the nozzle is over-expanded. Figure 2 displays the gradual decrease in base pressure from NPR 1.5 until L/D = 4. (a). At NPR = 2, the value required to produce sonic flow, the findings confirm that the passive flow management is effective when duct lengths L = 2D and longer, as depicted in Fig. 2. (b). The impact of the cavity's presence on the ground pressure is greatest when NPR = 2. Therefore, when a control mechanism is employed the base pressure is more than that without cavity. What's further, it holds for the correct expansion. Figure 2 displays the simulation outcomes for NPR = 3. (c). For L/D 3, the existence of the cavity lessens pressure at the recirculation area; for L/D > 3, the results are approximately the same with and without the cavity, indicating the necessity of a minimum duct length. In contrast, as shown in Fig. 2, (d) for favorable pressure at the nozzle, NPRs 4 and 5 were chosen for the current investigation (e).

The base pressure is significantly affected by the cavity in these situations. For L/D values between 1.5 and 5, the pressure at the wake corner is essentially identical. This demonstrates that the pressure at the wake pressure was unaffected by the cavity aspect ratio.

Figure 2 can be used to talk about the impact of cavity geometry (a–e). In Fig. 2a, c, and d, the orange and yellow lines denoting two different aspect ratios are superimposed (e). Figure 2 b demonstrates, however, that a higher trend of base pressure can be seen with a cavity aspect ratio (ASR) of 0.5 compared to that of a cavity aspect ratio (ASR) of 1 or no cavity at all. Therefore, in both the overexpanded and correct expansion cases, the base pressure was unaffected by the geometrical changes in the cavity.

3.2.2 Base Pressure Variance with NPR at Fixed Duct Size

The base pressure ratio changes with NPR at a constant duct length, as depicted in Fig. 3. In this section, NPRs were chosen for this study, which has a base pressure that is altered by the occurrence of the regulation at the duct. The wake pressure is high when the nozzle is not correctly expanded and stays that way until it reaches the perfect expansion at NPR 2; it begins to drop and gradually increases at the underexpanded nozzle flow. Figure 3 demonstrates an increase from NPR 1.5 to NPR 2 in the base pressure. Base pressure dropped sharply through NPR 3, then steadily rose through NPR 5.

3.3 *Consequence of Cavity Presence in the Base on the Wake Pressure*

3.3.1 At Constant Nozzle Pressure Ratio (NPR)

Base pressure findings with and without cavities from L/D 1 to 6 are shown in Fig. 4. The blue line represents the base pressure, and it is lower for the plain duct than for the enlarged duct with control. However, the results from the base are the main topic of discussion. Figure 4 shows that the base pressure ratio for NPR 1.5 is highest for the green line (a). As a result, for NPR 1.5, the optimal cavity distance from the base is 0.5D for L/D values between 1 and 6. In Fig. 4b, it is found that the optimal cavity placement for NPR 2 varies with L/D. The pressure at the base drops from L/D 1 to 2. The yellow line depicts the optimal 1D position away from the base. The base pressure is then lowered across the duct at all control points when the L/D ratio is greater than 3. Base pressure for NPR 3 dropped from L/D 1 to 2 as shown in Fig. 4c. When duct length is more than 2, the resulting base pressure is nearly constant. This demonstrates that L = 2D is the bare minimum duct size required for the attached flow. As shown by the green line, 0.5D with ASR 1 commencing the base is the

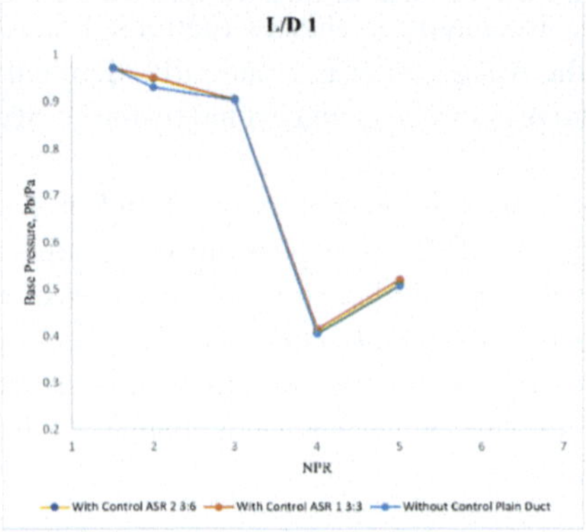

(a) L = 1D

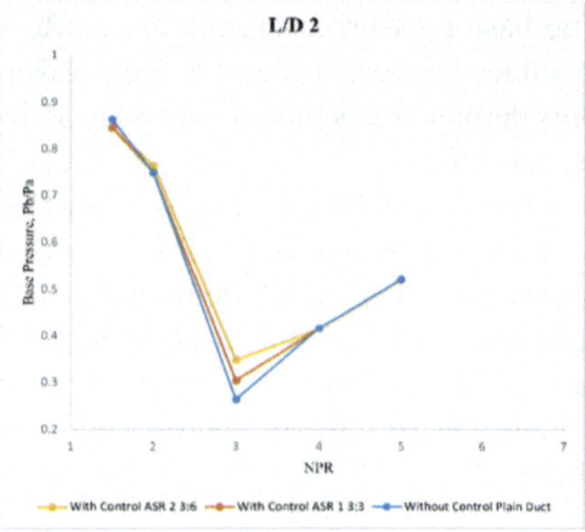

(b) L = 2D

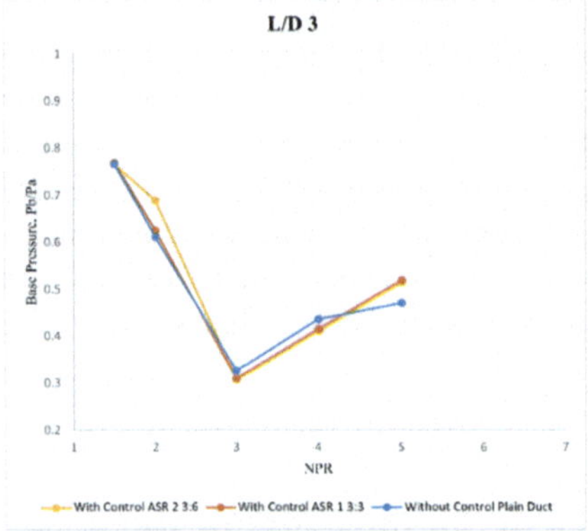

(c) L = 3D

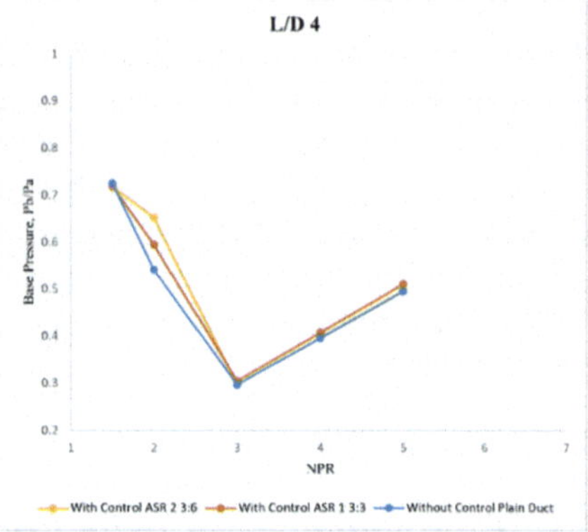

(d) L = 4D

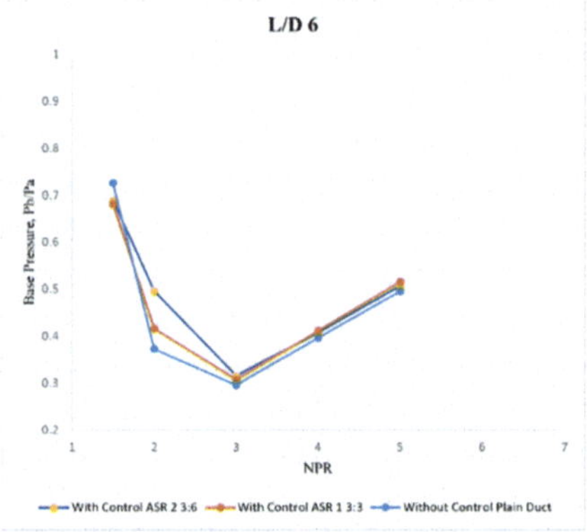

(e) L = 6D

Fig. 3 The base pressure ratio variation with NPR for plain duct and duct with control optimal cavity position for NPR 3. The pressure at the base shifts marginally from L = 1D to 6D between NPR 4 and NPR 6. Figure 4 shows that the green line represents a higher baseline pressure. Therefore, 0.5D with ASR 1 is optimal for these NPR cavities.

3.3.2 At Fixed Duct Length

Figure 5 depicts the pressure in the base area for abruptly expanded ducts with control at a variety of locations and expansion levels. The blue color line denotes the pressure in the wake of the duct, while the other colors reflect the various degrees of control. It is obscure to verify that has the optimal cavity placement for achieving the largest base pressure, as displayed in Fig. 5, and flow from the converging nozzle is facing an adverse pressure gradient for NPR values below two. These results show that dynamic or reactive control is ineffective when the flow-accelerating devices are operated under the presence of adverse pressure. It is also noted that the L/Ds in this investigation all had different base pressure distributions. Figure 5a shows that the pressure of wake fell significantly for NPR from 1.5 to 2 for L/Ds 1 through 6. This drop in the lowest pressure is proportional to the ratio of the nozzle pressure at the lip with that of atmospheric pressure. The pressure gradient is favorable above NPR = 3 as the NPR and consequently the magnitude of the pressure increase gradually.

In comparison to NPR, whose base pressure rose gradually for NPR more than 3, this shift in direction is due to the expansion rates changing. Successfully raising wake pressure when flow-accelerating devices are operated nozzle exit plane pressure is larger than back pressure. The duct model has the least base pressure as long as they are seeing adverse pressure at the nozzle exit. Figure 5 shows that the most effective model is the one with the larger pipe at a cavity position of 0.5D for the pressure of the wake. The green lines in the diagrams indicate the nozzles with the highest base pressure. Figure 5 shows that overall, NPR increases when a cavity is present.

4 Conclusion

When Mach (M) is equal to one and a duct diameter of 15 mm, the base pressure will be impacted by the flow control in the duct. To comprehend the shifts of the wake pressure due to the change of the expansion level at the lip of the flow accelerating device and the duct length, the NPRs at various expansions and L/Ds are examined and shown. The difference between the pressure in a duct at the base corner with a cavity and a plain duct is large. However, adjusting the cavity's geometry has little influence on the wake pressure. Altering the wake pressure at the convergent nozzle's base wall and moving the cavity around can improve performance by a small margin. Passive control with a cavity is effective when the converging flow accelerating devices are operated under the impact of under-expansion.

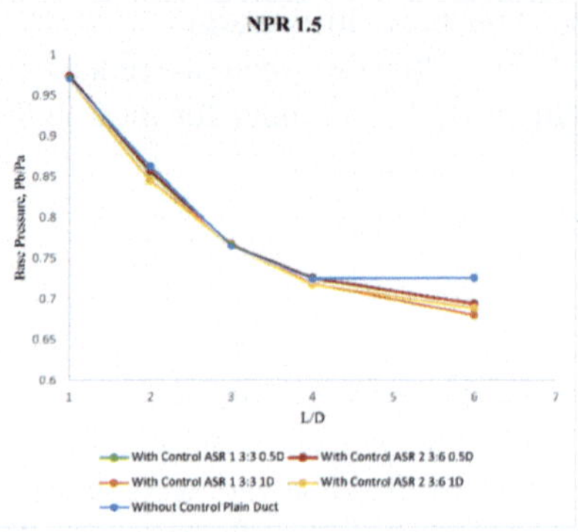

(a) NPR = 1.5

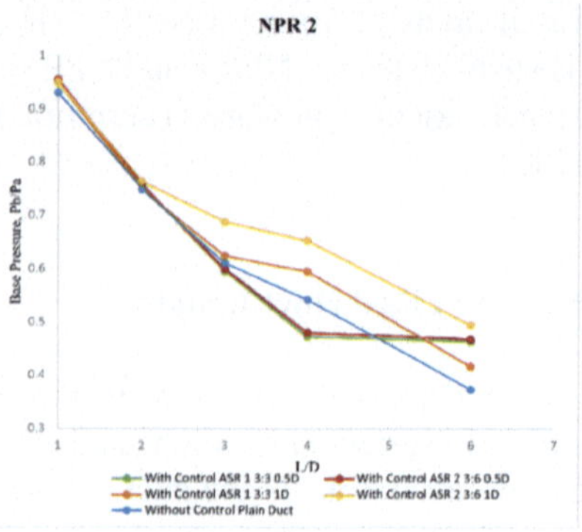

(b) NPR = 2

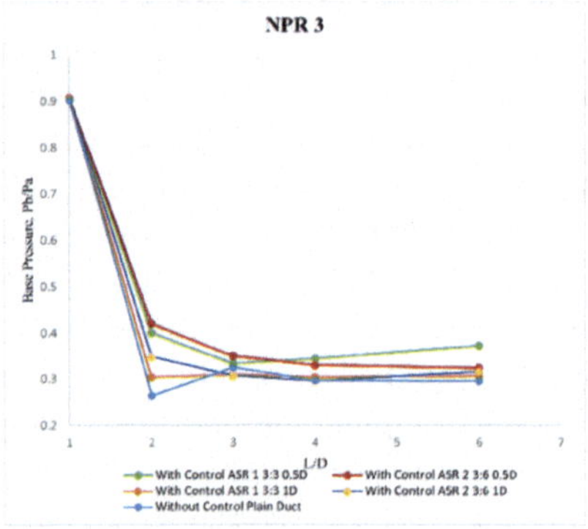

(c) NPR = 3

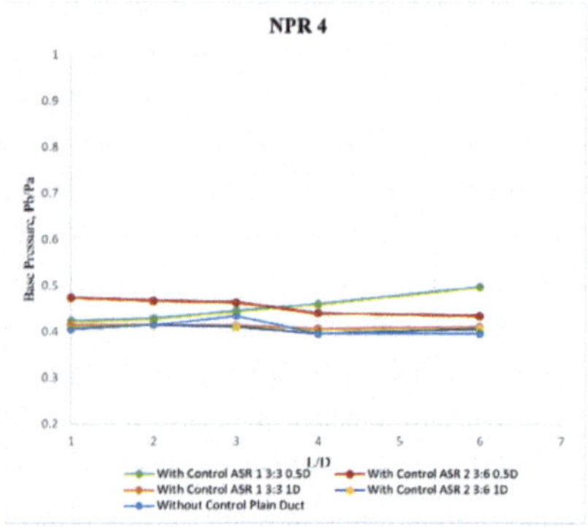

(d) NPR = 4

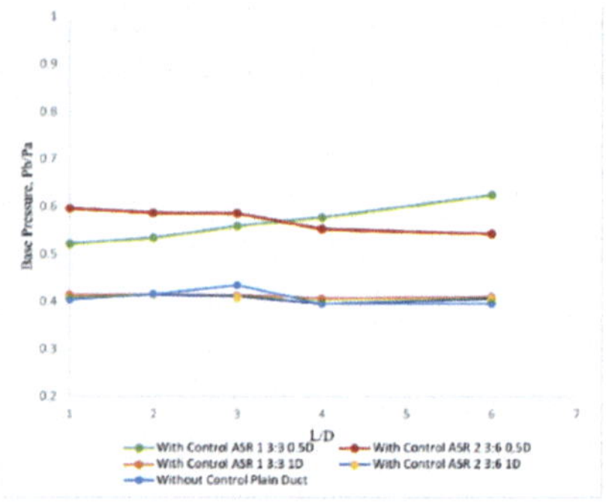

(e) NPR = 6

Fig. 4 The base pressure ratio variation with L/D for plain duct and duct with a cavity as control at numerous locations

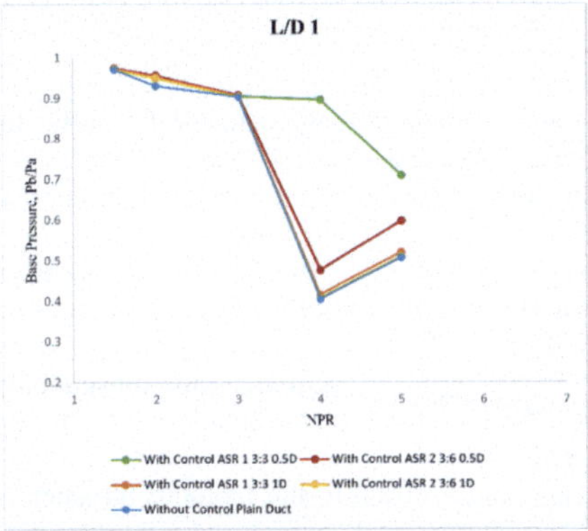

(a) L= 1D

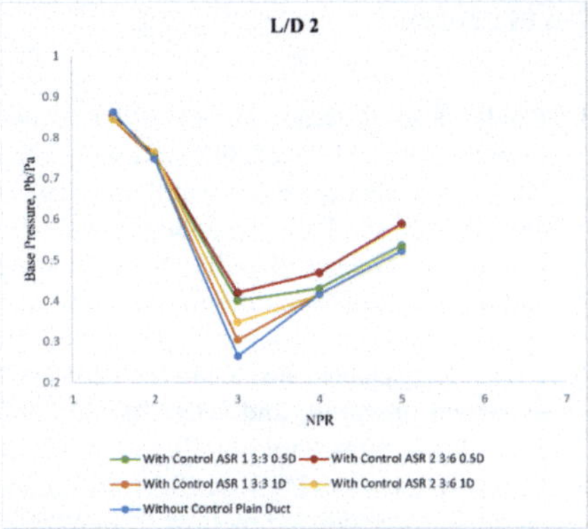

(b) L = 2D

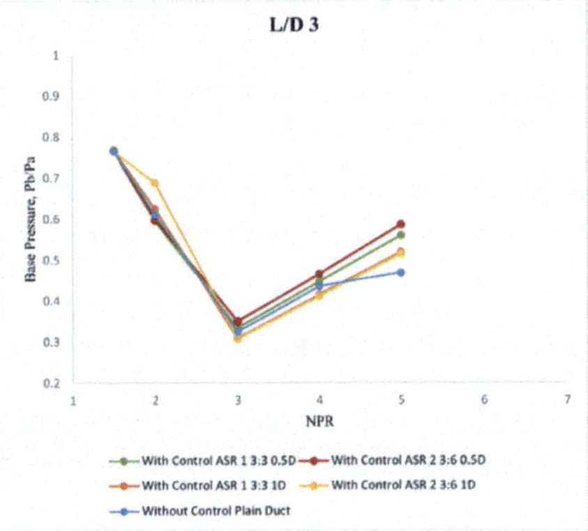

(c) L= 3D

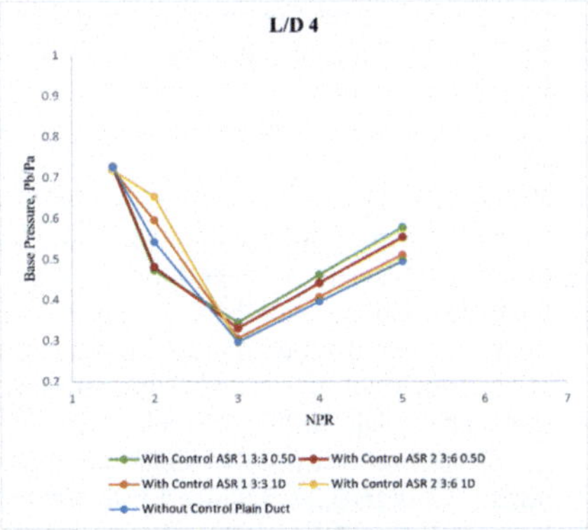

(d) L = 4D

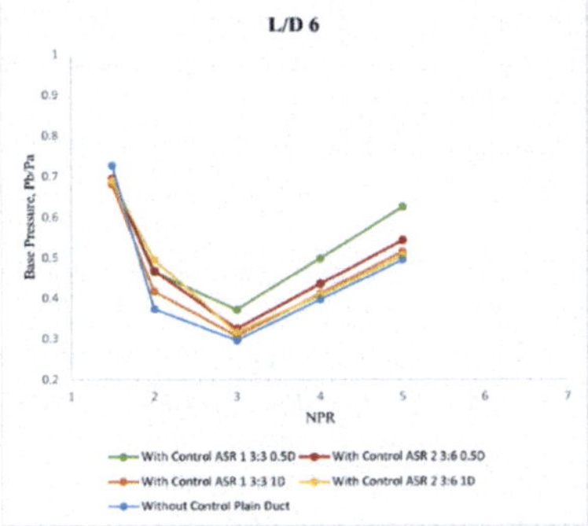

(e) L = 6D

Fig. 5 The base pressure ratio variation with NPR for plain duct and duct control at numerous locations

References

1. Pandey KM, Rathakrishnan E (2006) Annular cavities for base flow control. Int J Turbo Jet Engines 23(2):113–127. https://doi.org/10.1515/TJJ.2006.23.2.113
2. Pathan KA, Dabeer PS, Khan SA (2019) Investigation of base pressure variations in internal and external suddenly expanded flows using CFD analysis. CFD Lett 11(4):32–40
3. Pathan KA, Dabeer PS, Khan SA (2019) An investigation of the effect of control jets location and blowing pressure ratio to control base pressure in suddenly expanded flows. J Therm Eng 6(2):15–23. https://doi.org/10.18186/thermal.726106
4. Pathan KA, Dabeer PS, Khan SA (2017) CFD analysis of the effect of area ratio on suddenly expanded flows. In: 2nd International conference for convergence in technology (I2CT), pp 1192–1198. https://doi.org/10.1109/I2CT.2017.8226315
5. Pathan KA, Dabeer PS, Khan SA (2017) CFD analysis of effect of flow and geometry parameters on thrust force created by flow from nozzle. In: 2nd International conference for convergence in technology (I2CT), pp 1–5
6. Pathan KA, Dabeer PS, Khan SA (2017) CFD analysis of effect of mach number, area ratio, and nozzle pressure ratio on velocity for suddenly expanded flows. In: 2nd International conference for convergence in technology (I2CT), pp 1–7
7. Pathan KA, Dabeer PS, Khan SA (2018) An investigation to control base pressure in suddenly expanded flows. Int Rev Aerosp Eng 11(4):162–169. https://doi.org/10.15866/irease.v11i4.14675
8. Pathan KA, Dabeer PS, Khan SA (2018) Optimization of area ratio and thrust in suddenly expanded flow at supersonic Mach numbers. Case Stud Therm Eng 12:696–700. https://doi.org/10.1016/j.csite.2018.09.006
9. Khan SA, Fatepurwala MA, Pathan KN, Dabeer PS, Ali Baig MA (2018) CFD analysis of human powered submarine to minimize drag. Int J Mech Prod Eng Res Develop 8(3):1057–1066. https://doi.org/10.24247/ijmperdjun2018111
10. Pathan KA, Dabeer PS, Khan SA (2019) Effect of nozzle pressure ratio and control jets location to control base pressure in suddenly expanded flows. J Appl Fluid Mech 12(4):1127–1135. https://doi.org/10.29252/jafm.13.02.30049
11. Pathan KA, Dabeer PS, Khan SA (2020) Enlarge duct length optimization for suddenly expanded flows. Adv Aircr Spacecr Sci 7(3):203–214. https://doi.org/10.12989/aas.2020.7.3.203
12. Aabid A, Murtuza MA, Shushrutha A (2015) Experimental study of mechanical properties of tellurium nanowires reinforced epoxy composites. Am J Mater Sci 5(April):194–197. https://doi.org/10.5923/c.materials.201502.37
13. Khan SA, Aabid A, Ali Baig MA (2018) CFD analysis of CD nozzle and effect of nozzle pressure ratio on pressure and velocity for suddenly expanded flows. Int J Mech Prod Eng Res Dev 8(3):1147–1158
14. Khan A, Aabid A, Khan SA (2018) CFD analysis of convergent-divergent nozzle flow and base pressure control using micro-JETS. Int J Eng Technol 7(3):232–235
15. Khan SA, Aabid A (2018) CFD analysis of CD nozzle and effect of nozzle pressure ratio on pressure and velocity for suddenly expanded flows. Int J Mech Prod Eng Res Dev 8(June):1147–1158
16. Khan SA, Mohamed O, Aabid A (2021) CFD analysis of compressible flows in a convergent-divergent nozzle. Mater. Today Proc (xxxx). https://doi.org/10.1016/j.matpr.2021.03.074
17. Shen Z, Yu S, Zheng S, Nofal TA, Musa A, Li Z (2022) Numerical study of multi-jet with upstream divergent ramp at supersonic cross flow. Aerosp Sci Technol 127:107689. https://doi.org/10.1016/j.ast.2022.107689
18. Bai C, Wang M, Wu Z (2022) Symmetric Mach reflection configuration with asymmetric unsteady solution symmetric mach reflection configuration with asymmetric unsteady solution. Chin J Aeronaut. https://doi.org/10.1016/j.cja.2022.06.012

19. Shamitha, Crasta A, Pathan KA, Khan SA (2023) Analytical and Numerical Simulation of Surface Pressure of an Oscillating Wedge at Hypersonic Mach Numbers and Application of Taguchi's Method. J Adv Res in A Scie and Eng Tech 30(1) 15:30. https://doi.org/10.37934/araset.30.1.1530
20. Samimy M, Kim JH, Kastner J, Adamovich I, Utkin Y (2007) Active control of high-speed and high-Reynolds-number jets using plasma actuators. J Fluid Mech 578:305–330. https://doi.org/10.1017/S0022112007004867
21. Gaitonde DV, Samimy M (2011) Coherent structures in plasma-actuator controlled supersonic jets: axisymmetric and mixed azimuthal modes. Phys Fluids 23(9). https://doi.org/10.1063/1.3627215
22. Pathan KA, Khan SA, Shaikh AN, Pathan AA, Khan SA (2021) An investigation of boat tail helmet to reduce drag. Adv Aircr Spacecr 8(3):239–250. https://doi.org/10.12989/aas.2021.8.3.239
23. Li H, Choi J, Li B, Kim I, Heo J (2016) Numerical analysis on the gas flow dynamics from a rectangular slot-nozzle for pulse cleaning of filter unit. Powder Technol 297:330–339. https://doi.org/10.1016/j.powtec.2016.04.040
24. Shaikh JS, Pathan KA, Kumar K, Khan SA (2023) Effectiveness of Cone Angle on Surface Pressure Distribution along Slant Length of a Cone at Hypersonic Mach Numbers. J Adv Res in F Mech Ther Scie 104(1): 185–203. https://doi.org/10.37934/arfmts.104.1.185203
25. Wang PC, McGuirk JJ (2013) Large eddy simulation of supersonic jet plumes from rectangular con-di nozzles. Int J Heat Fluid Flow 43:62–73. https://doi.org/10.1016/j.ijheatfluidflow.2013.06.002
26. Mousavi SM, Roohi E (2014) Three-dimensional investigation of the shock train structure in a convergent-divergent nozzle. Acta Astronaut 105(1):117–127. https://doi.org/10.1016/j.actaastro.2014.09.002
27. Najar NA, Dandotiya D, Najar FA (2013) Comparative analysis of K-epsilon and Spalart-Allmaras turbulence models for compressible flow through a convergent-divergent nozzle. Int J Eng Sci 2(8):8–17
28. Pathan KA, Ashfaq S, Dabeer PS, Khan SA (2022) Analysis of parameters affecting thrust and base pressure in suddenly expanded flow from nozzle. J Adv Res Fluid Mech Therm Sci 64(1):1–18
29. Shaikh JS, Kumar K, Pathan KA, Khan SA (2022) Analytical and computational analysis of pressure at the nose of a 2D wedge in high-speed flow. Adv Aircr Spacecr Sci 9(2):119–130. https://doi.org/10.12989/aas.2022.9.2.119
30. Salvador FJ, Jaramillo D, Romero J-V, Roselló M-D (2017) Using a homogeneous equilibrium model for the study of the inner nozzle flow and cavitation pattern in convergent-divergent nozzles of diesel injectors. J Comput Appl Math 309:630–641. https://doi.org/10.1016/j.cam.2016.04.010
31. Shamitha, Crasta SA, Pathan KA, Khan SA (2023) Numerical simulation of surface pressure of a wedge at supersonic Mach numbers and application of design of experiments. J Adv Res Appl Mech 101(1):1–18. https://doi.org/10.37934/aram.101.1.118
32. Shaikh JS, Kumar K, Pathan KA, Khan SA (2023) Computational analysis of surface pressure distribution over a 2d wedge in the supersonic and hypersonic flow regimes. Fluid Dyn Mater Process 19(6):1637–1653. https://doi.org/10.32604/fdmp.2023.025113
33. Shaikh SK, Pathan KA, Chaudhary ZI, Khan SA (2020) CFD analysis of an automobile catalytic converter to obtain flow uniformity and to minimize pressure drop across the monolith. CFD Lett 12(9):116–128
34. Shaikh SK, Pathan KA, Chaudhary ZI, Marlpalle BG, Khan SA (2020) An investigation of three-way catalytic converter for various inlet cone angles using CFD. CFD Lett 12(9):76–90
35. Pathan KA, Dabeer PS, Khan SA (2019) Influence of expansion level on base pressure and reattachment length. CFD Lett 11(5):22–36

Numerical Study of Large-Scale Control in Compressible Turbulent Channel Flows

Moghees Ahmad, M. F. Baig, and S. F. Anwer

Abstract Direct numerical simulations in a fully developed compressible turbulent channel flows have been carried out using the large-scale control of large vortical structures along streamwise direction. Effectiveness of the control at Mach number, $Ma = 1.5$ and Reynolds number $Re_b = 3000$ is studied for different amplitudes and wavelengths of forcing. The large-scale control concentrates the turbulent kinetic energy in certain regions near the walls and it can be observed in $y - z$ plane of the channel. These high turbulent kinetic energy regions are also the regions of high streamwise vorticity as well. The control reduces the wall-normal velocity fluctuations near the walls for all the cases but, on the other hand, Reynolds shear stresses are reduced for drag reduction cases and are increased for the case where drag increases. The skin-friction drag reduction(DR) of around 7% is achieved in the present study while, for the case with high strength vortical structures, the skin-friction drag increases.

1 Introduction

The turbulence is an inescapable phenomenon in fluid dynamics and to be precise in our daily life as well. The manipulation of turbulent flows to achieve skin-friction drag reduction is desired, to achieve the benefits related to environmental and economic concerns. Various methods and controls are proposed to achieve this target of taming turbulence. The control methods to contain turbulence are characterized into passive and active control methods based on their input. Passive control techniques require no power input, while active control methods require continuous power input for

M. Ahmad (✉) · M. F. Baig · S. F. Anwer
Department of Mechanical Engineering, Aligarh Muslim University, Aligarh 202002, India
e-mail: mahmad37@myamu.ac.in

M. F. Baig
e-mail: mfbaig.me@amu.ac.in

S. F. Anwer
e-mail: sfahadanwer@zhcet.ac.in

M. A. Siddiqui et al. (eds.), *Advances in Heat Transfer and Fluid Dynamics*, Lecture Notes in Mechanical Engineering, https://doi.org/10.1007/978-981-99-7213-5_2

maintaining the desired changes in the flow. Passive methods such as riblets, coatings, splitter plate, etc., are rather easy but give limited drag reduction of up to 10%. Considerable research has been performed on passive devices like compliant surfaces, wavy walls and riblets [1].

Schoppa and Hussain [2] proposed a new method to control the drag which they named large-scale control. In this control strategy, large-scale vortices are generated along the flow in the streamwise direction. These large-scale vortices affect the near-wall coherent structures responsible for turbulence generation; by mitigating wall-normal vorticity and streaks formation. Schoppa reported 20% drag for certain wavelengths and strengths of the vortices. Canton et al. [3] further investigated the large-scale control by carrying out DNS study at different friction Reynolds numbers (Re_τ) of 104 and 180 and for five different wavelengths. Canton et al. [3] found that the control strategy by Schoppa and Hussain [2] is only provisionally effective and hence suggested a modification to sustain drag reduction and reported drag reduction of around 18%.

This paper aims to study large-scale control in compressible turbulent flows in order to see the efficacy of the control for supersonic flows and to analyse related physics, as this problem has not been explored yet. The results can be applied to transport of compressible gases in rectangular/circular duct flows.

2 Numerical Setup

The dimensional governing equations used for compressible channel flows are listed:

$$\frac{\partial \rho}{\partial t} + \frac{\partial (\rho u_i)}{\partial x_i} = 0 \tag{1}$$

$$\frac{\partial (\rho u_i)}{\partial t} + \frac{\partial (\rho u_i u_j)}{\partial x_j} = -\frac{\partial p}{\partial x_i} + \frac{\partial \tau_{ij}}{\partial x_j} + \rho f_i \tag{2}$$

$$\frac{\partial (\rho e)}{\partial t} + \frac{\partial (\rho e + p) u_i}{\partial x_i} = \frac{\partial (\tau_{ij} u_j)}{\partial x_i} + \rho f_i u_i + \frac{\partial (q_i)}{\partial x_i} \tag{3}$$

where ρ is the density of fluid, t is the time, u_i is the velocity component along x_i direction, e is the total specific energy of the fluid, p is the pressure in the flow field, τ_{ij} is stress tensor and f_i is the driving force.

The present numerical study is performed with an in-house, finite difference code developed for compressible turbulent channel flows and parallelized in all three directions using Message Passing Interface(MPI). A hybrid scheme is used for discretization of the convective or non-linear terms of the governing equations, in which third-order upwinding [4] or fourth-order central differencing is used. Cell-based Peclet number (Pe), $Pe = \rho u_i \Delta x_i / \mu$ is the parameter which decides the discretization of the convective terms, where Δx_i is the step size of the computational domain and u_i is the velocity of the ith direction. For streamwise and spanwise directions, if the $| Pe | < 2$, fourth-order central differencing is used and for $| Pe | \geq 2$, upwinding is used. While, for the wall-normal direction, Peclet number value which decides between upwinding and central differencing is $Pe = 0.2$, if $| Pe | \geq 0.2$ then upwinding is used otherwise central differencing is used. The cell Peclet number-based discretization had been studied and tested by previous studies as well [5–8]. The diffusive and pressure terms are discretized using fourth-order central difference discretization. The temporal integration is done using second-order Runge–Kutta method due to its efficiency and accuracy.

The directions x, y and z are along streamwise, spanwise and wall-normal, respectively. The channel length $\mathsf{L_x}$, $\mathsf{L_y}$ and $\mathsf{L_z}$ in the x, y and z directions are $4\pi\delta \times 4\pi\delta/3 \times 2\delta$, where δ is the half-channel height. The flow through channel is periodic in spanwise and streamwise directions, respectively, while no-slip condition is applied on the channel walls. The mesh is uniform in streamwise and spanwise directions, while non-uniform mesh is used in the wall-normal direction to capture the small-scale structures close to the channel wall. The number of mesh points considered for x, y and z directions are $192 \times 130 \times 160$, respectively. The bulk Reynolds number is taken $Re_b = 3000$ at Mach number, $Ma = 1.5$ for the present study. The grid spacing for x and y directions in viscous scale is $\Delta x^+ \approx 14$ and $\Delta y^+ \approx 7$, while, for wall-normal direction 'z', it varies from $\Delta z^+_{min} \approx 0.04$ to $\Delta z^+_{max} \approx 4.2$. The resolution in wall-normal direction is finer than that of Kolmogorov scale, which is 0.53 in viscous scale. Two-point correlations both in streamwise and spanwise directions confirm that the dimensions of the channel are sufficient for the computations. The code is validated with results from Coleman et al. (CKM) [9].

The large-scale vortices are produced by using the body force method as proposed by Canton et al. [3]. The dimensional equations for the body forces are

$$f_{x_{con}} = 0 \tag{4}$$

$$f_{y_{con}}(y, z) = -A\beta\ cos(\beta y)(1 + \ cos(\pi(\frac{z}{h} - 1))) \tag{5}$$

$$f_{z_{con}}(y, z) = -A\pi\ sin(\beta y)\ sin(\pi(\frac{z}{h} - 1)) \tag{6}$$

where $f_{x_{con}}$, $f_{y_{con}}$ and $f_{z_{con}}$ are the control on the body forces in x, y and z directions, respectively. A is the amplitude of the force applied for the generation vortices and β is the wavenumber of the spanwise length of the vortices.

3 Results and Discussion

With the aim to gain insight into the large-scale control and to study the changes it brings to the physics of the flow, a number of numerical simulations are carried out by varying forcing amplitude and spanwise wavelength of the vortices for compressible turbulent channel flows. The wavelength of the vortices is denoted by $\Lambda^+ = 2\pi\beta^+$. The control is applied to a fully developed compressible turbulent channel flow field at $Re_b = 3000$ and $Ma = 1.5$. The Dirichlet boundary condition is used on the turbulent channel walls. Table 1 shows the parametric details of the simulations performed. The wavelength $\Lambda^+ = 900$ is selected for the simulations and one case of $\Lambda^+ = 450$ is also simulated, where $\Lambda^+ = 900$ corresponds to one pair of vortices, while $\Lambda^+ = 450$ corresponds to two pairs of vortices in the $y - z$ plane. The strength of the vortices produced is determined by quantity $\max|\langle w\rangle_{(y,z)}|$ which is the maximum value of wall-normal velocity component in $y - z$ plane obtained by averaging in streamwise direction x and time t. The skin-friction drag reduction DR is calculated by the mathematical equation Eq. (7).

$$DR\% = \frac{\tau_w|_{unc} - \tau_w|_{con}}{\tau_w|_{unc}} \times 100 \tag{7}$$

where subscript '*unc*' refers to uncontrolled case and '*con*' to controlled case.

The control is applied throughout the axial length of the channel and to quantify the effect of the control, the temporal variation of the wall shear stress is recorded. The averaging of the variables is done in streamwise direction x and time t, hence depicted in $y - z$ plane and the averaged quantity is denoted by $\langle * \rangle_{y,z}$. Figure 1 shows contour plots of vortices produced by large-scale control in $y - z$ plane of the channel. Here, Fig. 1a shows two pairs of vortices produced for Case 3 and Fig. 1b shows one pair of vortices produced for case 4. These plots show the cross-flow velocity with velocity vectors superimposed onto the contours of cross-flow velocity to show the rotation of the vortices. It can be observed from Fig. 1b that the large-scale control aggregates the streaks at the bottom wall from where the fluid lifts up from the bottom to top wall and on outer sides at the top wall. Mean streamwise vorticity $\overline{\omega}_x$ are shown in $y - z$ plane, Fig. 2. These plots show the high vorticity concentration areas in the $y - z$ plane. The mean streamwise vorticity $\overline{\omega}_x$ has high values in regions where the fluid lifts up and settles or spreads out onto the wall due

Table 1 Simulations details

Case	Ma	Re_b	A	Λ^+	$\max\lvert\langle w\rangle_{(y,z)}\rvert$	$DR\%$
1	1.5	3000	0.00025	900	3.3	2.92
2	1.5	3000	0.0005	900	11.03	2.89
3	1.5	3000	0.0005	450	7.7	7.2
4	1.5	3000	0.00075	900	13.8	−4.5

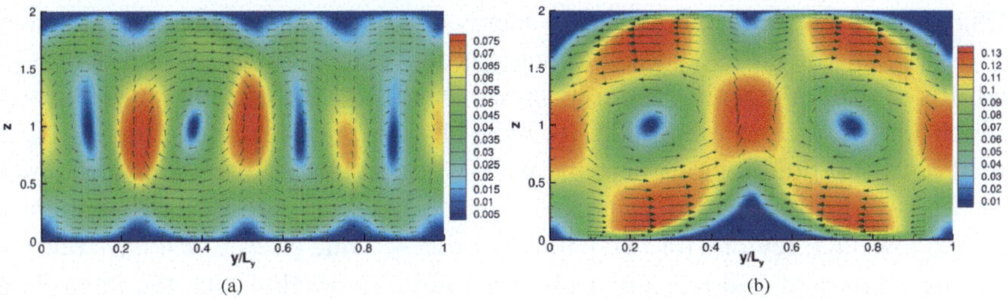

Fig. 1 Cross-flow velocity plots averaged in x & t for $y-z$ plane **a** Case 3, **b** Case 4

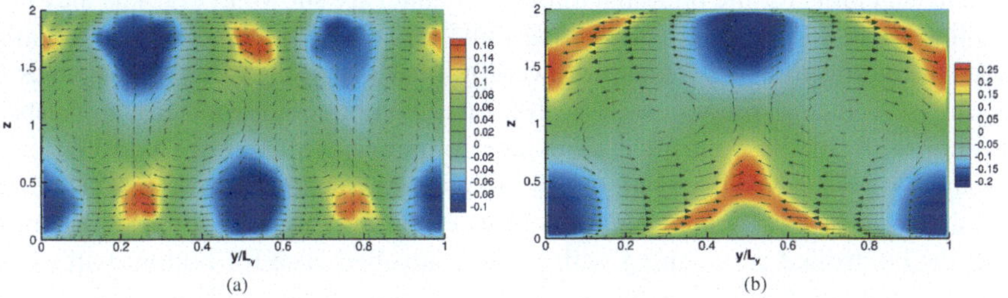

Fig. 2 Mean streamwise vorticity $\overline{\omega}_x$ plots averaged in x & t for $y-z$ plane **a** Case 3, **b** Case 4

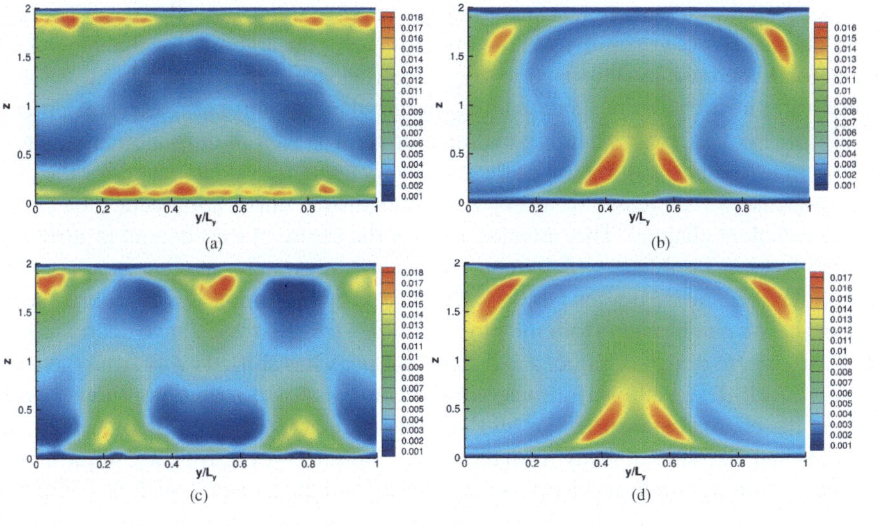

Fig. 3 Comparison of Turbulent kinetic energy (TKE) in y-z plane **a** Case 1, **b** Case 2, **a** Case 3, **b** Case 4

to high shear rate. As can be seen from $\overline{\omega}_x$ contour plots, that for drag reduction case, i.e. case 3, the values of $\overline{\omega}_x$ are less than that for drag increase case, i.e. case 4.

Figure 3 depicts the turbulent kinetic energy (TKE) distribution across the channel in $y - z$ plane. The turbulent kinetic energy is concentrated in the regions where the fluid lifts up from the wall. Figure 3a shows regions of low and high TKE for case 1 which has low forcing amplitude. It can be observed that regions of high TKE are spread in lift-up fluid regions near the wall, whereas low TKE regions are loosely spread across the channel as compared to other cases. Figure 3c shows pattern of high and low TKE regions for two pairs of vortices, while cases 2 & 4 have one pair of strong vortices and show similar plots as in Fig. 3b, d. However, the intensity of high TKE regions is considerably more for drag increase (DI) case, i.e. case 4. The spread of low TKE regions in cases 2 & 4 leads to horseshoe pattern.

Statistical plots for the bottom half of the channel are shown in Fig. 4 as they help visualize mean variation in wall-normal direction. These have been obtained after planar averaging in $x - y$ plane followed by time averaging. The RMS values of streamwise velocity component u', wall-normal component w' and Reynolds shear stresses $\overline{\rho u'w'}$ of the controlled cases are compared with the uncontrolled case. To magnify the near-wall variation of these plots log-scale is used for wall-normal distance. Figure 4a compares the RMS plots of streamwise velocity component u' of all the controlled cases along with the uncontrolled case for bottom half of the turbulent channel. Case 1 has the lowest forcing amplitude and hence its effect on the flow is minimal. For cases 1 & 3, u' is less than that of the uncontrolled flow in the near-wall region which signifies reduction in skin-friction drag, whereas, for Cases 2 & 4, the quantity u' has increased as compared to uncontrolled case. The case with drag increase, i.e. case 4 has the highest u' plot throughout the channel, which leads to an increase in skin-friction drag. The RMS values of wall-normal velocity component w' are an important quantity for turbulence control and they are depicted in Fig. 4b. For all cases pertaining to drag reduction as well as drag increase, there is a drop in w' in near-wall region of the turbulent channel flow. However, for the drag increase case, w' increases to a very high value as the curve approaches the centre of the turbulent channel. This increase in w' at the centre of the channel is attributed to high forcing amplitude which in turn increases the quantity $\max|\langle w\rangle_{(y,z)}|$ as well. The Reynolds shear stresses ($\overline{\rho u'w'}$) averaged in wall-normal direction are shown in Fig. 4c. As can be seen from the plot, Reynolds shear stresses have decreased for the cases where drag has reduced and also their peak has shifted away from the wall. Case 1 has little variation from uncontrolled case as low forcing amplitude hardly affects the channel statistics. For all the controlled cases except Case 4, $\overline{\rho u'w'}$ has reduced and also shifted away from the wall as well. However, for Case 4, $\overline{\rho u'w'}$ has increased in magnitude and its peak has also moved closer to the wall as compared to the uncontrolled case, signifying an increase in the skin-friction drag.

The vortical structures which signify the intensity of the turbulence are shown in Fig. 5 for all the cases studied, along with the vortical structures for uncontrolled turbulent channel flow. The Q-criterion which was proposed by Hunt [10] is used to visualize the iso-surfaces of the vortical structures. The streamwise velocity component u is superimposed on these vortical structures as well. The vortical structures for only the bottom half of the channel are shown in these plots. It can be observed from Fig. 5 that, for all the controlled cases, there is concentration of the vortical

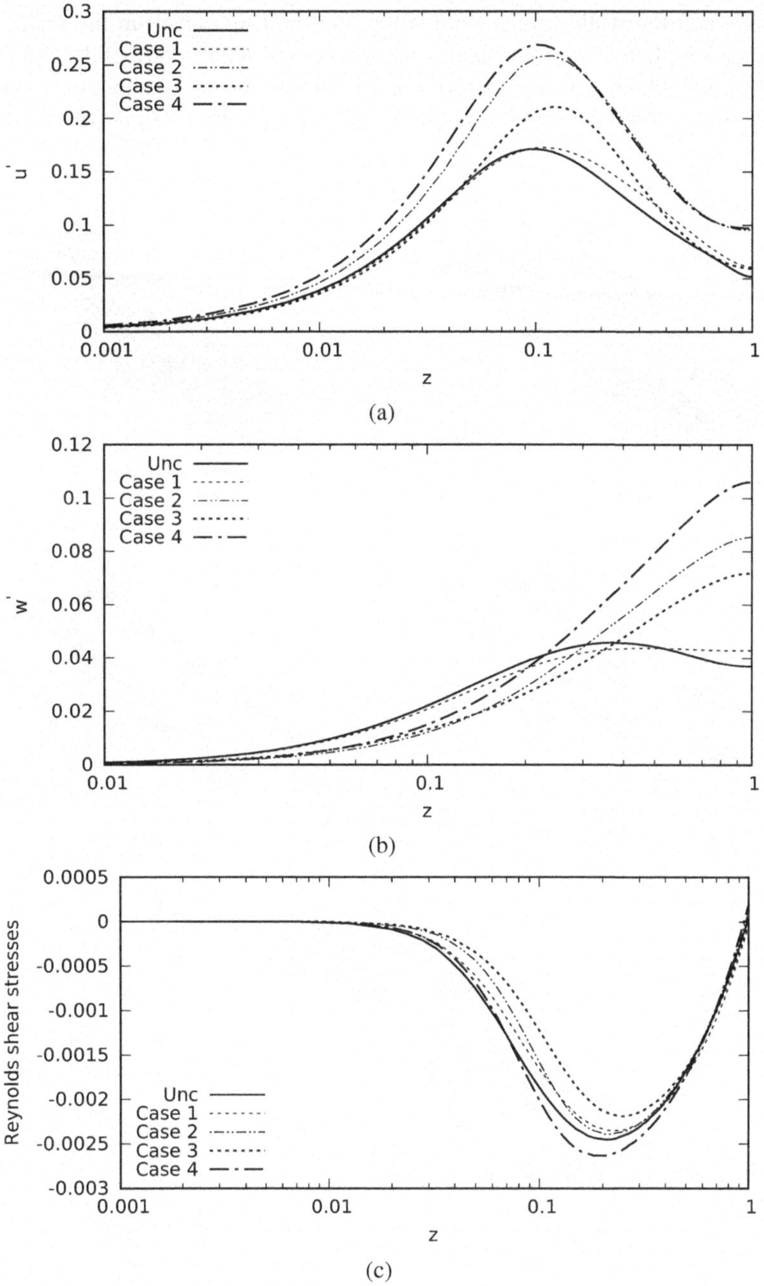

Fig. 4 Comparison of RMS quantities for uncontrolled and controlled cases averaged in wall-normal direction, bottom half of channel is shown **a** RMS of Streamwise velocity component, u', **b** RMS of Wall-normal velocity component, w', **c** Reynolds shear stresses, $\overline{\rho u' w'}$

structures near the wall surface from where the fluid lifts up from the wall. At the bottom wall, high density of vortical structures occurs at the centre of the channel for Cases 1, 2 & 4 which have one pair of vortices, while, for case 3, high concentration of vortical structures is observed at two locations, as it has two pair of vortices. At

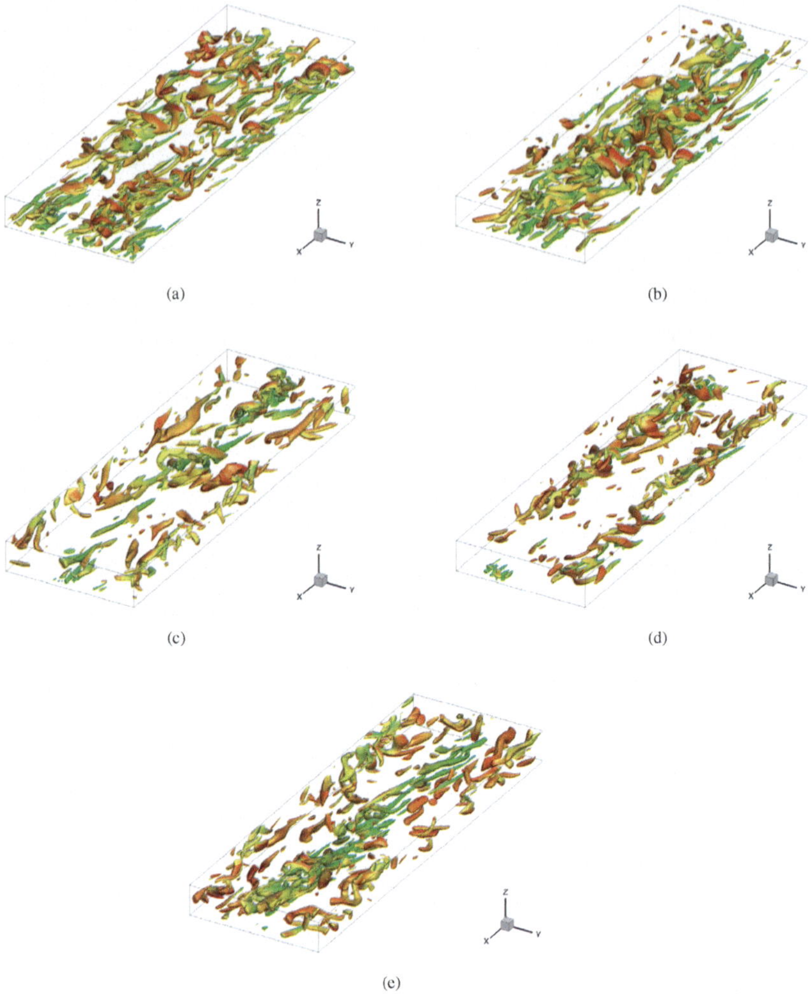

Fig. 5 Iso-surfaces visualization using Q-invariant criterion at a value of Q=0.28, **a** Uncontrolled case, **b** Case 1, **c** Case 2, **d** Case 3, **e** Case 4

the top wall of the channel, these structures are found near the outer sides of the wall. For high-strength cases, the small-scale vortical structures are being moved in a circulatory fashion by the large-scale vortical structures.

4 Conclusions

Numerical simulations of a fully developed compressible turbulent channel flow have been carried out by generating large-scale vortical structures along the streamwise direction. The effectiveness of the control for Mach number, $Ma = 1.5$ and Reynolds number $Re_b = 3000$ is studied at different amplitudes of forcing and wavelength. The large-scale control leads to concentration of vortical structures on the bottom and top walls, leading to zones of high and low turbulent kinetic energy. In the near-wall region, the control decreases the wall-normal velocity fluctuations monotonically; however, Reynolds shear stresses decrease for drag reduction cases and increases for drag increase cases. The drag reduction of $\approx 7\%$ is achieved in the present study, while, for high forcing amplitude of the large-scale vortices, the drag increases in the channel flow.

References

1. Gad-El-Hak M (2000) Flow control
2. Schoppa W, Hussain F (1998) A large-scale control strategy for drag reduction in turbulent boundary layers. Phys Fluids 10:1049–1051. https://doi.org/10.1063/1.869789
3. Canton J, Örlü R, Chin C, Hutchins N, Monty J, Schlatter P (2016) On large-scale friction control in turbulent wall flow in low Reynolds number channels. Flow Turbul Combust 97:811–827
4. Veersteg H, Malalasekera W (1995) An introduction to computational fluid dynamics
5. Ahmad H, Baig MF, Fuaad PA (2015) Numerical investigation of turbulent-drag reduction induced by active control of streamwise travelling waves of wall-normal velocity. Eur J Mech B Fluids 49:250–263. https://doi.org/10.1016/j.euromechflu.2014.09.004
6. Fuaad PA, Baig MF, Ahmad H (2016) Drag-reduction in buoyant and neutrally-buoyant turbulent flows over super-hydrophobic surfaces in transverse orientation. Int J Heat Mass Transfer 93:1020–1033. https://doi.org/10.1016/j.ijheatmasstransfer.2015.10.068
7. Khawar O, Baig MF, Sanghi S (2021) Taylor-couette flows undergoing orthogonal rotation subject to thermal stratification. Phys Fluids 33:035107. https://doi.org/10.1063/5.0035546
8. Naim MS, Baig MF (2019) Turbulent drag reduction in taylor-couette flows using different super-hydrophobic surface configurations. Phys Fluids 31:095108. https://doi.org/10.1063/1.5116316
9. Coleman GN, Kim J, Moser RD (1995) A numerical study of turbulent supersonic isothermal-wall channel flow. J Fluid Mech 305:159–183. https://doi.org/10.1017/S0022112095004587
10. Hunt JCR, Wray AA, Moin P (1988) Eddies, streams, and convergence zones in turbulent flows. Publisher Centre for Turbulence Research

Drag Reduction in Turbulent Compressible Channel Flows Using Spanwise Velocity Waves

Moghees Ahmad, M. F. Baig, and S. F. Anwer

Abstract Numerical simulations in compressible turbulent channel flows are carried out by applying the active control technique of spanwise velocity waves traveling in streamwise direction. The sinusoidal waves traveling in upstream direction are selected for the present study. For Mach number $Ma = 1.5$ and bulk Reynolds number $Re_b = 3000$, the effects of varying the amplitude of control velocity while keeping other parameters constant are analyzed with the aim to achieve maximum turbulent drag reduction. The application of this active control technique for all the amplitudes studied successfully modifies the near-wall structures by reducing high-speed streaks and vortical structures. The maximum turbulent drag reduction, $\text{DR} = 37.01\,\%$ is observed at threshold amplitude $Amp = 1.5$ for a selected wave number κ and oscillation frequency ω. The spanwise vorticity near the walls is suppressed and hence, the turbulent drag produced by eddies is reduced as well.

1 Introduction

Since the advent of study of fluids, turbulence is the riddle yet to be overcome. The understanding of turbulence is important, as almost all natural and engineering flows around us are turbulent in nature and there is no escape to it. Turbulent flows are commonly encountered and the drag associated with turbulent flows is also present in all natural and engineering flows. There are flows where turbulence is desired, but in engineering flows where energy input is to be reduced, turbulent drag is unwanted and should be reduced. The economical and environmental costs associated with the turbulent drag for the real-life flows and problems are high. The near-wall turbulent

M. Ahmad (✉) · M. F. Baig · S. F. Anwer
Department of Mechanical Engineering, Aligarh Muslim University, Aligarh 202002, India
e-mail: mahmad37@myamu.ac.in

M. F. Baig
e-mail: mfbaig.me@amu.ac.in

S. F. Anwer
e-mail: sfahadanwer@zhcet.ac.in

M. A. Siddiqui et al. (eds.), *Advances in Heat Transfer and Fluid Dynamics*, Lecture Notes in Mechanical Engineering, https://doi.org/10.1007/978-981-99-7213-5_3

structures are the culprit for this drag and hence, the efforts should be such as to reduce these near-wall structures. The methods to modify these near-wall structures can be broadly divided into two categories; passive and active control methods [1]. Passive techniques such as riblets, coatings, splitter plates, etc., are considered simple, but limited drag reduction is achieved by it. Hence, to hunt for higher drag reductions, the active control methods which are rather difficult to implement are used.

Jung et al. [2] proposed one such active open-loop control technique of spanwise wall oscillation. Jung et al. [2] carried out Direct Numerical Simulations(DNS) for channel flows. In this control technique, the channel wall is oscillated using a sinusoidal wave, which successfully altered near-wall structures, which led to turbulent drag reduction. Kim and Sung [3] carried out unsteady blowing into the turbulent boundary layer and reported the changes in the near-wall structures. Tardu and Doche [4] confirmed experimentally that a maximum drag reduction of 50% can be achieved by periodic blowing through a spanwise slot. Ahmad and Baig [5] performed DNS simulations using active control of upstream and downstream traveling waves of wall-normal velocity and they computed a 10% reduction in skin-friction drag for turbulent channel flows at $Re_\tau = 180$.

Quadrio et al. [6] carried out DNS study of active control technique of applying spanwise velocity waves traveling in streamwise direction in turbulent channel flows and reported approximately 48% reduction in skin-friction drag. Auteri et al. [7] carried out experimental study of spanwise velocity waves traveling in streamwise direction for pipe flows and they reported a maximum of 33% reduction of skin-friction drag. The present study is carried out in a compressible turbulent channel for different amplitudes in order to find the effects of the amplitude variation and its related physics in the compressible channel flows. Physically, spanwise velocity waves traveling in streamwise direction can be produced by blowing and suction at the channel walls.

2 Numerical Setup

Governing equations for compressible channel flows in dimensional forms are listed as

$$\frac{\partial \rho}{\partial t} + \frac{\partial (\rho u_i)}{\partial x_i} = 0 \tag{1}$$

$$\frac{\partial (\rho u_i)}{\partial t} + \frac{\partial (\rho u_i u_j)}{\partial x_j} = -\frac{\partial p}{\partial x_i} + \frac{\partial \tau_{ij}}{\partial x_j} + \rho f_i \tag{2}$$

$$\frac{\partial (\rho e)}{\partial t} + \frac{\partial (\rho e + p) u_i}{\partial x_i} = \frac{\partial (\tau_{ij} u_j)}{\partial x_i} + \rho f_i u_i + \frac{\partial (q_i)}{\partial x_i} \tag{3}$$

where ρ is the fluid density, u_i is the velocity component, direction, the total specific energy of the fluid is e, p is the pressure in the flow field, τ_{ij} is stress tensor, t is the time, and f_i is the body force along x_i direction.

An in-house, finite difference code for supersonic channel flows is used for simulations. The code is parallel in all the three directions and uses Message Passing Interface (MPI) for parallelization. The code uses a hybrid scheme for discretization of the convective or non-linear terms of the governing equations. Cell-based Peclet number (Pe) is the parameter which acts as a switch and decides whether third-order upwinding [8] will be used or fourth-order central differencing. Peclet number is defined as $Pe = \frac{\rho u_i \Delta x_i}{\mu}$, where Δx_i is the step size and u_i is the velocity of the ith direction where $i = 1, 2, 3$. For streamwise and spanwise directions, if $| Pe | \geq 2$, then upwinding is used, else fourth-order central differencing. For wall-normal direction, if Peclet number $| Pe | \geq 0.2$, then upwinding is used, otherwise, central differencing. This hybrid method has been studied previously as well [5, 9]. Fourth-order central difference method is used for dicretization of other terms in governing equations, while second-order Runge–Kutta method is used for time integration.

The channel coordinates in streamwise, spanwise, and wall-normal are denoted by x, y, and z. The channel length in streamwise, spanwise, and wall-normal directions are taken as $4\pi\delta \times \frac{4\pi\delta}{3} \times 2\delta$, where δ is the half-channel height. The flow is periodic in x and y directions. The non-uniform mesh is used in the wall-normal direction, to capture high gradients in z direction while mesh is uniform in x and y directions, respectively. Mesh points in streamwise, spanwise, and wall-normal directions are $192 \times 130 \times 160$, respectively. The resolution in wall-normal direction is finer than that of the Kolmogorov scale, which is 0.53 in viscous scale, which is calculated using the following equation:

$$\frac{\eta}{l_o} \sim Re^{\frac{-3}{4}} \tag{4}$$

The code is validated from the results of Coleman et al. (CKM) [10] at bulk Reynolds number $Re_b = 3000$ and Mach number $Ma = 1.5$.

The dimensional equation for producing the streamwise traveling wave of spanwise velocity is

$$V_{(x,y,z_w,t)} = Amp \times sin(\kappa x + \omega t) \tag{5}$$

where $V_{(x,y,z_w,t)}$ is the spanwise velocity component at the wall, Amp is the oscillation amplitude, κ is the wavenumber in streamwise direction, x is the streamwise coordinate, ω is the oscillation frequency, and t is the time.

3 Results and Discussions

Table 1 shows the details of the simulations carried out by the present study. A total of four DNS are carried out by changing the amplitude of the velocity wave, for the same wavenumber and frequency. The variation of the amplitude of the spanwise velocity on the drag reduction is studied. The other parameters are kept same for all the cases studied, while the Dirichlet boundary condition is used on the channel walls. An uncontrolled fully developed turbulent channel flow field is taken as initial condition and the control is applied on both the walls of the turbulent channel. To quantify the effect of control in achieving turbulent drag reduction, the mean of the temporal variations of the wall shear stresses are recorded.

The Drag Reduction (DR) for the cases is calculated by the mathematical Eq. (6):

$$\mathrm{DR}\% = \frac{\tau_w|_{unc} - \tau_w|_{con}}{\tau_w|_{unc}} \times 100 \tag{6}$$

where $\tau_w|_{unc}$ is shear stresses for uncontrolled case and $\tau_w|_{con}$ are shear stresses for controlled cases.

Highest drag reduction DR= 37.01% is obtained for case 3, having an amplitude $Amp = 1.5$. Although the amplitude for case 4 ($Amp = 2.5$) is significantly larger than case 3 but gives similar DR. Figure 1 shows Reynolds shear stresses $\overline{\rho u'w'}$ obtained for all the controlled cases along with the uncontrolled case. The upper wall of the channel is not shown, since the control is applied on both the walls, therefore upper half-length variation is symmetric to the lower half. Since DR is achieved in all the cases, therefore $\overline{\rho u'w'}$ has reduced for all the cases studied. The control successfully reduces the maxima for all the controlled cases and also moves it away from the wall. Reduction in $\overline{\rho u'w'}$ signifies the reduction in turbulent fluctuations as well. Figure 2 compares the production P_k of controlled cases with that of uncontrolled cases. The plot shows a significant reduction in production P_k very near to the wall. From Fig. 2, it can be observed that stronger the control amplitude lesser is P_k maxima and farther it is from the wall as well. Case 4 has the lowest maxima even though case 3 has the highest drag reduction, since larger is the penetration of the control effect into the wall-normal direction.

Yoon et al. [11] gave relationship between the skin-friction coefficient C_f and spanwise vorticity Ω_y, it is expressed by mathematical equation(7):

Table 1 Simulations details

Case	Ma	Re_b	Amp	κ	ω	DR %
1	1.5	3000	0.25	1	0.5	20.21
2	1.5	3000	0.5	1	0.5	29.53
3	1.5	3000	1.5	1	0.5	37.01
4	1.5	3000	2.5	1	0.5	36.92

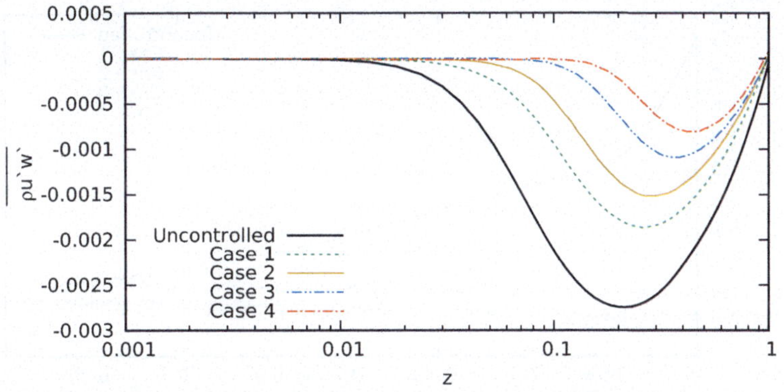

Fig. 1 Comparison of planar averaged Reynolds shear stresses $\overline{\rho u'w'}$ for uncontrolled and controlled cases

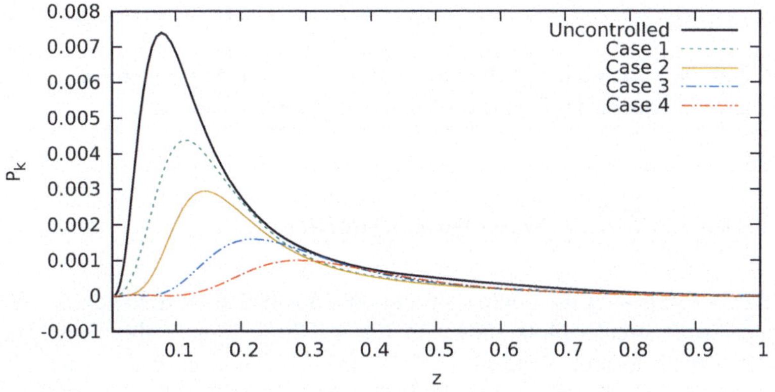

Fig. 2 Comparison of planar averaged Production P_k for uncontrolled and controlled cases

$$C_f = \frac{2\tau_w^*}{\frac{1}{2}\rho U^2} = \frac{2}{Re}\left(\left.\frac{\partial w}{\partial x}\right|_{z=0} - \Omega_y|_{z=0}\right) \tag{7}$$

However, for the applied control, $(\partial w/\partial x)|_{z=0}$ is zero. The vorticity $\boldsymbol{\Omega}$ signify the rotation of the fluids and eddies in the flow. Vorticity variation from the wall is an important parameter, it gets generated at the fluid boundary and thereafter, diffuses into the flow. The spanwise vorticity Ω_y also changes as the control is applied, the RMS fluctuations of spanwise vorticity $\Omega_y^{'}$ from the bottom wall are shown in Fig. 3. The only relevant near-wall region is shown in this figure. It can be observed from the plot that, for all the controlled cases, the fluctuations of $\Omega_y^{'}$ are less than that for uncontrolled case for all the wall-normal distance. The applied control suppresses the intensity of $\Omega_y^{'}$, which signifies the reduction of the skin-friction drag. For uncontrolled case, the maxima is at the walls of the channel, while, for controlled case 1, the maxima at the wall has reduced. For cases 2, 3, & 4, $\Omega_y^{'}$ has reduced at

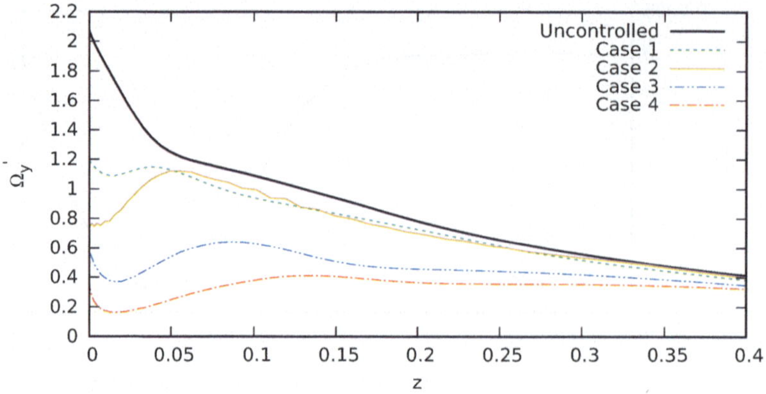

Fig. 3 Comparison of planar averaged RMS fluctuations of spanwise vorticity Ω_y' for uncontrolled and controlled cases

the wall and the maxima has shifted away from the wall. Moreover, for cases 3 & 4, the Ω_y' curve is almost flat for most of the wall-normal distance.

3.1 Visualization of Near-Wall Structures

The effects of applying the control on the uncontrolled flow field can be observed in the near-wall streaks plots shown in Fig. 4. These figures show instantaneous streaks at a wall-normal plane of $z^+ \approx 30$ from the bottom wall of the channel. Figure 4a shows streaks for uncontrolled case, where high and low-speed streaks are alternately arranged one after another. As compared to low-speed streaks, high-speed streaks contain a large amount of turbulent kinetic energy. Figure 4b–e show streaks for the controlled cases. All the controlled cases show a marked change from the uncontrolled case. For cases 1 & 2, high-speed streaks are reduced in magnitude and their spatial density is much lesser than for uncontrolled case. For high-amplitude cases, i.e., cases 3 & 4, the penetration of effect of the control is deep into the flow field, as can be observed from Fig. 4d and e. High-speed streaks for both the high-amplitude cases are reduced markedly with more low-speed regions. Even at a height of $z^+ \approx 30$, the low-speed streaks are predominant and hence, effectively reduces turbulence and turbulent kinetic energy as well.

To get an insight into the flow characteristics, the iso-surfaces of vortical structures are also visualized using the Q-criterion proposed by Hunt et al. [12], which is based on the principle that the vortical structures are the regions in the flow field, where rotation rate is greater than strain rate. At low Reynolds number flows, these vortical structures have large sizes, while for high Reynolds number flows, they tend to get smaller in proportion to increase in Reynolds number. As shown in Fig. 5, vortical

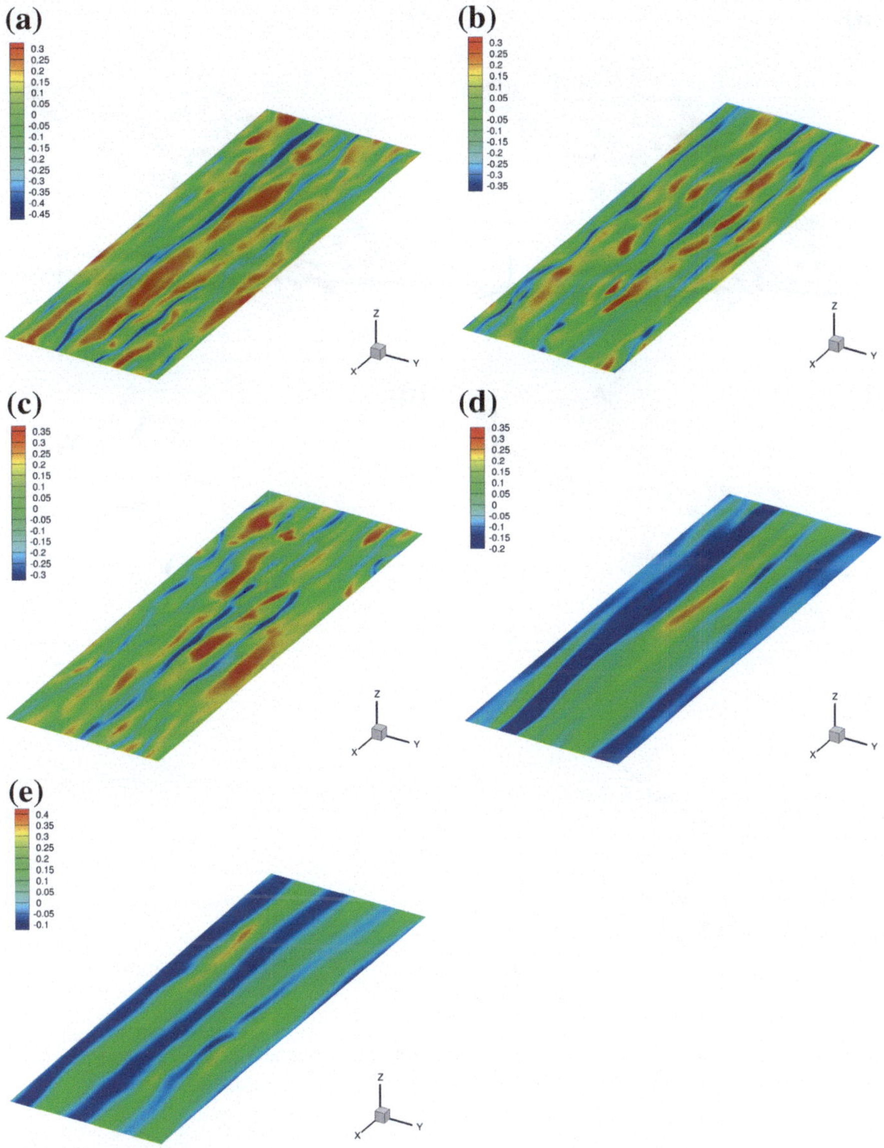

Fig. 4 Comparison of near-wall streaks at wall-normal distance of $z^+ = 30$ from the bottom wall: **a** Uncontrolled Case, **b** Case 1, **c** Case 2, **d** Case 3, **e** Case 4

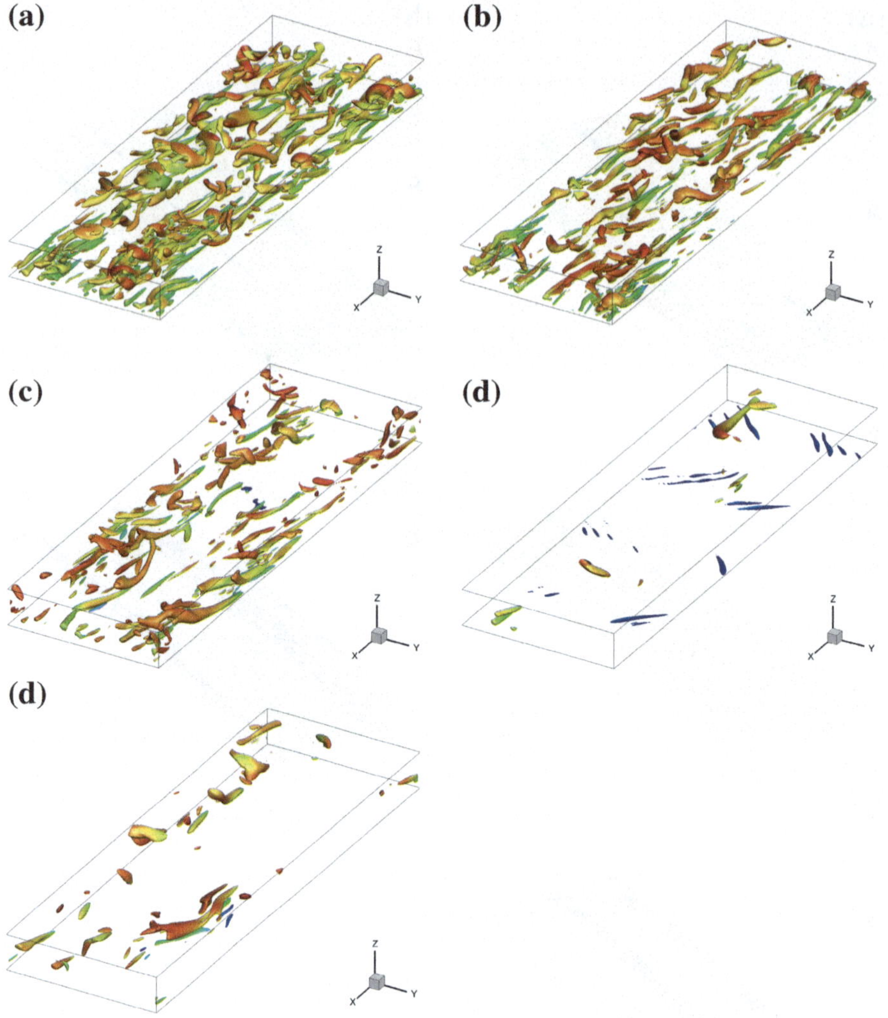

Fig. 5 Comparison of vortical structures using Q-criterion shown at a value $Q = 0.28$ in bottom half of channel: **a** Uncontrolled Case, **b** Case 1, **c** Case 2, **d** Case 3, **e** Case 4

structures are plotted for bottom half of the channel for a carefully chosen positive value of $Q = 0.28$ and streamwise velocity u is superimposed onto the vortical structures. High density of vortices in flow denotes high skin-friction drag. Therefore, the concentration of these vortical structures should reduce on positive application of the control. The vortical structures for the uncontrolled case are shown in Fig. 5a. For all the controlled cases, as shown in Fig. 5b–e, the coherent vortical structures have reduced in the order of drag reduction achieved, as compared to uncontrolled case and hence, signifies reduction in TKE. As can be seen in Fig. 5d, the vortical structures are least for case 3 which has achieved the highest reduction in drag. However, despite other quantities like Ω_y', $\overline{\rho u' w'}$ being less for case 4, it has higher vortical structures than case 3.

4 Conclusions

In the present paper, DNS of compressible channel flows has been carried out by applying the active control technique of spanwise velocity waves traveling in streamwise direction. The sinusoidal upstream traveling waves at Mach number, $Ma = 1.5$ and Reynolds number $Re_b = 3000$ are selected for the study. The effect of varying the amplitude of control velocity while keeping other parameters constant is analyzed with the aim to achieve maximum turbulent drag reduction. The maximum turbulent drag reduction, DR$= 37.01\%$ is observed at an amplitude, $Amp = 1.5$ of the spanwise velocity waves.

The effects of the control on the spanwise vorticity, production, Reynolds shear stress, streaks, and vortical structures are analyzed. The control successfully alters spanwise vorticity in near-wall region. The high-speed streaks at wall-normal plane, $z^+ = 30$ have reduced and also reduced the turbulence in the flow. The present paper studied only four cases; however, a thorough study can be carried out for a carefully selected range of amplitude and other parameters as well for turbulent compressible channel flows.

References

1. Gad-El-Hak M (2000) Flow control
2. Jung WJ, Mangiavacchi N, Akhavan R (1992) Suppression of turbulence in wall-bounded flows by high-frequency spanwise oscillations. Phys Fluids A Fluid Dyn 4:1605–1607. https://doi.org/10.1063/1.858381
3. Kim K, Sung HJ (2006) Effects of unsteady blowing through a spanwise slot on a turbulent boundary layer. J Fluid Mech 557:423–450
4. Tardu SF, Doche O (2009) Active control of the turbulent drag by a localized periodical blowing dissymmetric in time. Exp Fluids 47:19–26
5. Ahmad H, Baig M, Fuaad P (2015) Numerical investigation of turbulent-drag reduction induced by active control of streamwise travelling waves of wall-normal velocity. Eur J Mech B Fluids 49:250–263. https://doi.org/10.1016/j.euromechflu.2014.09.004
6. Quadrio M, Ricco P, Viotti C (2009) Streamwise-travelling waves of spanwise wall velocity for turbulent drag reduction. J Fluid Mech 627:161–178. https://doi.org/10.1017/S0022112009006077
7. Auteri F, Baron A, Belan M, Campanardi G, Quadrio M (2017) Experimental assessment of drag reduction by traveling waves in a turbulent pipe flow. Phys Fluids 22:117
8. Veersteg H, Malalasekera W (1995) An introduction to computational fluid dynamics
9. Fuaad P, Baig M, Ahmad H (2016) Drag-reduction in buoyant and neutrally-buoyant turbulent flows over super-hydrophobic surfaces in transverse orientation. Int J Heat Mass Transfer 93:1020–1033. https://doi.org/10.1016/j.ijheatmasstransfer.2015.10.068

10. Coleman GN, Kim J, Moser RD (1995) A numerical study of turbulent supersonic isothermal-wall channel flow. J Fluid Mech 305:159–183. https://doi.org/10.1017/S0022112095004587
11. Yoon M, Ahn J, Hwang J, Sung HJ (2016) Contribution of velocity-vorticity correlations to the frictional drag in wall-bounded turbulent flows. Phys Fluids 28:081702. https://doi.org/10.1063/1.4961331
12. Hunt JCR, Wray AA, Moin P (1988) Eddies, streams, and convergence zones in turbulent flows. Centre for Turbulence Research

Passive Control of Base Flow at Supersonic Mach Number for Area Ratio 4

Nur Aqilah, Khizar Ahmed Pathan, and Sher Afghan Khan

Abstract This paper presents simulation results to assess a cavity's impact on base pressure at a supersonic Mach number of 1.4. The flow field in a sudden expansion flow is disrupted due to the presence of a cavity to control the base pressure and reduce base drag. This study also focuses on understanding the recirculation region's flow pattern and the flow's behavior through a converging–diverging nozzle. C-D nozzles are used in many aerospace industries. This research is carried out by the numerical approach using Computational Fluid Dynamics (CFD) Analysis. The simulation part, considering the Mach number, area ratio of 4, and the nozzle pressure ratio in the range of 2 to 10, was gathered at various length-to-diameter (L/D) ratios and cavity aspect ratios when conducting this analysis. The cavity aspect ratio used in this research is 3:3, 3:6, 6:3, and 6:6, which will show the results of the base pressure. Throughout this investigation, the cavity is located at 1D. The work applied the K-standard wall function turbulence model with the commercial computational fluid dynamics (CFD) and verified it. The C-D nozzle was created and modeled. The results demonstrate that the expansion level (Nozzle Pressure Ratio), the cavity aspect ratio, and the cavity's location significantly influenced the base pressure.

Keywords Computational fluid dynamics (CFD) · Passive control · Nozzle pressure ratio (NPR) · Base pressure

N. Aqilah · S. A. Khan
Mechanical and Aerospace Engineering Department, Faculty of Engineering, International Islamic University, 53100 Kuala Lumpur, Malaysia
e-mail: sakhan@iium.edu.my

K. A. Pathan (✉)
Department of Mechanical Engineering, CSMSS Chh. Shahu College of Engineering, Aurangabad 431011, India
e-mail: kapathan@csmssengg.org

M. A. Siddiqui et al. (eds.), *Advances in Heat Transfer and Fluid Dynamics*, Lecture Notes in Mechanical Engineering, https://doi.org/10.1007/978-981-99-7213-5_4

1 Introduction

High-speed aerodynamics is an important topic among engineers and researchers today. Knowledge of high-speed aerodynamics may be utilized in everyday technology, particularly vehicle and aircraft technology. Reduced drag in aerodynamic vehicles is one of the critical problems. When the Mach number (M) is more than one, an increase in the area leads to an increase in velocity and vice versa. In a divergent duct, velocity increases, whereas in a converging duct, velocity decreases. This is inverse to how subsonic flow behaves in divergent and convergent ducts. The abrupt axisymmetric expansion flow field is a complex phenomenon characterized by flow separation, recirculation, and reattachment [1]. A shear layer can separate a flow field into two main zones, one for flow recirculation and the other for main flow [2]. When the dividing streamline reaches the wall, the reattachment line is formed. Due to a quick shift in the rear geometry, a massive division of the shear layer arises at the base of the baseflow. Split flow in the base area causes two distinct issues: base instability and the more significant drag. In this study, the base pressure is the parameter affected by the recirculation zone at the base region. This research aims to manage the base pressure to reduce base drag. Active and passive control are the two sorts of methods for optimizing the base pressure [3]. These two methods are critical because controlling the low pressure in the base region and bringing it closer to air pressure is critical for achieving near-zero base drag.

This study is to improve the aerodynamic efficiency of flight vehicles by reducing the various forms of drag is of considerable interest to many researchers. In a practical situation, an increment in drag affects the flow separation. The flow separation can be controlled by two methods which are active and passive control. The base pressure is regulated in passive control by manipulating the geometric changes of the structures utilizing ribs, cavities, boattails, splitter plates, and locked vortex devices. The distinction between the active and passive control may be summarized as follows: dynamic control requires an external energy source, whereas passive control does not.

Converging–diverging (CD) nozzle is used to accelerate compressible flows to supersonic speeds and is commonly used in the propulsion systems of rockets and jet engines. The nozzle contains three sections: a converging section, a throat (with the smallest cross-sectional area), and a diverging part. Depending on the nozzle geometry and inlet–outlet pressure ratios, the flow at the nozzle exit is either subsonic or supersonic. This simulation analyzes the compressible flow through a 2D axisymmetric CD nozzle and the flow behavior at two different conditions.

Passive control techniques are used in various applications, including flow over blunt projectiles, missiles, jet engines, vernier rockets, internal combustion engine ports, and many more. Passive flow control approaches are potential methods for directly manipulating the boundary layer structure, vortices, and wake flows. The research will look at the aerodynamic efficacy of various passive and active flow control techniques in aerodynamic flows for drag reduction, separation, and wake control.

According to authors in [3], various passive control techniques, such as cavities, ribs, dimples, static cylinders, spikes, and others, focus on managing the base pressure and controlling the drag force. It is discovered that the passive control device might change the base pressure, resulting in higher and lower base drag. Furthermore, passive control is successful when a favorable pressure gradient is near the nozzle exit.

At transonic speeds, the efficacy of different base modifications or passive devices for lowering the base and total afterbody drag is examined. Several base modifications were explored, including cavities, ventilated cavities with varied ventilation geometries, and two vortex suppression devices, as well as the effects of these devices on boat-tailed and flared bases [4].

According to [5], using a passive control method for manipulating base drag includes locking vortex after-bodies, simple cavities multi-step after-bodies, and simple after-bodies using non-axis-symmetric boat-tailing in various speed regimes.

2 Modeling and Meshing

The geometry of the 2D converging–diverging nozzle and the enlarged duct is shown in Fig. 1. The enlarged duct with an annular cavity is attached to the converging–diverging nozzle. The annular cavity is provided in the dilated duct to control the base pressure. The geometries are created in ANSYS Workbench software for various combinations of geometric parameters. The dimensions of the converging–diverging nozzle for Mach number 1.4 are givens in Table 1. When the geometries are symmetric about the axis, the analysis is conducted by solving the 2-D axisymmetric problem, according to a study [2–16]. Various authors have studied the flow through the nozzle using computational fluid dynamics and found that the CFD analysis results agree with the experimental results [17–35].

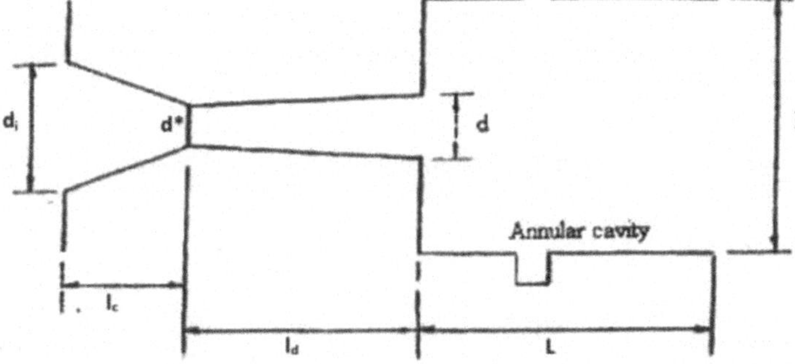

Fig. 1 Schematic of an experimental model for validation part

Table 1 The dimensions of the converging–diverging nozzle at Mach No. 1.4

Characteristics	Dimension
Mach number	1.4
Inlet diameter (di)	40 mm
Throat diameter (d*)	9.4706 mm
Outlet diameter (d)	10 mm
Convergent length (lc)	26.44 mm
Divergent length (ld)	15.18 mm
Area ratio	4
Duct diameter, D	20 mm
Duct length, L	It depends on the L/D ratio

Table 2 Cavity aspect ratio with width and height dimensions

Cavity aspect ratio, ASR	Width, w (mm)	Height, h (mm)
1	3	3
2	3	6
1	6	3
2	6	6

2.1 Cavity Dimensions

This present work is to study the effect of the cavity and its location on base pressure. Four cavity geometries have been used in this work, and Table 2 shows the cavity dimension and its location, respectively.

2.2 Cavity Locations

In addition, this study used 1D as the cavity position from the base wall. Equation 1 was used to determine cavity location. Further, Table 3 shows the cavity locations from the base wall for the 1D location.

$$\text{Cavity Location} = \text{D} - \frac{w}{2} \tag{1}$$

Table 3 The location of the cavity from the base wall of the duct at the 1D location

Parameter (location)	Width (mm)	Location (mm)
1D	3	18.5
1D	6	17

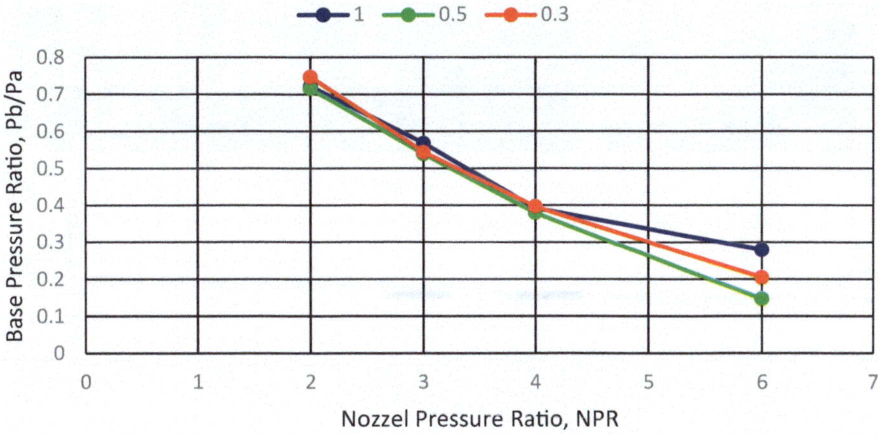

Fig. 2 The base pressure versus NPR for the mesh independence test

2.3 *Mesh Independence Test*

The mesh independence check is used to determine the appropriate element size. More repetitions and computing time are required for a higher NPR. By selecting the proper element size, higher accuracy may be achieved with less computation time. The same geometry is utilized for the mesh independence check. Simulated element sizes include three distinct sizes. The NPR for each element size varied between 2.00, 3.00, 4.00, and 6.00. Three element sizes, 1 mm, 0.5 mm, and 0.3 mm, are considered for the grid independence test. Figure 2 shows the base pressure Vs. nozzle pressure ratios that are plotted for various element sizes.

The appropriate element size is 0.3 mm, with about 5764 nodes and 5511 elements. This is because the base pressure ratio difference between the element sizes of 0.5 mm and 0.3 mm is less than 10%. The number of iterations will be greater than 10,000 if the element size is lowered to 1 mm, and the computational time will be longer. As a result, the most appropriate element size is 0.3 mm. The structures meshed model with an element size of 0.3 mm is shown in Fig. 3.

3 CFD Analysis

Ansys FLUENT was used for all simulations and analyses in this study. This software can comprehensively validate across a wide range of applications. The computational fluid dynamics analysis is carried out for all the combinations of the parameters. The density-based solver is considered for the analysis. The pressure ratio required for correct expansion at Mach number 1.4 is 3.18. The design condition for the model was set at this value. The values selected were lower than the design condition for the

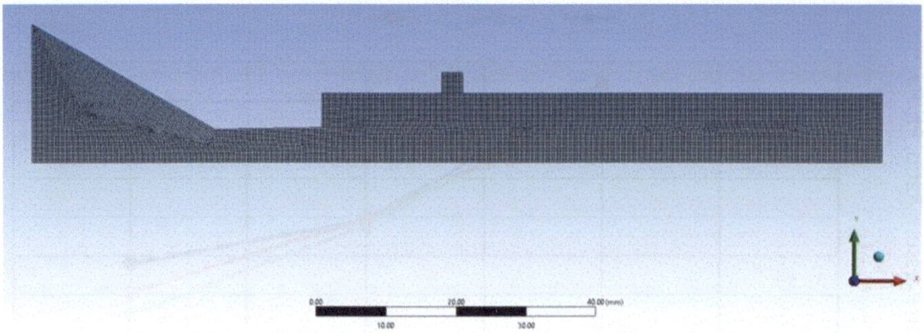

Fig. 3 The structured mesh with 0.3 element size

overexpanded cases 2 and 3. The values chosen for the under-expanded situations were higher than the design condition, such as 5, 6, and 7. Simulations were conducted for aspect ratios of the cavity as 3:3, 3:6, 6:3, and 6:6 for NPRs ranging from 2 to 7. Results are compared with and without cavities at a fixed cavity location equal to the diameter of the duct (1D).

4 Results and Discussion

Base pressure results for duct length (L) equal to two times the duct diameter (2D) are shown in Fig. 4. The NPR needed for correct expansion is 3.18. For any NPR less than 3.18, the flow exiting from the nozzle will be over-expanded. It is seen that there is a gradual decrease in the base pressure up to NPR = 4 later due to a reduction in the level of over-expansion. The decline in the base pressure has increased till NPR = 5. With a further increase in NPR, the drop in base pressure is significant. However, from NPR = 6 and above, this decline in the base pressure is arrested, and later at NPR = 7, there is a slight rise in the base pressure. However, for a cavity with geometry 6:6, the base pressure values are higher than for the rest of the cavity geometries.

When the duct length is three times the diameter (3D), the base pressure results are shown in Fig. 5. In Fig. 4, when the duct length was 2D, the effectiveness of the cavity was not seen till NPR 4, whereas in this case, the variation in base pressure values is observed from NPR 3 onwards. For the plain duct, the declining trend continues till NPR 5, and there is an increase in the base pressure; later, at NPR 7 again, there is a decline in the base pressure. This change in the base pressure is attributed to the shock structure and impact of the ambient pressure. Various cavity geometries behave differently. For instance for a cavity with dimension 3:3, there is a continuous decrease in the base pressure till NPR 6, and later there is a marginal change in the base pressure. When the cavity has dimension 3: 6, this downward trend is arrested at NPR = 5, and the base pressure increases till NPR 6, and at NPR 7, there is a

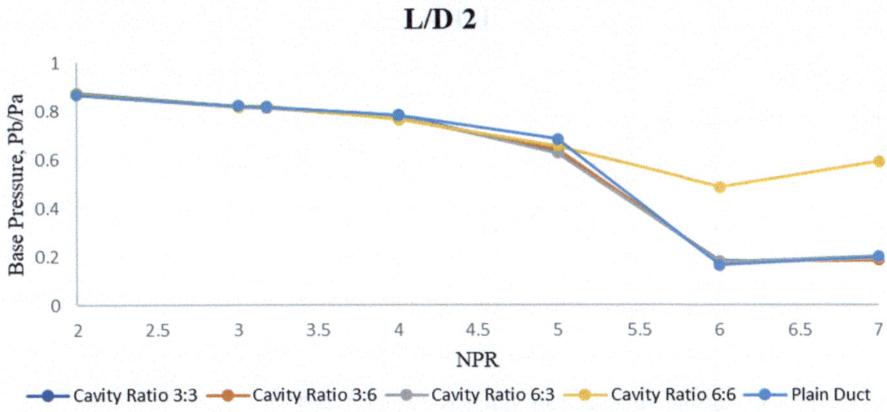

Fig. 4 The base pressure variation with NPR at constant L/D = 2

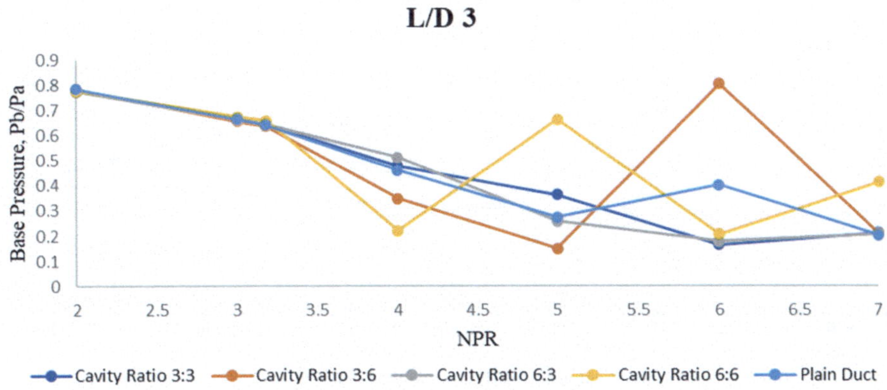

Fig. 5 The base pressure variation with NPR at constant L/D = 3

decline in the base pressure. For a cavity with a 6:3 dimension, the declining trend of the base pressure continues till NPR 6, and later there is a marginal change in the base pressure. Finally, for Cavity size 6:6, there is a mixed trend. As usual, the base pressure decreases until NPR = 4, then increases at NPR 5, but delining and increasing trend revisits at NPR 6 and 7. Hence, we can say that it is a case sensitive to duct length, NPR, Mach number, cavity dimension, and cavity location.

Figure 6 shows the base pressure variations when the duct length is four times the diameter (4D). The increased duct length influences the base pressure results. Base pressure values for plane duct and cavity 3:3 show a similar trend, except when a cavity is employed, which decreases base pressure till NPR 4 and later up to NPR 5, resulting in a rise and again declines at NPR 6 and 7. For cavity 3:6, with the usual decline, it increases at NPR 5 and decreases at NPR 6 and 7. Cavities with

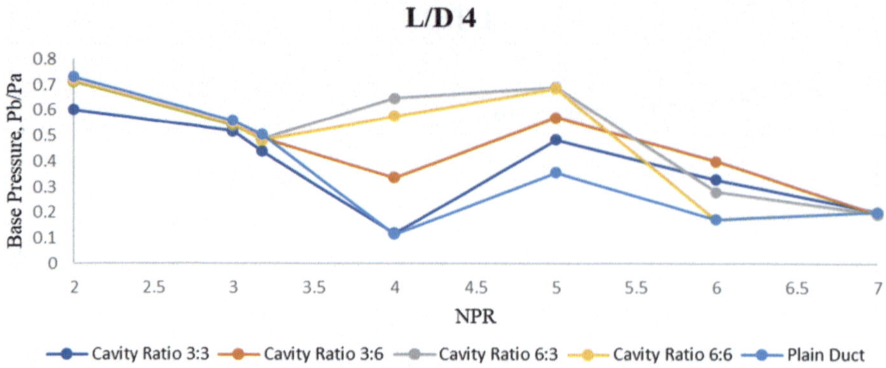

Fig. 6 The base pressure variation with NPR at constant L/D = 4

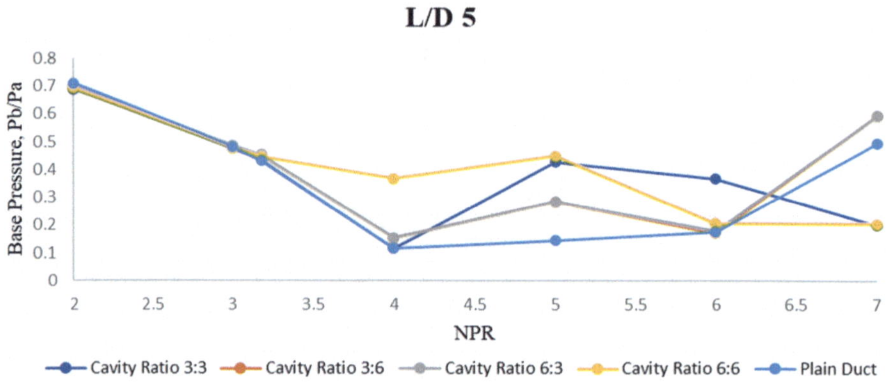

Fig. 7 The base pressure variation with NPR at constant L/D = 5

dimensions 6:3 and 6:6 show similar trends where base pressure is on the lower side for the 6:3 cavity, and the other geometry remains at higher values compared to 6:3.

For duct length, L is five times the duct diameter (5D), and the base pressure results are displayed in Fig. 7. Due to the increased duct length, fluctuations in the base pressure at various NPRs are suppressed. The cavity having dimension 6:6 attains higher base pressure till NPR 5; any increase in NPR results in a decrease in base pressure. For plain duct, the declining trend continues till NPR 4. Later, there is a gradual increase in the values at NPR 7 and a considerable increase in the base pressure. For cavity 6:3, the maximum growth is attained at NPR 5; For other values of NPR, it continues to achieve low base pressure values.

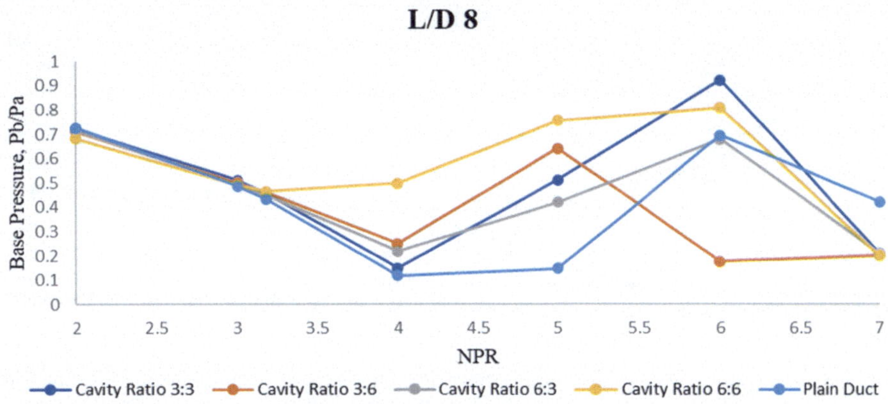

Fig. 8 The base pressure variation with NPR at constant L/D = 8

Figure 8 exhibits that the finding at duct length (L) is eight times the duct diameter (8D). The cavity with 6:6 performed better till NPR 6. The following geometry of the cavity is 3:6 and is also up to NPR 6. Then cavity with 3:3 is effective in regulating the base pressure, and the last is 6:3 in control effectiveness.

At a fixed NPR = 2, the base pressure results with duct length are shown in Fig. 9. This indicates that the base pressure becomes independent of duct length from L = 4D. In most cases, the control decreases the base pressure as the jets are over-expanded. The marginal increase in the NPR as 3 shows a decline in the over-expansion level, as shown in Fig. 10. Control effectiveness is seen at L/D 2 to 6, and at the Duct length L/D 6, the flow parameters attain a steady state. The results of base pressure variations at NPR = 3.18 show a similar trend as at NPR 3, as shown in Fig. 11. Results at NPR = 4. Figure 12 shows the variations of base pressure at NPR 4. Based on the results, it can be seen that due to the rise in NPR, the requirement of minimum duct length to reattach the flow is reduced to L/D 3. The fluctuations in the base pressure exist for various aspect ratios of the cavity. The cavity with dimensions 6:3, 6:6, and 3:6 have become effective.

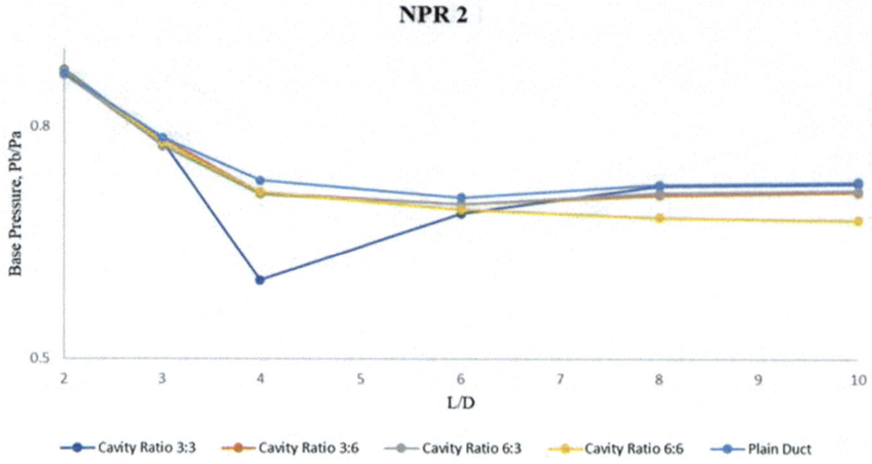

Fig. 9 The base pressure variation with L/D at constant NPR = 2

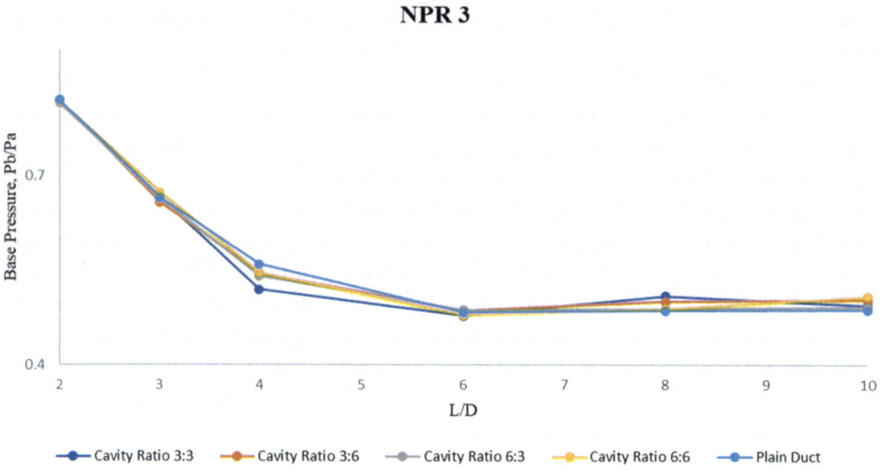

Fig. 10 The base pressure variation with L/D at constant NPR = 3

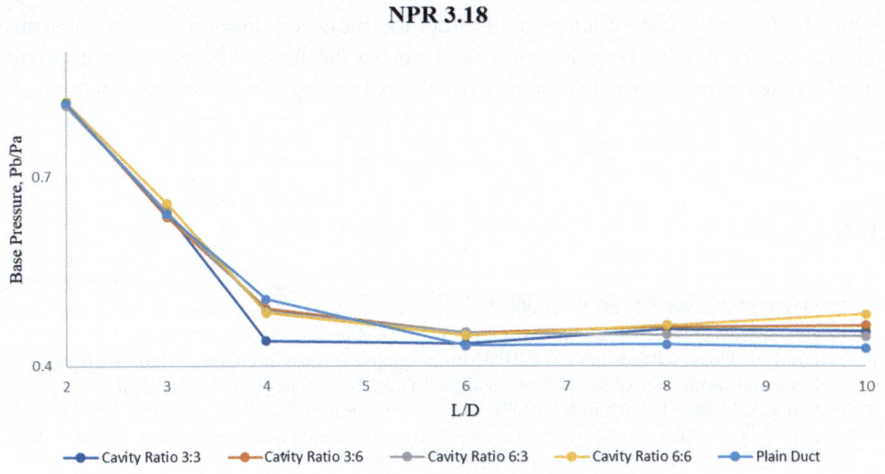

Fig. 11 The base pressure variation with L/D at constant NPR = 3.18

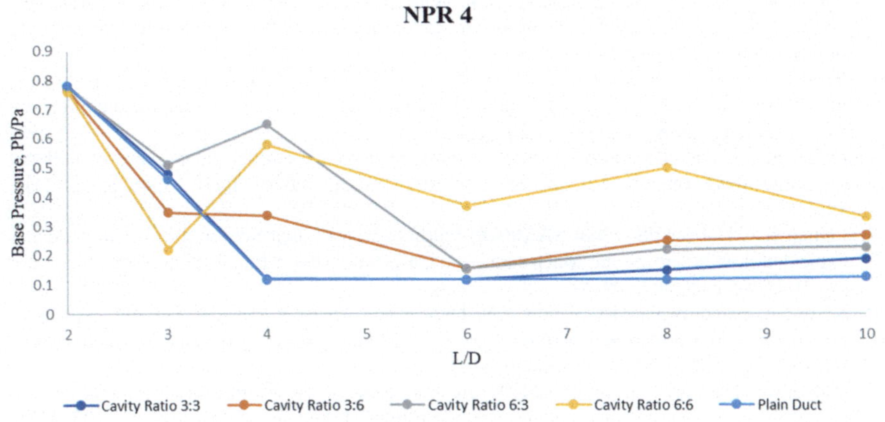

Fig. 12 The base pressure variation with L/D at constant NPR = 4

5 Conclusion

In conclusion, the presence of cavities impacts base pressure and is one of the effective ways to reduce net drag. The method that will be applied for future research is effectively illustrated. With a constant cavity location, a cavity in the duct affected the base pressure at Mach 1.4 and an area ratio of 4. The base pressure is only slightly impacted by raising the cavity aspect ratio. The cavity's height has been increased in this investigation. It is clear from the data that any increase in cavity height is negligible and has no appreciable impact on base pressure. But if the cavity is widened, the base pressure will undoubtedly vary since the altered cavity width can be inside the reattachment length and affect the base flow field. The base pressure may be higher if the cavity location is closer to the base. The pressure gradient based on the results shows that the passive control has no adverse effects on the wall pressure.

References

1. Pandey KM, Rathakrishnan E (2006) Annular cavities for base flow control. Int J Turbo Jet Engines 23(2):113–127. https://doi.org/10.1515/TJJ.2006.23.2.113
2. Pathan KA, Dabeer PS, Khan SA (2019) Investigation of base pressure variations in internal and external suddenly expanded flows using CFD analysis. CFD Lett 11(4):32–40
3. Pathan KA, Dabeer PS, Khan SA (2020) An investigation of the effect of control jets location and blowing pressure ratio to control base pressure in suddenly expanded flows. J Therm Eng 6(2):15–23. https://doi.org/10.18186/thermal.726106
4. Pathan KA, Dabeer PS, Khan SA (2017) CFD analysis of the effect of area ratio on suddenly expanded flows. In 2nd International conference for convergence in technology (I2CT), pp 1192–1198. https://doi.org/10.1109/I2CT.2017.8226315
5. Pathan KA, Dabeer PS, Khan SA (2017) CFD analysis of effect of flow and geometry parameters on thrust force created by flow from nozzle. In 2nd International conference for convergence in technology (I2CT), pp 1–5
6. Pathan KA, Dabeer PS, Khan SA (2017) CFD Analysis of effect of mach number, area ratio, and nozzle pressure ratio on velocity for suddenly expanded flows. In 2nd International conference for convergence in technology (I2CT), pp 1–7
7. Pathan KA, Dabeer PS, Khan SA (2018) An investigation to control base pressure in suddenly expanded flows. Int Rev Aerosp Eng 11(4):162–169. https://doi.org/10.15866/irease.v11i4.14675
8. Pathan KA, Dabeer PS, Khan SA (2018) Optimization of area ratio and thrust in suddenly expanded flow at supersonic Mach numbers. Case Stud Therm Eng 12:696–700. https://doi.org/10.1016/j.csite.2018.09.006
9. Khan SA, Fatepurwala MA, Pathan KN, Dabeer PS, Ali Baig MA (2018) CFD analysis of human powered submarine to minimize drag. Int J Mech Prod Eng Res Dev 8(3):1057–1066. https://doi.org/10.24247/ijmperdjun2018111
10. Pathan KA, Dabeer PS, Khan SA (2019) Effect of nozzle pressure ratio and control jets location to control base pressure in suddenly expanded flows. J Appl Fluid Mech 12(4):1127–1135. https://doi.org/10.29252/jafm.13.02.30049
11. Pathan KA, Dabeer PS, Khan SA (2020) Enlarge duct length optimization for suddenly expanded flows. Adv Aircr Spacecr Sci 7(3):203–214. https://doi.org/10.12989/aas.2020.7.3.203
12. Aabid A, Murtuza MA, Shushrutha A (2015) Experimental study of mechanical properties of tellurium nanowires reinforced epoxy composites. Am J Mater Sci 5(April):194–197. https://doi.org/10.5923/c.materials.201502.37
13. Khan SA, Aabid A, Ali Baig MA (2018) CFD analysis of CD nozzle and effect of nozzle pressure ratio on pressure and velocity for suddenly expanded flows. Int J Mech Prod Eng Res Dev 8(3):1147–1158
14. Khan A, Aabid A, Khan SA CFD analysis of convergent-divergent nozzle flow and base pressure control using micro-JETS. Int J Eng Technol 7(3):232–235
15. Khan SA, Aabid A (2018) CFD analysis of CD nozzle and effect of nozzle pressure ratio on pressure and velocity for suddenly expanded flows. Int J Mech Prod Eng Res Dev 8(June):1147–1158
16. Khan SA, Mohamed O, Aabid A (2021) CFD analysis of compressible flows in a convergent-divergent nozzle. Mater Today Proc xxxx. https://doi.org/10.1016/j.matpr.2021.03.074

17. Shen Z, Yu S, Zheng S, Nofal TA, Musa A, Li Z (2022) Numerical study of multi-jet with upstream divergent ramp at supersonic cross flow. Aerosp Sci Technol 127:107689. https://doi.org/10.1016/j.ast.2022.107689
18. Bai C, Wang M, Wu Z (2022) Symmetric Mach reflection configuration with asymmetric unsteady solution symmetric mach reflection configuration with asymmetric unsteady solution. Chinese J Aeronaut. https://doi.org/10.1016/j.cja.2022.06.012
19. You A, Be MAY, In I (2022) Control of supersonic compression corner flow using a plasma actuator. Phys Fluids
20. Samimy M, Kim JH, Kastner J, Adamovich I, Utkin Y (2007) Active control of high-speed and high-Reynolds-number jets using plasma actuators. J Fluid Mech 578:305–330. https://doi.org/10.1017/S0022112007004867
21. Gaitonde DV, Samimy M (2011) coherent structures in plasma-actuator controlled supersonic jets: axisymmetric and mixed azimuthal modes. Phys Fluids 23(9). https://doi.org/10.1063/1.3627215
22. Pathan KA, Khan SA, Shaikh AN, Pathan AA, Khan SA (2021) An investigation of boat tail helmet to reduce drag. Adv Aircr Spacecr 8(3):239–250. https://doi.org/10.12989/aas.2021.8.3.239
23. Li H, Choi J, Li B, Kim I, Heo J (2016) Numerical analysis on the gas flow dynamics from a rectangular slot-nozzle for pulse cleaning of filter unit. Powder Technol 297:330–339. https://doi.org/10.1016/j.powtec.2016.04.040
24. Shaikh JS, Pathan KA, Kumar K, Khan SA (2023) Effectiveness of Cone Angle on Surface Pressure Distribution along Slant Length of a Cone at Hypersonic Mach Numbers. J of Adv Res in F Mech and Ther Scie 104(1):185–203. https://doi.org/10.37934/arfmts.104.1.185203
25. Wang PC, McGuirk JJ (2013) Large Eddy Simulation of supersonic jet plumes from rectangular con-di nozzles. Int J Heat Fluid Flow 43:62–73. https://doi.org/10.1016/j.ijheatfluidflow.2013.06.002
26. Mousavi SM, Roohi E (2014) Three-dimensional investigation of the shock train structure in a convergent-divergent nozzle. Acta Astronaut 105(1):117–127. https://doi.org/10.1016/j.actaastro.2014.09.002
27. Najar NA, Dandotiya D, Najar FA (2013) Comparative analysis of K-epsilon and Spalart-Allmaras turbulence models for compressible flow through a convergent-divergent nozzle. Int J Eng Sci 2(8):8–17
28. Pathan KA, Ashfaq S, Dabeer PS, Khan SA (2019) Analysis of parameters affecting thrust and base pressure in suddenly expanded flow from nozzle. J Adv Res Fluid Mech Therm Sci 64(1):1–18
29. Shaikh JS, Kumar K, Pathan KA, Khan SA (2022) Analytical and computational analysis of pressure at the nose of a 2D wedge in high-speed flow. Adv Airc Spacecr Sci 9(2):119–130. https://doi.org/10.12989/aas.2022.9.2.119
30. Salvador FJ, Jaramillo D, Romero J-V, Roselló M-D (2017) Using a homogeneous equilibrium model for the study of the inner nozzle flow and cavitation pattern in convergent-divergent nozzles of diesel injectors. J Comput Appl Math 309:630–641. https://doi.org/10.1016/j.cam.2016.04.010
31. Shamitha AC, Pathan KA, Khan SA (2023) Numerical simulation of surface pressure of a wedge at supersonic Mach numbers and application of design of experiments. J Adv Res Appl Mech 101(1):1–18. https://doi.org/10.37934/aram.101.1.118
32. Shaikh JS, Kumar K, Pathan KA, Khan SA (2023) Computational analysis of surface pressure distribution over a 2d wedge in the supersonic and hypersonic flow regimes. Fluid Dyn Mater Process 19(6):1637–1653. https://doi.org/10.32604/fdmp.2023.025113
33. Shaikh SK, Pathan KA, Chaudhary ZI, Khan SA (2020) CFD analysis of an automobile catalytic converter to obtain flow uniformity and to minimize pressure drop across the monolith. CFD Lett 12(9):116–128
34. Shaikh SK, Pathan KA, Chaudhary ZI, Marlpalle BG, Khan SA (2020) An investigation of three-way catalytic converter for various inlet cone angles using CFD. CFD Lett 12(9):76–90

35. Pathan KA, Dabeer PS, Khan SA (2019) Influence of expansion level on base pressure and reattachment length. CFD Lett 11(5):22–36

Numerical Analysis on the Effect of Constriction on the Mixing of Fluids in Serpentine Microchannels

Kamran Rasheed, Sameen Mustafa, Mubashshir Ahmad Ansari, and Shahnawaz Alam

Abstract Fluid mixing at micro level is a key function in microfluidic systems for the homogenization of fluid samples. Extensive work has been done by many researchers in the designing of micromixers for achieving efficient mixing. The impact of constrictions on fluid flow and mixing caused by the rectangular bend microchannel is numerically analyzed in this work. The micromixer has two aligned inlet channels and a main mixing channel which has constrictions in its rectangular bend section. Numerical analysis of mixing has been carried out for three different Reynolds number viz. 10, 30 and 60 and at two constriction values ($s = 0\ \mu m$ and $s = 50\ \mu m$). The findings suggest that the blending efficiency of the micromixer having constriction is much better than the mixing performance exhibited by the channel having no constriction ($s = 0\ \mu m$).

Keywords Micromixer · Numerical analysis · Mixing index · Serpentine microchannel · Constrictions

K. Rasheed (✉) · S. Alam
Department of Mechanical Engineering, Integral University, Lucknow, India
e-mail: kamran@iul.ac.in

S. Alam
e-mail: alamshah@iul.ac.in

S. Mustafa
Faculty of Science and Technology, Libera Università di Bolzano, Bolzano, Italy

M. A. Ansari
Department of Mechanical Engineering, Faculty of Engineering and Technology ZHCET, Aligarh Muslim University, Aligarh, India
e-mail: mubashshir@zhcet.ac.in

M. A. Siddiqui et al. (eds.), *Advances in Heat Transfer and Fluid Dynamics*, Lecture Notes in Mechanical Engineering, https://doi.org/10.1007/978-981-99-7213-5_5

1 Introduction

Microfluidics is the complex confluence of science and technology which offers the handling and investigation of fluid flow behaviour in a microchannel having length less than a millimetre [1, 2]. Microfluidics system provides better opportunity to deal with very low volume of fluid sample and for short reaction time [3]. The very first microfluidic device was developed with photolithography on glass or silicon wafers which was further used for electroosmotic pumping and electrophoretic separation [4]. Microfluidic devices are used to perform various chemical and biological analyses in microscale chips [5, 6]. Microfluidic systems have made their strong presence in the biomedical industry, biochemical analysis, and pharma industry in last few decades [7].

One of the basic operations of these microfluidics devices is the intermixing of fluid samples which carried out the required biochemical reaction for high yield results. Micromixers are the essential part of the microfluidics system, and performance of these micromixers depends upon the design of the micromixers [8, 9]. At microscale level, achieving fast mixing is a tough task due to low Reynold's number flow where mixing occurs because of molecular diffusion, which is very sluggish and the less effective method. Expeditious blending of the fluid sample in a micromixer is key factor for several process executions [5, 10].

Micromixers are classified as active micromixers and passive micromixers. An active micromixer can be operated by using an external source of energy for better mixing performance. Pressure field, electric field, or magnetic fields are all possible external sources of power. Although they are very efficient in blending of fluids but their manufacturing and assembly with the microfluidic system is difficult [11–13]. In comparison to active micromixers, passive micromixers are simple, inexpensive and no external source of energy is required for their operation. They can be easily incorporate with microfluidic systems. Effective mixing of fluids occurs because of exceptional geometry of the microchannel which causes unique flow field. There are two additional categories for passive micromixers that are defined according to their mixing methodology: laminar micromixers and convective micromixers [14]. Lie et al. carried out a computational and experimental study for fluid merging in a three dimensional serpentine microchannel having geometry of square wave and straight microchannel. From experimental results they observed that a 3D serpentine microchannel have better mixing performance than the straight mixing channel and square wave mixing channel for the scale of 6–70 Reynolds number. They also observed that the ability of mixing of 3D serpentine mixing channel is 1.6 times higher in comparison to square wave mixing channel and ten times higher relative to the straight channel [15]. Tsai et al. investigated a multidirectional vortex-based planar micromixer with radial baffles in a curved channel. The combination of radial baffles and microchannel's curvature created vortices in multiple directions and converging–diverging flow and they both work together to improve mixing. Their study reveals that mixing performance of the basic C-shaped micromixer can be enhanced by attaching the first baffle to the internal cylinder and the second baffle to the external cylinder [16]. Hossain et al. conducted a computational cal analysis on fluid flow and their mixing in microchannel with various configurations like curved, square wave and zigzag. They analysed the mixing of water and ethanol for a wide range of Reynolds number that vary from 0.267 to 267. They claimed that the microchannel having square wave geometry exhibit high mixing performance among the above. They also studied the effect of pressure drop for each geometry and found that the trend of the pressure decrease in each channel is almost identical at Re = 50. The pressure reductions in the zigzag and square-wave microchannels are almost equal for Reynolds numbers greater than 50, however the curved shape mixing channel exhibits a significantly lesser drop in pressure in comparison to the above two mixing channels since it does not have an abrupt bend like the other two channels [17]. Ansari et al. proposed a modified form of simple planar serpentine square wave bend microchannel to upgrade the mixing performance. Square wave serpentine channel carries two layers of equal heights. For conducting the numerical simulations at different Reynolds number range from 1–80, three different offset positions i.e. (s/w = ¼, ½, ¾) were created by the lateral movement of the upper layer keeping the bottom layer fixed. The results show that a bilayer channel is more efficient than a simple channel [18]. Kuo et al. reported a very unique attempt to enhance the abilities of mixing for three microfluidic mixers with zigzag, square-wave and curved mixing channels by placing them on a rotating Compact Disc platform. As the disk rotates three separate forces viz. centrifugal force because of system revolutions, Coriolis force and centrifugal force developed in the microchannel due to its curvature within the microchannel. By performing the number of computational and experimental research they found that the microchannel having square-wave geometry delivers the best mixing performance in comparison to aforementioned devices because of the robust chaotic stirring and extended holding time [19]. Yang et al. developed a dean vortices based three dimensional micromixer having two layers of spiral channels overlapped together in the vertical direction, by using a smooth channel twisted into double-layer spiral geometry. Their numerical analysis shows that the pair of spiral channels overlapped in transverse direction were constituted with three semicircles, demonstrated a high level of mixing efficiency. Additionally, they discover that by raising the vertical channel's height and changing its form from cuboid to cylinder, the high mixing efficiency of the vertical channel can be further enhanced [20]. Chen et al. conducted an in-depth experimental and computational simulation for $0.1 \leq Re \leq 100$, to explore and analyze the performance of mixing for serpentine micromixers. Six designs of micromixers underwent mixing studies using blue and yellow ink, and three improved configurations (zigzag, multi-wave, and square-wave) were chosen for study. The micromixer with square wave geometry was found highly effective in fluid sample mixing than both the multi-wave and zigzag micromixers due to its sharper turns and longer flow path but it also reflects higher pressure drop as compared to both micromixers due to sharp turns. Mixing efficiency rises linearly from the first segment to the last portion at lower Reynolds number and further mixing efficiency rises with increase in Re [21]. Another experiment on serpentine micromixers with five different geometries (zig-zag, loop, mouth, square-wave, and multi-wave patterns) was carried out by Chen et al. as part of a numerical analysis for assessing the ability of blending of serpentine micromixers. The objective of the research was to find the influence of the microchannel's shape on the mixing index. From the simulation results they concluded that the micromixer with square-wave microchannels can achieve a higher mixing index than others. Additional numerical study on square wave units demonstrates that modifying the proportion of microchannels to square-wave units is an efficient and easy way to improve mixing performance [22]. Gidde et al. conducted a numerical simulation study to examine the impact of design parameters of geometry (i.e., constriction channel's width) on the performance of mixing for passive micromixers that included circular and square mixing enclosures using microfluidic recirculation to improve the mixing ability for Reynolds values between 0.1 and 75. Mixing and flow field studies were performed in two phases for Reynolds values lying in between 0.1 and 75. In the first phase two distinct arrangements of the micromixer consisting eight chambers for mixing (4- square shaped and 4- circular shaped) are analyzed. In the second phase of analysis, comparison of mixing performances of the suggested micromixers was carried out for Reynolds numbers ranging from 0.1 – 75 on the basis of the mixing chamber design. At Re = 0.1, 1, and 5, both micromixers had almost similar mixing efficiency, additionally for Re > 5, where the square chamber micromixer demonstrated improved mixing performance. In comparison to micromixers with circular chambers, their investigation shows that square chamber micromixers function better in terms of performance index and shows less pressure drop [23]. In addition to a parametric analysis on geometrical factors, Solehati et al. evaluated the mixing performance of the standard T-junction and compared it with that of wavy channels and complex channel designs with the help of Taguchi statistical approach for analyzing the effect of all design variables (figure of merit, mixing efficiency and pumping potential). It was found that Taguchi technique is reliable for determining the ideal arrangement of design elements with a highest relative deviation of less than 7%. They concluded that high wavy amplitude, high wavy frequency, narrow width, and small cross-sectional channel are the ideal design criteria for high value chemical products on the other hand for low cost and for bulk manufacturing, pumping power based product optimization is recommended. For intermediate values, the shape of the micromixer can also be optimize on the basis of both mixing performance and pumping power [24].

2 Numerical Analysis

Numerical analysis conducted to examine the effect of constriction on fluid blending and flow of fluid in a three-dimensional serpentine channel at different Reynolds number. ANSYS CFX was used for simulation, and to simplify the Navier–Stokes equations along with a coupled solver by employing the finite volume method. Mass conservation equation along with continuity and momentum equations are used for flow analysis. The following mathematical equations are used for analysis [25]:

$$\nabla.\mathbf{u} = 0 \tag{1}$$

$$\rho(\mathbf{u}.\nabla)\mathbf{u} = -\nabla P + \mu\nabla^2\mathrm{u} \tag{2}$$

Advection–diffusion equation is used to explain the mass transport of the mixed fluid, which is written below:

$$\mathbf{u}.(\nabla c) = D\nabla^2 c \tag{3}$$

where c represents the concentration of species in $\mathrm{mol/m^3}$ and D represents the diffusivity in $\mathrm{m^2/s}$.

To get reliable results for mixing analyses, a high quality mesh is essential. A hexahedral grid for the complete model was created. For multicomponent fluid flow analysis few assumption are made i.e. fluids are blend at molecular scale, having same pressure, velocity etc., mass movements occurs through convection and diffusion [18].

Zero static pressure at outlet, normal velocities at two inlets and no-slip condition at wall has been chosen as the limiting conditions for the model. The wall has been assigned no-slip boundary condition. For the calculation of the micromixer's level of mixing, variance of mass fraction is used on the perpendicular plane to the fluid flow direction and is defined as [26]:

$$\sigma = \sqrt{1/\mathrm{N}} \sum (\mathrm{c_i} - \mathrm{c_m})^2 \tag{4}$$

where N refers to the number of sampling points on the cross sectional plane, c_i refers to the mass fraction at the point of sampling i.e. c_i and c_m is the optimal mixing mass fraction. Interpolations are used to acquire the values at the sample locations from values at neighboring computing points. The mixing index at axial position is used to assess the level of mixing of sample fluids on an orthogonal plane in the direction of fluid flow and expresses as [25]:

$$M = 1 - \sqrt{\sigma/\sigma^2_{\mathrm{max}}} \tag{5}$$

where σ_{max} is the maximum variance over the data range. The variance is highest and lowest, respectively, for a fully unmixed and fully mixed fluid.

The working fluids for the mixing process are chosen as water and ethanol. The density of water is 998 $\mathrm{kg/m^3}$ and density of ethanol is 789 $\mathrm{kg/m^3}$ are used as fluid characteristics at 20 °C. The investigation was done for Reynolds numbers lying in between 10 and 60 and formula for the calculation of Reynolds number is given:

$$Re = \rho\, u \mathrm{D_h} / \mu \tag{6}$$

where, ρ is the density of fluid, u is the velocity of the fluid in the channel, D_h is the hydraulic diameter of the channel and μ is the average absolute viscosity of the fluids.

3 Micromixer Design

The main purpose of the present study is to propose an effective micromixer design for quick and effective mixing by increasing the local advection in the fluid flow. The proposed micromixer has five bend rectangular mixing chambers in a serpentine shape with constriction in the form of converging channels as shown in Fig. 1. The inner walls of the rectangular chambers are tapered along the flow direction, and the micromixer's constriction channel's width varies from the inner wall to outer wall. The inlet length, the mixing length and the width of the micromixer is kept constant. Length (l), height (h) and width (w) of the inlet channel are 1500 μm, 200 μm and 200 μm respectively, shown in Fig. 1. The dimensions of the main mixing channel are as follows: (Channel Height) $H = 2 \times 10^2$ μm, (Channel Width) $W = 2 \times 10^2$ μm and (Axial Channel Length) $L = 5.6 \times 10^3$ μm. Two different constriction values (s = 0 μm & s = 50 μm) are selected for numerical investigation. For the mathematical analysis width/depth ratio of mixing channel is taken as 1, which is known as the aspect ratio of the channel.

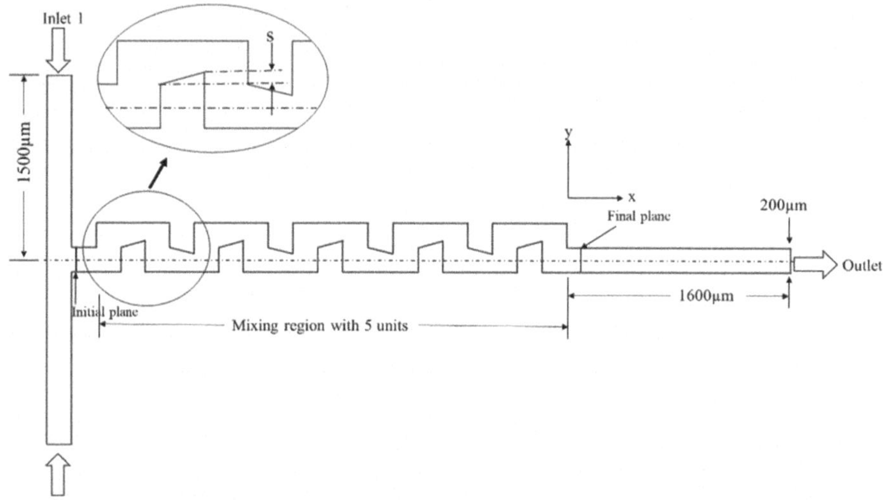

Fig. 1 Design concept for proposed micromixer

4 Results and Discussion

The mixing performance of a three-dimensional rectangular serpentine micromixer having tapered constrictions in bend has investigated. Steady continuity, Navier–Stokes, advection–diffusion equations are used in this numerical analysis. For several grid systems, the distribution of the mixing index has been assessed on planes across the flow direction along the x-axis. An isometric view of a section of the serpentine micromixer is shown in Fig. 2a, and b shows the meshing at constriction values (s) = 50 μm.

The mixing capabilities of the suggested micromixer is explained in detail for various Reynolds numbers (10, 30 & 60) and for different values of constriction (s = 0 μm and 50 μm). Figure 3a and b represent the comparison of streamlines distribution in the microchannel at s = 0 μm and s = 50 μm (aspect ratio = 1, Re = 10), Fig. 4a and b represent the comparison of streamlines distribution in microchannel at s = 0 μm and s = 50 μm (aspect ratio = 1, Re = 30) and similarly Fig. 5a and b represent the comparison of streamlines distribution in the microchannel at s = 0 μm and s = 50 μm (aspect ratio = 1, Re = 60).The fluid flow lines are originated from two separate inlets and merged in the main mixing channel and after collision they proceed along the path on their own side. The projected streamlines represent the development of flow lines in the main microchannel at s = 0 μm and at s = 50 μm.

Figures 6 and 7 illustrates the numerical outcomes for the formation of the ethanol mass-fraction distribution on the YZ-plane along the micromixer's axial length for s = 0 m and s = 50 m for Re = 10, 30, and 60, respectively.

The mixing ability of the micromixer is evaluated for different Reynolds numbers by conducting numerical simulations. Figure 8a, b and c show the comparison of mixing index for Reynolds numbers 10, 30 & 60 respectively at s = 0 μm and 50 μm. The graphs are plotted in between the mixing index and number of plane along the length of the mixing channel. The above three graphs show very interesting results, where the mixing index decreases as the Reynolds number increases. In Fig. 8c it is clearly visible that at Re = 60, the value of the mixing index is minimum in

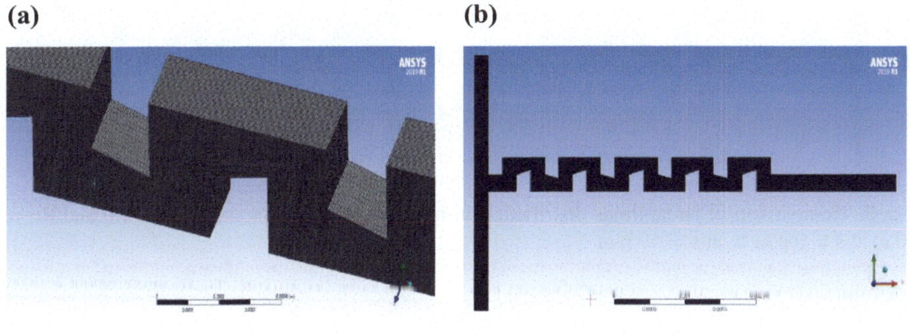

Fig. 2 **a** Isometric view of a section of the micromixer for s = 50 μm, **b** Grid system for s = 50 μm

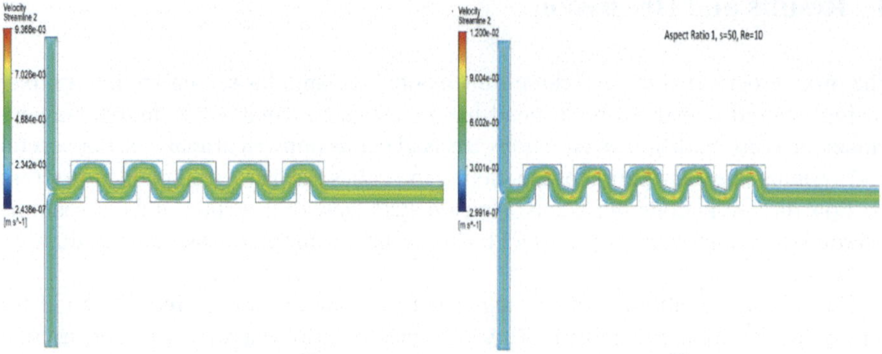

(a) (b)

Fig. 3 Comparison of streamlines distribution in microchannel for Aspect ratio = 1.0 and Re = 10 **a**: at s = 0 μm **b**: at s = 50 μm

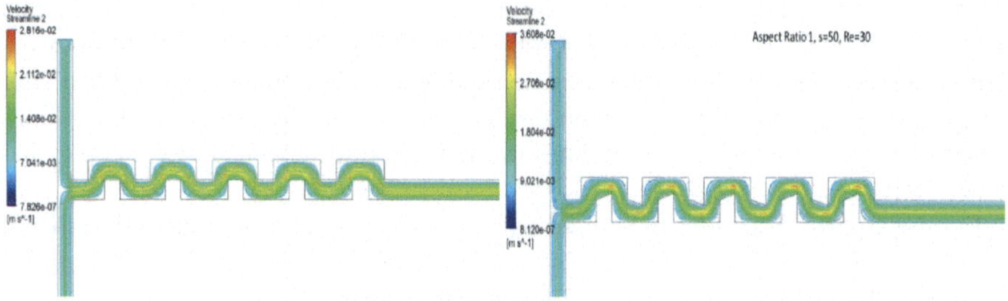

(a) (b)

Fig. 4 Comparison of streamlines distribution in microchannel for Aspect ratio = 1.0 and Re = 30 **a**: at s = 0 μm **b**: at s = 50 μm

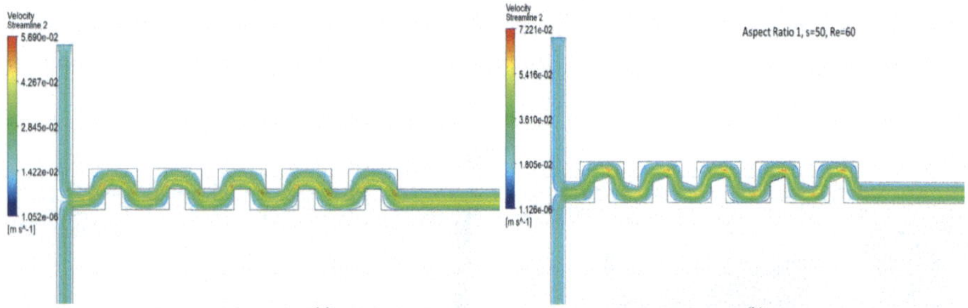

(a) (b)

Fig. 5 Comparison of streamlines distribution in microchannel for Aspect ratio = 1.0 and Re = 60 **a**: at s = 0 μm **b**: at s = 50 μm

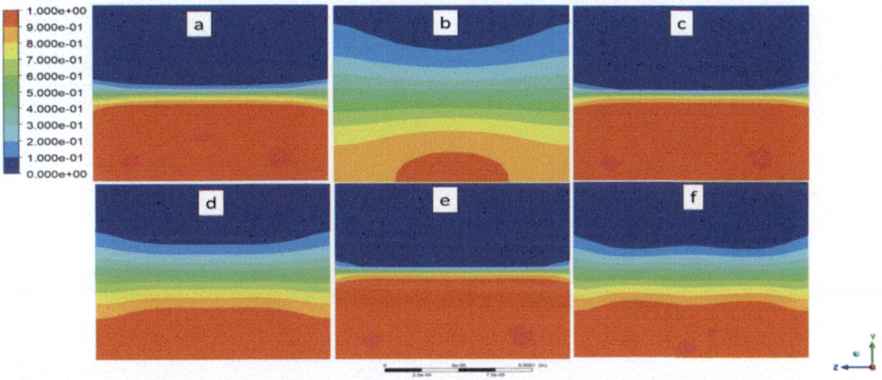

Fig. 6 Mass fraction contour plots for **a** initial plane, s = 0 μm, Re = 10, **b** final plane, s = 0 μm, Re = 10, **c** initial plane, s = 0 μm, Re = 30, **d** final plane, s = 0 μm, Re = 30, **e** initial plane, s = 0 μm, Re = 60, **f** final plane, s = 0 μm, Re = 60

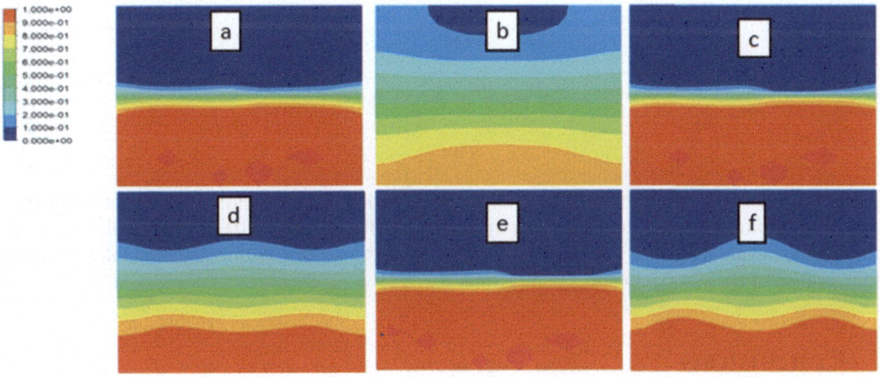

Fig. 7 Mass fraction contour plots for **a** initial plane, s = 50 μm, Re = 10, **b** final plane, s = 50 μm, Re = 10, **c** initial plane, s = 50 μm, Re = 30, **d** final plane, s = 50 μm, Re = 30, **e** initial plane, s = 50 μm, Re = 60, **f** final plane, s = 50 μm, Re = 60

comparison to the value of the mixing index at Re = 10 & 30 with constriction and without constriction. Low mixing index results from a decrease in residence time in the mixing channel, which is due to an increase in fluid velocity as the Reynolds number increases. The main purpose of this study is to select the suitable design of the micromixer for improved mixing performance. However at low, moderate and higher Reynolds number (10, 30 and 60) the mixing channel having constrictions (s = 50 μm) shows better mixing performance in comparison to non-constricted channel (s = 0 μm).

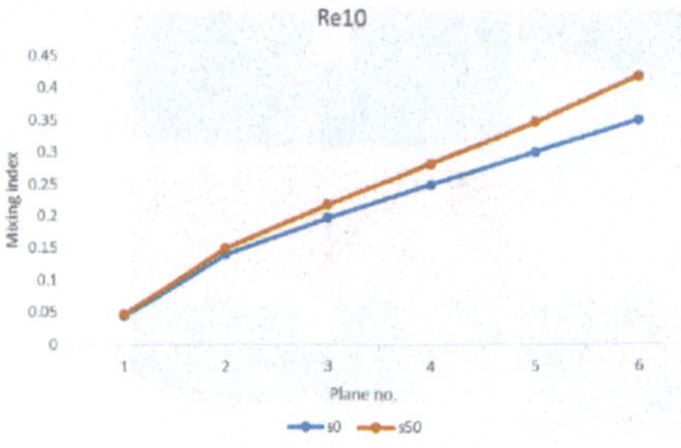

(a)

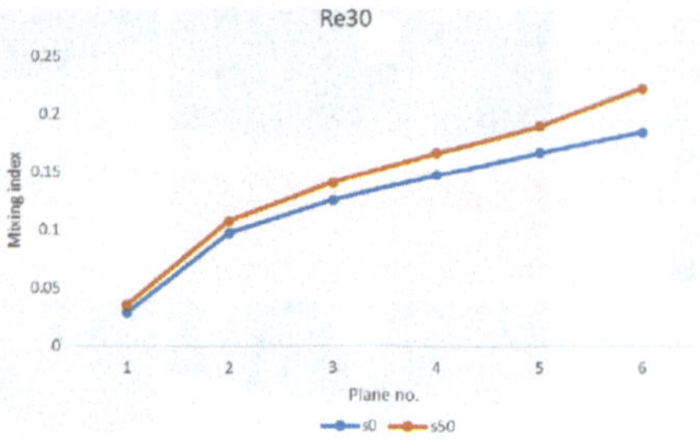

(b)

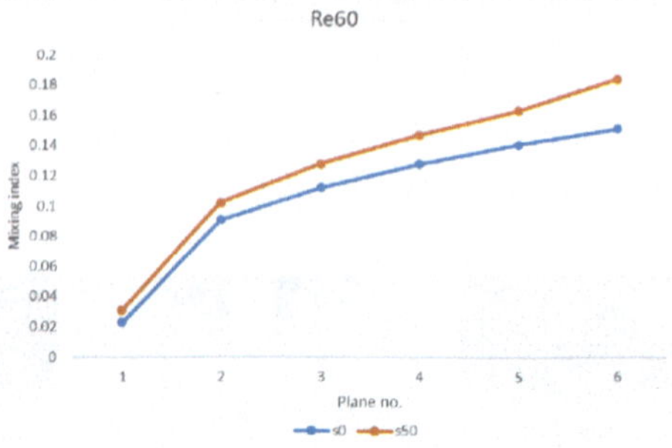

(c)

Fig. 8 Plots of mixing index vs Plane no. (Along the length of the micromixer). **a** Numerical simulation: mixing index evaluated for s = 0 μm and s = 50 μm at Re = 10. **b** Numerical simulation: mixing index evaluated for s = 0 μm and s = 50 μm at Re = 30. **c** Numerical simulation: mixing index evaluated for s = 0 μm and s = 50 μm at Re = 60

5 Conclusions

In the present paper, we report an investigation on the effect of constriction on mixing and fluid flow characteristics in a rectangular serpentine-type micromixer. The bends in a microchannel generates secondary flows that enhance mixing of fluids, while the constriction locally enhances the advection flows. The effect of the constriction values on mixing ability of the micromixer is presented for distinct Reynolds number ranges from 10–60. The results from numerical simulations show that the increase in constriction value for various Reynolds number converted the laminar flow of working fluids into the engulfment flow at the constriction area which enhances the mixing performance of the micromixer.

Acknowledgements Manuscript communication number (MCN): IU/R&D/ 2022-MCN0001771 Office of Research and Development Cell, Integral University, Lucknow, India.

References

1. Manz A, Graber N, Widmer HM (1990) Miniaturized total chemical analysis systems: a novel concept for chemical sensing. Sens Actuat B Chem 1(1–6):244–248. https://doi.org/10.1016/0925-4005(90)80209-I
2. Stone HA, Stroock AD, Ajdari A (2004) Engineering flows in small devices: microfluidics toward a lab-on-a-chip. Annu Rev Fluid Mech 36:381–411. https://doi.org/10.1146/annurev.fluid.36.050802.122124
3. Beebe DJ, Mensing GA, Walker GM (2002) Physics and applications of microfluidics in biology. Annu Rev Biomed Eng 4:261–286. https://doi.org/10.1146/annurev.bioeng.4.112601.125916
4. Figeys D, Pinto D (2000) Lab-on-a-chip: a revolution in biological and medical sciences. Anal Chem 72(9):330–335. https://doi.org/10.1021/ac002800y
5. Ansari MA, Kim KY, Anwar K, Kim SM (2010) A novel passive micromixer based on unbalanced splits and collisions of fluid streams. J. Micromech Microeng 20(5). https://doi.org/10.1088/0960-1317/20/5/055007
6. D EW, Jia S, Sibbitts J, Sadeghi J, Sellens K, Culbertson CT (2016) Micro total analysis systems: fundamental advances and applications. Anal Chem 88(1):320–338. https://doi.org/10.1021/acs.analchem.5b04310
7. Zhang J, Luo X (2018) Mixing performance of a 3D micro T-mixer with swirl-inducing inlets and rectangular constriction. Micromachines 9(5). https://doi.org/10.3390/mi9050199
8. Lin CH, Tsai CH, Fu LM (2005) A rapid three-dimensional vortex micromixer utilizing self-rotation effects under low Reynolds number conditions. J. Micromech Microeng 15(5):935–943. https://doi.org/10.1088/0960-1317/15/5/006
9. Ward K, Fan ZH (2015) Mixing in microfluidic devices and enhancement methods. J Micromech Microeng 25(9). https://doi.org/10.1088/0960-1317/25/9/094001
10. Soleymani A, Kolehmainen E, Turunen I (2008) Numerical and experimental investigations of liquid mixing in T-type micromixers. Chem Eng J 135(1). https://doi.org/10.1016/j.cej.2007.07.048
11. Nguyen NT, Wu Z (2005) Micromixers-a review. J Micromech Microeng 15(2):1–16. https://doi.org/10.1088/0960-1317/15/2/R01
12. Long M, Sprague MA, Grimes AA, Rich BD, Khine M (2009) A simple three-dimensional vortex micromixer. Appl Phys Lett 94(13):13–16. https://doi.org/10.1063/1.3089816
13. Hessel V, Löwe H, Schönfeld F (2005) Micromixers-a review on passive and active mixing principles. Chem Eng Sci 60(8–9):2479–2501. SPEC. ISS. https://doi.org/10.1016/j.ces.2004.11.033
14. Raza W, Hossain S, Kim KY (2020) A review of passive micromixers with a comparative analysis. Micromachines 11(5). MDPI AG. https://doi.org/10.3390/MI11050455
15. Liu RH et al (2000) Passive mixing in a three-dimensional serpentine microchannel. J Microelectromechanical Syst 9(2):190–197. https://doi.org/10.1109/84.846699
16. Tsai RT, Wu CY (2011) An efficient micromixer based on multidirectional vortices due to baffles and channel curvature. Biomicrofluidics 5(1). https://doi.org/10.1063/1.3552992
17. Hossain S, Ansari MA, Kim KY (2009) Evaluation of the mixing performance of three passive micromixers. Chem Eng J 150(2–3):492–501. https://doi.org/10.1016/j.cej.2009.02.033
18. Ansari MA, Park CW, Hur N, Kim D (2013) Non-aligned bilayer square-wave bend microchannel for mixing. J Mech Sci Technol 27(12):3851–3859. https://doi.org/10.1007/s12206-013-0929-6
19. Kuo JN, Jiang LR (2014) Design optimization of micromixer with square-wave microchannel on compact disk microfluidic platform. Microsyst Technol 20(1):91–99. https://doi.org/10.1007/s00542-013-1769-0
20. Yang J, Qi L, Chen Y, Ma H (2013) Design and fabrication of a three dimensional spiral micromixer. Chinese J Chem 31(2):209–214. https://doi.org/10.1002/cjoc.201200922

21. Chen X, Li T, Zeng H, Hu Z, Fu B (2016) Numerical and experimental investigation on micromixers with serpentine microchannels. Int J Heat Mass Transf 98:131–140. https://doi.org/10.1016/j.ijheatmasstransfer.2016.03.041
22. Chen X, Li T, Hu Z (2017) A novel research on serpentine microchannels of passive micromixers. Microsyst Technol 23(7):2649–2656. https://doi.org/10.1007/s00542-016-3060-7
23. Gidde RR, Pawar PM, Ronge BP, Misal ND, Kapurkar RB, Parkhe AK (2018) Evaluation of the mixing performance in a planar passive micromixer with circular and square mixing chambers. Microsyst Technol 24(6):2599–2610. https://doi.org/10.1007/s00542-017-3686-0
24. Solehati N, Bae J, Sasmito AP (2018) Optimization of wavy-channel micromixer geometry using Taguchi method. Micromachines 9(2):1–14. https://doi.org/10.3390/mi9020070
25. Ansari MA, kim KY (2009) Parametric study on mixing of two fluids in a three-dimensional serpentine microchannel. Chem Eng J 146(3):439–448. https://doi.org/10.1016/j.cej.2008.10.006
26. Ansari MA, Kim KY (2009) A numerical study of mixing in a microchannel with circular mixing chambers. AIChE J 55(9):2217–2225. https://doi.org/10.1002/aic.11833

Investigation on Fluid Flow in Biomimetic Microchannel

Mohd Amir Khan, Arees Qamreen, and Mubashshir Ahmad Ansari

Abstract In this work, a numerical investigation on fluid flow and mixing has been performed on split and recombined microchannels based on Murray's law. Murray's law is derived from nature-inspired tree-like structures that are commonly found in living and non-living things. The human cardiovascular system and plant water transport systems are the most famous and investigated examples of such a type of structure. The same concept is introduced to design a micromixer having asymmetric bifurcation in order to solve the mixing problem at the microscale. Mixing at a small scale is a very complicated phenomenon, and due to its integration with bioengineering and chemical investigations, homogeneous and quick mixing has long been a prerequisite for microfluidic devices. In this paper, asymmetric and symmetric bifurcation micromixers were investigated with Reynold's number ranging from 0.01 to 300. The results demonstrate that almost all the Reynolds number ranging from 0.01 to 300 are better handled by the symmetric bifurcated micromixer. However, for the asymmetric bifurcated micromixer, the increase in mixing index is only seen at Reynold number 300; below that, its efficiency is consistently lower than that of the symmetric micromixer. This is due to the larger daughter vessel dominating the smaller daughter vessel. The ability to take part in the mixing of smaller daughter vessel was obstructed by the bulk flow of the larger daughter vessel. This seems to be the main cause of inadequate mixing. The research reveals that symmetric bifurcated micromixers with ratio $d_1/d_2 = 1.0$ channels are much more effective in design than asymmetric bifurcated micromixers with ratio $d_1/d_2 = 0.6$ channels.

M. A. Khan (✉) · A. Qamreen · M. A. Ansari
Department of Mechanical Engineering, Zakir Husain College of Engineering and Technology, Aligarh Muslim University, Aligarh, India
e-mail: gh3021@myamu.ac.in

M. A. Ansari
e-mail: mubashshir@zhcet.ac.in

M. A. Siddiqui et al. (eds.), *Advances in Heat Transfer and Fluid Dynamics*, Lecture Notes in Mechanical Engineering, https://doi.org/10.1007/978-981-99-7213-5_6

1 Introduction

The practice of replicating natural design has increased over the last several decades. The primary reason for this attention is that Nature never harms the environment and always performs accordingly to the concepts of budget and effectiveness. The natural mechanisms are capable of self-healing and purification [1]. The activity of imitating naturally occurring substances in human production is known as "biomimicry" [2]. Currently, many natures inspired designs have been produced in many sectors and found they are flawlessly accomplished. Examples of technologies that operate well include flight inspired by birds, Velcro, hydrophobia based on the lotus, [1] and many more. Similarly, microfluidic systems are also significantly impacted by this field [3]. The term "microfluidics" refers to both the science and the technology used to create extremely small devices with compartments and passages by which fluids can either flow freely or be constrained. Microfluidics, as the name suggests, deals with handling extremely small fluid volumes. The mixing operation in microfluidic devices faces several difficulties because of the small dimensions and laminar character of the flow. While turbulent flow propagates eddies dissipate, and only diffusive mixing between the two flow streams remains active. Due to this, creating flow conditions that encourage mixing fluid in the microfluidic devices is the main objective in the design of a micromixer. A nature-inspired bifurcation microchannel is utilized in this paper. The design and optimization of bifurcating fluid distribution microchannel networks is one major application of biomimetic ideas [4]. There are several bifurcation channels in living and nonliving systems, They are involved in many activities that are typically in charge of managing fluids with sophisticated rheological behavior. The vascular system in the human body, which distributes blood and other essential fluids to the different organs, the human lungs' system for transferring oxygen, and the water-transporting system xylem in plants and trees are examples of such bifurcating system [5, 6].

Murrays in 1926 investigated the mammalian circulatory system and found that there is a correlation between the parent vessels diameter and the diameter of the distal branches, i.e., the cube of the maternal vascular radius equals the sum of the cubes of the daughter vascular radius [5]. In this study, we use Murray's law to develop an Asymmetric bifurcated micromixer. Before that, multiple types of research on the same concept had previously been conducted [7–11], but they had changed the fundamental Murray's law to take into account the feasibility of fabricating microchannel devices. The various manufacturing methods include wet etching, lithographic techniques, and micro-milling. A majority of the fabrication utilized a constant depth strategy to fabricate the microfluidic device to reduce cost and complexity. Krzysztof et al. adopted Murray's law to develop a symmetric fractal-based micromixer with constant height and numerically analyzed its mixing effectiveness. The results showed that the micromixer achieves excellent mixing at low Reynold numbers (from 0.01 to 1) [7]. Lim et al. developed a fractal-based micromixer with non-uniform depth. They adjusted the channel radius to ensure that the structure followed Murray's law [8]. McCulloh et al. confirmed that the water transport network of plants also agreed with Murray's law [6]. Liu et al. enhanced Murray's law to comprise the entire vascular arterial tree, instead of focusing on just one vessel. Emerson et al. demonstrated the use of Murray's Laws in the design of microfluidic systems by extending it to uniform-height applications on various microfluidic systems [10]. Xu et al. developed the Y, T, and semicircular type micromixers based on a modified version of Murray's rule and investigated the effects of structural factors on the degree of mixing using CFD. They came to the conclusion that the basic T-shaped micromixer had a greater degree of mixing [11].

1.1 Theoretical Background

Nature has a highly challenging and intricate bifurcation system. However, the majority of either living or nonliving systems are related mathematically in some way [12]. Murray proposed a fundamental law after examining earlier studies [5], which is now known as Murray's law. It is regarded as one of the fundamental ideas of bifurcation networks. The cardiovascular and respiratory systems of humans, river split, and the water transportation system in plants and trees are a few illustrations of existing bifurcating networks on which research has already been conducted [5, 6]. In the human circulatory system, Murray recognized that two separate energies needed to be mixed to keep the vessel's minimal operating work. First, energy is necessary to overcome a fluid's viscous action p_f, and then energy is essential metabolically to retain the fluid volume constant p_m.

The expression that keeps blood power flowing through arterial vessels.

$$p_f = \Delta P Q \tag{1}$$

where, ΔP represents the blood vessel pressure difference and Q is the volumetric flow rate. The metabolic rate of the vessel i.e., power necessary to restore the blood in the vessel is p_m.

$$p_m = m.v = m\Pi r^2 l \tag{2}$$

where, m and v are the metabolic coefficient and volume of fluid respectively.

Therefore, the total power required is:

$$p_t = p_f + p_m = \Delta P Q + m\Pi r^2 l \tag{3}$$

According to Hagen Poiseuille law,

$$\Delta P = \frac{8\mu l}{\Pi r^4} Q \tag{4}$$

Rearranging,

$$p_t = \frac{8\mu l}{\Pi r^4} Q^2 + m\Pi r^2 l \tag{5}$$

Hence, a minimal value of p_t exists where, $\partial p_t / \partial r = 0$

$$\frac{\partial p_t}{\partial r} = \frac{\partial}{\partial r}\left(\frac{8\mu l}{\Pi r^4} Q^2 + m\Pi r^2 l\right) \tag{6}$$

$$\frac{-4.8\mu}{\Pi r^5} Q^2 + m\Pi r^2 = 0 \tag{7}$$

$$Q^2 = \frac{m\Pi^2 r^6}{16\mu} \tag{8}$$

So, the correlation between arteries and capillaries is as follows.

$$Q = N r^3 \tag{9}$$

where N consider to be constant which is equal to $\frac{\pi}{4}\sqrt{\frac{m}{u}}$. Based on the conservation law. In one branch of a blood artery, the relationship is as follows:

$$N r_0^3 = N r_1^3 + N r_2^3 \tag{10}$$

Also,

$$r_0^3 = r_1^3 + r_2^3 \tag{11}$$

If the bifurcation is symmetric then the radii of both the daughter vessels are equal. i.e. $r_1 = r_2$

$$r_0^3 = 2 r_1^3 \tag{12}$$

All the aforementioned derivations are of Murray's laws regardless of the fact that the formulation of this law involves various assumptions, Despite this it is also suitable for several other biological bifurcation systems, including plant water transport bifurcation systems. Murray's law is also applicable to nonliving systems as originally it was found for biological systems.

2 Micromixer Design and Numerical Methods

2.1 Micromixer Design Concept

In this research, we investigated networks with varied widths and four successive generations of constant depth. The channel design is supposed to have a depth of $h = 125$ μm which was taken constant for the entire domain and a variable width to allow Murray's law to be followed. At the outlet, the rectangular cross-section was kept to be 250 μm × 125 μm, resulting in an output aspect ratio of $\alpha = 0.5$ and the aspect ratio is variable at each inlet. In the present situation, the hydraulic diameter at each daughter's vessel is different, thus the larger one of each generation is considered to produce the length of the micromixer. The length of each generation is designed to be proportionate to its hydraulic diameter i.e., $l = kd_h$. Figure 1 demonstrates that the Asymmetric bifurcated three-dimensional geometric structure satisfied Murray's law. The micromixer structure has been developed with four generations., eight inputs, and one output. For all four generations, the relationship between the daughter vessels is taken to be 0.6 i.e., $d1 = 0.6d2$.

2.2 Mathematical Model

Computational Fluid Dynamics (CFD) is a significantly affordable approach for carrying out research. The computational fluid dynamics software Ansys CFX was used to carry out the numerical simulations on the mixing operation on the model (Fig. 1). It solves the steady continuity and Navier Stokes along with species convection–diffusion equation to address the mixing problem. The assumptions made in this problem are that the flow is laminar, steady, isothermal and incompressible.

$$\nabla.\vec{V} = 0 \tag{13}$$

$$\rho\vec{V}.\nabla\vec{V} = -\nabla P + \mu\nabla^2\vec{V} \tag{14}$$

$$\left(\vec{V}.\nabla\right)C = D\nabla^2 C \tag{15}$$

where, C represents species concentration coefficient and D is the diffusion coefficient. μ and ρ denote fluid dynamic viscosity and density of the fluid, and $\vec{V}$ and P are the velocity vector, the pressure respectively.

The computational fluid dynamics program Ansys CFX is used to simulate the operation on the micromixer. To analyze the mixing of two fluids, the convection–diffusion equation was solved. The design modular was used to generate the symmetric and asymmetric bifurcated micromixer designs. Figures 1a & b show

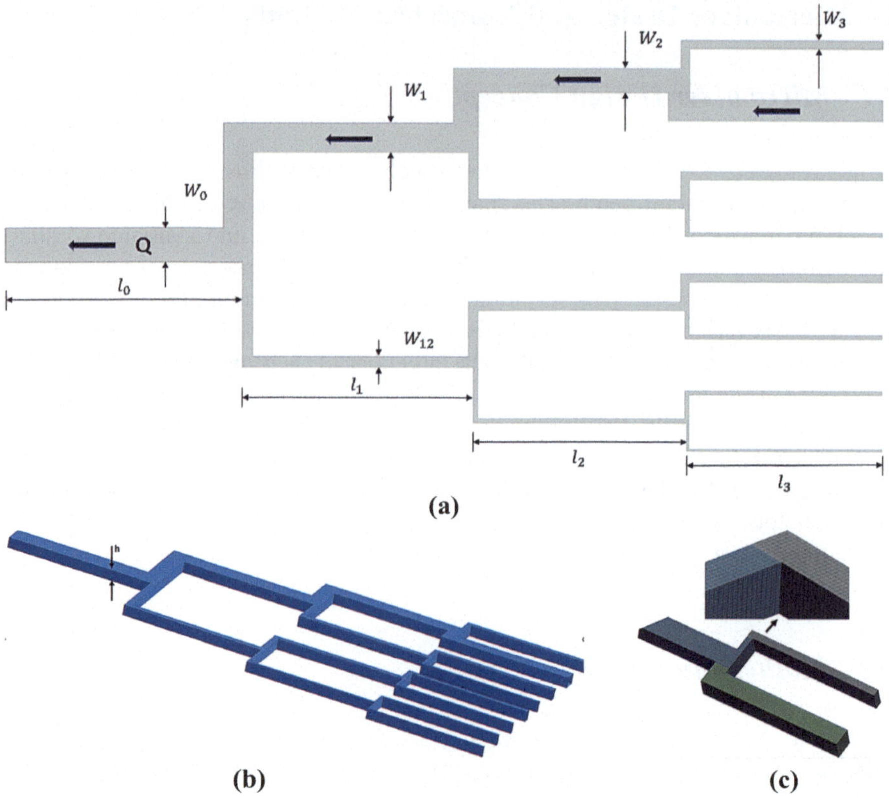

Fig. 1 Bifurcating microfluidic network with constant depth and variable width having four generation ($i = 0, 1, 2, 3$) with aspect ratio $\alpha = 0.5$. **a** Systematic arrangement of width along the length **b** Three-dimensional view with constant height(h) **c** Meshing along with magnifying view

the mixer's schematic design. As can be seen in Fig. 1c, the geometries of both the mixers are meshed with structured hexahedral elements. The hexahedral elements are very well oriented with flow directions which results in reduced numerical diffusion. Another important feature of hexahedral meshing is that it produces fewer elements than other meshing systems, which further drastically reduces the computational time required to solve the problem. All simulations were conducted with dual precisions.

Grid independence test was executed on a particular problem in order to ensure the effect of grid density on the results. A fine mesh structure is employed on the problem to reduce diffusion error as much as possible. Although there are several strategies to minimize numerical diffusion, it is highly rare that it can be eliminated entirely from the result.

The simulation's initial setup is as follows. The variation in the velocity at the microchannel wall is assumed to be zero, Water and ethanol were chosen as the working fluids for the analysis at temperature 20 °C. The densities of water and ethanol are considered to be 9.96×10^2 Kg/m^3 and 7.9×10^2 Kg/m^3 respectively.

The velocity at all eight inlets are set by the local Reynolds number $Re_0 = V_0 D_{h_0}/v$ defined at the outlet trunk of the channel where it is maximum. The outlet was assigned as zero static pressure. The convergence criterion that was taken into account. when the solution approached the root mean square residual value of 10^{-12}.

2.3 *Quantification of Mixing*

The quantification strategy for mixing used in this paper is by calculating the variance of the mass fraction of one of the two working fluids that will be mixed [13]. The quantification of mixing was carried out by taking the mass fraction at number of node on the cross-sectional plane. In this context, the water mass fraction is used to evaluate the mixing index. The average fluid fraction value which is ($\overline{c}_i = 0.5$) is then used to adjust the standard deviation of the fluid mass fraction to be mixed. The formula definition is as follows:

$$MI = 1 - \left[\frac{\sqrt{\frac{1}{n-1} \sum_{i=1}^{n} \left(C_i - \overline{C}_i\right)^2}}{\overline{C}_i} \right] \tag{14}$$

where, C_i here used to show mass fraction at a particular sample point on the plane, $\overline{C}_i$ is the average value of mass fraction, and n is the number of sample point on the selected plane. To ensure precision, a significant range of sample points are taken and kept perfectly symmetrical on the cross-sectional plane (Fig. 2).

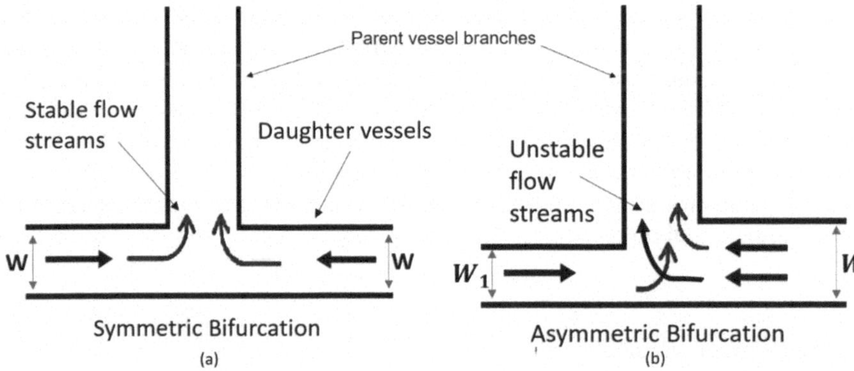

Fig. 2 Schematic representation of the bifurcated micromixer based on Murray's law: **a)** Symmetric bifurcated micromixer and **b)** Asymmetric bifurcated micromixer

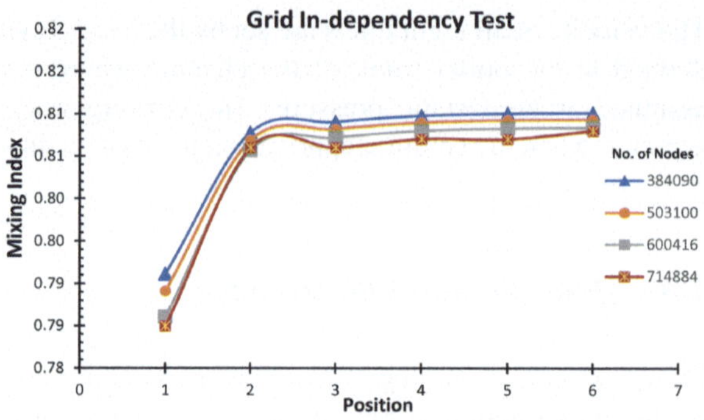

Fig. 3 Grid independency test at Re = 30, Mixing Index at six planes in the main channel

3 Result and Discussion

3.1 Grid Independence Test

A numerical analysis of the symmetric and asymmetric bifurcated micromixer at various Reynolds numbers (0.1 to 300) has been performed. To assure the results accuracy, grid independency test and validation study have been performed simultaneously to conduct this analysis. Four different grids with different mesh density mesh sizes-Mesh 1(3.84×10^5 lakh), Mesh 2(5.03×10^5 lakh), Mesh 3(6.0×10^5 lakh) and Mesh 4(7.14×10^5 lakh) are taken into consideration. In the preliminary situation, Mesh 1 is used and is considered as a coarse mesh, and overall domain is discretized by the element size, which is 10 μm for the first case. The meshes are made fine in the subsequent cases by reducing the element size.

It was observed that the mixing index decreases as the mesh density increases (see Fig. 3). At Re 30, the percentage difference in Mixing Index between two successive meshes was calculated. The grids with Mesh 1 and Mesh 2 were found to have a MI difference of 3.53%. Between Mesh 2 and Mesh 3, it was 0.12%, and for Mesh 4, a finer mesh, it was further lowered to 0.04%. Consequently, a mesh with Mesh 3 and 4 provides a good approximation, and no additional mesh refinement was carried out. Hence, we could choose Mesh 3 to carry out the simulation investigation in order to save the computation time.

3.2 Validation Study

The symmetric bifurcated micromixer obeying Murray law is used to validate the current model [7]. Figure 4 shows a plot of mixing index of both mixers along the axial length. The largest difference in the mixing quality values between the two mixers is about 5%. Since the results are in agreement with the previous research, the current simulation is considered to be reliable (Figs. 5, 6 and 7).

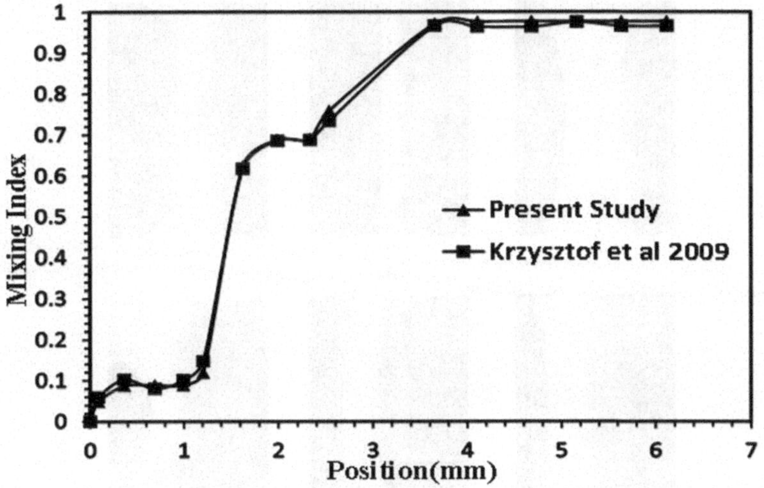

Fig. 4 Validation of present numerical results of symmetric bifurcated micromixer model with the result of Cieslicki K, Piechna A. [7]

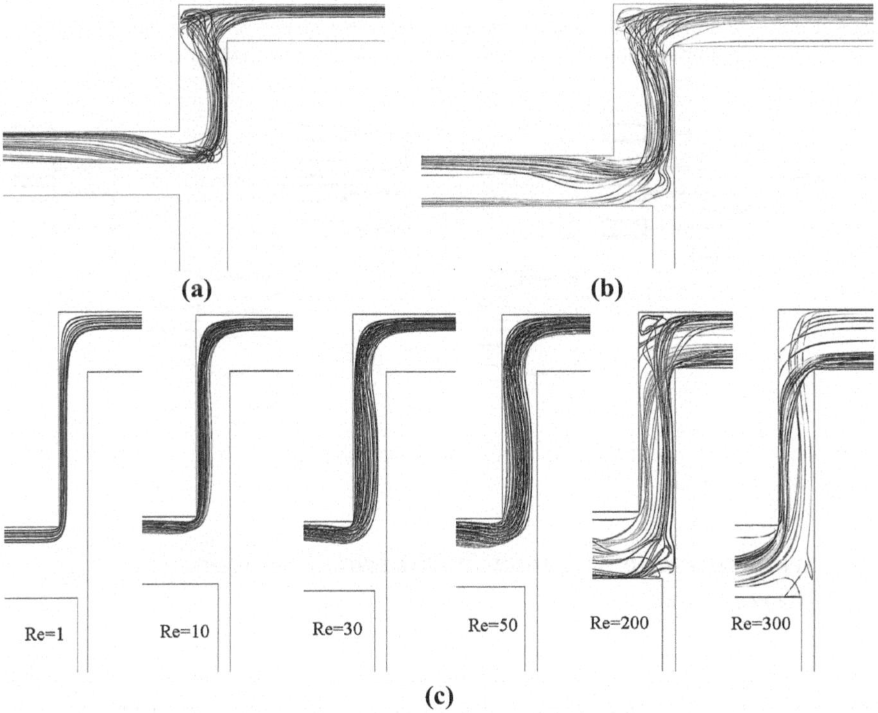

Fig. 5 Projected streamlines at Reynold number 100 for **a)** symmetric bifurcated micromixer **b)** Asymmetric bifurcated micromixer and **c** represents the behavior of streamlines of asymmetric bifurcated micromixer with rise in Reynold numbers

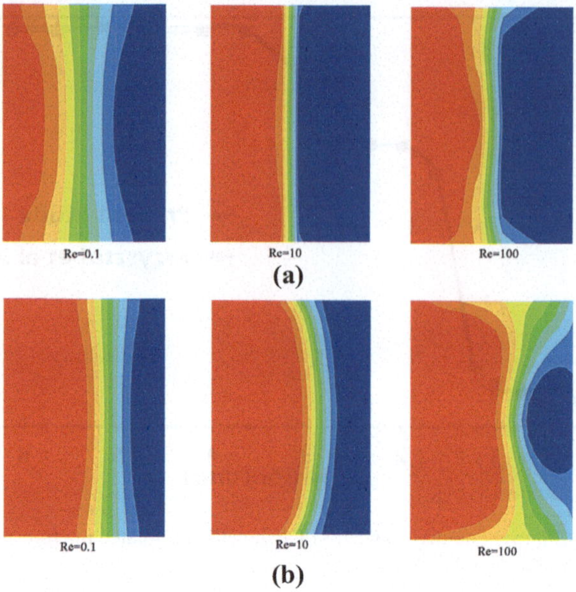

Fig. 6 Comparison of the numerical results showing the distribution of the ethanol mass fraction on y–z plane at first bifurcation **a)** symmetric bifurcation **b)** asymmetric bifurcation

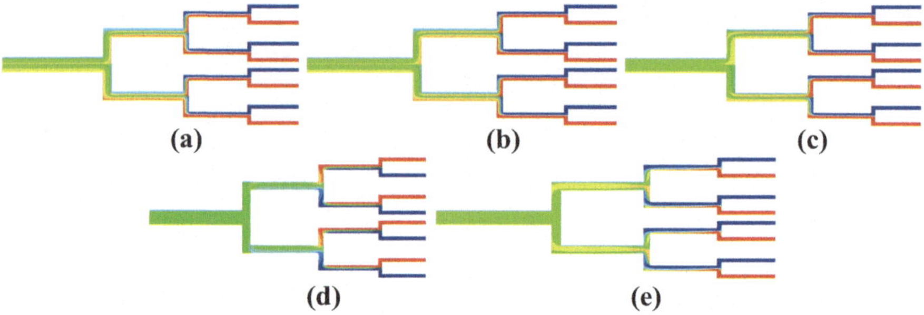

Fig. 7 Mass fraction distribution; **a)** Re = 1 **b)** Re = 10 **c)** Re = 30 **d)** Re = 100 **e)** Re = 300 Cieslicki K, Piechna A [7]

3.3 Symmetric and Asymmetric Bifurcated Biomimetic Micromixer

Figure 5 represents the streamline for the fluid flow of the symmetric and asymmetric bifurcated micromixer from input 1 to outlet. It can be seen clearly from the differences that symmetric bifurcated micromixer streamlines are stable and flow in a controlled manner, whereas asymmetric bifurcated micromixer streamlines spread out across the entire width of the channel and promotes instability in the fluid flow.

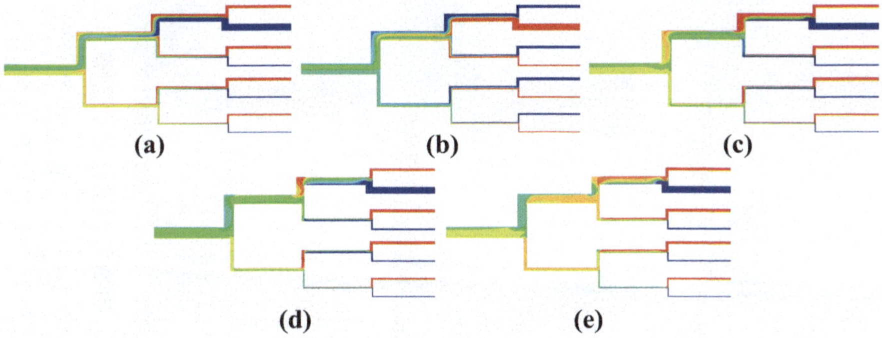

Fig. 8 Mass fraction distribution; **a)** Re = 1 **b)** Re = 10 **c)** Re = 30 **d)** Re = 100 e) Re = 300

The evolution of symmetrical flow in the main channel is represented by the projected streamline for various Reynolds numbers. For Re $\leq$ 10, the entering fluid streamlines formed identical flow structures, but as the Reynolds number increases, the crossover of one streamline to other starts and disrupted the developed symmetric flow pattern. The imbalance in the stream can be clearly observed at Re = 100 to Re = 300, which demonstrated that at high Reynolds numbers, the mixing efficiency significantly increases. However, symmetric micromixers produce higher mixing efficiency than asymmetric micromixers for low and intermediate Reynolds numbers. An asymmetric micromixer have not showed noticeable impact on mixing performance at low Reynold's numbers. The top view of the streamline is displayed at various Reynolds numbers (1, 10, 30, 50, 100, 200, and 300), which is beneficial for visualizing the progress of the flow field and for highlighting the justification behind improving mixing performance.

With increase in Reynold's number, the flow inside an asymmetric bifurcated micromixer becomes more unbalanced, directly increasing the interface area (see Fig. 8). Extending the symmetric bifurcation to an asymmetric bifurcation was carried out with the intention of causing an imbalance between the flows, which is in good agreement with the primary purpose. However, the objective on the enhancement of mixing is not achieved for which the present studywas anticipated.

3.4 *Flow Analysis*

This part describes the micromixer's flow pattern for symmetric and Asymmetric bifurcated micromixers that promotes to mixing. (Fig. 9) illustrates the numerical findings for the velocity vector on the X–Y plane at the fixed location at different Reynolds numbers (0.02, 0.1, 10, 100, 200 and 300). The degree of the disproportion between the larger daughter vessel and the smaller daughter vessel varies with an increase in Reynolds number, where the intensity becomes more erratic. However,

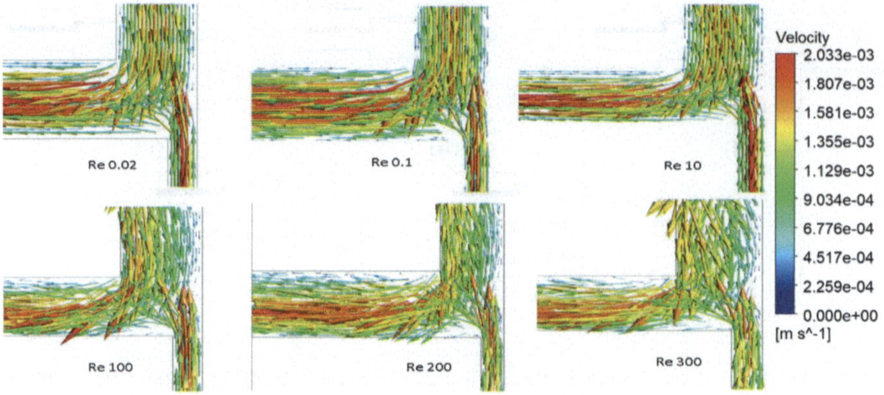

Fig. 9 Velocity vectors on the XY-plane with different Reynolds numbers

the collision of the two fluids stream are significant and disturbs the fluid interface due to the disparity in inertial forces among the two fluids in the channel.

Figure 10 provides an overview of the overall effect of asymmetric and symmetric bifurcated micromixers on mixing performance at different Reynolds numbers (Re 0.01 to 300). For Re 0.02, the mixing of the both mixers is considerably high, after that it declines until Re 0.1. Beyond Re 100, saturation in the mixing can be observed. The production of extremely prominent imbalances between the flows causes the mixing to jump from Re 100 to Re 300. When compared to other Reynolds numbers, the mixing phenomenon at Re 0.02 is rather dissimilar. Both cases show pretty high mixing at Re 0.02; this occurs because the fluid had sufficient time for molecular diffusion to take place at a larger residence time in the microchannel.

Fig. 10 Mixing indexes with Reynolds number for symmetric and asymmetric bifurcated micromixer

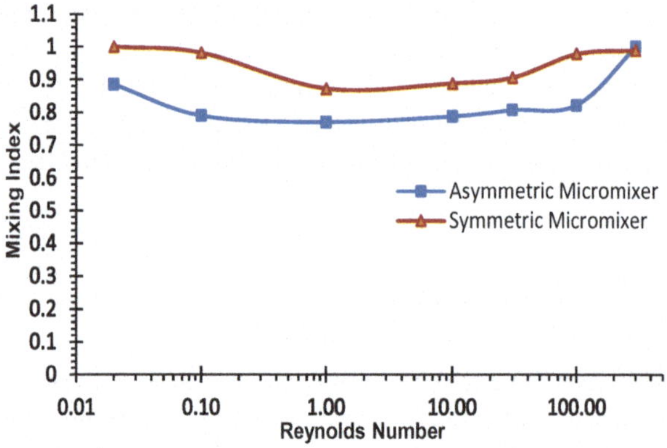

4 Conclusion

A numerical analysis was carried out on a biomimetic-inspired bifurcated micromixer design thats follow Murray's law. Murray law was essentially found for the biological system. The concept behind this law is that structures that follow this law will experience a reduction in overall friction losses in laminar flow due to viscous effect. In this paper, a numerical analysis has been conducted to figure out the effects of symmetric and asymmetric bifurcated micromixers on mixing and fluid flow for a wide range of Reynolds numbers (Re 0.01 to 300). The depth of the both micromixers were taken same in order to examine the flow behavior due to change in configuration. It is found that the index of mixing for the symmetric micromixer is shown to be high at lower Re (0.01 to 1), whereas the indices are lower for the asymmetric bifurcated micromixer. Consecutively the results of symmetric bifurcated micromixer are in good agreement at high Reynold numbers ranging from 1 to 300 as compared to the asymmetric bifurcated micromixer. The Asyxmmetric micromixer performed well at only Reynolds number 300. Asymmetric micromixers were designed with the intention of disrupting the meeting of the two fluid streams as they approached each other for mixing. As intended, it caused some disturbances however it is not capable of creating good mixing at low Reynold numbers. This is due to the fact that the shorter daughter vessel's fluid entrance was interrupted by the larger daughter vessel's bulk flow, preventing the shorter daughter from participating in mixing. According to the overall analysis, symmetric micromixers continue to be the most effective for mixing applications, and asymmetric bifurcated micromixers require additional modifications to improve the mixing index.

References

1. Shahda MA, Elmokadem A, Abd Elhafeez MM Biomimicry Levels as an approach to the architectural sustainability. Port Said Eng Res J. researchgate.net/publication/328726155
2. Vincent JFV et al (2006) Biomimetics: its practice and theory. J R Soc Interface 3(9):471–482. https://doi.org/10.1098/rsif.2006.0127.PMC1664643.PMID16849244
3. Inamdar NK, Borenstein JT (2011) Microfluidic cell culture models for tissue engineering. Curr Opin Biotechnol 22(5):681–689
4. Yang S, Undar A, Zahn JD (2006) A microfluidic device for continuous, real time blood plasma separation. Lab Chip 6:871–880
5. Murray CD (1926) The physiological principle of minimum work: I. The vascular system and the cost of blood volume. Proc Natl Acad Sci USA 12:207–214
6. McCulloh KA, Sperry JS, Adler FR (2003) Water transport in plants obeys Murray's Law. Nature 421:939–942
7. Cieslicki K, Piechna A (2009) Investigations of mixing process in microfluidic manifold designed according to biomimetic rule. Lab Chip 9(5):726–732
8. Khan MA, Suhaib M, Ansari MA (2023) Investigations on fluid flow and mixing in fractal tree like biomimetic microchannel based on Murray's law. Chemical Engineering and Processing-Process Intensification, 194, p.109564.https://doi.org/10.1016/j.cep.2023.109564
9. Liu Y, Kassab GS (2007) Vascular metabolic dissipation in Murray's law. Am J Physiol Heart Circulat Physiol 292 (3):H1336–H1339

10. Barber RW, Emerson DR (2008) Optimal design of microfluidic networks using biologically inspired principles. Microfluid Nanofluid 4:179–191
11. Xu J, Chen X (2019) Mixing performance of a fractal-like tree network micromixer based on Murray's law. Xu J, Chen X. Int J Heat Mass Transfer 141:346–352
12. Bassingthwaighte JB, Liebovitch LS, West BJ (1994) Fractal physiology. Oxford University Press, New York
13. Danckwerts PV (1952) The definition and measurement of some characteristics of mixtures. Appl Sci Res 3(4):279–296. https://doi.org/10.1007/BF03184936

Numerical Analysis of Transient Induced Flow Through Open Ended Tall Vertical Concentric Annulus

Jawed Mustafa, Saeed Alqaed, Shahid Husain, and M. Altamush Siddiqui

Abstract In this work, thermally induced water flow behavior in an open, tall, vertical annulus is carefully characterized using numerical analysis. The annuli in this study had aspect ratios of 1.184 and 352 (length to annular gap) (outer radius to inner radius). When constructing the machinery for nuclear reactor systems, the circular test sections with such a high aspect ratio are essential. The proprietary code employs a tried-and-true semi-implicit finite difference method as the numerical approach, which is based on the SMAC technique. The induced flow is brought on by a step change in the inner wall heat flux even while the exterior wall is assumed to be adiabatic. The Raleigh number (Ra) ranges from 1.1×10^4 to 6.6×10^5, although the Prandtl number (Pr) is 6.43. The research examines the impact of heating on the mass flow rate, pressure drop, and heat transfer coefficient among other design parameters. Along with the outcomes, a time- and location-based temperature and velocity profile is given. Additionally taken into account are the temporal fluctuations in induced flow and pressure distribution along the annulus.

Keywords Natural convection · Tall annulus · Closed loop thermo-siphon · Radius ratio · High aspect ratio · Constant heat flux

J. Mustafa (✉) · S. Alqaed
Mechanical Engineering Department, College of Engineering, Najran University, P.O. Box (1988), Najran 61441, Saudi Arabia
e-mail: jmmustafa@nu.edu.sa

S. Husain · M. A. Siddiqui
Department of Mechanical Engineering, Zakir Husain College of Engineering and Technology, Aligarh Muslim University, Aligarh 202002, India

M. A. Siddiqui et al. (eds.), *Advances in Heat Transfer and Fluid Dynamics*, Lecture Notes in Mechanical Engineering, https://doi.org/10.1007/978-981-99-7213-5_7

1 Introduction

In an annular cylindrical cavity with a closed end, both the flow of fluid and the transmission of heat share all the fundamental ideas that apply to energy-confined natural convection flows. In heated vertical open-ended annuli, such as annular chambers, parallel plates, and circular tubes, convective heat and fluid mass transfer naturally occurs. The fuel components of nuclear reactors, chemical distillery operations, channel-type solar energy collectors, heat protection systems, and gas-cooled electrical cables are a few examples of practical applications for such systems. These issues with natural convection heat transmission have gotten a lot of attention lately.

Thomas [1] statistically examined the fluid flow in the closed annular cavity formed by two concentric vertical cylinders and two horizontal planes. It was found that the motion is caused by a radial density gradient caused by a thermal boundary condition that maintains the inner cylinder heated and the outside cylinder cold. In addition, just one cell moves for low to intermediate Rayleigh numbers; nevertheless, for high Rayleigh numbers, the beginning of multicellular motion is seen. Temperature, stream function, and vorticity are broken down into Fourier components of a body-fitted curvilinear coordinate in Mochimaru's [2] approach for improving the solution of transient natural convection heat transfer in enclosures. Ho and Lin [3] investigated cold water natural convection heat transfer in an eccentric horizontal annulus made up of two isothermal cylinders and a density inversion using the finite difference method in a computer research. The results show that the positioning of the annulus' inner cylinder and the combined impact of the water density inversion significantly affect the flow pattern and heat transmission properties. In order to solve the boundary layer equations regulating the unstable laminar free convection flow in open-ended vertical concentric annuli, Shaarawi and Aattas [4] created a finite-difference method. The annulus height and the induced flow rate were suggested to be correlated over a range of time parameter values, commencing with quiescence and ending with the ultimate steady state. Statistics were used to examine the transient behavior of laminar free convection in a vertical annulus by El-Shaarawi and Al-Attas [5]. It has been discovered that when the dimensionless annulus height lowers, or as modified Gr rises, the temperature overshot phenomenon—which is caused by the dominance of the conduction heat transfer mode—becomes more prominent at the small-time values. In conjugate (combined conduction and free convection) thermal boundary conditions, where one cylinder is heated isothermally and the other is maintained at the temperature of the inlet fluid, El-Shaarawi et al. [6] investigated the impact of annulus geometrical parameters on the induced flow rate and heat transfer. When the numerical findings for a fluid with a Prandtl number of 0.7 were published, it was discovered that the geometry parameters (the fluid-annulus radius ratio and the eccentricity) had a significant influence on the outcomes. In a vertical cylindrical annulus, Sankar and Younghae [7] looked at how discrete heating affected convective heat transmission. The numerical findings demonstrate that the heat transfer rate varied with radii and aspect ratios and was always greater at the bottom heater. Desrayaud et al. [8] created a benchmark solution for each boundary condition in order to examine the sensitivity of the natural convection that takes place in a vertical asymmetric heated channel for four sets of open boundary conditions. According to their findings, the channel's exit has a return flow, and changing the flow patterns has no effect on how quickly fluid leaves the channel. Mustafa et al. [9–11] conducted an experimental and numerical analysis on an open-ended vertical concentric annulus. This work will next conduct a numerical analysis to ascertain the exact behavior of thermally induced water flow and the results of partial heating in an open-ended tall vertical annulus. For different heat fluxes, the fluid flow's dynamic behavior is also investigated. The aspect ratios of the annuli in this investigation were 1.184 and 352 (length to annular gap) (outer radius to inner radius). We place a strong emphasis on assessing the impact of heating for extremely high aspect ratios on design elements like pressure distribution and heat transfer coefficient. We also search for flow dynamics, such the flow pattern in annuli.

2 Governing Equations

A pictorial representation of the problem that served as the foundation for the numerical analysis is shown in Fig. 1. The inner tube of the annulus consists of a heated zone in the center and unheated (adiabatic) zones at the entry and exit ends of the test liquid, while the outer tube that makes up the annulus is intended to be adiabatic. The numerical analysis employed a cylindrical coordinate system and made the assumption that the flow was axisymmetric. The fluid was also modeled using the Boussinesq supposition. The relevant non-dimensional parameters and the non-dimensional governing equations are listed below.

$$\nabla \cdot \vec{U} = 0 \tag{1}$$

$$\frac{\partial \vec{U}}{\partial \tau} + \left(\vec{U} \cdot \vec{\nabla}\right)\vec{U} = -\vec{\nabla} P + \Pr\left(\nabla^2 \vec{U}\right) + \mathrm{Ra}\,\Pr\,\theta \hat{e}_z \tag{2}$$

$$\frac{\partial \theta}{\partial \tau} + \left(\vec{U} \cdot \vec{\nabla}\right)\theta = \nabla^2 \theta \tag{3}$$

Prandtl number $\Pr = \frac{\nu}{\alpha}$, Reylaigh number $\mathrm{Ra} = \frac{g\beta|q_w|b^4}{k\nu\alpha}$

The governing equations for the free convection processes in a vertically heated concentric cylindrical annulus must be simultaneously solved in accordance with the provided boundary conditions. The non-dimensional boundary conditions for the geometry of the issue are shown in Fig. 2. The heating zone is located in the axial non-dimensional height between $Z = 0$ and $Z = A = 352$. Specifically, $R_i = r_i/b$ and $R_o = r_o/b$.

(a) Wall (Inner) ($R = R_i$): $U = W = 0$ and $\frac{\partial \theta}{\partial R} = -1$ for $0 \le z \le A$

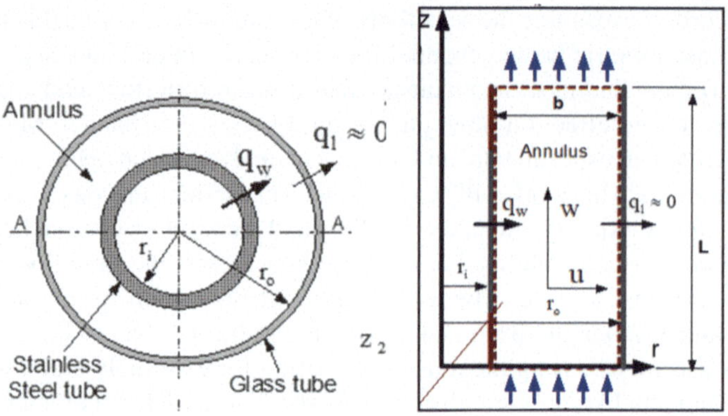

Fig. 1 Schematic view of the test section showing the vertical annulus (not drawn to scale)

Fig. 2 Non-dimensional boundary conditions for numerical solution of the problem

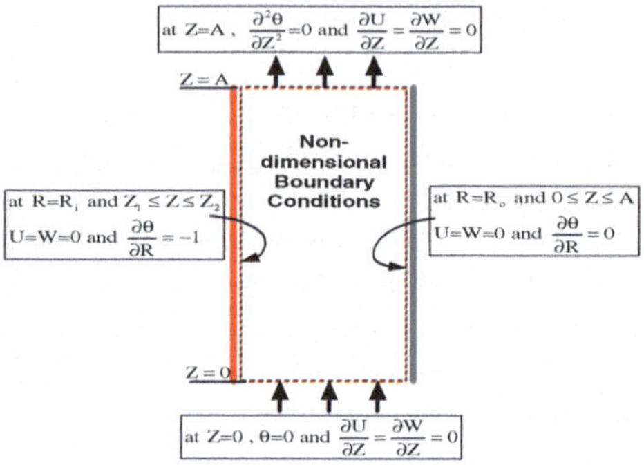

(b) Wall (Outer) ($R = R_o$): $U = W = 0$ and $\frac{\partial\theta}{\partial R} = 0$ for $0 \leq Z \leq A$
(c) Inlet (inflow) condition ($Z = 0$): $\frac{\partial U}{\partial Z} = \frac{\partial W}{\partial Z} = 0$ and $\theta = 0$ for $R_i < R < R_o$
(d) Outlet (outflow) condition ($Z = A$): $\frac{\partial U}{\partial Z} = \frac{\partial W}{\partial Z} = 0$ and $\frac{\partial^2\theta}{\partial Z^2} = 0$ for $R_i < R < R_o$.

3 Modeling and Simulation

The plan uses a predictor-corrector method in two steps. The diffusion components in Eqs. (2) and (3) are treated implicitly in the prediction step to produce tentative estimates of the temperature and velocity fields at the new time level. This is demonstrated mathematically as follows:

$$U_i^* - \Pr\left[\delta\tau\left(\frac{\partial^2 U_i^*}{\partial X_j^2}\right)\right] = U_i^n - \delta\tau\left(U_j^n\frac{\partial U_i^n}{\partial X_j} + \frac{\partial P^n}{\partial X_i} - \mathrm{Ra}\Pr\theta^n\right) \tag{4}$$

$$\theta^{n+1} - \delta\tau\left(\frac{\partial^2\theta^{n+1}}{\partial X_j^2}\right) = \theta^n - \delta\tau\left(U_j^n\frac{\partial\theta^n}{\partial X_j}\right) \tag{5}$$

The non-solenoidal preliminary estimate of the velocity field is adjusted in the corrector stage using an irrotational pressure correction field that preserves vorticity. The corresponding pressure fields, $P^{(n+1)}$, and the divergence-free velocity field, $U_i^{(n+1)}$, are made to fulfill the momentum equation as follows:

$$U_i^{n+1} - \Pr\left[\delta\tau\left(\frac{\partial^2 U_i^*}{\partial X_j^2}\right)\right] = U_i^n - \delta\tau\left(U_j^n\frac{\partial U_i^n}{\partial X_j} + \frac{\partial P^{n+1}}{\partial X_i} - \mathrm{Ra}\,\Pr\,\theta^n\right) \tag{6}$$

$$\nabla^2 P' = \frac{\partial U_i^*/\partial X_i}{\delta\tau} \tag{7}$$

Consequently, the Poisson equation is solved to yield the pressure adjustment (7). According to the boundary conditions proposed by Saad and Schultz [12], Cheng and Armfield [13], and published in Hasan et al. [14], the Eq. (7) is solved for the inner nodes using the GMRES solver.

The corrected velocity field at current–time will be recovered as,

$$U_i^{n+1} = U_i^* - \delta\tau\frac{\partial P'}{\partial X_i} \tag{8}$$

3.1 Domain Decomposition and Parallelization

On distributed memory devices, the current code has been parallelized using MPI (Message Passing Interface) [15]. Domain decomposition is the simple parallelization technique [16]. whereby the n processors each get an equal share of the calculation. Utilizing the MPI commands MPI SEND, MPI WAIT, and MPI RECV, an explicit point-to-point communication is established. When the actions on the data set are to be carried out in two distinct sections of the code, communication between processors is required. Figure 3 depicts a schematic representation of the domain decomposition used.

Non-equidistant meshes are required for the majority of near-wall flow simulations because local mesh rinement makes it simpler to resolve steep flow gradients. The discretization of the physical space is transferred onto the computational space in the r-direction (normal to the wall) as a result, and a mapping-function is constructed (R). As one approaches the left and right margins of the domain, Fig. 4 demonstrates that the horizontal mesh spacing in the radial direction (normal to the wall) obtains a significant degree of non-uniformity.

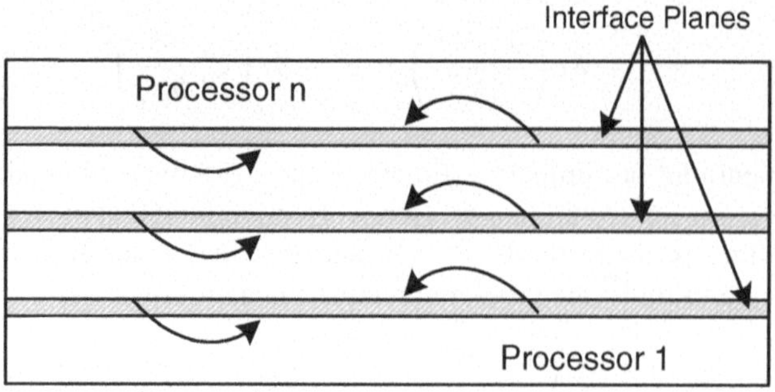

Fig. 3 1*D* 'slab' decomposition of a 2*D* domain

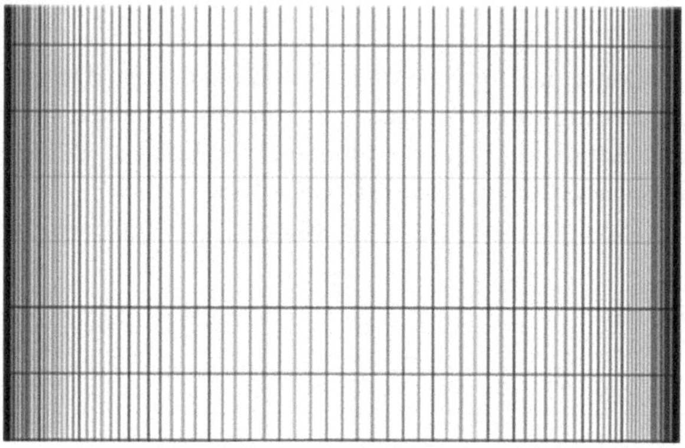

Fig. 4 Representative Numerical grid of 100 × 2500

$$\frac{\partial \theta}{\partial R} = \frac{1}{\left(\frac{\partial R}{\partial \varepsilon}\right)} \frac{\partial \theta}{\partial \varepsilon} \tag{9}$$

4 Validation

Numerical simulations of 2-D, laminar, continuous flow in a vertical, open-ended channel produced by natural convection were performed by Desrayaud et al. [8] and Amine et al. [17]. Their findings were used to support the approach used in the present inquiry. The benchmark solution for natural convection flow in a vertical channel has been given by Desrayaud et al. [8] using the thermally driven vertical channel issue as a model. They looked at the air convection that naturally takes place between two vertical walls, one of which has insulation and is partly heated by a homogenous heat flow. This kind of problem leads to a phenomena called flow reversal, which occurs when channel walls are heated unevenly. The study of Amine et al. [17] on air convection between two vertical walls is relevant. The Mustafa et al. [9] study published the findings of the validation.

5 Results and Discussion

Figure 5 depicts the transient behavior of the dimensionless temperature profile over the annulus gap at mid-height (a). In contrast, as shown in Fig. 5, the non-dimensional value of temperature increases with time, reaches a maximum, and then declines to attain its steady state value at any radial point (a). The transient temperature, in other words, may be said to be greater than the steady-state temperature. Temperature overrun is the term for this occurrence. This is a result of conduction heat transmission's early supremacy. In this instance, the convection component on the left side of the energy equation is subordinate to the diffusion term on the right side of the equation. Conduction is the pattern that the heat transmission process follows, and as time passes, the heat transfer coefficient deviates from steady-state convection to a lower value. As a result, the temperature rises above the corresponding steady-state convection temperature. Convection, on the other hand, causes the temperature to continue to drop while increasing the heat transfer coefficient until it achieves its steady-state value. As can be shown, until non-dimensional time hits 0.5, conduction mode outperforms convection mode. Convection then begins to prevail over conduction, reaching steady state at a non-dimensional time equal to 1.6. Dimensionless temperature drops radially from the heated inner wall to the outside adiabatic wall at any given axial location. The temperature distribution in the radial direction is shown in Fig. 6b. The heated wall has the maximum temperature at various axial points. A minimum is reached at the adiabatic wall after which it steadily declines in the radial direction. It illustrates the formation of a thermal barrier layer. Any location along the radial axis results in an axial increase in the dimensionless wall temperature. This has happened as a result of the steady heat flux given at time t.

Figures 6 and 7 show the isotherms formed in the annuli for the length $Z = 0$ to 9, up until a significant temperature change. When the temperature starts to rise from conduction, it does so quickly before peaking at $t = 0.5$. When the convection mode becomes significant and axial flow takes place, the temperature decreases and the width of the thermal boundary layer formed decreases. As the Rayleigh number increases, the transient period decreases and a steady state is quickly reached. This is because an increase in buoyancy force coincides with an increase in the Rayleigh number. In Fig. 6, different Rayleigh numbers for the thermal boundary layer at the intake of the fully heated annulus are compared. In the case of water, the Rayleigh number increases as the thermal boundary layer thickness decreases. This is because water loses some of its viscosity at higher temperatures. In comparison to low Rayleigh numbers, convection is stronger at high Rayleigh numbers. Figure 8 depicts the location from which the flow is fully established. Since this is where the flow nearly fully develops, the thermal entry length lengthens as the Rayleigh number increases.

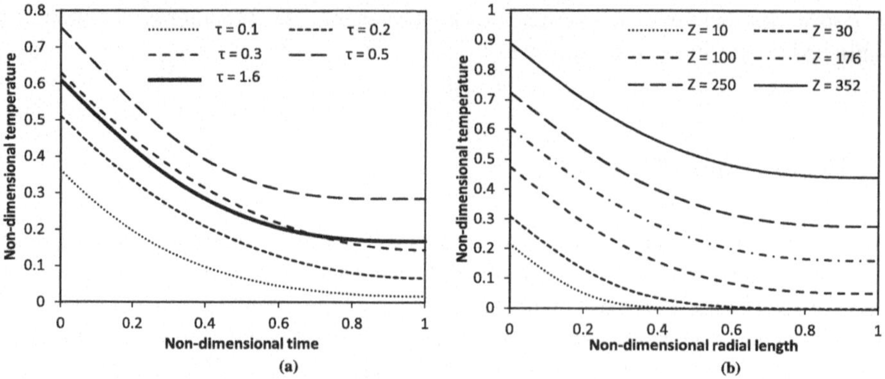

Fig. 5 Variation of temperature **a** along the radial direction at mid height with time **b** along the radial direction at different axial length ($Ra = 4.4 \times 10^4$)

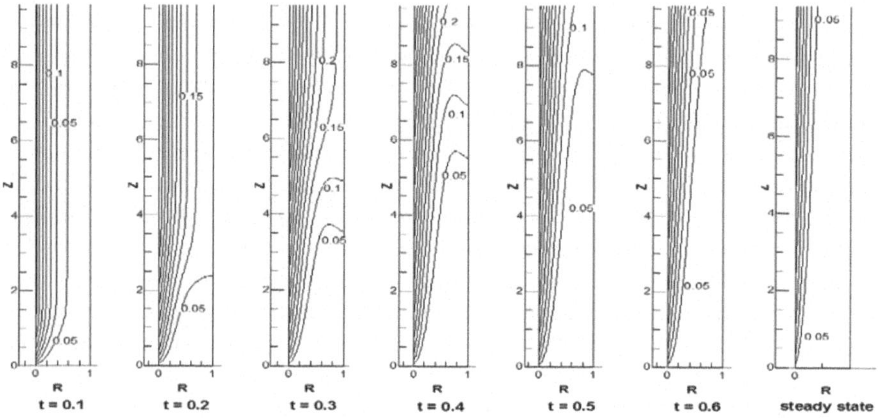

Fig. 6 Temperature contours at the inlet of the annulus with time for fully heated case

Figure 9a through d display the temporal evaluation of radial velocity at various locations (c). While Fig. 9b shows the radial velocity at various radial locations over time, Fig. 9a shows the radial velocity at the midsection at various time instances. Figure 9c shows the average radial velocity in the axial direction over time, while Fig. 9d shows steady state radial velocities at various planes at a fixed $Ra = 4.4 \times 10^4$. As heat is applied, the heated surface starts to develop radial velocity, which is positive near the inside wall and negative toward the outside surface. As soon as the heating process starts, fluid close to the heated surface moves away, while fluid close to the outer surface moves toward the inner wall. While the negative radial velocity increases toward the outside wall, the positive radial velocity near the heated surface grows in magnitude over time. The magnitude of the positive velocity close to the heated wall and the negative radial velocity close to the outer wall both decrease as

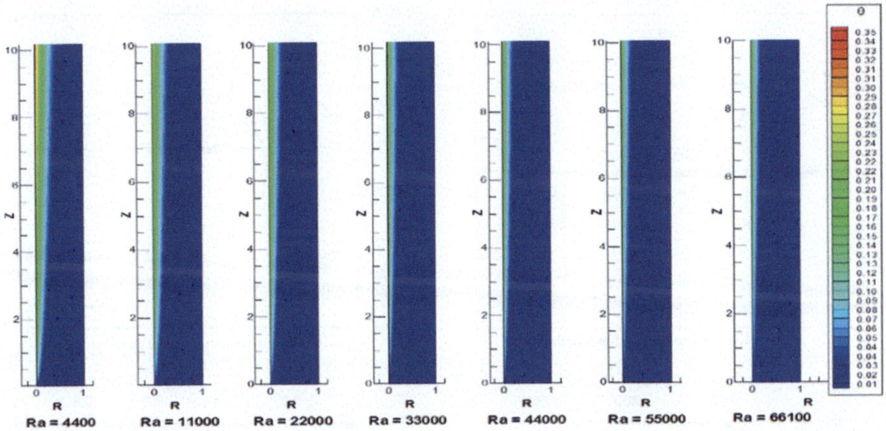

Fig. 7 Comparison of thermal boundary layer along the axial length for fully heated annulus at different Rayleigh numbers at steady state

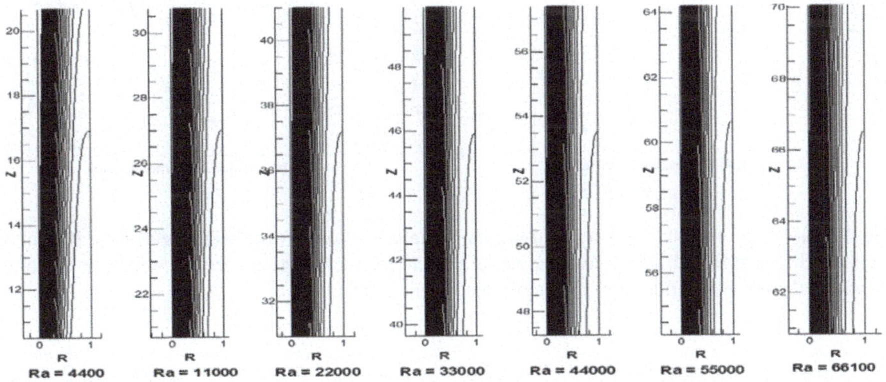

Fig. 8 Comparison of thermal entrance length full heated annulus at different Rayleigh numbers at steady state

the period is prolonged, signifying the emergence of buoyancy and the start of flow. As the flow develops, the radial velocity decreases until it reaches virtually zero at a steady state. The radial velocity time history in Fig. 9b demonstrates that the flow eventually vanishes as steady state approaches and the flow reaches mid height.

Radial velocity decreases from the bottom part of the vertical annulus to the top part. The radial velocity component must vanish at the axial site where the steady or parallel flow completely develops. Figure 9 depicts the variation in radial velocity along different axial lengths (d). The intake of the annulus has a radial velocity that is almost zero. There is little velocity close to the heated area because there is only axial flow at the entrance. The highest radial velocity, shown in this diagram to be at

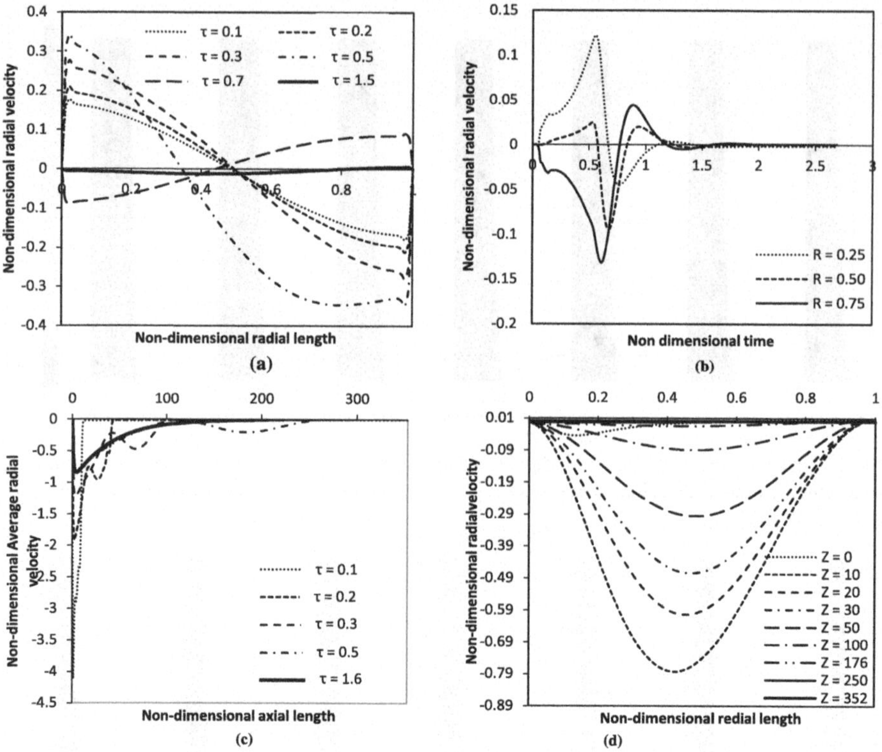

Fig. 9 Variation of radial velocity **a** along the radial direction at mid height with time **b** at mid height at different R with time **c** average radial velocity along the axial direction with time **d** along the radial direction at different axial length ($Ra = 4.4 \times 10^4$)

the bottom of the annulus (Z = 20), virtually vanishes at the point where the flow is fully formed.

Additionally, a negative value indicates a flow reversal. As a result, fluid is forced into the heated interior cylinder from regions close to the outer border. The radial velocity maxima are located inside the heated cylinder. The radial velocity vanishes as the flow stabilizes and develops to its full potential. Radial velocity is not possible because the flow becomes parallel in the fully formed zone. In addition, if the annulus is not high enough at a steady state, flow may ascend without fully forming.

When the heat supply first starts, there are more oscillations in the lower zone. The bottom zone does not exhibit vanishing radial velocity in steady state because the fluctuations have finally been reduced. But if we move in the axial direction, the magnitude of the radial velocity decreases until it reaches zero when the flow has fully developed in that direction.

Figure 10a and b show the transient increase of the radial velocity profile with time at the annulus' inlet for different Rayleigh numbers (b). As the heating process gets going, negative high-intensity radial velocity contours start to appear close to the intake. As time goes on, the contours lengthen toward the axial length, the radial velocity intensity decreases, and they stabilize at a steady state. Conduction's early dominance in this situation is once again obvious. But as time goes on, convection starts because of the buoyancy force and ultimately triumphs over conduction, as shown in Fig. 10b. Figure 10a–b show that when the Rayleigh number grows, the transient response of radial velocity increases with large magnitude (b). The growth of the negative radial velocity is restricted to the lower area at the dimensionless time (t = 0.1) when the Rayleigh number is low, but it is wider in the axial length when the Rayleigh number is high. It suggests that as the axial length approaches, the radial velocity-negative outlines expand more quickly. So with high Rayleigh numbers compared to low Ra flows, buoyant convection grows more rapidly. Figure 11 shows the radial velocity profile at the annulus inlet for different Rayleigh numbers. While the radial velocity's amplitude is negligibly small for smaller Rayleigh numbers, it grows and peaks in the axial direction for larger Rayleigh numbers. The thermal entrance length increases as a result.

The axial velocity contours for the fully heated annulus are shown changing over time in Fig. 12. The rise in axial velocity over time is clearly seen in Fig. 12. Axial velocity increases initially near the heated wall. It then gradually spreads over the whole annulus as time goes on. The intensity of the axial velocity increases as well.

Figure 13 depicts the axial velocity variation for different Rayleigh numbers at the midway of the annulus. The trend in axial velocity is steady. Only the magnitude rises as Rayleigh numbers increase.

Heat transmission has been described using local and average Nusselt numbers, calculated at the inner wall's heated area. The local Nusselt number Nu of the heated cylindrical inner wall is given as follows:

$$\text{Local Nusselt number, } \mathrm{Nu}_z = \frac{qb}{k(T_{w,z} - T_b)} = \frac{1}{(\theta_{w,z} - \theta_b)} \tag{10}$$

where T_b is the dimensional bulk liquid temperature and is defined as:

$$T_b = \frac{\int \rho C_p w T dA}{\int \rho C_p w dA} \tag{11}$$

where the integrals are evaluated over the heated test section of the annulus,

And average Nusselt number ($\overline{Nu}$) is define as:

$$\overline{Nu} = \frac{1}{(z_2 - z_1)} \int_{z_1}^{z_2} Nu_z dz \tag{12}$$

where z_1 and z_2 are the positions where heating start and end, respectively.

Figure 14a shows, for the fully heated and partially heated cases, the variation of the local Nusselt number along the heated axial length for different Rayleigh numbers. The local Nusselt number decreases from high values near the annular entry and then gradually becomes constant over the whole axial length of the annulus. The increasing

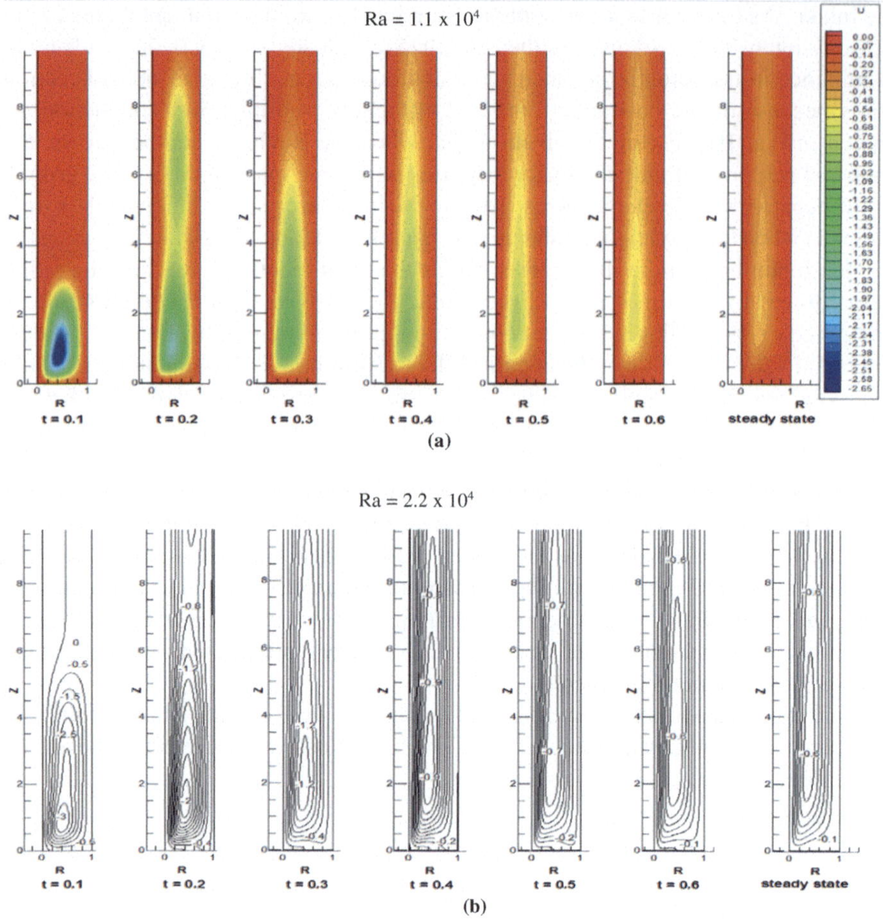

Fig. 10 Transient growth of radial velocity at the inlet of the annulus **(a)** Ra $= 1.1 \times 10^4$ **(b)** Ra $= 4.4 \times 10^4$

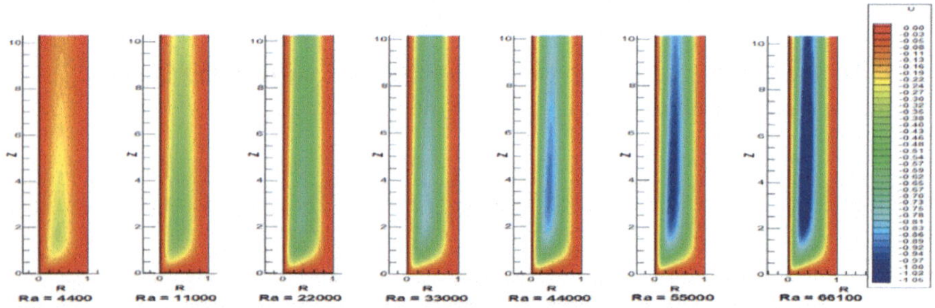

Fig. 11 Steady state contours of radial velocity at the inlet along the axial length for fully heated inner wall of the annulus at different Rayleigh numbers

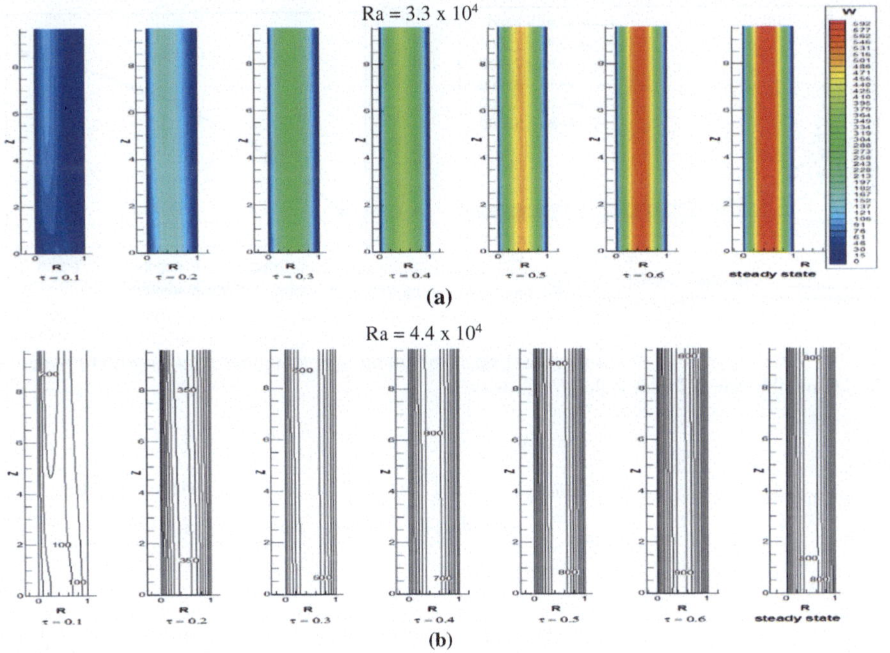

Fig. 12 Contours of axial velocity component at the inlet with time **(a)** Ra $= 3.3 \times 10^4$ **(b)** Ra $= 4.4 \times 10^4$

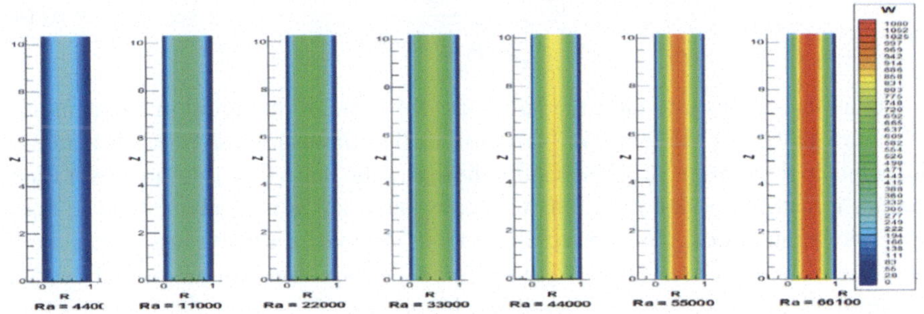

Fig. 13 Comparison of axial velocity along the axial length for full heated annulus at different Rayleigh numbers

boundary layer at the annulus' entrance and fully formed flow over the remainder of its length are therefore represented by these Nusselt number variations. As the Rayleigh number increases, so does the Nusselt number. Figure 14 illustrates the variation between the typical Nusselt and Rayleigh numbers (b). The average Rayleigh number may be expressed in terms of the following relationships, as shown in Table 1. Rayleigh number increases steadily as the average Nusselt number increases.

The non-dimensional volume flow rate $\overline{Q}$ through the annulus is expressed as:

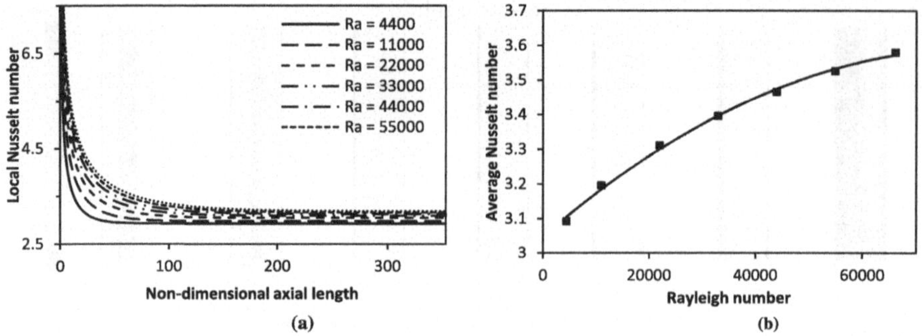

Fig. 14 Variation of Nusselt numbers **(a)** Local Nusselt number along the axial length **(b)** Average Nusselt number for different Raleigh numbers

Table 1 Correlations of Nusselt number

Nu_a	Mean deviation
$3.049 + 0.00001Ra\text{-}8 \times 10^{-11}Ra^2$	±3.15%

$$\overline{Q} = \int_{R_i}^{R_o} 2\pi RWdR \tag{13}$$

Similarly, mean bulk axial velocity is define as:

$$\overline{W} = \frac{\overline{Q}}{\int_{R_i}^{R_o} 2\pi RdR} \tag{14}$$

Equation 14, thus evaluates the bulk mass flow rate of the liquid in the annulus.

Figure 15 shows the temporal variation of the dimensionless mass flow rate at different axial positions (a). The results show that the mass flow rate rises over time, reaches a maximum, and then falls to a constant value at a steady state. The mass flow rate being almost constant at all points in the axial direction in steady state is evidence of mass conservation. Figure depicts the non-dimensional mass flow rate variation for different Rayleigh numbers. Figure 15b. In both cases, the mass flow rate increases along with the Rayleigh number. Convection heat transfer is therefore greater for larger Rayleigh numbers than for lower Rayleigh numbers.

Figure 16 depicts the correlation between the Rayleigh number and the non-dimensional bulk axial velocity. The buoyancy force increases the bulk axial velocity with the Rayleigh number, as seen in Fig. 16.

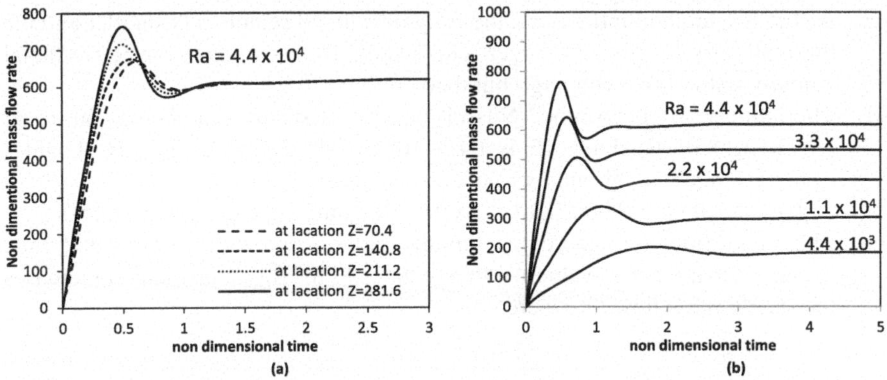

Fig. 15 Variation of mass flow rate with time **(a)** At different axial locations **(b)** For different Rayleigh numbers

Fig. 16 Variation of bulk axial velocity for different Rayleigh numbers

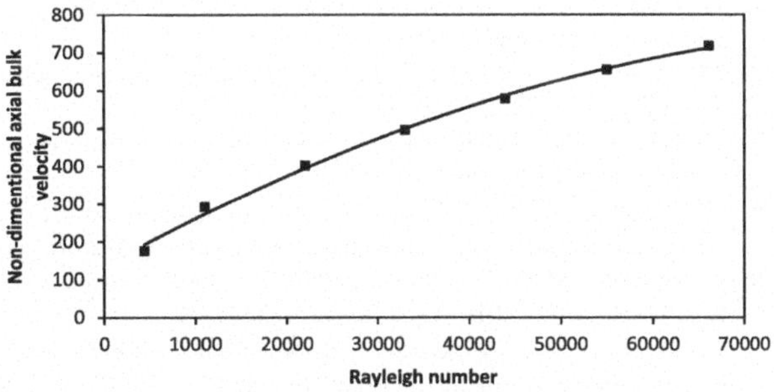

6 Conclusions

The current work examines the fluid flow characteristics and heat transfer for laminar free convection of liquid water in a vertical concentric annular thermosiphon subjected to continuous wall heat flux (Prandtl number = 6.43). Studies on Rayleigh numbers ranging from 4.4×10^3 to 6.61×10^4 have been conducted. For totally hot occurrences, numerical computations were made. The parallel finite difference numerical algorithm was validated using a variety of numerical results. The research's conclusions are as follows.

(i) It was observed that temperature rose quickly and reached a maximum for every radial position owing to conduction. Axial flow then develops as the convection increases in strength, and steady state is shown. The obtained steady state temperature is lower than the transient phase maxima. The term "temperature overshoot" refers to this. In every instance, the aforementioned phenomena was seen.

(ii) As the Rayleigh number rises, the fully developed region gets smaller and the thermal entry length gets longer. This results from stronger convective radial velocity with higher Rayleigh numbers.

(iii) The numerically determined Nusselt number at steady state comes out to be 3.09 to 3.58 for Rayleigh, Ra is 4.4×10^3 to 4.4×10^4. The numerical values of the average Nusselt number get correlated as:

$$Nu_a = 3.049 + 0.00001Ra - 8x10^{-11}Ra^2, \text{mean deviation} = \pm 3.15\%$$

(iv) The mass flow rate increases initially, attains maximum value and then decreases to a constant value at a steady state. As the Rayleigh number increases the steady mass flow rate increases.

References

1. De Vahl Davis G, Thomas RW (1969) Natural convection between concentric vertical cylinder. Phy Fluids 12(suppl II):198–207
2. Mochimaru Y (1987) New way for simulation of transient natural convection heat transfer. J Fluid Mech 8(3):235–239
3. Ho CJ, Lin YH (1988) Natural convection heat transfer of cold within an eccentric horizontal cylinder annulus. J Heat Transf ASME 110:894–901
4. El-Shaarawi MAI, Al-Attas MA (1992) Unsteady natural convection in open-ended vertical concentric annuli. Int J Num Meth Heat Fluid Flow 2:503–516
5. El-Shaarawi MAI, Al-Attas M (1993) Transient induced flow through a vertical annulus. Japan Soc Mech Eng (JSME) Int J Ser B 36:156–165
6. El-Shaarawi MAI, Mokheimer EMA, Jamal A (2007) Geometry effects on conjugate natural convection heat transfer in vertical eccentric annuli. Int J Numer Methods Heat Fluid Flow 17(5):461–493
7. Sankar M, Do Y (2010) Numerical simulation of free convection heat transfer in a vertical annular cavity with discrete heating. Int Commun Heat Mass Transf 37:600–606
8. Desrayaud G, Chénier E, Joulin A, Bastide A, Brangeon B, Cherif Y, Eymard R, Garnier C, Giroux-Julien S, Harnane Y, Joubert P, Lassue S, Le Quéré P, Xin S, Zoubir A (2013) Benchmark solutions for natural convection flows in vertical channels submitted to different open boundary conditions. Int J Therm Sci 72:18–33
9. Mustafa J (2014) Experimental and numerical analysis of heat transfer in a vertical annular thermo-siphon. Ph D thesis, AMU, Aligarh
10. Mustafa J, Altamush Siddiqui M, Fahad Anwer S (2018) Experimental and numerical analysis of heat transfer in a tall vertical concentric annular thermo-siphon at constant heat flux condition. Heat Transf Eng 40(11). https://doi.org/10.1080/01457632.2018.1446867
11. Mustafa J, Husain S, Altamush Siddiqui M (2017) Experimental studies on natural convection of water in a closed loop vertical annulus. Exp Heat Transf 30(1):25–45. Taylor & Francis
12. Saad Y (2011) Numerical methods for large eigenvalue problems. Society for Industrial and Applied Mathematics, Philadephia
13. Cheng L, Armfield S (1995) A simplified marker and cell method for unsteady flows on non-staggered grids. Int J Numer Meth Fluids 21:15–34
14. Hasan N, Anwer SF, Sanghi S (2005) On the outflow boundary condition for external incompressible flows: a new approach. J Comput Phys 206(2):661–683
15. MPI Forum (2009) MPI: a message-passing interface standard. Version 2.2. http://www.mpi-forum.org (March.2011)

16. Gropp W, Lusk E, Skjellum A (1999) Using MPI: portable parallel programming with the message-passing interface. MIT Press, Cambridge, MA, USA, 2 edn. ISBN 0–262–57132–3
17. Amine Z, Daverat C, Xin S, Giroux-Julien S, Pabiou H, Ménézo C (2013) Natural convection in a vertical open-ended channel: comparison between experimental and numerical results. J Energy Power Eng 7:1265–1276

Numerical Solutions of 2D Riemann Problems of Gas Dynamics Using a Hybrid PVU-M+ Scheme

Altaf Ahmed and Nadeem Hasan

Abstract The present work consists of numerical solution of two-dimensional Riemann problems for gas dynamics using PVU-M+ scheme (Hasan et al. Comput Fluid 119:58–86 2015 [1]). The test cases and their initial data are taken from Lax and Liu (SIAM J Sci Comput 19:319–340, 1998 [2]). The flow domain consists of a square geometry divided in 4 quadrants such that there exists only one planar wave (shock, rarefaction or slip line) between each two quadrants. The solutions are analyzed through density contours and line plots at a suitable location in X–Y plane. Although the PVU-M+ scheme resolved most flow features, it showed difficulty in capturing pressure field in a weak velocity field behind a discontinuity (rarefaction/shock). To overcome this, pressure field at cell interface is calculated based on acoustic speeds and a pressure gradient at cell center is obtained using this pressure field which is then blended with forward/backward approximation of pressure gradient using a mean hybrid weight function, $\overline{W}_h = \frac{1}{3}\sum_{k=1}^{3}(W_h)_{i+2-k}$, where the hybrid weight function defined as, $W_h = Ae^{-\left(\frac{M}{B}\right)^2}$ has two adjustable constants ("A" and "B") which can be adjusted according to the flow problems and 'M' being the local Mach number based on convective flux velocity (u or v). This modified PVU-M+ scheme is named "**PVU-M+H**" where "H" stands for hybrid. Recommended values of "A" and "B" are 0.2 and 0.12, respectively, for all type of flow problems. All test cases were solved keeping these values of "A" and "B".

A. Ahmed (✉) · N. Hasan
Department of Mechanical Engineering, AMU, Aligarh 202002, India
e-mail: altaf.ahmed@am.iitd.ac.in

A. Ahmed
Department of Applied Mechanics, Indian Institute of Technology Delhi, New Delhi 110016, India

M. A. Siddiqui et al. (eds.), *Advances in Heat Transfer and Fluid Dynamics*, Lecture Notes in Mechanical Engineering, https://doi.org/10.1007/978-981-99-7213-5_8

1 Introduction

Any newly developed numerical scheme is required to pass a number of standard tests. These test cases involve 1D advection equation, 1D Riemann problems 1D Burger's equation, 1D inviscid Burger's equation, forward facing step, inviscid flow past circular cylinder, 2D Riemann Problems [2–5]. Riemann problems are generally an initial value problem in which initial data is discontinuous. 2D Riemann Problems (RPs) for gas dynamics are very interesting benchmark test cases for a compressible flow numerical scheme. RPs for gas dynamics are governed by Euler Equations. Testing of numerical schemes on flow governed by Euler equations has several advantages such as discontinuities in the flow act as true discontinuities, numerical viscosity of the scheme can be assessed accurately. Furthermore, it is also believed that the discretization of non-linear term (convection term) is the most formidable task for any computational fluid dynamists. To accurately capture pressure field in low velocity region for a compressible flow numerical scheme is a challenging task. Euler equations contain both these terms in its formulation.

2 Literature Survey

In the year 2006, Qamar et al. [6] developed a flux based finite difference numerical scheme for computation of compressible fluid flow problems. It was based on two-step predictor–corrector method which divided the flux of Navier–Stokes equation into two parts namely convective and non-convective. Pressure term, viscous term, conduction term, source term and body forces were part of the non-convective part and convective term was included in the convective part. For discretization of the convective term, it was divided into two parts namely convective velocity and convective property. The convective velocity at cell interface was evaluated using averaging of the neighboring cell value and convective property at cell interface was calculated using the first order upwind scheme. The non-convective part was approximated at grid points using forward difference in the predictor step and backward difference in the corrector step. The major advantage of this scheme was that it was able to capture discontinuities without any spurious oscillations and the major disadvantage of this scheme was that it uses 1st order upwind scheme for approximating convective properties at cell interfaces in the entire flow domain, hence compromising accuracy. It was named "PVU scheme" i.e., Particle Velocity Upwind.

In the year 2010, again Qamar et al. [7] proposed a new method for approximating the convective flux at the cell interface to increase the overall accuracy of the earlier proposed PVU scheme [6]. The strategy was to use low order accurate approximation near steep gradients/discontinuities and higher order accurate approximation near smoothly varying solution. To detect region of discontinuities they have used a smoothness indicator function suggested by Jiang and Shu [8] and accordingly changed 1st order upwind approximation near discontinuities and 2nd order upwind approximation in smooth region for convective properties at cell interface. Convective velocity at cell interface was determined by quadratic interpolation. The main advantage of this scheme over the previous PVU scheme was that it was more accurate in the smooth region. The transition from higher order approximation to lower order approximation was not smooth. The switch worked as binary either first order or higher order depending upon the threshold value of smoothness indicator. To remove this abrupt change in approximation of convective flux near the threshold value of the smoothness indicator, some kind of blending was required between higher order and lower order approximation which can increase the accuracy near discontinuities without any spurious oscillations.

In the year 2015, Hasan et al. [1] modified the earlier proposed PVU scheme [6] and presented a newly developed PVU-M+ scheme [1]. This scheme was based on blending of lower order and higher order approximation for convective property and convective velocity at cell interface. The scheme was tested against different types of 1D/2D benchmark viscous/inviscid test problems. It has performed well in these test cases as compared to state of art schemes such as Van Leer, HLL and AUSMPW+. Although the scheme has been tested against a number of benchmark test problems, the 2D Riemann test problems have been left out.

3 Governing Equations

The non-dimensional Euler equations for a perfect gas in 2D, in the absence of body forces, can be given as,

$$\frac{\partial \boldsymbol{U}}{\partial t} + \frac{\partial \boldsymbol{F}}{\partial x} + \frac{\partial \boldsymbol{G}}{\partial y} = 0 \tag{1}$$

$$\boldsymbol{U} = \begin{Bmatrix} \rho \\ \rho u \\ \rho v \\ \rho E_t \end{Bmatrix} \boldsymbol{F} = \begin{Bmatrix} \rho u \\ \rho u u + p \\ \rho v u \\ \rho h_t u \end{Bmatrix} \boldsymbol{G} = \begin{Bmatrix} \rho v \\ \rho u v \\ \rho v v + p \\ \rho h_t v \end{Bmatrix} \tag{2}$$

where ρ is the non-dimensional density, u is the non-dimensional x-velocity, v is non-dimensional the y-velocity, $E_t = e + \frac{1}{2}(u^2 + v^2)$ is the non-dimensional total energy, e is the non-dimensional internal energy of the gas, $p = (\gamma - 1)\rho e$ is the non-dimensional pressure, and γ is the ratio of specific heats, $\gamma = 1.4$. A reference density ρ_{ref} and a reference pressure p_{ref} are chosen to scale density and pressure. These reference values are based on the type of gas selected for the simulation and only required to convert the non-dimensional initial condition into dimensional while solving dimensional Euler equations. The x and y velocities are scaled with a common reference velocity U_{ref} which is defined as $U_{ref} = \sqrt{\frac{p_{ref}}{\rho_{ref}}}$. Here U_{ref} is not equal to speed of sound at (ρ_{ref}, p_{ref}) but is $\frac{1}{\sqrt{\gamma}}$ times the speed of sound. The total energy "E$_t$" is non dimensionalized using U_{ref}^2. *These non-dimensional parameters are chosen in such a way that non-dimensional Euler equations remains same as the dimensional Euler equations.*

4 Two-Dimensional Riemann Problems

The 2D Riemann problem is the Euler equation with discontinuous initial data such that only one planar wave (shock, rarefaction or slip line) can exist at an interface between two neighboring discontinuous data. The computational domain is a square geometry in Cartesian coordinate system ranging $(x, y) = (0, 1) \times (0, 1)$ and is divided in four quadrants such that

$$(\rho, u, v, p)(x, y, 0) = \begin{cases} (\rho_1, u_1, v_1, p_1), x > 0.5, y > 0.5 \\ (\rho_2, u_2, v_2, p_2), x < 0.5, y > 0.5 \\ (\rho_3, u_3, v_3, p_3), x < 0.5, y < 0.5 \\ (\rho_4, u_4, v_4, p_4), x > 0.5, y < 0.5 \end{cases}$$

where indices k = 1, 2, 3, 4 represent the quadrant of the computational domain and (ρ_k, u_k, v_k, p_k) are constant in each quadrant (Fig. 1).

Classification of the Riemann problem is based on the combination of four elementary planar waves that define it. Suppose w and $w\prime$ represent normal and

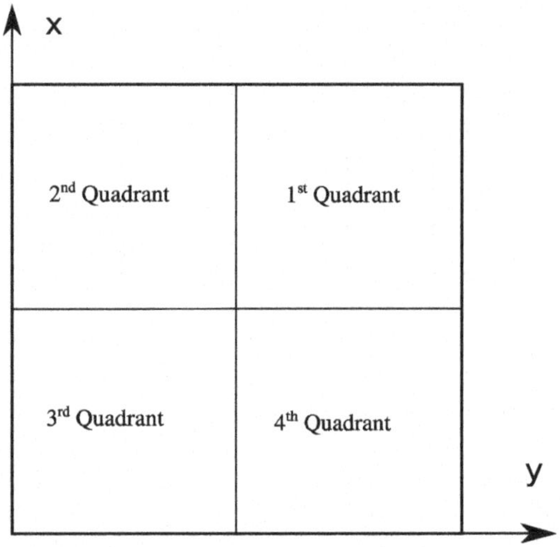

Fig. 1 Flow domain

tangential velocity components to the line of discontinuity. Define (l, r) such that $(l, r) \in \{(2, 1), (3, 2), (3, 4), (4, 1)\}$, A forward rarefaction wave $\overrightarrow{R_{lr}}$, or backward rarefaction wave $\overleftarrow{R_{lr}}$ is calculated by the formula,

$$\overrightarrow{R_{lr}} : w_l - w_r = \frac{2\sqrt{\gamma}}{\gamma - 1}\left(\sqrt{\frac{p_l}{\rho_l}} - \sqrt{\frac{p_r}{\rho_r}}\right), \; w_l\prime = w_r\prime$$

$$\overleftarrow{R_{lr}} : w_l - w_r = \frac{2\sqrt{\gamma}}{\gamma - 1}\left(\sqrt{\frac{p_r}{\rho_r}} - \sqrt{\frac{p_l}{\rho_l}}\right), \; w_l\prime = w_r\prime$$

For polytropic gas,

$$\overrightarrow{R_{lr}} \text{ and } \overleftarrow{R_{lr}} : \frac{p_l}{p_r} = \left(\frac{\rho_l}{\rho_r}\right)^{\gamma}$$

A forward shock wave $\overrightarrow{S_{lr}}$ or a backward shock $\overleftarrow{S_{lr}}$ is calculated by the formula,

$$\overrightarrow{S_{lr}} : \frac{w_r - w_l}{\rho_r - \rho_l} = +\sqrt{\frac{1}{\rho_r \rho_l}\frac{p_r - p_l}{\rho_r - \rho_l}} \; w_l\prime = w_r\prime$$

$$\overleftarrow{S_{lr}} : \frac{w_r - w_l}{\rho_r - \rho_l} = -\sqrt{\frac{1}{\rho_r \rho_l}\frac{p_r - p_l}{\rho_r - \rho_l}} \; w_l\prime = w_r\prime$$

For polytropic gas,

$$\overrightarrow{S_{lr}} \text{ and } \overleftarrow{S_{lr}} : \frac{\rho_l}{\rho_r} = \left(\frac{p_l}{p_r} + \frac{\gamma - 1}{\gamma + 1}\right) / \left(1 + \frac{(\gamma - 1)p_l}{(\gamma + 1)p_r}\right)$$

In 2D a positive slip line J_{lr}^{+} and a negative slip line J_{lr}^{-} is given by (Table 1),

$$J_{lr}^{+} : w_l = w_r \, p_l = p_r w_l\prime \le w_r\prime$$

$$J_{lr}^{-} : w_l = w_r \, p_l = p_r w_l\prime \ge w_r\prime$$

5 Numerical Methodology

PVU-M+ scheme [1] is essentially a modified version of an earlier PVU scheme [6]. It is a flux based finite difference scheme for compressible flow which is based on direct discretization of conservation equations (Mass, Momentum and Energy) on

Table 1 Classification of 2D Riemann problems

4R	Test case 01			Test case 02											
		$\overrightarrow{R_{21}}$			$\overrightarrow{R_{21}}$										
	$\overrightarrow{R_{32}}$		$\overrightarrow{R_{41}}$	$\overleftarrow{R_{32}}$		$\overrightarrow{R_{41}}$									
		$\overrightarrow{R_{34}}$			$\overleftarrow{R_{34}}$										
4S	Test case 03			Test case 04											
		$\overleftarrow{S_{21}}$			$\overleftarrow{S_{21}}$										
	$\overleftarrow{S_{32}}$		$\overleftarrow{S_{41}}$	$\overrightarrow{S_{32}}$		$\overleftarrow{S_{41}}$									
		$\overleftarrow{S_{34}}$			$\overrightarrow{S_{34}}$										
4J	Test case 05			Test case 06											
		J_{21}^-			J_{21}^-										
	J_{32}^-		J_{41}^-	J_{32}^+		J_{41}^+									
		J_{34}^-			J_{34}^-										
2J + 2R	Test case 07			Test case 08			Test case 09			Test case 10					
		$\overrightarrow{R_{21}}$			$\overleftarrow{R_{21}}$			J_{21}^+			J_{21}^-				
	J_{32}^-		$\overrightarrow{R_{41}}$	J_{32}^-		$\overleftarrow{R_{41}}$	$\overrightarrow{R_{32}}$		$\overrightarrow{R_{41}}$	$\overrightarrow{R_{32}}$		$\overrightarrow{R_{41}}$			
		J_{34}^-			J_{34}^-			J_{34}^+			J_{34}^+				
2J + 2S	Test case 11			Test case 12			Test case 13			Test case 14					
		$\overleftarrow{S_{21}}$			$\overrightarrow{S_{21}}$			J_{21}^-			J_{21}^+				
	J_{32}^+		$\overleftarrow{S_{41}}$	J_{32}^+		$\overrightarrow{S_{41}}$	$\overleftarrow{S_{32}}$		$\overleftarrow{S_{41}}$	$\overleftarrow{S_{32}}$		$\overleftarrow{S_{41}}$			
		J_{34}^+			J_{34}^+			J_{34}^-			J_{34}^-				
2J + R + S	Test case 15			Test case 16			Test case 17			Test case 18			Test case 19		
		$\overrightarrow{R_{21}}$			$\overleftarrow{R_{21}}$			J_{21}^-			J_{21}^+			J_{21}^+	
	J_{32}^-		$\overleftarrow{S_{41}}$	J_{32}^-		$\overrightarrow{S_{41}}$	$\overleftarrow{S_{32}}$		$\overrightarrow{R_{41}}$	$\overleftarrow{S_{32}}$		$\overrightarrow{R_{41}}$	$\overleftarrow{S_{32}}$		$\overrightarrow{R_{41}}$
		J_{34}^+			J_{34}^+			J_{34}^-			J_{34}^+			J_{34}^-	

numerical grid. Before applying PVU-M+ scheme, the 2D non-dimensional Euler equations can be written in a little different way,

$$\frac{\partial \boldsymbol{U}}{\partial t} + \frac{\partial(\boldsymbol{F}^c + \boldsymbol{F}^{nc})}{\partial x} + \frac{\partial(\boldsymbol{G}^n + \boldsymbol{G}^{nc})}{\partial y} = 0$$

where,

$$\boldsymbol{U} = \begin{Bmatrix} \rho \\ \rho u \\ \rho v \\ \rho E_t \end{Bmatrix} \boldsymbol{F}^c = \begin{Bmatrix} \rho u \\ \rho uu \\ \rho vu \\ \rho h_t u \end{Bmatrix}, \boldsymbol{G}^c = \begin{Bmatrix} \rho v \\ \rho uv \\ \rho vv \\ \rho h_t v \end{Bmatrix}, \boldsymbol{F}^{nc} = \begin{Bmatrix} 0 \\ p \\ 0 \\ 0 \end{Bmatrix}, \boldsymbol{G}^{nc} = \begin{Bmatrix} 0 \\ 0 \\ p \\ 0 \end{Bmatrix}.$$

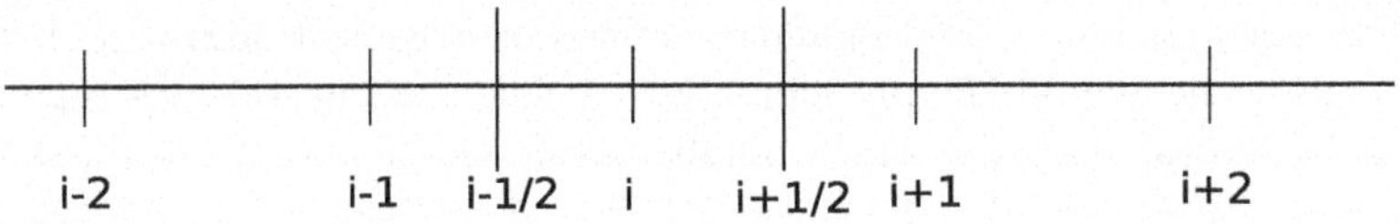

Fig. 2 Computational molecule for ith node

In 1-D framework PVU-M+ scheme can be described with the help of Fig. 2 which shows a typical one-dimensional mesh and a computational segment surrounding the ith node.

PVU-M+ scheme is basically a two-step predictor–corrector based numerical algorithm, to obtain solution at the new time level (n + 1) the predictor and corrector steps for the solution vector at the ith grid point are given as,

Predictor step:

$$U^* = U^n - \Delta t\left\{\frac{F^c_{i+1/2}(U^n) - F^c_{i-1/2}(U^n)}{x_{i+1/2} - x_{i-1/2}} + \frac{F^{nc}_{i+1}(U^n) - F^{nc}_{i}(U^n)}{x_{i+1} - x_i}\right\},$$

Corrector step:

$$U^{n+1} = \frac{(U^* + U^n)}{2} - \frac{\Delta t}{2}\left\{\frac{F^c_{i+1/2}(U^*) - F^c_{i-1/2}(U^*)}{x_{i+1/2} - x_{i-1/2}} + \frac{F^{nc}_{i}(U^*) - F^{nc}_{i-1}(U^*)}{x_i - x_{i-1}}\right\}.$$

The time integration is 2nd order accurate. The detail analysis of the scheme is given in [1].

6 Modification in PVU-M+ Scheme

The estimation of pressure gradient is changed in new modification. It is found that forward/backward estimation of the pressure term does not work in the low velocity region in some cases of 2D Riemann problems. It is proposed that in the low velocity region characteristics based pressure splitting should be used to capture the pressure waves for estimation of pressure gradients. The pressure gradient at the ith grid point in 1D frame is then given as:

$$\left.\frac{\partial p}{\partial x}\right|_i = \overline{W}_h \left.\frac{\partial p}{\partial x}\right|_i^s + (1 - \overline{W}_h)\left.\frac{\partial p}{\partial x}\right|_i^{f/b}$$

$\overline{W}_h$ is mean hybrid weight function which, at ith grid point, is defined as:

$$\overline{W}_h = \frac{\left(W_h|_{i-1} + W_h|_i + W_h|_{i+1}\right)}{3}$$

Hybrid weight function "W_h" should be defined such that its value ranges between 0 and 1. Its value should sharply increase to 1 near the low velocity region and decrease to zero as the local Mach number increases beyond a threshold value. A Gaussian distribution type function for W_h can deliver such characteristics. Therefore, we set:

$$W_h = Ae^{-\left(\frac{M}{B}\right)^2}$$

where "M" is local Mach number and "A" and "B" are adjustable constants.

$\left.\frac{\partial p}{\partial x}\right|_i^{f/b}$ is due to the original PVU-M+ scheme and $\left.\frac{\partial p}{\partial x}\right|_i^{s}$ is defined as,

$$\left.\frac{\partial p}{\partial x}\right|_i^{s} = \frac{p_{i+\frac{1}{2}} - p_{i-\frac{1}{2}}}{x_{i+\frac{1}{2}} - x_{i-\frac{1}{2}}}$$

where superscript "s" stands for "split" based on wave-based pressure splitting. The intercell value of pressure $p_{i\pm\frac{1}{2}}$ can be determined based on well-known splitting techniques [9] as,

$$\begin{aligned} p_{i+\frac{1}{2}} &= f^{+}(M_i) \times p_i + f^{-}(M_{i+1}) \times p_{i+1}, \\ p_{i-\frac{1}{2}} &= f^{+}(M_{i-1}) \times p_{i-1} + f^{-}(M_i) \times p_i \end{aligned}$$

$f^{\pm}$ are pressure weight functions which are functions of Mach number 'M'. f^{+} is related to the right running wave at ith grid point and f^{-} is related to the left running wave at the (i + 1)th grid point for intercell value of pressure at (i + 1/2) i.e., $p_{i+\frac{1}{2}}$ and similarly for $p_{i-\frac{1}{2}}$. $f^{\pm}$ should be defined in such a manner that at supersonic speed one weight function should be zero and other should be 1 and at subsonic speed contribution from both sides (left and right) should be used to determine intercell pressure value. Values of weight functions $f^{\pm}$ should be proportional to the magnitude of characteristic speeds ($(u \pm a)$ or $(M \pm 1)$) to which they are related in subsonic flow. Following the above arguments, the simplest function for $f^{\pm}$ which can be thought of is a 1st order polynomial function of characteristic speed $(M \pm 1)$ given [9, 10] as,

$$f^{\pm}(M) = \begin{cases} \frac{1 \pm M}{2}, & |M| \le 1 \\ \frac{1 \pm sgn(M)}{2}, & |M| > 1 \end{cases}$$

A third order polynomial function was also given by [8] which is differential at Mach number 1 as,

$$f^{\pm}(M) = \begin{cases} \frac{(M \pm 1)^2 (2 \mp M)}{4}, & |M| \le 1 \\ \frac{1 \pm sgn(M)}{2}, & |M| > 1 \end{cases}$$

Observation of special cases:

$$M = +1 \quad f^+(M) = 1.0 \quad f^-(M) = 0$$
$$M > 1 \quad f^+(M) = 1.0 \quad f^-(M) = 0$$
$$M \le -1 \quad f^+(M) = 0 \quad f^-(M) = 1.0$$

The new modified scheme was named '**PVU-M+H**' where '**H**' stands for 'hybrid'.

7 Performance of the Proposed PVU-M+H Scheme

To simulate 2D Riemann problems for gas dynamics, we have chosen a square geometry in the Cartesian coordinate system ranging from (0, 0) to (1, 1) as shown in Fig. 1. The geometry is divided into 400 × 400 uniform grid points in X and Y directions, respectively. The governing equation (Eq. 1) is then coded in ForTran language using PVU-M+ and PVU-M+H schemes. Time step 'Δt' and kappa 'κ' are taken to be 10^{-05} and 0.1 respectively for all 19 test cases. All four sides of the square domain are open boundary type i.e., the square domain can be seen as a part of larger square domain cut around the center. The boundary values are extrapolated from interior points by putting one sided second order accurate approximation of gradient of variables equal to zero. The Initial conditions are taken from [2]. Figure 3 shows the zoomed version of discretized flow domain showing the grid shape and size.

We will compare the both PVU-M+ and PVU-M+H schemes using density contours of the test cases. Since we can visualize contact discontinuity, shock waves and rarefaction waves using density, hence it is very important variable for comparison purposes in compressible flow. 2D contours can only show us the qualitative aspect of the schemes. To assess the schemes on quantitative basis, we will also compare the profile of different variables at suitable locations. Every contour plot

Fig. 3 Zoomed view of the discretized domain

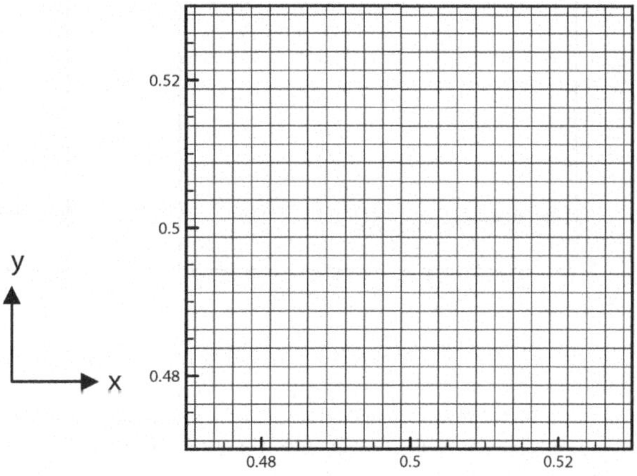

has been supplemented with streamlines (black color lines with arrows on it) which will help us in visualizing flow directions. Only test cases 01, 02, 03 and 06 which contain only one type of elementary waves at interfaces are discussed here.

Test case 01: Test case 01 involves four forward moving rarefaction waves at initial discontinuities (quadrant interfaces). PVU-M+ scheme (Fig. 4a) captures rarefaction waves very well. Direction of movement of all the waves is opposite to the flow direction. The scheme was able to resolve flow features without any difficulties and the contours conform to Lax and Liu [2]. PVU-M+H (Fig. 4b) resolves rarefaction

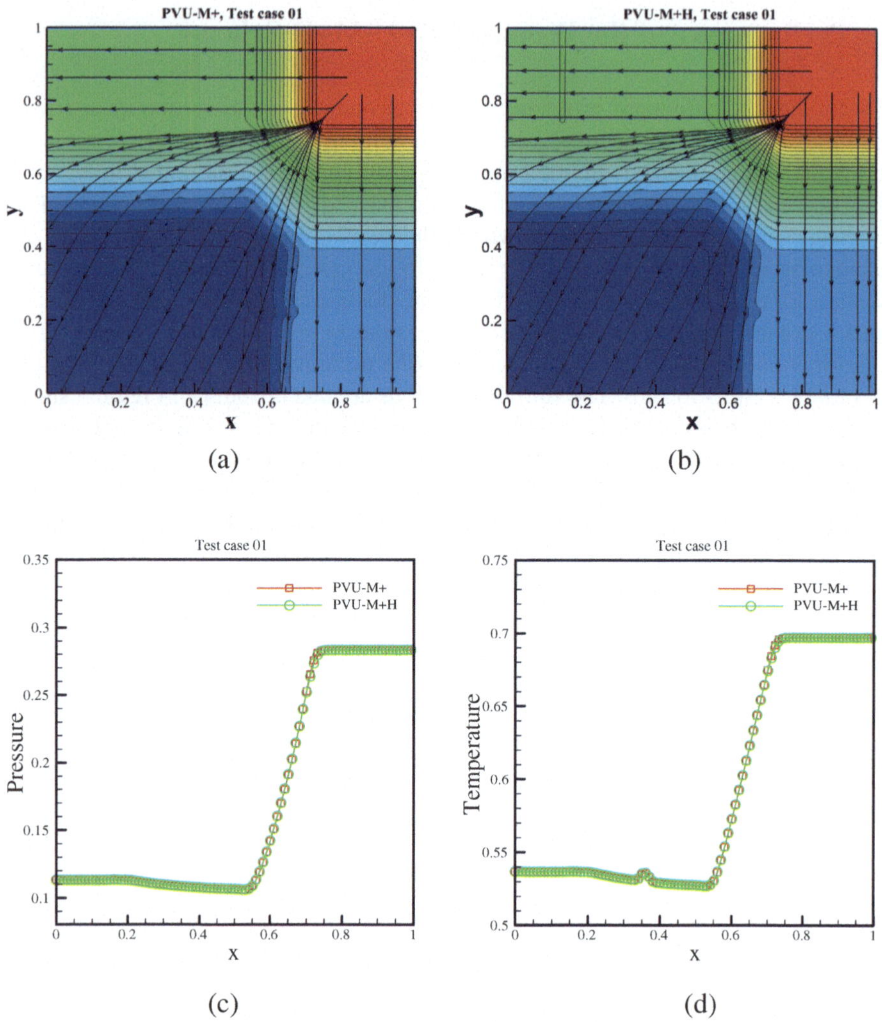

Fig. 4 **a** & **b** Show density contours of test case 01 for PVU-M+ & PVU-M+H respectively at time t = 0.2 and **c** & **d** show profiles for PVU-M+ and PVU-M+H at y = 0.5

waves same as PVU-M+. Profiles of temperature (Fig. 4d) by both the schemes show overheating at approx. x = 0.35. These artifacts (overheating) can be seen in S-E and N-W regions with more visibility near the S-E region (Figs. 4a–b). These artifacts start building from initial discontinuities at time t = 0 and move with the flow.

Test case 02: Test case 02 too involves 4 rarefaction waves but this time two forward and two backward running waves. Figure 5 shows comparison of PVU-M+ and PVU-M+H. Very large oscillations can be seen near S-E and N-W regions in density contour of PVU-M+ (Fig. 5a) which is not desirable if we look at Lax and Liu [2] whereas PVU-M+H (Fig. 5b) performs way better than PVU-M+ removing

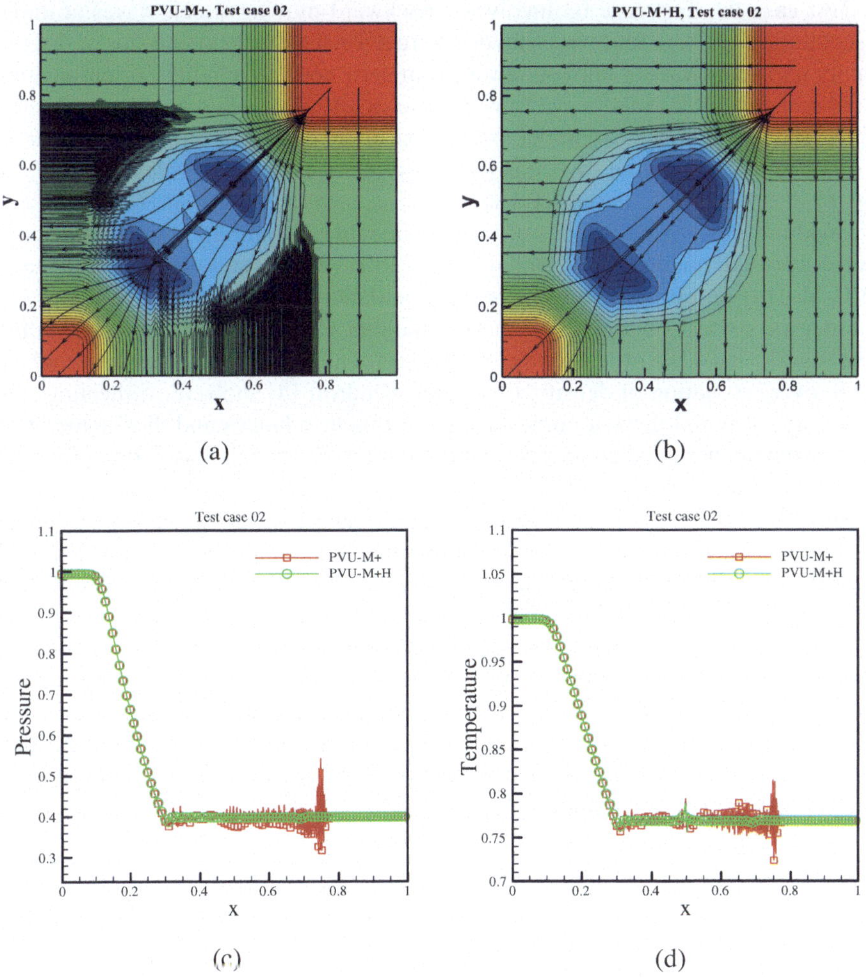

Fig. 5 **a** & **b** Show density contours of test case 02 for PVU-M + & PVU-M+H respectively at time t = 0.2 and **c** & **d** show profiles for PVU-M+ and PVU-M+H at y = 0.1

large oscillations. Take a look at oscillations between quadrant 3 and 4, these are oscillating in the y-direction and if we look at initial data between these two quadrants, we find that velocity normal to the wave front ($\overleftarrow{R_{34}}$) i.e., x-velocity, is zero in the 4th quadrant and the wave is moving away from the zero-velocity zone (dead zone). The overheating phenomenon can also be found at initial discontinuity (not visible in PVU-M+) that does not move with flow as the normal velocity (x-velocity) is zero around that region. Same can be said about oscillations between quadrant 2 and 3 with normal velocity (y-velocity) being zero in the 2nd quadrant. Figure 5b still have some oscillations at initial discontinuities of 2–3 and 3–4 quadrants. But the overall effects of PVU-M+H are positive. Two shock waves are created due to interaction of rarefaction waves near N-E and S-W regions.

Test case 03: Test case 03 involves 4 backward moving shock waves at initial discontinuities. These four waves involve zero normal velocity at one side of discontinuity and each wave is moving away from zero normal velocity region creating oscillations parallel to the wave front in case of PVU-M+ (Fig. 6a). In quadrant 2 of Fig. 6a, oscillations are parallel to x-axis since y-velocity is zero and in quadrant 4 oscillations are parallel to y-axis since x-velocity is zero (not clearly visible in Fig. 6a due to chosen contour level).

Quadrant 1 is having patches of contour line due to combine effect of discontinuities between quadrants 1–2 and 4–1 as both x-velocity and y-velocity are zero. PVU-M+H (Fig. 6b) able to suppress the oscillations away from shock waves but fail to resolve shocks without spurious oscillations. Contrary to our previous finding (test case 01 and 02) a little cooling (Fig. 6d) is found at initial discontinuities which led to over estimation of density. Liska and Wendroff [4] showed overheating (dip in density) at initial discontinuities using 8 different schemes and they argued that the scheme having good contact resolution don't have enough dissipation to reduce these type of errors in density.

Test case 06: Test case 06 involves 2 positive and 2 negative slip lines at initial discontinuities. The solution (Fig. 7a) shows the formation of vortex (low pressure region) in the middle of the computational domain. The PVU-M+ performed well in capturing the slip lines and its other flow features.

All slip lines are moving in anti-clockwise direction leading to the formation of some type of shock (compressive wave) just above and below the central region. The entire flow domain is divided in two parts (a) subsonic: area near center and (b) supersonic as move away from center. Different profiles along centerline parallel to x-axis are shown in Fig. 7c and d. The pressure profile (Fig. 7c) confirms the low-pressure region in the middle.

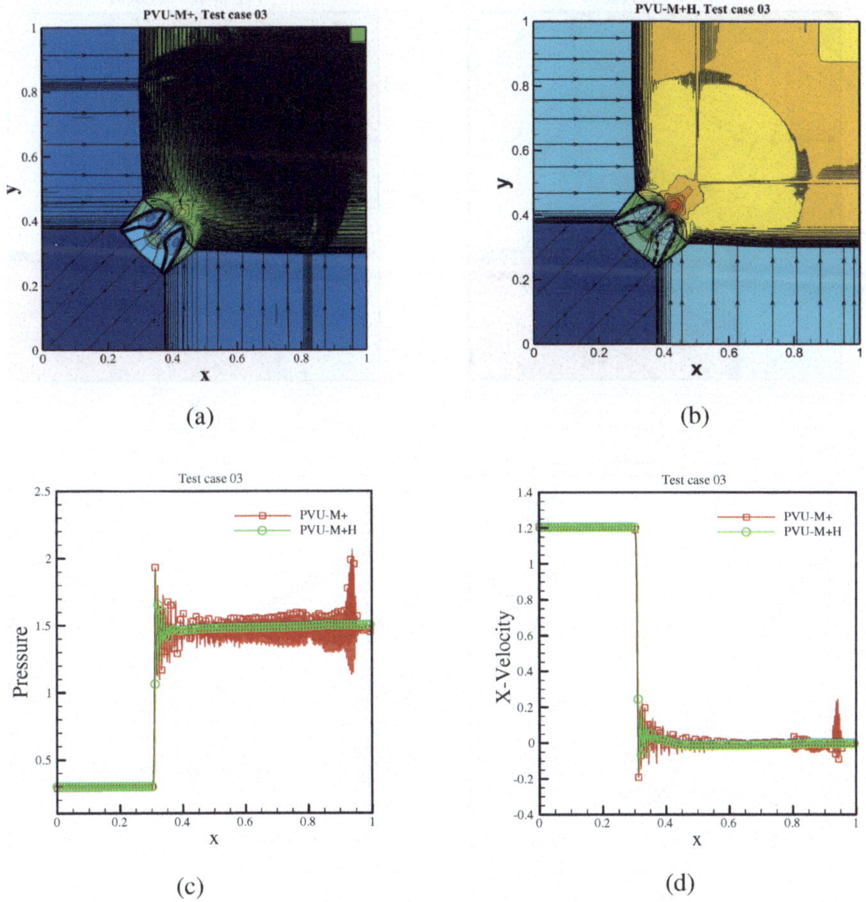

Fig. 6 **a** & **b** Show density contours of test case 03 for PVU-M + & PVU-M+H respectively at time t = 0.3 and **c** & **d** show profiles for PVU-M+ and PVU-M+H at y = 0.6

8 Conclusion

The PVU-M+ scheme was tested for all the 19 test cases. Of all the 19 test problems, we found that there were two regions where PVU-M+ was not good. (1) Region with zero normal (with respect to discontinuity) velocity behind a shock/rarefaction and (2) Vicinity of a stationary slip line when there are rarefaction waves interacting with the slip lines. In the first region the amplitude of the spurious oscillations was very high as compared to the second region. The new PVU-M+H scheme was tasked to address the first problem and it has done well in suppressing the grid scale oscillations in the dead zone behind the discontinuities (Figs. 5 and 6). The adjustable constant for W_h gives a wide control for the blending of two types of pressure approximation. It should

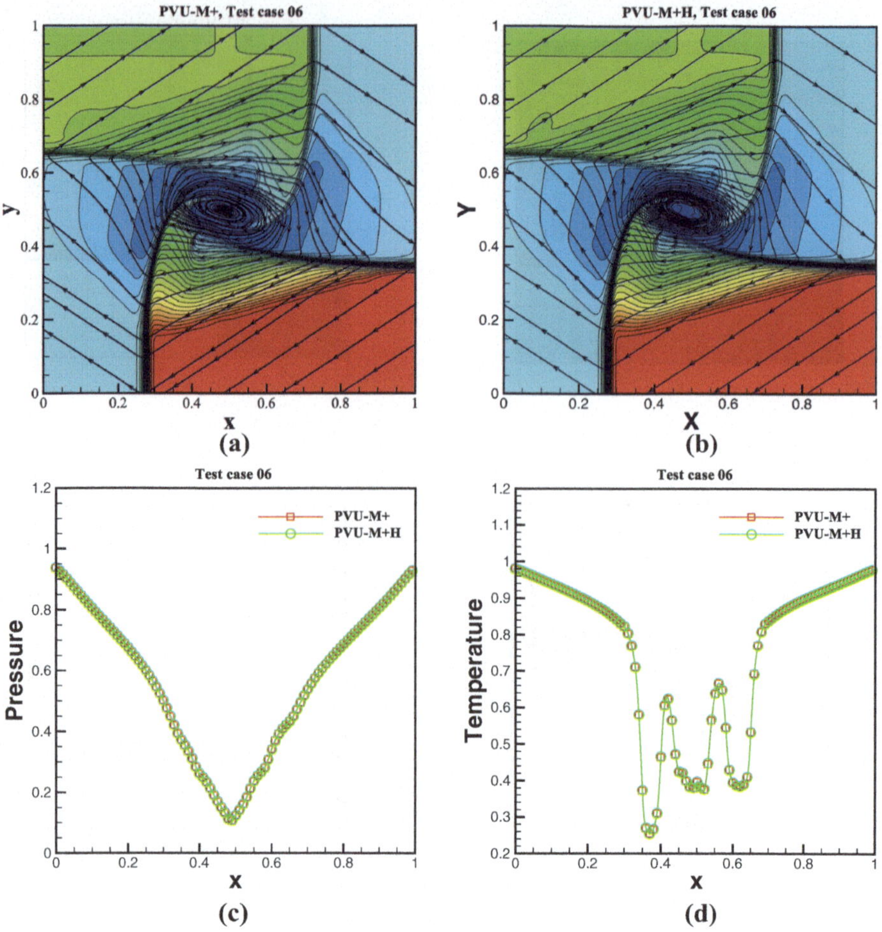

Fig. 7 **a** & **b** Show density contours of test case 06 for PVU-M + & PVU-M+H respectively at time t = 0.3 and **c** & **d** show profiles for PVU-M+ and PVU-M+H at y = 0.5

be noted that grid scale oscillations behind the discontinuities can be completely removed by PVU-M+ with fine tuning of W_h. The recommended upper limit of W_h is 0.2 for any general flow problem. Going beyond this limit starts creating problem for rarefaction waves containing sonic points.

The rarefaction wave with sonic point and discontinuities (shock or rarefaction waves) moving away from a low velocity region or dead zone provide two extreme flow scenarios for the PVU-M+H scheme to test. The highest amplitude of W_h (A = 1.0) results in sharp capturing of shock waves and noise free dead zone whereas rarefaction wave with sonic point gets distorted. The lowest amplitude of W_h (A = 0.0) results in shock waves with spurious oscillations behind it whereas rarefaction wave with sonic point does not pose any difficulty.

Nomenclature

$\overrightarrow{R_{lr}}/\overleftarrow{R_{lr}}$	Forward/backward rarefaction waves
$\overrightarrow{S_{lr}}/\overleftarrow{S_{lr}}$	Forward/backward shock waves
J_{lr}^{+}/J_{lr}^{-}	Positive/negative slip lines
$W_h/\overline{W}_h$	Hybrid/mean hybrid weight function
A/B	Adjustable parameters
M	Local Mach number
$f^{\pm}$	Pressure weight functions
PVU	Particle Velocity Upwind
PVU-M+/PVU-M+H	Original/modified numerical schemes
x, y	Cartesian coordinate directions
u/v	Non-dimensional x/y-direction velocities
ρ	Non-dimensional density
P	Non-dimensional pressure
e	Non-dimensional internal energy
E_t	Non-dimensional total energy
γ	Ratio of specific heats
U	Solution vector
F/G	Flux vectors
$\mathbf{F}^c/\mathbf{G}^c$	Convective flux vectors
$\mathbf{F}^{nc}/\mathbf{G}^{nc}$	Non-convective flux vectors
S-E	South-East
N-E	North-East
S-W	South-West
N-W	North-West

References

1. Hasan N, Mujaheed Khan S, Shameem F (2015) A new flux-based scheme for compressible flows. Comput Fluids 119:58–86. https://doi.org/10.1016/j.compfluid.2015.06.026
2. Lax PD, Liu X (1998) Solution of two-dimensional Riemann problems of gas dynamics by positive schemes. SIAM J Sci Comput 19(2):319–340
3. Kurganov A, Tadmor E (2002) Solution of two-dimensional Riemann problems for gas dynamics without Riemann problem solvers. Numer Methods Partial Differ Equ 18(5):584–608. https://doi.org/10.1002/num.10025
4. Liska R, Wendroff B (2003) Comparison of several difference schemes on ID and 2D test problems for the euler equations. SIAM J Sci Comput 25(3):995–1017. https://doi.org/10.1137/S1064827502402120
5. Jung CY, Nguyen TB (2018) Fine structures for the solutions of the two-dimensional Riemann problems by high-order WENO schemes. Adv Comput Math 44(1):147–174. https://doi.org/10.1007/s10444-017-9538-8
6. Qamar A, Hasan N, Sanghi S (2006) New scheme for the computation of compressible flows. AIAA J 44(5):1025–1039. https://doi.org/10.2514/1.14793

7. Qamar A, Hasan N, Sanghi S (2010) A new spatial discretization strategy of the convective flux term for the hyperbolic conservation laws. Eng Appl Comput Fluid Mech 4(4):593–611. https://doi.org/10.1080/19942060.2010.11015344
8. Jiang G-S, Shu C-W (1996) Efficient implementation of weighted ENO schemes. J Comput Phys 126(1):202–228. https://doi.org/10.1006/jcph.1996.0130
9. Liou MS, Van Leer B, Shuen JS (1990) Splitting of inviscid fluxes for real gases. J Comput Phys 87(1):1–24. https://doi.org/10.1016/0021-9991(90)90222-M
10. Liou M-S, Steffen CJ (1993) A new flux splitting scheme. J Comput Phys 107(1):23–39. https://doi.org/10.1006/jcph.1993.1122

Long Time Evolution of Optimally Perturbed Wing-Tip Vortices

Mohd Suhail Naim and Navrose

Abstract We numerically investigate the effectiveness of optimal perturbations in exciting instabilities in the wake of flow past a finite aspect ratio wing at Re=1000. Analysis of the long-term evolution of perturbed wing-tip vortices is done in a frame of reference that moves roughly at the speed of the perturbation ($\sim U_\infty$; free stream speed). We simulate the scenarios in which the optimal perturbation is periodically injected into the wake. The wing-tip vortices displacement in response to the perturbation serves as a model for the actual perturbation. The maximum displacement experienced by perturbed wing-tip vortices as a function of time is analysed. The maximum displacement displays short time growth (up to 10-time units) after initial decay for the scenario where a single perturbation is present over a length of ~8b (where 'b' is the separation between the wing-tip vortices at $x = 50c$(c: chord length) from the leading edge of the finite wing in the wake) along the wing-tip vortices. Additionally, the impact of altering the spacing between the wing-tip vortices has been investigated. When the separation is reduced to ~0.37b, the initial rate of decay of maximum displacement slows down by $\sim 40\%$. Increased perturbations and further narrowing of the space between the wing-tip vortices are predicted to produce intriguing findings.

1 Introduction

The trailing vortex system, which consists of a pair of counter-rotating vortices, is one of the most prominent characteristics of an aircraft's wake. In the near wake, there may be additional vortices, but in the far wake, the trailing vortex system consists primarily of a pair of counter-rotating vortices, also known as wing-tip vortices. The

M. S. Naim (✉) · Navrose
Department of Aerospace Engineering, Indian Institute of Technology Kanpur, Kanpur 208 016, India
e-mail: mshaim@iitk.ac.in

Navrose
e-mail: navrose@iitk.ac.in

M. A. Siddiqui et al. (eds.), *Advances in Heat Transfer and Fluid Dynamics*, Lecture Notes in Mechanical Engineering, https://doi.org/10.1007/978-981-99-7213-5_9

pair of vortices is a robust flow structure. The average lifespan of trailing vortices of a commercial aircraft is between 1 and 3 min. For a heavy airliner, the flow around these vortices is extremely energetic and poses a risk of Wake Vortex Encounter (WVE: where the following aircraft interacts with the wake of a leading aircraft) for the following aircraft, resulting in a sudden loss of lift and violent vibrations. Due to the lack of altitude to recover, WVE can be extremely dangerous, especially during landing and takeoff. As part of the regulations, a minimum distance is maintained between the leading and trailing aircraft to prevent WVE, which limits an airport's operational capacity. The lifespan of 1 to 3 min suggests that the trailing vortices do not simply decay just by diffusion. They undergo an instability that is sinusoidal and symmetrical about the midplane, later stages of evolution include reconnection at regular intervals to form a series of vortex rings, and eventually flow transforms into a less harmful turbulent state. It is highly desirable to mitigate the lifespan of the trailing vortices. Widnall [1], Spalart [2], and Leweke et al. [3] review articles give a full picture of all the work that has been done on vortex systems.

Significant work by Crow [4] in which a linear stability analysis of the aircraft wake modelled as two line vortices determined that the pair is linearly unstable. The most unstable mode results in sinusoidal and symmetrical (relative to the midplane) displacement of vortices with a wavelength $\sim$8.6 times the gap between the vortices. The study was modal (asymptotic analysis), lacking transient flow processes, and restricted to early stages of instability development. Several methods have been proposed in an effort to accelerate the decay of trailing vortices. Farrell [5] and Trefethen et al. [6] demonstrate that flow systems can support transient growth, which can stimulate non-linear processes that may result in an earlier onset of turbulence (Schmid [7] and Kerswell [8]).

Recently, Navrose et al. [9] performed the transient growth analysis of flow perturbation past a fine-span wing. They determined the optimal perturbation for varying *Re*(Reynolds number), *AR*(aspect ratio), α(angle of attack), the geometry of wing cross-section, and Time horizon. The optimal perturbation was identified as chordwise periodic structures near the surface of the wing. For the investigated parameters, the general shape of perturbation remained unchanged. It was discovered that the near-wake supported the transient growth of the optimal perturbation. In the wake, the evolution of optimal perturbation and perturbed wing-tip vortices were examined until *60c* from the wing leading edge. During its motion, the perturbation moves as a wavepacket at a speed of $\sim U_\infty$ and displaces the tip vortices helically about the unperturbed location.

In the present work, we intend to investigate further the evolution of perturbed tip vortices up to $\sim$*370c* from the finite wing leading edge. Full-scale simulations for such a large domain in the streamwise direction and for different parameter values, such as *Re, AR,* and frequency of perturbation injection into wake are highly computationally expensive. We propose a perturbation model in which the actual perturbation is represented solely by the displacements it induces in the wing-tip vortices. To study the evolution of perturbed wing-tip vortices, the '*model perturbation*' is superimposed on the wing-tip vortices and the governing equations are integrated over time. To investigate the effect of multiple perturbations on the dynamics of wing-tip vortices, the number of perturbations may be varied along the streamwise extent. We varied the distance between the wing-tip vortices because reducing the distance between the wing-tip vortices encourages the development of instability Crow [4].

2 Governing Equations

The flow (u,p) is governed by the incompressible Navier–Stokes equations expressed in non-dimensional form as:

$$\nabla \cdot \mathbf{u} = 0. \tag{1}$$

$$\frac{\partial u}{\partial t} + (\mathbf{u} \cdot \nabla)u = -\nabla p + \frac{1}{Re}\nabla^2 u. \tag{2}$$

Here, *u, p,* and *Re* represent velocity, pressure, and Reynolds number, respectively. The physical length and velocity scales have been non-dimensionalized by employing the chord length (c) and free stream velocity (U_∞), respectively. The equations are accompanied by initial and boundary conditions.

3 Perturbation and Baseflow Description

Figure 1a shows the unperturbed wake of the flow past a finite flat plate wing at Re = 1000. The wake is comprised of a pair of wing-tip vortices with opposite circulations, as depicted by a two-colour scheme. The linear optimal perturbation Navrose et al.[9] computed for this fully developed steady flow is depicted in Fig. 1b for T = 10(time horizon: time for which optimization has been carried out). The linear optimal perturbation evolves as a travelling vorticity structure with nearly the same velocity as the free stream speed. Figure 1c depicts the shape of the perturbation when it reaches x = 54c. When this perturbation is added to the flow depicted in Fig. 1a, it displaces the vortices at its location in a helical manner about the unperturbed location of wing-tip vortices (see Fig. 1d).

To carry out our study, we first extract the vertical and spanwise displacement (see Fig. 2a) of the perturbed wing-tip vortices as a function of streamwise distance corresponding to flow instant shown in Fig. 1d. Secondly, we propose a '*model perturbation*' that is collectively represented by the displacements extracted in the first step. The third step involves defining baseflow for conducting our perturbation study, for which we construct a Batchelor vortex corresponding to the vortex circulation and radius at x = 50*c* of the flow shown in Fig. 1a and extend it in the streamwise direction up to $\sim 8b$(Crow[4]). In the final step, the extracted displacements from the

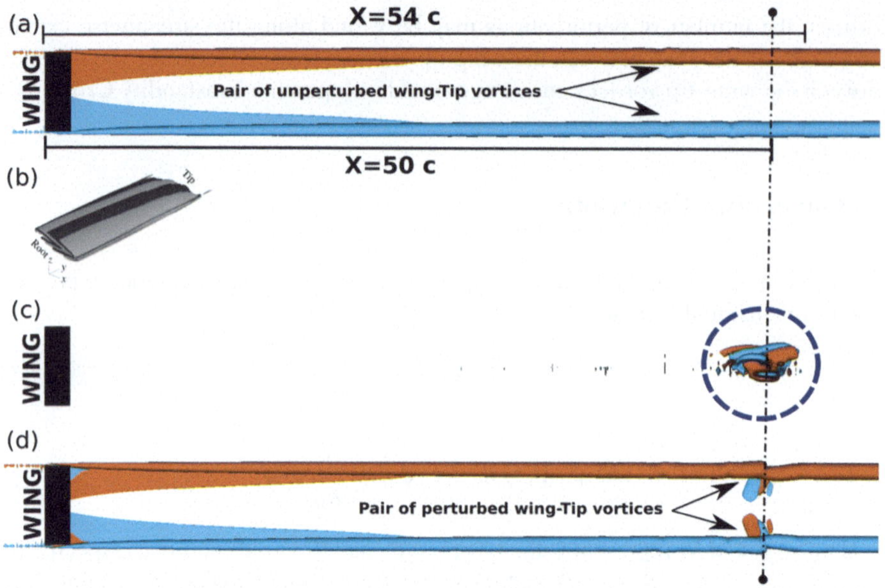

Fig. 1 Re = 1000 flow past a flat plate wing with $\alpha = 5°$ and AR = 6: **a** isosurfaces of streamwise vorticity showing pair of unperturbed wing-tip vortices, **b** linear optimal perturbation for T = 10 shown by the spanwise vorticity Navrose et al. [9], **c** isosurfaces of streamwise vorticity shows perturbation shape when it has reached x = 54c from the leading edge in the far wake and **d** corresponding perturbed wing-tip vortices near x = 50c

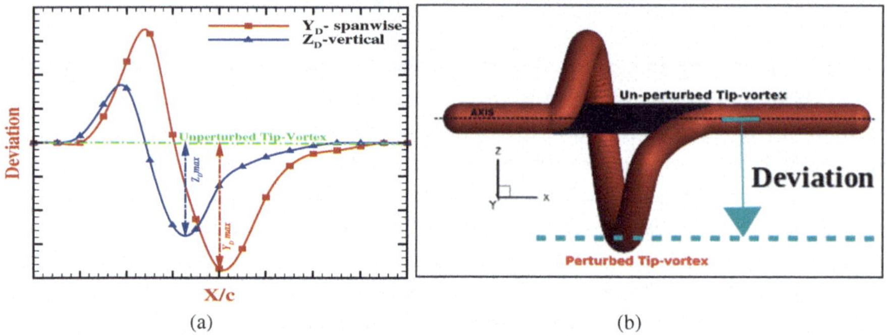

Fig. 2 **a** Displacement of the wing-tip vortices in the spanwise and vertical direction by the perturbation located at x = 54c and **b** model representation of perturbed wing-tip vortices corresponding to instant shown in Fig. 1d; here the displacement has been superimposed on the Batchelor vortex equivalent to velocity field at x = 50c

first step are superimposed on the baseflow obtained in the third step. This concludes our model description(see the magnified view in Fig. 2b) corresponding to the flow situation shown in Fig. 1d. In the axial direction, periodic boundary conditions have been applied. As we are only interested in the most unstable mode; long-wavelength

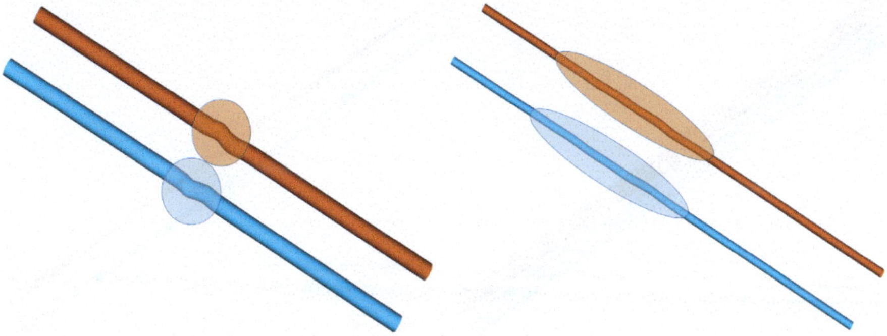

(a) (b)

Fig. 3 Stages of evolving Wing-tip vortices perturbed with a single perturbation located in the middle of the axial extent highlighted by shaded region: **a** initial condition and **b** at time = 330, shown by isocontours of axial vorticity

Crow instability, which happens to be symmetric about the midplane, only one vortex has been simulated by imposing a symmetric boundary condition on the midplane. Outflow boundary conditions are used in other directions.

The calculations are performed using the open source spectral element solver NEK5000 (*Fischer, Lottes, and Kerkemeier*)[10]. Simulations are conducted on a mesh of extent $L_x[0,43.0]$, $L_y[0,8.0]$ and $L_z[-10,10]$. There are 128,000 spectral elements within the computational domain. Each spectral element is discretized further with $N_{GLL} = 8$ Gauss-Lobato-Legendre (GLL) points in each direction. In the axial(X), spanwise(Y), and vertical(Z) directions there are 50, 80, and 32 uniformly distributed elements, respectively. The largest divisions in each direction are $\Delta x_{max}/c \approx 0.18$, $\Delta y_{max}/c \approx 0.0524$ and $\Delta z_{max}/c \approx 0.0524$. The time step for each simulation is 10^{-3}, which corresponds to a CFL number of 0.003. For computing the flow up to 320-time units, around 23000 CPU hours are spent per case.

4 Results

4.1 Effect of No. of Perturbations and Gap Between the Wing-Tip Vortices

Figure 3 shows the case when a single perturbation is present over the axial extent of wing-tip vortices. Initial conditions as well as the evolved perturbed wing-tip vortices are illustrated. Figure 4 shows the impact of a smaller gap between the wing-tip vortices. Detailed investigation of the evolved perturbed wing-tip vortices can be seen if we plot the variation of maximum amplitude of displacement as a function of time as in Fig. 5. For the time span of about 320-time units, the evolution appears to occur in two stages. Region-A extends from t = 0 to t = 170 and Region-B

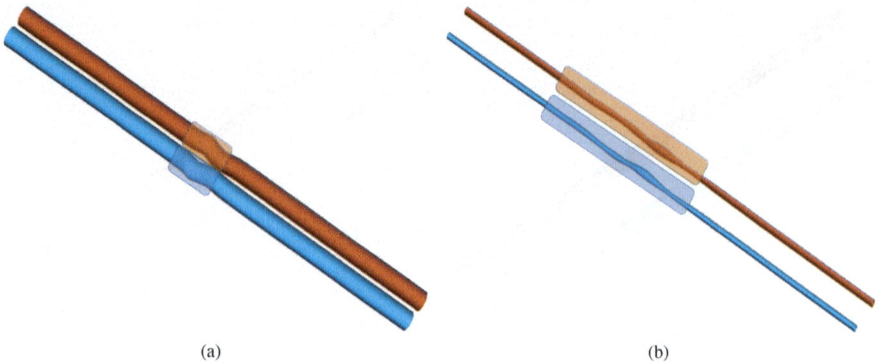

Fig. 4 Stages of evolving wing-tip vortices perturbed with a single perturbation located in the middle of the axial extent highlighted by shaded region with gap between them reduced to 37% of b:**a** initial condition and **b** at time = 320 shown by isocontours of axial vorticity

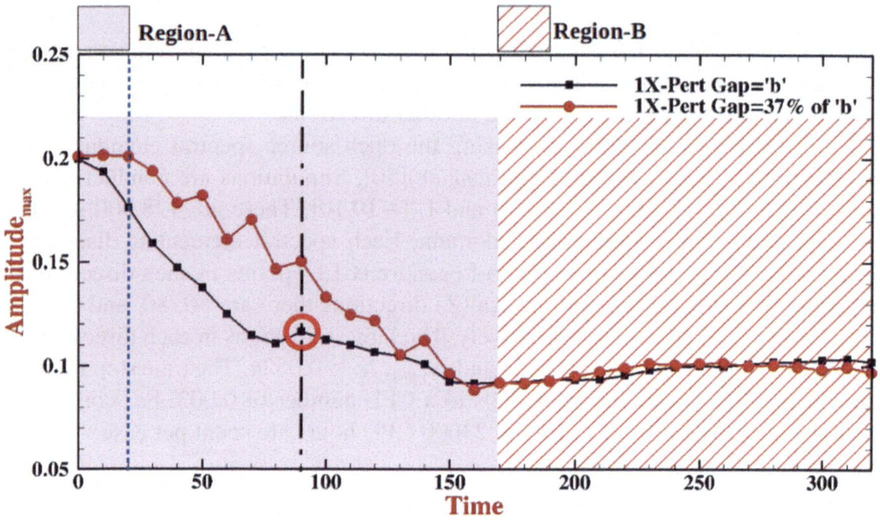

Fig. 5 Evolution of the maximum displacement of the perturbed wing-tip vortices about the unperturbed location

extends from t = 170 onwards till t = 320. Region-A shows transient behaviour in which a gradual reduction of the maximum amplitude of displacement takes place. Region-B represents the stage of evolution in which the displacement does not show any significant decay or growth. Case '1X-pert Gap=b' demonstrate an interestingly momentarily growth at t = 90 before following the gradual decline. Interestingly the gap reduction slows down the amplitude decay in Region-A.

5 Conclusion

We have simulated the evolution of perturbed wing-tip vortices in the far wake up to *370c* from the leading edge in a frame of reference moving roughly with the speed of perturbation. To investigate the evolution of perturbed wing-tip vortices, a large streamwise domain for the flow past a finite wing must be taken, which has a high computational cost. Instead, we define a model perturbation and baseflow corresponding to the flow situation shown in Fig. 1(d). The approach gives us the freedom to study the effect of different parameters like the Reynolds number, the gap between the vortices, and the number of perturbations on the axial domain. The present study considers single perturbation and the effect of reducing the gap on the evolution of perturbed wing-tip vortices. Evolution was analysed in terms of the maximum amplitude of displacement. Reduction of the gap remarkably decreases the decay rate of maximum displacement in Region-A, whereas the later evolution in Region-B involves almost no growth or decay. An increase in the number of perturbations, the interactions between them and the impact of gap reduction is expected to give interesting results.

Acknowledgements The authors acknowledge the use of computational resources at High-Performance Computing(HPC) and PARAM Sanganak facilities set up under the aegis of the Department of Science and Technology (DST) and the National Supercomputing Mission(NSM), Government of India at the Indian Institute of Technology, Kanpur.

References

1. Widnall SE (1975) The structure and dynamics of vortex filaments. Annu Rev Fluid Mech 7:141–165
2. Spalart PR (1998) Airplane trailing vortices. Ann Rev Fluid Mech 30:107
3. Leweke T, Le Dizes S, Williamson CH (2016) Dynamics and instabilities of vortex pairs. Ann Rev Fluid Mech 48:507–541
4. Crow SC (1970) Stability theory for a pair of trailing vortices 8:2172–2179
5. Farrell BF (1988) Optimal excitation of perturbations in viscous shear flow. Phys Fluids31:2093–2102
6. Trefethen LN, Trefethen AE, Reddy SC, Driscoll TA (1993) Hydrodynamic stability without eigenvalues 261:578–584
7. Schmid PJ (2007) Nonmodal stability theory. Ann Rev Fluid Mech 39:129–162
8. Kerswell R (2018) Nonlinear nonmodal stability theory. Ann Rev Fluid Mech 50:319–345
9. Brion V, Jacquin L (2019) Transient growth in the near wake region of the flow past a finite span wing. J Fluid Mech 866:399–430
10. Fischer P, Lottes J, Kerkemeier S (2008) Nek5000. Web Page http://nek5000.mcs.anl.gov

Comparison of Various RANS Turbulence Models for Dry Bed Simulation of Rotating Packed Bed (RPB)

Gaurav Kumar, Dheeraj Singh, Shweta Gole, and D. S. Murthy

Abstract The potential of rotating packed beds (RPB) is currently being explored by researchers to intensify heat transfer in thermal systems, primarily in cooling towers. The RPB has originally been implemented in chemical industries as a replacement for giant columns for carrying processes like distillation, separation, and absorption. In a rotating packed bed, the water flows radially outwards by the centrifugal force (100–1000 times the gravitational force in conventional columns) generated by a spinning porous cylindrical rotor, and the air is directed to flow radially inwards in a counter-current direction by a blower. To close the gap in fundamental knowledge of RPB and promote research further to achieve process intensification in heat transfer systems, deeper comprehension of fluid dynamics is essential. In the present work, the dry bed flow of air was simulated in ANSYS Fluent software and various RANS turbulence models have been explored and compared. The CFD simulation has been performed for a wide range of gas flow rates and two rotational speeds 800 rpm and 1600 rpm. The simulation results have been validated with the results from experiments.

1 Introduction

The rotating packed beds are established as gas–liquid contactors in chemical engineering fields where they are used to intensify the mass transfer process. The ability to generate 100–1000 times the gravitational force gives them a clear advantage over the conventionally used columns [1]. The conventional columns were being used to promote mass transfer during processes such as distillation, adsorption, absorption, separation, etc. These columns contain a porous packing where the working fluids interact with each other in a counter-current way. The liquid falls freely from the top of the column to the floor, and the gas rises from the bottom up. Since the gravity is

G. Kumar (✉) · D. Singh · S. Gole · D. S. Murthy
Department of Mechanical Engineering, College of Technology, G.B. Pant University of Agriculture and Technology, Pantnagar 263145, Uttarakhand, India
e-mail: 50904_gauravkumar@gbpuat-tech.ac.in

M. A. Siddiqui et al. (eds.), *Advances in Heat Transfer and Fluid Dynamics*, Lecture Notes in Mechanical Engineering, https://doi.org/10.1007/978-981-99-7213-5_10

only driving force for liquid movement, the problems such as flooding at top of the packing arise which limits the water flow rate and becomes a cause of these rotors' enormous size. The work of rotating packed beds is revolutionary in replacing these columns. Apart from intensifying the mass transfer process, it substantially reduced the size and weight of equipment. This almost doubles the mass-transfer rate and throughput, allowing for up to 10 times size reduction for the same process when compared to conventional columns [2].

In some recent research, the heat transfer aspects of the RPB have also been explored which stimulated the idea of process intensification in thermal systems using RPB [3–5]. Thermal systems such as cooling towers work on the same principle as conventional columns. In cooling towers, the water enters as a spray at the top of the tower and flows down through the tower and ambient air drawn into the tower interacts with the water stream in a counter or cross-current direction. The interior of the tower is filled with filling or packing material. Apart from being humungous in size, the proper water distribution and flooding of the inlet at a high water flow rate are some of the problems in such towers. These problems can be addressed by RPB as it has complete control over the water flow but a thorough investigation conducted in a stepwise manner is required to replace the cooling towers by rotating packed beds. Sandeep et al. [5] calculated the similar thermal performance parameters for an RPB that are utilized to evaluate the cooling tower performance, The results obtained for rate of water evaporation, rate of heat transfer, cooling range, effectiveness, and Merkel number for RPB suggested that the RPB is a feasible alternative of cooling towers. Saurabh and Murthy [6] developed the correlation for heat exchange between participating fluids in a rotating packed bed and observed that the heat transfer in RPB strongly depends on the gas flow rate, liquid flow rate, air inlet temperature, water inlet temperature, and rotational speed. Saurabh et al. [3] presented the various surface contours for air pressure drop, velocity, water temperature, and rate of water–air heat transfer in the rotating packed bed.

The researchers have been studying mass transfer, pressure drop, and flow behavior inside RPB using both experimental [7] and numerical methods [8]. The understanding of basic airflow behavior inside a rotating packed bed is crucial to understand the underlying physics involved in heat and mass transport processes. This understanding would be further helpful when we consider the scale-up of the RPB system for actual thermal applications. The most economic, practical, and widely used tool for this purpose is computational fluid dynamics. Using conventional experimental methodologies, it is difficult to precisely assess the flow field in RPBs since the flow pattern is extremely complicated and turbulent. In an RPB, the primary flow features that are needed to be modelled are flow through porous media, rotation of the rotor, and turbulence modeling. For the modeling of flow through porous media, the authors have mainly used two approaches. One is the obstacle simplification approach, where the porous media is modeled by placing simplified obstacles of different shapes in both two dimensions and three dimensions [9–12]. The other approach is the porous media simplification approach where the influence of porous media on the flow field is mathematically modelled by using viscous and inertial resistance formulations [13, 14]. The former approach gives better visualization of the flow field but is less accurate and more time-consuming due to poor quality and a large number of mesh elements. In the latter, it is possible to create better quality elements, and is less time-consuming. For the modeling of rotation, two approaches used by the researchers are the sliding mesh method (SM) and the multiple reference frame method (MRF). Out of these two, the MRF method is robust and less time-consuming. CFD has a number of models to perform turbulent flow modeling. The selection of a suitable turbulence model is crucial for correctly estimating the flow and turbulence characteristics in RPBs. In CFD simulations, Reynolds-averaged-Navier–Stokes (RANS) turbulence models are frequently used to assess turbulent flow behavior in a variety of applications. To simulate turbulent viscosity and eddy diffusivities, RANS models solve the time-averaged equations of motion and incorporate turbulence closure models. RANS turbulence models' accuracy varies depending on the application and the model of choice. The most commonly used turbulent models in the simulation of flow inside the rotating packed bed are realizable $k - \varepsilon$, standard $k - \varepsilon$, and SST $k - \omega$ model [15].

Chavez and Larachi [13] simulated the three-dimensional unsteady-state turbulence in an RPB having only one phase. Wojtasik et al. [16] performed a CFD analysis on dry bed gas pressure drop in RPB CO_2 absorbers using the standard $k - \varepsilon$ model and found it to be effectual in analyzing the behavior of gas flow in RPB setups having a porous packing. In the present work, the turbulence flow inside the rotating packed bed has been modelled using various turbulent models i.e., Spalart–Allmaras model, standard $k - \varepsilon$, realizable $k - \varepsilon$, and SST $k - \omega$ model and the results obtained by them have been compared. The validity of the simulation results is confirmed by comparing them with the results obtained from experiments. The results of this investigation will shed important light on the reliability and applicability of different RANS turbulence models for estimating the flow and turbulence characteristics in RPBs.

2 Rotating Packed Bed: Setup and Working

In a rotating packed bed, a cylindrical porous material is fixed between two annular discs. The combined assembly of discs and packing is called the rotor. The selection of design and material for packing have a crucial role on how efficiently RPBs work. In the present work, the packing was made by using wire mesh which is mostly used by researchers due to its low cost, ease of fabrication and high porosity. The packing had a porosity of 0.81. Higher porosity is widely recognized to improve mass transfer and reduce pressure drop. The packing had a diameter of 80 mm on the inside and 240 mm on the outside. The axial thickness of the packing was 45 mm. The water was inserted at the eye of the rotor through a liquid distributor and the air was inserted from the top of the casing. The rotor was rotated with the help of a motor to generate the required centrifugal force for water flow. The casing enclosed and secured the whole rotor assembly and prevented the water from splashing out. The water collected

Fig. 1 Schematic view of RPB unit

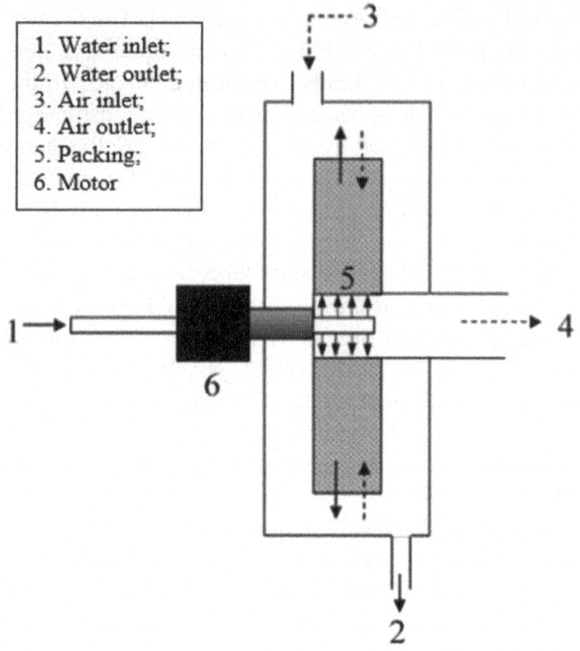

inside the casing was drained out from the bottom of the casing. After travelling radially inwards through the packing, the air exits through an outlet duct connected to the eye of the packing. Figure 1 illustrates the experimental configuration through a schematic diagram. The differential pressure across the rotating packed bed was measured by placing probes at the top of the casing and the end of the air outlet duct. The results of experiment were used for validating the CFD simulations.

3 Computational Domain and Discretization

To accurately simulate the fluid flow in the rotary packed bed (RPB), a computational domain and grid discretization were established. The geometry of the RPB was simplified for the CFD analysis by considering only three sections as shown in Fig. 2a. Section I contained the peripheral portion having a radial thickness of 35 mm, section II contained the packing portion where the porous media model was applied, and section III contained the eye of the rotor and the air outlet duct. The flow inside the casing was neglected in the simulation. The air inlet was provided at the circumference of section I and the outlet was provided at the end of section III. The packing was rotated counterclockwise about the positive z-axis. The model used for rotation was the multiple reference frame (MRF) model. The RPB geometry was then discretized to generate a computational mesh.

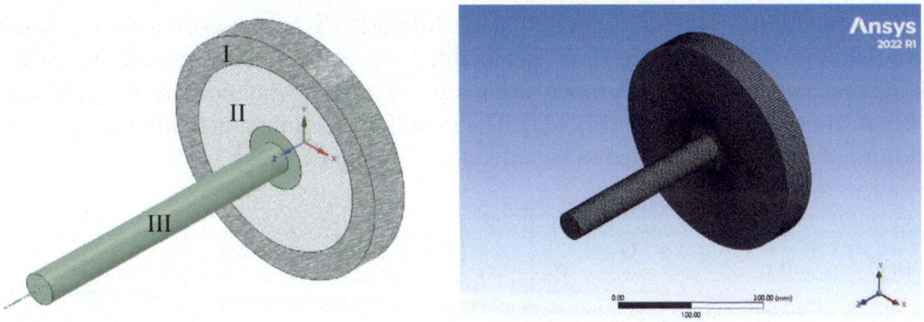

(a) (b)

Fig. 2 **a** Simplified Geometrical model of RPB used for CFD simulations, **b** Discretized geometry

The meshing was done in ANSYS meshing using structured hexahedron elements for the whole region as shown in Fig. 2b. After performing the grid sensitivity analysis, the number of elements selected for the simulation was about 1 million with average element quality of 0.83 and average skewness of 0.11. This level of mesh refinement was deemed sufficient to accurately capture the details of the fluid flow within the RPB. With this well-constructed computational domain and grid discretization, we are able to conduct simulations of the RPB that accurately capture the physical processes occurring within it.

4 CFD Modeling

The governing equations were averaged continuity and momentum equations with added source terms in the momentum equation for porous media involving viscous resistance $\left(150(1-e)^2/D_P^2e^3\right)$ and inertial resistance $\left(3.5(1-e)/D_Pe^3\right)$. Here e is porosity and D_p is equivalent particle diameter. It is important to note that due to the complexity of the turbulent flow, choosing the right turbulence model for a particular application might be difficult. As a result, it is essential to thoroughly assess the precision and computational effectiveness of various turbulence models before including them into a CFD simulation. Since the computational domain contains both stationary and rotating frames, a multiple reference frame model with relative velocity formulation was applied. By utilizing this approach, two further acceleration factors, namely Coriolis acceleration and centripetal acceleration, were identified in the momentum equation. The effect of these accelerations on the pressure drop across the RPB has been mentioned by some authors previously. The turbulence was analyzed by using various RANS (Reynolds Averaged Navier–Stokes) models. The air was assumed incompressible and the effect of gravity was also neglected as compared to the centrifugal force generated by the rotor.

One equation turbulence model (Spalart–Allmaras (S-A) model) and two-equation turbulence models (Standard $k-\varepsilon$, Realizable $k-\varepsilon$, and SST $k-\omega$ model) were used in the present work. The S-A model addresses a transport equation for the kinematic eddy (turbulent) viscosity, $\tilde{v}$ (Eq. (1)). The model is less time-consuming as only one additional equation is needed to be solved for turbulence.

$$\frac{\partial}{\partial t}(\rho\tilde{v}) + \frac{\partial}{\partial x_i}(\rho\tilde{v}u_i) = G_v + \frac{1}{\sigma_{\tilde{v}}}\left[\frac{\partial}{\partial x_j}\left\{(\mu + \rho\tilde{v})\frac{\partial\tilde{v}}{\partial x_j}\right\} + C_{b_2}\rho\left(\frac{\partial\tilde{v}}{\partial x_j}\right)^2\right] - Y_v \tag{1}$$

Here, G_v and Y_v are the production and destruction of turbulent viscosity, respectively.

The standard $k - \varepsilon$ and realizable $k - \varepsilon$ have similar forms as both contain the transport equations for turbulent kinetic energy, k and dissipation rate, ε, however, they differ in modeling of turbulent viscosity and generation and destruction terms in the dissipation equation. Equation (2) represents the transport equation for turbulent kinetic energy, and Eqs. (3) and (4) represent the transport of viscous dissipation rate for standard and realizable $k - \varepsilon$ model, respectively.

$$\frac{\partial}{\partial t}(\rho k) + \frac{\partial}{\partial x_i}(\rho k u_i) = \frac{\partial}{\partial x_j}\left\{\left(\mu + \frac{\mu_t}{\sigma_k}\right)\frac{\partial k}{\partial x_j}\right\} + G_k - \rho\varepsilon \tag{2}$$

$$\frac{\partial}{\partial t}(\rho\varepsilon) + \frac{\partial}{\partial x_i}(\rho\varepsilon u_i) = \frac{\partial}{\partial x_j}\left\{\left(\mu + \frac{\mu_t}{\sigma_\varepsilon}\right)\frac{\partial\varepsilon}{\partial x_j}\right\} + C_{1\varepsilon}\frac{\varepsilon}{k}G_k - C_{2\varepsilon}\rho\frac{\varepsilon^2}{k} \tag{3}$$

$$\frac{\partial}{\partial t}(\rho\varepsilon) + \frac{\partial}{\partial x_i}(\rho\varepsilon u_i) = \frac{\partial}{\partial x_j}\left\{\left(\mu + \frac{\mu_t}{\sigma_\varepsilon}\right)\frac{\partial\varepsilon}{\partial x_j}\right\} + \rho C_1 S\varepsilon - \rho C_2\frac{\varepsilon^2}{k + \sqrt{v\varepsilon}} \tag{4}$$

Here, G_k is the production of turbulent kinetic energy.

The SST $k-\omega$ model obtains the solution for the transport equation for turbulent kinetic energy and specific dissipation rate, ω.

$$\frac{\partial}{\partial t}(\rho\omega) + \frac{\partial}{\partial x_i}(\rho\omega u_i) = \frac{\partial}{\partial x_j}\left(\left(\mu + \frac{\mu_t}{\sigma_\omega}\right)\frac{\partial\omega}{\partial x_j}\right) + G_\omega - Y_\omega \tag{5}$$

Here, G_ω represents the generation of ω and Y_ω represents its dissipation.

The model constants for each method are shown in Table. 1. The conditions at the boundary of the model equations are as follows:

- Constant mass flow rate of air at the air inlet boundary $\left(29.35\,\mathrm{m^3/h} \le \dot{m} \le 58.70\,\mathrm{m^3/h}\right)$.
- No slip condition at the wall.
- Zero-gauge pressure at the air outlet (101.325 kPa).

Table 1 Values of model constant for each turbulence model

Model	Model constants
Spalart–Allmaras model	$C_{b_1} = 0.136,\ C_{b_2} = 0.622$, and $\sigma_{\tilde{\nu}} = 0.667$
Standard $k - \varepsilon$ model	$C_{1_\varepsilon} = 1.44,\ C_{2_\varepsilon} = 1.92,\ \sigma_k = 1.0$, and $\sigma_\varepsilon = 1.3$
Realizable $k - \varepsilon$ model	$C_{1\varepsilon} = 1.44,\ C_2 = 1.9,\ \sigma_k = 1.0$, and $\sigma_\varepsilon = 1.2$
SST $k - \omega$ model	$\sigma_{k,1} = 1.176,\ \sigma_{\omega,1} = 2.0,\ \sigma_{k,2} = 1.0$, and $\sigma_{\omega,2} = 2.0$

5 Results and Discussion

The validation of the drop in air pressure across the rotating packed bed using different turbulent modes is an important step in understanding the behavior of the system. The pressure drop is a key factor in determining the efficiency of the process and can have a significant impact on the overall performance. The drop in air pressure across the rotating packed bed obtained by using different turbulent modes was validated with the results obtained from experiments. The pressure drop has been plotted along with the gas flow rate at 800 rpm and at 1600 rpm as shown in Fig. 3a, b, respectively. The error bars show ±20% of the experimental values and it can be seen that all the values obtained from simulation using each turbulence model are lying in that range. The pressure drop obtained by the Spalart–Allmaras (S-A) model and SST $k - \omega$ model was found closer to the experimental results than the standard $k - \varepsilon$ model and realizable $k - \varepsilon$ for all the gas flow rates in both the figures.

The percentage error in the simulations using different turbulent models for the RPB rotating at 800 rpm and 1600 rpm is shown in Table 2. It can be seen from the table that the error in pressure drop results obtained from the Spalart–Allmaras model is the least among all the methods for most of the gas flow rates at both 800 and 1600 rpm. This suggests that the S-A and SST $k - \omega$ models may be more accurate in predicting the pressure drop across the rotating packed bed.

The analysis of the flow field characteristics can be further assessed by examining the turbulence kinetic energy and eddy viscosity. Figure 4a, c show the contours of turbulent kinetic energy on a plane at the center of the packing in the y–z axis obtained by the Standard $k - \varepsilon$, Realizable $k - \varepsilon$, and SST $k - \omega$ model, respectively. Figure 5a, d show the contours of eddy viscosity on a plane at the center of the packing in the y–z axis obtained by the Spalart–Allmaras, realizable $k - \varepsilon$, standard $k - \varepsilon$, and SST $k - \omega$ model, respectively.

It was found that the air starts rotating prior to entering the rotor; starts spinning with the rotor as it enters the rotor; and finally exits from the outlet pipe in a spiral path. Due to the rotation of the packing, the air travelling through the rotor experienced motion in two different directions. One is the radially inward motion developed due to the pressure gradient generated by the blower and the other is the rotational motion developed due to the drag between packing and air. This gives rise to Coriolis acceleration. The effect of Coriolis acceleration inside the packing is neutralized by the resistance offered by packing however as the air enters the eye section, in the absence of packing the effect of Coriolis acceleration becomes significant and the

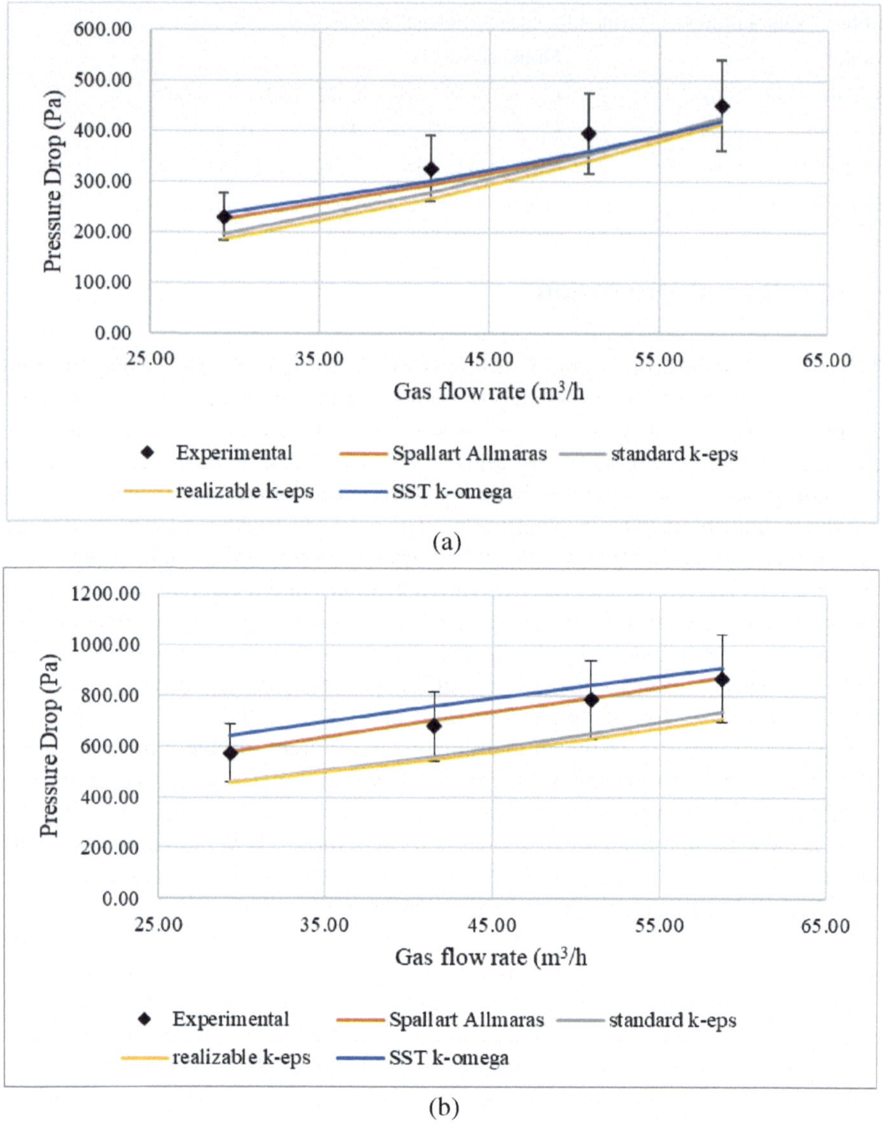

Fig. 3 A comparison between experimental and simulated pressure drop by varying gas flow rate at rotor speed **a** 800 rpm, and **b** 1600 rpm

velocity of air suddenly increases inside the eye section. This certainly increases the turbulence in that section. Also, the air from the eye section changes its direction towards the gas outlet pipe by taking a right-angle turn. Figures 4 and 5 show that the realizable $k-\varepsilon$, and standard $k-\varepsilon$ model capture turbulence in the eye section better than the other models. Figure 6a, d show the velocity vectors obtained by different

Table 2 Percentage error in the simulations using different turbulence models at 800 rpm and 1600 rpm

Gas flow rate (in m^3/hr)	Spalart–Allmaras		Standard $k-\varepsilon$		Realizable $k-\varepsilon$		SST $k-\omega$	
	800 rpm (%)	1600 rpm (%)	800 rpm (%)	1600 rpm (%)	800 rpm (%)	1600 rpm (%)	800 rpm (%)	1600 rpm (%)
29.35	2.10	1.11	15.13	19.72	18.32	19.48	3.37	12.39
41.51	9.86	4.09	14.55	17.37	17.86	18.93	7.34	11.84
50.84	9.94	1.19	10.59	16.82	13.68	19.10	8.75	7.37
58.70	7.21	0.63	5.60	14.84	8.03	17.97	7.01	5.13

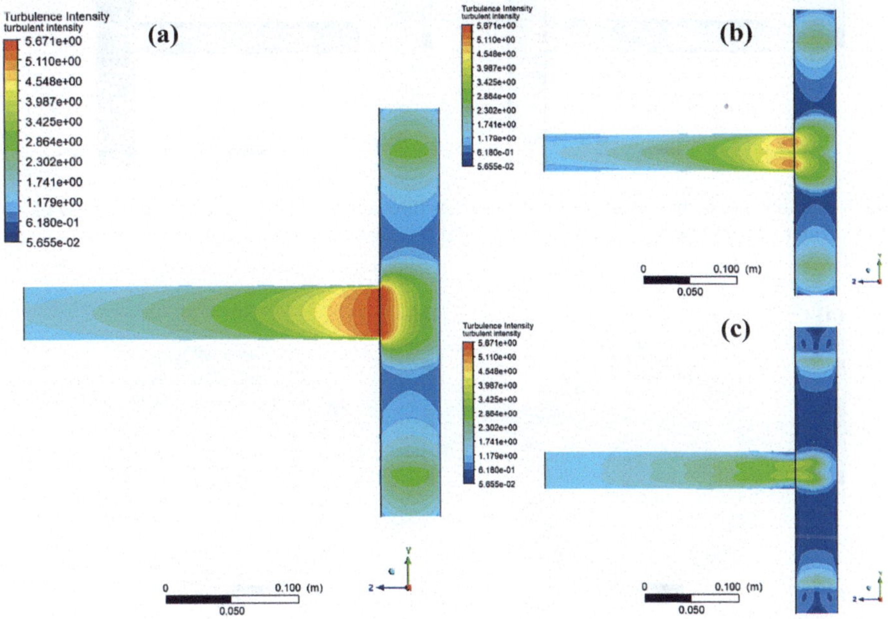

Fig. 4 Contours of turbulent kinetic energy for gas flow rate 58.70 m^3/h and 800 rpm on y–z plane at the center of packing obtained by **a** standard $k-\varepsilon$, **b** SST $k-\omega$ model, and **c** realizable $k-\varepsilon$ model

turbulence models in the x–y plane cut from the center of the packing. It can be seen from the figure that the rotation of air as it comes to the proximity of packing is better visible by the realizable $k-\varepsilon$, and standard $k-\varepsilon$ model.

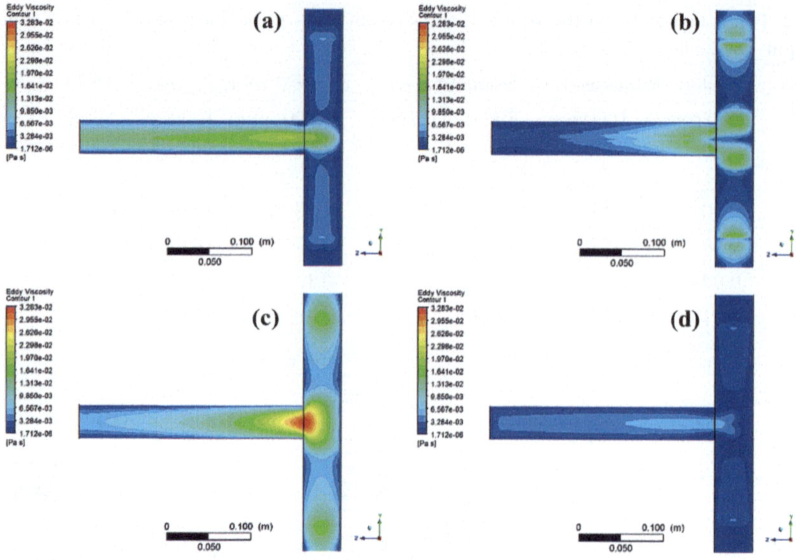

Fig. 5 Contours of turbulent eddy viscosity for gas flow rate 58.70 m^3/h and 800 rpm on y–z plane at the center of packing obtained by **a** S-A model **b** realizable $k-\varepsilon$ model, **c** standard $k-\varepsilon$ and **d** SST $k-\omega$ model

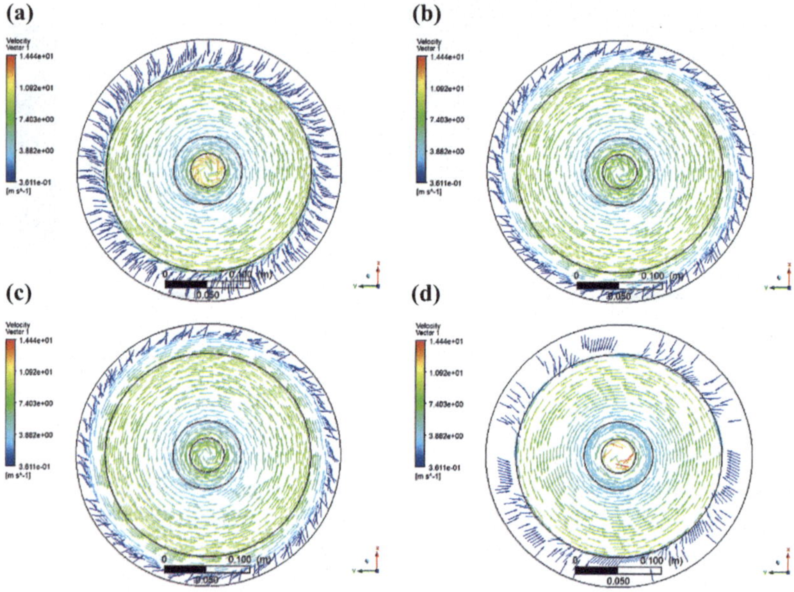

Fig. 6 Velocity vectors on x–y plane for gas flow rate 58.70 m^3/h and 800 rpm at the center of packing obtained by **a** S-A model **b** realizable $k-\varepsilon$ model, **c** standard $k-\varepsilon$, and **d** SST $k-\omega$ model

6 Conclusion

The CFD simulations on the RPB were performed using various turbulence models for airflow only. The simulations were performed in the ANSYS fluent software using the S-A model, standard and realizable $k - \varepsilon$ model, and SST $k - \omega$ model for turbulence modeling and the results were validated with the experiments. All the results obtained from simulation were under 20% of that obtained from experiments. The S-A and SST $k - \omega$ model gave the closest results to experimental data in terms of pressure drop however the turbulent behavior of the flow inside the RPB was better explained by the standard $k - \varepsilon$ model and realizable $k - \varepsilon$ model. The pressure drop results obtained by the Spalart–Allmaras model gave the least error among all the methods for most of the gas flow rates at both 800 and 1600 rpm.

Acknowledgements This research work was funded by AICTE under their Research Promotion Scheme.

References

1. Rao DP, Bhowal A, Goswami PS (2004) Process intensification in rotating packed beds (HIGEE): an appraisal. Ind Eng Chem Res 43(4):1150–1162. https://doi.org/10.1021/ie030630k
2. Kumar MP, Rao DP (1990) Studies on a high-gravity gas-liquid contactor. Ind Eng Chem Res 29(5):917–920. https://doi.org/10.1021/ie00101a031
3. Saurabh SK, Murthy DS (2022) Insights into thermal transactions of a novel rotating packed bed. J Therm Sci Eng Appl 14(1):011003–1, 2022. https://doi.org/10.1115/1.4050836
4. Saurabh DDSM, Kumar S (2019) CFD analysis of multiphase flow hydrodynamics in rotating packed bed. Twelfth Int Conf Therm Eng Theory Appl, 5–7, 2019, [Online]. Available: https://journals.library.ryerson.ca/index.php/ictea/article/download/1184/1217
5. Kumar S, Kumar G, Murthy DS (2022) Experimental investigation on thermal performance characteristics of rotating packed bed. Exp Heat Transf 35:1–13. https://doi.org/10.1080/08916152.2022.2027576
6. Saurabh, Murthy DS (2020) Analysis and optimization of thermal characteristics in a rotating packed bed. Appl Therm Eng 165:114533. https://doi.org/10.1016/j.applthermaleng.2019.114533
7. Sandilya P, Rao DP, Sharma A, Biswas G (2001) Gas-phase mass transfer in a centrifugal contactor. Ind Eng Chem Res 40(1):384–392. https://doi.org/10.1021/ie0000818
8. Yang YC, Ouyang Y, Zhang N, Yu QJ, Arowo M (2019) A review on computational fluid dynamic simulation for rotating packed beds. J Chem Technol Biotechnol 94(4):1017–1031. https://doi.org/10.1002/jctb.5880
9. Guo T-Y, Shi X, Chu G-W, Xiang Y, Wen L-X, Chen J-F (2016) Computational fluid dynamics analysis of the micromixing efficiency in a rotating-packed-bed reactor. Ind Eng Chem Res 55(17):4856–4866. https://doi.org/10.1021/acs.iecr.6b00213
10. Xie P, Lu X, Yang X, Ingham D, Ma L, Pourkashanian M (2017) Characteristics of liquid flow in a rotating packed bed for CO_2 capture: a CFD analysis. Chem Eng Sci 172:216–229. https://doi.org/10.1016/j.ces.2017.06.040
11. Shi X, Xiang Y, Wen LX, Chen JF (2013) CFD analysis of liquid phase flow in a rotating packed bed reactor. Chem Eng J 228:1040–1049. https://doi.org/10.1016/j.cej.2013.05.081

12. Yang W, Wang Y, Chen J, Fei W (2010) Computational fluid dynamic simulation of fluid flow in a rotating packed bed. Chem Eng J 156(3):582–587. https://doi.org/10.1016/j.cej.2009.04.013
13. Llerena-Chavez H, Larachi F (2009) Analysis of flow in rotating packed beds via CFD simulations-Dry pressure drop and gas flow maldistribution. Chem Eng Sci 64(9):2113–2126. https://doi.org/10.1016/j.ces.2009.01.019
14. Martínez EL, Jaimes R, Gomez JL, Filho RM (2012) CFD simulation of three-dimensional multiphase flow in a rotating packed bed. Comput Aided Chem Eng 30:1158–1162. https://doi.org/10.1016/B978-0-444-59520-1.50090-7
15. Ouyang Y et al. (2019) Evaluation of various turbulence models for numerical simulation of a multiphase system in a rotating packed bed. Comput Fluids, 194. https://doi.org/10.1016/j.compfluid.2019.104296
16. Wojtasik J, Maciej M, Marcin J (2022) CFD simulation of gas pressure drop in porous packing for rotating packed beds (RPB) C. Environ Sci Pollut Res, 0123456789. https://doi.org/10.1007/s11356-022-20859-x

Effect of Buoyancy on Three-Dimensional Flow Around a Heated Square Cylinder in Mixed Convection

Mohd Perwez Ali, Nadeem Hasan, and Sanjeev Sanghi

Abstract Three-dimensional flow around a heated square cylinder is studied using direct numerical simulation (DNS) in a mixed convective flow regime. In this study, an infinite square cylinder is immersed in horizontal free-stream crossflow (air, Prandtl number $Pr = 0.7$) at right angles to gravity. Numerical results are shown with different heat levels defined by the over-heat ratio $\epsilon = (T_w - T_\infty)/T_\infty$, where T_w and T_∞ are equal to the surface and surrounding temperature. At large-scale heating $\epsilon \sim O(1)$, the thermal straining and transport properties of the fluid particles are varied. To capture this variation, an in-house solver of the compressible flow model is employed. The compressible flow governing equations in Cartesian coordinates are transformed into a body-fitted coordinate system and solved using the flux-based particle velocity upwind-modified+ (PVU-M+) scheme. The low Mach number $M = 0.1$ is used for all computations. The results obtained using the in-house solver are validated with the values reported in the literature achieved by experimental, DNS, and Floquet methods. The disordered vortical structure of the mode B transition changes its shape and spanwise wavelength with the increase of the heating level from $\epsilon = 0$ to $\epsilon = 1$ at the Reynolds number $Re = 500$. In addition, a very significant change in the force coefficient and the vortex shedding frequency is observed upon increasing the heating level. Variations in the frequency spectra of the lift coefficient are observed with varying heating.

M. P. Ali (✉) · S. Sanghi
Indian Institute of Technology Delhi, New Delhi 110 016, India
e-mail: perwezali7@gmail.com

S. Sanghi
e-mail: sanghi@am.iitd.ac.in

N. Hasan
Aligarh Muslim University, Aligarh 202 002, Uttar Pradesh, India
e-mail: nadhasan@gmail.com

M. A. Siddiqui et al. (eds.), *Advances in Heat Transfer and Fluid Dynamics*, Lecture Notes in Mechanical Engineering, https://doi.org/10.1007/978-981-99-7213-5_11

1 Introduction

The dynamics of the three-dimensional vortex structure behind a bluff-body wake is a very interesting phenomenon. Many experimental and numerical studies have been conducted to understand the two-dimensional and three-dimensional wake dynamics behind a bluff body. Three-dimensional flow studies mostly focus on the isothermal wake behind unheated bluff bodies, especially circular and square cylinders. Only a few three-dimensional studies have been conducted around a heated cylinder in the mixed convective flow regime. However, in mixed convection, the three-dimensional flow around a heated bluff body is of great importance in engineering and industries based on applications such as electronics cooling, chemical reactors, and compact heat exchangers.

The sequence of three-dimensional transition modes with an increase in Reynolds number Re has been studied in the literature using various methods. These three-dimensional transition modes are recognized by the shape and spanwise wavelength λ_z of the vortical structure. The first transition, i.e., mode A, with longer wavelengths,$\lambda_z = 5D - 5.8D$ (where D is the side length of the square cylinder), is observed in the Reynolds number range between 150 and 200 [1–5]. For $Re = 190 - 250$, the second transition, represented by mode B, appears with a shorter wavelength $\lambda_z = 1D - 1.2D$ [1, 2, 5]. The third transition known as mode C with intermediate wavelength $\lambda_z = 2.6D - 2.8D$ between is detected using Floquet stability analysis for $Re = 200 - 215$ [2, 6]. At higher Reynolds numbers ($Re \geqslant 250$), an irregular three-dimensional instability is observed in the cylinder wake [7].

In mixed convective flow regimes, experimental and numerical studies have been performed on three-dimensional flow transitions behind a heated circular cylinder [8, 9]. In these studies, water flows perpendicular to gravity within the range of Reynolds number $75 < Re < 117$ and Richardson number range $0.35 < Ri < 2.5$. Based on these studies, it can be concluded that the three-dimensional transition with $\lambda_z = 2D$ in the heated cylinder wake, denoted as mode E, occurs at a lower Reynolds number than the mode A transition of the unheated cylinder. Arif and Hasan [10] studied buoyancy effects on the three-dimensional transition around a square cylinder, which is subjected to horizontal crossflow using air and water as working fluids. In this investigation, buoyancy effects on flow properties such as force coefficient, Strouhal number, and Nusselt number are investigated for $Re = 50 - 250$ and $Ri = 0 - 2$. In literature, three-dimensional transitions around a heated cylinder have only been studied using the Boussinesq approximation (small-scale heating scenario). In large-scale heating in mixed convection, only two-dimensional wake transitions behind a heated square cylinder have been demonstrated at $Re = 100$ using the non-Oberbeck Boussinesq model [11–13]. These studies describe the effects of buoyancy on vortex shedding and flow parameters.

In the present investigation, an infinite square cylinder is immersed in a uniform free-stream crossflow (with fluid velocity U_∞ and temperature T_∞) perpendicular to gravity (see Fig. 1a). Since the spanwise length is infinite, the three-dimensional

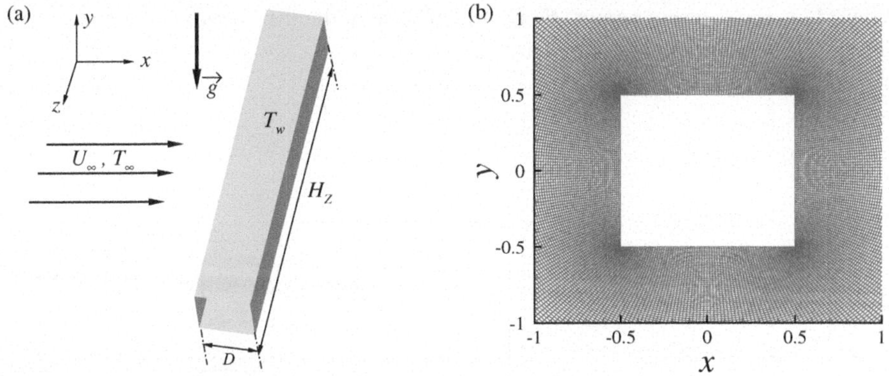

Fig. 1 (**a**) Horizontal flow past a square cylinder, and (**b**) Close view of two-dimensional mesh in $\xi - \eta$ plane

modes having definite spanwise wavelength can be captured by restricting the spanwise domain with a periodic boundary condition, over a finite length $H_z = 6D$ (see Fig. 1a). The flow is described by Cartesian coordinates in which the $x-$, $y-$, and $z-$coordinates are in the streamwise, negative gravity, and spanwise directions, respectively. A circular domain of radius $R = 60D$ is used around a square cylinder whose center is at the same position as the cylinder center. An $O-$type body-fitted mesh is generated with 281, 309, and 61 grid points in the ξ, η, and z directions, respectively. The mesh is initially generated in two-dimensional $\xi - \eta$ coordinates and then uniformly replicated in the $z-$direction. Figure 1b shows a two-dimensional magnified grid with a minimum dimensionless grid size of 1.7×10^{-3}. To capture the flow field on this fine mesh, the dimensionless time-step $\Delta t = 5 \times 10^{-5}$ is employed. In current computations, the Reynolds number, Mach number, and Froude number are kept fixed at $Re = 500$, $M = 0.1$, and $Fr = 1.0$, while the heating level varies from $\epsilon = 0$ to $\epsilon = 1$.

2 Numerical Model

2.1 Governing Equations

The three-dimensional governing equations of compressible gas flow in the strong conservative form are given in non-dimensional form:

$$\frac{\partial \boldsymbol{U}}{\partial t} + \frac{\partial \boldsymbol{F}}{\partial x} + \frac{\partial \boldsymbol{G}}{\partial y} + \frac{\partial \boldsymbol{H}}{\partial z} = \boldsymbol{J}, \tag{1}$$

where $\boldsymbol{U}$ is the solution vector, $\boldsymbol{F}$, $\boldsymbol{G}$, and $\boldsymbol{H}$ denote the flux vectors and $\boldsymbol{J}$ is source vector. These vectors are defined as

$$\boldsymbol{U} = [\rho \quad \rho u \quad \rho v \quad \rho w \quad \rho E]', \tag{2}$$

$$\boldsymbol{F} = \begin{bmatrix} \rho u \\ \rho u u + p - \frac{2\mu}{Re}\left\{\frac{\partial u}{\partial x} - \frac{1}{3}(\nabla\cdot\mathbf{V})\right\} \\ \rho u v - \frac{\mu}{Re}\left(\frac{\partial v}{\partial x} + \frac{\partial u}{\partial y}\right) \\ \rho u w - \frac{\mu}{Re}\left(\frac{\partial u}{\partial z} + \frac{\partial w}{\partial x}\right) \\ \rho u E - \frac{\gamma\kappa}{RePr}\frac{\partial T}{\partial x} + \Phi^F \end{bmatrix}, \quad \boldsymbol{G} = \begin{bmatrix} \rho v \\ \rho v u - \frac{\mu}{Re}\left(\frac{\partial v}{\partial x} + \frac{\partial u}{\partial y}\right) \\ \rho v v + p - \frac{2\mu}{Re}\left\{\frac{\partial v}{\partial y} - \frac{1}{3}(\nabla\cdot\mathbf{V})\right\} \\ \rho v w - \frac{\mu}{Re}\left(\frac{\partial v}{\partial z} + \frac{\partial w}{\partial y}\right) \\ \rho v E - \frac{\gamma\kappa}{RePr}\frac{\partial T}{\partial y} + \Phi^G \end{bmatrix}, \tag{3}$$

$$\boldsymbol{H} = \begin{bmatrix} \rho w \\ \rho w u - \frac{\mu}{Re}\left(\frac{\partial w}{\partial x} + \frac{\partial u}{\partial z}\right) \\ \rho w v - \frac{\mu}{Re}\left(\frac{\partial w}{\partial y} + \frac{\partial v}{\partial z}\right) \\ \rho w w + p - \frac{2\mu}{Re}\left\{\frac{\partial w}{\partial z} - \frac{1}{3}(\nabla\cdot\mathbf{V})\right\} \\ \rho w E - \frac{\gamma\kappa}{RePr}\frac{\partial T}{\partial z} + \Phi^H \end{bmatrix}, \quad \boldsymbol{J} = \begin{bmatrix} 0 \\ 0 \\ (1-\rho)/Fr^2 \\ 0 \\ (\gamma - 1)\left\{\gamma(1-\rho)\left(\frac{M}{Fr}\right)^2 v - (\nabla\cdot\mathbf{V})\right\} \end{bmatrix}. \tag{4}$$

The terms Φ^F, Φ^G, and Φ^H are given as

$$\Phi^F = \gamma(\gamma - 1)M^2\left(pu + \frac{\mu}{Re}D_F\right), \tag{5}$$

$$\Phi^G = \gamma(\gamma - 1)M^2\left(pv + \frac{\mu}{Re}D_G\right), \tag{6}$$

$$\Phi^H = \gamma(\gamma - 1)M^2\left(pw + \frac{\mu}{Re}D_H\right). \tag{7}$$

The quantities D_F, D_G, and D_H in Eqs. 5–7 are given as

$$D_F = \left[\frac{2}{3}u\left(-2\frac{\partial u}{\partial x} + \frac{\partial v}{\partial y} + \frac{\partial w}{\partial z}\right) - v\left(\frac{\partial v}{\partial x} + \frac{\partial u}{\partial y}\right) - w\left(\frac{\partial w}{\partial x} + \frac{\partial u}{\partial z}\right)\right], \tag{8}$$

$$D_G = \left[\frac{2}{3}v\left(\frac{\partial u}{\partial x} - 2\frac{\partial v}{\partial y} + \frac{\partial w}{\partial z}\right) - u\left(\frac{\partial v}{\partial x} + \frac{\partial u}{\partial y}\right) - w\left(\frac{\partial w}{\partial y} + \frac{\partial v}{\partial z}\right)\right], \tag{9}$$

$$D_H = \left[\frac{2}{3}w\left(\frac{\partial u}{\partial x} + \frac{\partial v}{\partial y} - 2\frac{\partial w}{\partial z}\right) - u\left(\frac{\partial w}{\partial x} + \frac{\partial u}{\partial z}\right) - v\left(\frac{\partial w}{\partial y} + \frac{\partial v}{\partial z}\right)\right]. \tag{10}$$

The above governing equations are closed by thermodynamics state relations which are expressed in dimensionless form:

$$\rho = \frac{1 + \gamma M^2 p}{T}, \qquad e = \int_1^T C_v(T)dT + e_\infty, \qquad E = e + \frac{\gamma(\gamma - 1)}{2} M^2(u^2 + v^2 + w^2). \tag{11}$$

The symbols T, ρ, p, κ, μ, e and E represent the temperature, density, thermodynamic pressure, thermal conductivity, viscosity, specific internal energy, and total specific energy, respectively. In the above governing equations, the dimensionless fluid velocities u, v, w are the components of $\mathbf{V}$ in x, y, z directions and t denoted as non-dimensional time. In Eq. 11, the value of e_∞ is taken as unity. The values of κ, C_v are determined using the least squares curve-fitting technique [11, 14, 15], while Sutherland's law is used to calculate the molecular viscosity relation.

2.2 *Initial and Boundary Conditions*

In the present computations, free-stream values that exist at an infinitely large distance from the cylinder are used to initiate the simulation. At the cylinder surface, a uniform elevated temperature is given and no-slip, no-penetration conditions are used for the velocity. Pressure is determined using normal momentum equations, whereas density is obtained by equation of state. In the spanwise direction, periodic boundary conditions are employed. In order to induce transition to three-dimensionality, a spanwise random perturbation with an order of 10^{-7} is added at the initial condition of the density in the cylinder wake.

At the artificial boundary, the characteristic numerical boundary conditions of the Euler equations based on wave speed along the local normal direction have been applied [16]. At the inflow, where the waves enter the flow region, the characteristic variables are fixed and equal to the free-stream conditions, whereas, at outflow, these attribute variables are extrapolated from the internal values of the domain. For the pressure value at the outflow, the characteristic boundary condition reported in Bayliss and Turkel [17] is employed, which is given as

$$\frac{\partial p}{\partial t} - \left(\frac{1}{M}\right)\frac{\partial V_N}{\partial t} = 0, \tag{12}$$

where V_N is the normal velocity component. The value of density is obtained using the equation of state at the inflow and outflow portion of the domain.

3 Numerical Results

3.1 Validation Study

An in-house compressible solver is validated by comparing the obtained numerical values of the time-averaged drag coefficient and the Strouhal number. The drag coefficient (C_D), the lift coefficient (C_L), and the Strouhal number (St) are defined as

$$C_D = \frac{2F_D}{\rho_\infty {U_\infty}^2 D H_z}, \qquad C_L = \frac{2F_L}{\rho_\infty {U_\infty}^2 D H_z}, \qquad St = \frac{fD}{U_\infty}, \tag{13}$$

where f represents the vortex shedding frequency and F_D, F_L are the integrated forces (drag and lift) on the cylinder. At $Re = 200 - 300$ and $\epsilon = 0$, the obtained numerical values of St and $\overline{C_D}$ are compared with the reported values obtained via DNS, experimental, and Floquet methods as listed in Tables 1 and 2. The St values in the present study are calculated by the dominant frequency obtained using FFT algorithm of time history of lift coefficient (C_L). Table 1 shows that the St values of the present study are very close to the values of the DNS study reported by Jiang et al. [5], with a maximum deviation of 4%. Similarly, the $\overline{C_D}$ values are well validated by the values reported by Jiang and Cheng [18] with deviations less than 3%, as listed in Table 2.

Table 1 Comparison of Strouhal number (St) for various Re at $\epsilon = 0$

Re	Present study DNS	Jiang et al. [5] DNS	Mahir [19] DNS	Okajima [20] Exp.	Luo et al. [1] Exp.
200	0.1445	0.1490	0.154	0.1395	0.159
220	0.1457	0.1504	0.151	–	0.159
230	0.1466	0.1510	0.154	–	0.161
250	0.1452	–	0.152	0.1421	0.159
300	0.1391	0.1453	–	0.1399	–

Table 2 Comparison of time-averaged drag coefficients ($\overline{C_D}$) for various Re at $\epsilon = 0$

Re	Present study DNS	Jiang and Cheng [18] DNS	Sohankar et al. [3] DNS	Mahir [19] DNS	Robichaux et al. [2] Floquet
200	1.405	1.393	1.459	1.518	1.642
220	1.432	1.41	–	1.554	1.671
250	1.451	–	1.491	1.567	1.73
300	1.479	1.438	1.561	–	1.873

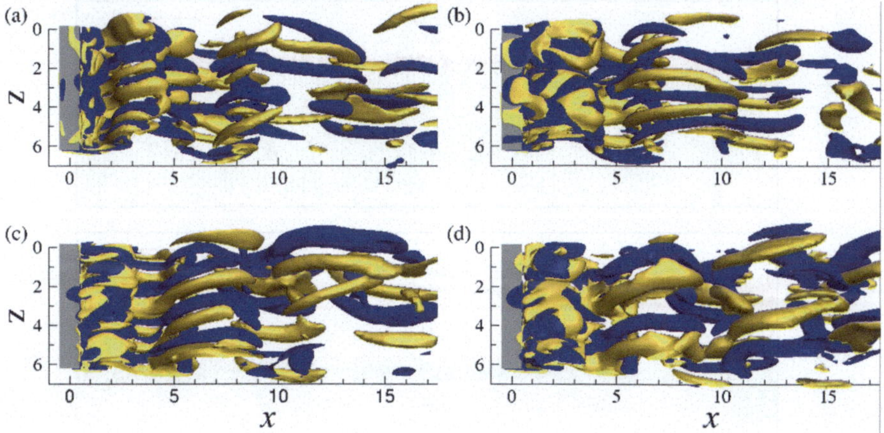

Fig. 2 Iso-surfaces of $\omega_x = \pm 0.5$ in the square cylinder wake at $Re = 500$ and $t = 900$ for (**a**) $\epsilon = 0.0$, (**b**) $\epsilon = 0.2$, (**c**) $\epsilon = 0.6$, and (**d**) $\epsilon = 1.0$. Yellow and blue represent positive and negative values, respectively

3.2 *Vortical Structure*

The three-dimensional flow in a bluff body wake depends on several factors such as bluff body shape, incoming flow, and buoyancy effect. The strong buoyancy causes the flow field to become asymmetric and significantly changes the shape of the vortical structure in the cylinder wake. Figure 2 shows the vortical structures of streamwise vorticity (ω_x) in a square cylinder wake with different heating at $Re =$ 500. The disordered vortical structure of the mode B transition is observed in square cylinder wake at $\epsilon = 0$ as shown in Fig. 2a. On increasing the heating level from $\epsilon = 0$ to $\epsilon = 0.6$, the vortical structures with $\lambda_z = 1.2D - 1.5D$ (depending on the vortex pair in the cylinder wake) do not change significantly (see Fig. 2a–c). Further increasing the heating value to $\epsilon = 1.0$, the strong effect of buoyancy transformed the disordered mode B structure into the mode C structure with wavelength $\lambda_z \sim 3D$ (see Fig. 2d).

3.3 *Time Histories of Force Coefficients*

Figure 3 shows the time history of C_D and C_L for $\epsilon = 0.0$ and $\epsilon = 1.0$ at $Re = 500$. Based on this plot, the amplitude of both the drag and the force coefficient becomes smaller with increase in the heating of the cylinder. The mean value of C_D decreases slightly on large-scale heating ($\epsilon = 1.0$), while the average value of C_L changes significantly with heating and gives negative values (see in Fig. 3b). This negative value of C_L at strong buoyant force is observed due to unbalanced pressure and shear forces around the cylinder developed by the asymmetric flow field.

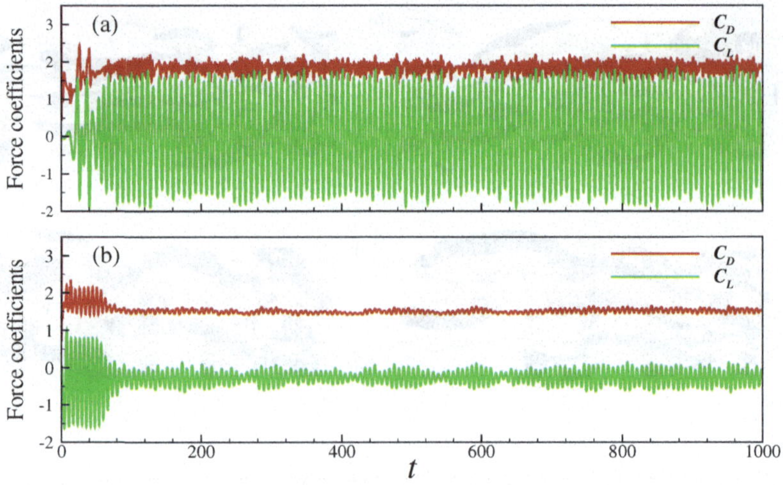

Fig. 3 Time histories of force coefficients (C_D and C_L) at $Re = 500$ for (**a**) $\epsilon = 0.0$ and (**b**) $\epsilon = 1.0$

3.4 Frequency Spectra of Lift Coefficient

Figure 4 shows the amplitude of C_L, with the St value at $Re = 500$ for $\epsilon = 0.0$ and $\epsilon = 1.0$. The vortex shedding frequency spectra are obtained from the Fast Fourier Transform (FFT) algorithm using the time history of the C_L values. At $\epsilon = 1.0$, the amplitude of C_L is about 10 times less than that of C_L amplitude of the unheated case (see Fig. 4). Additionally, the dominant peak frequency of the vortex shedding (i.e., St value) slightly increases upon heating of the cylinder surface ($\epsilon = 1.0$).

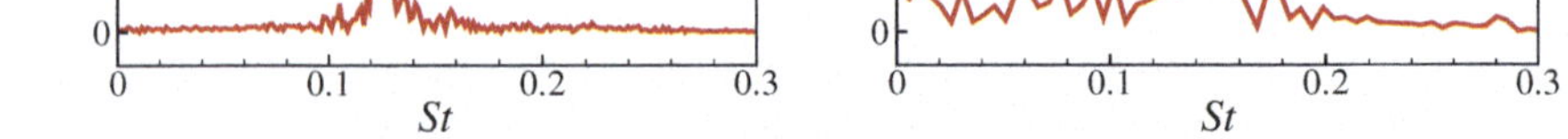

Fig. 4 At $Re = 500$, frequency spectra of lift coefficient (C_L) with heating (**a**) $\epsilon = 0.0$ and (**b**) $\epsilon = 1.0$

4 Conclusions

This paper presents a DNS study of the three-dimensional flow around a heated square cylinder subjected to a uniform free stream ($Pr = 0.7$) at right angle to gravity. At $Re = 500$, the disordered mode B vortical structure with $\lambda_z = 1.2D - 1.5D$ is observed in the isothermal wake behind the unheated square cylinder ($\epsilon = 0.0$). At $\epsilon = 0 - 0.6$, the vortical structure does not change significantly and the value of λ_z remains the same. At large-scale heating ($\epsilon = 1.0$), a significant changes is observed in the shape of the vortical structure with wavelength $\lambda_z \sim 3D$. The flow field around the heated cylinder becomes asymmetric due to the non-parallelism between the free-stream crossflow and buoyancy. The asymmetric flow field creates unbalanced pressure and shear forces that lead to the generation of negative lift. The amplitude of C_L becomes smaller as the buoyant force increases. Additionally, when the heating level is increased from $\epsilon = 0$ to $\epsilon = 1$, there is a very slight increase in the St value of the vortex shedding.

Acknowledgements The authors would like to acknowledge the computational support provided by the High-Performance Computing Facility at the Indian Institute of Technology, Delhi, India.

References

1. Luo S, Chew Y, Ng Y (2003) Characteristics of square cylinder wake transition flows. Phys Fluids 15:2549–2559
2. Robichaux J, Balachandar S, Vanka S (1999) Three-dimensional Floquet instability of the wake of square cylinder. Phys Fluids 11:560–578
3. Sohankar A, Norberg C, Davidson L (1999) Simulation of three-dimensional flow around a square cylinder at moderate Reynolds numbers. Phys Fluids 11:288–306
4. Agbaglah G, Mavriplis C (2017) Computational analysis of physical mechanisms at the onset of three-dimensionality in the wake of a square cylinder. J Fluid Mech 833:631–647
5. Jiang H, Cheng L, An H (2018) Three-dimensional wake transition of a square cylinder. J Fluid Mech 842:102–127
6. Sheard GJ, Fitzgerald MJ, Ryan K (2009) Cylinders with square cross-section: wake instabilities with incidence angle variation. J Fluid Mech 630:43–69
7. Roshko A (1954) On the development of turbulent wakes from vortex streets
8. Kieft RN, Rindt C, Van Steenhoven A, Van Heijst G (2003) On the wake structure behind a heated horizontal cylinder in cross-flow. J Fluid Mech 486:189–211
9. Ren M, Rindt CC, Van Steenhoven AA (2006) Three-dimensional transition of a water flow around a heated cylinder at $re = 85$ and $ri = 1.0$. J Fluid Mech 566:195–224
10. Mahir N, Altaç Z (2019) Numerical investigation of flow and combined natural-forced convection from an isothermal square cylinder in cross flow. Int J Heat Fluid Flow 75:103–121
11. Arif MR, Hasan N (2019) Vortex shedding suppression in mixed convective flow past a square cylinder subjected to large-scale heating using a non-Boussinesq model. Phys Fluids 31:023602
12. Arif MR, Hasan N (2020) Effect of thermal buoyancy on vortex-shedding and aerodynamic characteristics for fluid flow past an inclined square cylinder. Int J Heat Technol 38:463–471
13. Arif MR, Hasan N (2021) Effect of free-stream inclination and buoyancy on flow past a square cylinder in large-scale heating regime. Phys Fluids 33:073601

14. Ghoshdastidar PS (2012) Heat transfer. Oxford University Press. https://books.google.co.in/books?id=DxeKpwAACAAJ
15. Hasan N, Saeed A (2017) Effects of heating and free-stream orientation in two-dimensional forced convective flow of air past a square cylinder. Int J Therm Sci 112:1–30
16. Hirsch C (2007) Numerical computation of internal and external flows: the fundamentals of computational fluid dynamics. Elsevier
17. Bayliss A, Turkel E (1982) Far field boundary conditions for compressible flows. J Comput Phys 48:182–199
18. Jiang H, Cheng L (2018) Hydrodynamic characteristics of flow past a square cylinder at moderate Reynolds numbers. Phys Fluids 30:104107
19. Mahir N (2017) Three dimensional heat transfer from a square cylinder at low Reynolds numbers. Int J Therm Sci 119:37–50
20. Okajima A (1982) Strouhal numbers of rectangular cylinders. J Fluid Mech 123:379–398

Heat Transfer and Combustion

Effect of Divider Wall Shape on the Flow Development and Heat Transfer Characteristics in a Two-Pass Duct

Arun Chand, Nishab Ali, and Andallib Tariq

Abstract This paper presents the flow patterns and heat transfer distribution across the bend region in a two-pass rectangular duct for an inlet turbulent flow regime ($Re = 6500$). The results are presented at various vertical and horizontal planes located at different positions across the bend along the flow progression. Two divider wall configurations (in-between two-pass of the duct) are studied, i.e., (1) sharp-corner turn and (2) smooth curved turn. It has been noticed that flow gets completely modified because of divider wall shape. The numerical results reveal that primary and secondary flow in turn regime displays combined features of a bend-induced, Dean-type circulation. The flow dynamics and local heat transfer vary significantly with different divider wall configurations. For the duct with a sharp divider wall, a pair of counter-rotating Dean vortices induce after 90° near the end-wall of bend region. In contrast, for the duct with a smooth curved divider wall, vortices pair is induced close to the divider wall and stretched across the entire width. The results show that Dean vortices play an important role in enhancing localized heat transfer across the bend regime. Interestingly, the duct with a sharp divider wall exhibits higher localized heat transfer, whereas the duct with a smooth, curved divider wall exhibits lower localized heat transfer but more uniform distribution.

1 Introduction

The two-pass channels connected by a 180° turn are encountered in various engineering devices such as heat exchangers, ventilation piping systems, intake manifolds of engines, and the interior channels of gas turbine blades. The fluid experiences a strong curvature/bend in its path, which changes the flow characteristics. Bend regions in internal cooling passageways are associated with significant heat transfer

A. Chand (✉) · N. Ali · A. Tariq
AVTAR (Aerodynamics Visualization and Thermal Analysis Research) Lab, Mechanical and Industrial Engineering Department, Indian Institute of Technology Roorkee, Roorkee 247667, India
e-mail: arun_c@me.iitr.ac.in

M. A. Siddiqui et al. (eds.), *Advances in Heat Transfer and Fluid Dynamics*, Lecture Notes in Mechanical Engineering, https://doi.org/10.1007/978-981-99-7213-5_12

enhancement and a pressure drop penalty, which are directly related to flow characteristics at these turns. When fluid is passed through 180° turn, flow characteristics change due to the development of curvature-induced primary and secondary flows, thus making this an essential flow investigation problem [1–4]. Therefore, many researchers are motivated to investigate the flow features in curvature ducts, but the scope was limited to either sharp or smooth 180° turn configurations. A limited study is present on the effect of divider wall shape. Modern engineering requires higher heat transfer enhancement and lower pressure drops, demanding an optimum turn configuration. A thorough understanding of turbulent flow dynamics and its consequence on heat transfer at different turn configurations is required for this aerothermal problem and thus motivates the current study.

2 Literature Review and Objective

Wang and Chyu [2] investigated the flow behavior and heat transfer distribution in a two-pass square duct with various turn geometries. They reported flow separation and the formation of dean vortices at the bend region, which altered the flow behavior downstream in the second pass. The divider wall generated a swirl in the cross-plane flow downstream of the bend regime, increasing heat transfer, which is greatly influenced by turn geometry. Liou et al. [4] investigated the flow characteristics and the effect of divider wall thickness in a two-pass square duct with a sharp 180° bend using LDV. They observed flow separation as the mainstream flow detached from the sharp divider wall. Furthermore, increasing divider wall thickness shifts the separation region near the divider wall tip and decreases the average turbulent kinetic energy after the bend. Liou and Chen [5] investigated the secondary flow fields in the two-pass rectangular duct using laser Doppler velocimetry (LDV), with and without rotation. In the case of a stationary duct, they reported that flow behavior is affected up to 3 D_h before the turn and 11 D_h downstream of the turn. A small corner vortex along with a bigger recirculation zone near the divider wall is observed in the primary flow. A dean vortex is formed, which covers almost half of the cross-section in the secondary flow field. The size of the separation zone decreased by 75% with rotation compared to a stationary duct. Son et al. [6] investigated the flow dynamics in a two-pass square duct using Particle Image Velocimetry (PIV) experiments and correlated them with heat transfer enhancement. They reported that vortex characteristics and flow impingement behavior in both primary and secondary flows are strongly related to localized wall heat transfer at a sharp bend. Saha and Acharya [7] studied flow behavior and heat transfer with nine different bend geometries. Significant effects of bend geometries on the overall performance of the cooling channel were observed.

Erelli et al. [8] investigated different turn configurations in a two-pass square duct. They discovered that the flow behavior varies significantly with turn configurations and that a bend in a square duct induces secondary flow, which affects the local heat transfer distribution. Recently, Liu et al. [9] studied internal flow field characteristics in square channels with different turn geometries. Only primary flow results in the horizontal plane were reported. Yan et al. [10] varied the divider wall inclination angles in a two-pass duct and numerically investigated their effects on heat transfer characteristics and flow structure. Heat transfer was enhanced in the bend and upstream region, but decreased downstream of the bend as the positive divider inclination angle increased. With a negative divider inclination angle, the opposite heat transfer characteristics were observed. Reddy et al. [11] investigated the thermo-hydraulic performance of a novel curved serpentine coil and reported a 19% better performance compared to the conventional flat serpentine coil due to the enhanced chaotic nature of secondary flows. Wang et al. [12] investigated the evolution of secondary vortex structures and pressure loss for magneto-hydrodynamic flow in a square channel with a transverse magnetic field. The flow dynamics are reported to have a significant impact on convective heat transport in fusion reactor blankets.

The preceding discussion provides an overview of the complexities and importance associated with the aerothermal characterization of the bend regime. Available research emphasizes the overall flow field on a two-pass cooling channel and its effect on the temperature field. However, limited information is presented near the bend regime, which thus motivates the current study. This work focuses on studying flow structures and their dynamics at the 180° turn with two different divider shapes (sharp and smooth) at the bend region. The investigation aims to present a holistic insight into primary and secondary flow development and its consequences on heat transfer. Numerical simulation results are presented for a turbulent inlet condition at a Reynolds number of 6500. The results are expressed in terms of mean velocities, streamlines, and flow structures over primary and secondary flow regimes. Numerical simulation results are experimentally validated using Liquid Crystal thermography (LCT) and Particle Image Velocimetry (PIV) measurements. Although PIV technique can obtain velocity measurements with a high spatial resolution and accuracy, investigating such complex geometries where many measurement planes are required to understand the flow phenomena is quite challenging.

3 Computational Details

3.1 Computation Geometry and Grid Generation

The schematic and details of the two-pass rectangular duct geometries are shown in Fig. 1a, b. The computational domain dimensions are same as the experimental setup used for PIV measurement. The duct consists of an inlet and outlet section of length L(= 1180 mm) and hydraulic diameter D_h(=70.4 mm) on both passes of the duct. Constant heat flux is provided at the bottom wall of both duct passes. The turbulent flow profile was checked at plane P^1, which is at a section length of 12D_h from the inlet (Fig. 2). The uniform bend width W and clearance c were equal in both ducts. For the smooth divider wall duct, bend radius R_W was chosen to achieve a uniform bend width W (=2R_W) equal to that of the sharp divider wall. Hexahedral structured

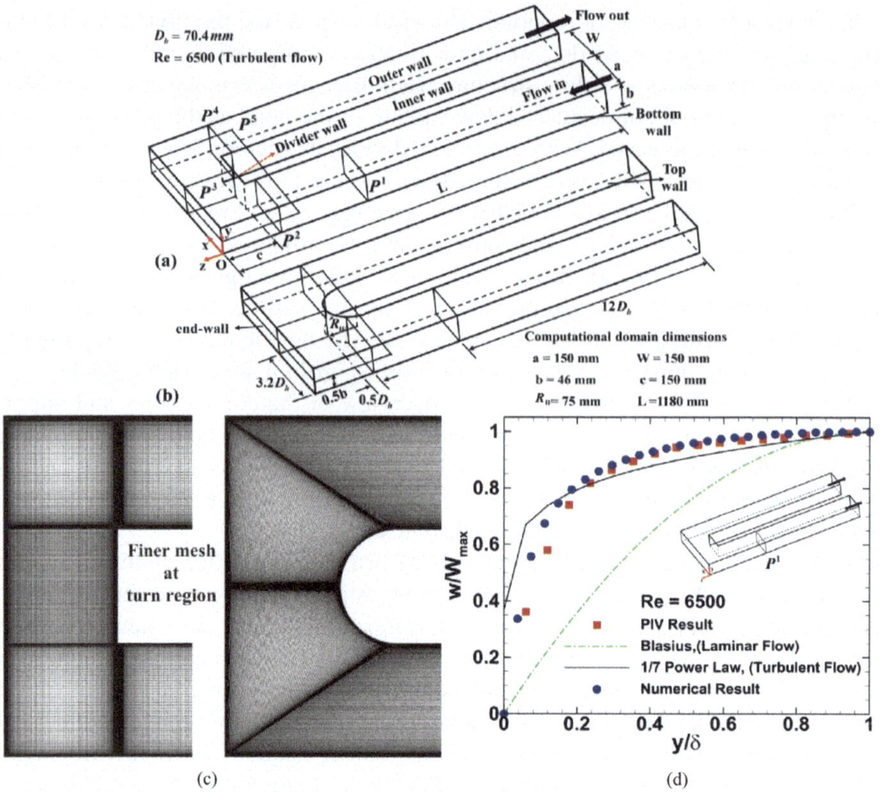

Fig. 1 Schematic diagram of computational domain with measurements planes (Two-pass duct with 180° bend) having **a** sharp divider wall, **b** smooth curved divider wall, **c** computational grid, **d** normalized inlet velocity profile at plane P^1

meshes were generated using ANSYS Meshing, where finer meshing is placed near the turn region, as shown in Fig. 2a. The turbulent flow was resolved using Realizable k-ε model with enhanced wall treatment, and a y^+ value of less than 1 was used for the first grid point near the wall. A total number of 7 million cells were used for the simulations.

3.2 Computational Setup and Boundary Conditions

The choice of a turbulence model is critical for turbulent flow involving three-dimensional flow phenomena, which demands accurate modelling. Numerous studies have demonstrated that the realizable k-ε turbulence model is capable of accurately predicting the flow structures and heat transfer distribution for flows involving strong adverse pressure gradients, recirculation, and separation, and it is widely used due to its shorter computational time and accuracy [7, 8, 10]. Hence, steady state numerical simulations are performed using the realizable k-ε turbulence model present in Ansys Fluent (version 21R1) software with enhanced wall treatment. The working fluid used is air with constant physical properties and is assumed incompressible. The Reynolds number ($Re = 6500$) calculated from the experiments in a two-pass duct with a sharp divider wall was used in numerical simulation for both the duct geometry. A separate flow simulation without energy equation was performed in a $20D_h$ long rectangular duct having an exact cross-section as the present geometry. The fully developed turbulent velocity profile was extracted from the simulation, which was then used at the inlet of present computational domain. The solution methods and boundary conditions used in numerical simulations are shown in Table 1. The generalized form of steady Reynolds-averaged Navier–Stokes (RANS) along with energy equations are presented by Eq. 1–6 [13]:

$$\text{Continuity equation}: \frac{\partial}{\partial x_i}(\rho u_i) = 0 \tag{1}$$

$$\text{Momentumequation}: \frac{\partial}{\partial x_i}(\rho u_j u_i) = -\frac{\partial p}{\partial x_i} + \frac{\partial}{\partial x_j}\left[\mu\left(\frac{\partial u_i}{\partial x_j} + \frac{\partial u_j}{\partial x_i} - \frac{2}{3}\delta_{i_j}\frac{\partial u_i}{\partial x_i}\right)\right] + \frac{\partial}{\partial x_j}\left(-\rho\overline{u'_i u'_j}\right) \tag{2}$$

$$\text{Energy equation}: \frac{\partial}{\partial x_i}(\rho u_i T) = \frac{\partial}{\partial x_j}\left[\left(\frac{\mu}{\mathrm{Pr}} + \frac{\mu_t}{\mathrm{Pr}_t}\right)\frac{\partial T}{\partial x_j} + u_i(\tau_{i_j})_{eff}\right] \tag{3}$$

where, $\left(-\rho\overline{u'_i u'_j}\right) = \mu_t\left(\frac{\partial u_i}{\partial x_j} + \frac{\partial u_j}{\partial x_i}\right) - \frac{2}{3}\left(\rho k + \mu_t\frac{\partial u_k}{\partial x_k}\right)\delta_{i_j}$, and $(\tau_{ij})_{eff} = \mu_{eff}\left(\frac{\partial u_i}{\partial x_j} + \frac{\partial u_j}{\partial x_i}\right) - \frac{2}{3}\left(\mu_{eff}\frac{\partial u_k}{\partial x}\right)\delta_{ij}$

$$\text{Turbulentkineticenergy (k)}: \nabla.\left(\rho k\bar{V}\right) = \nabla.\left[\left(\mu + \frac{\mu_t}{\sigma_k}\right)\nabla k\right] + G_k + G_b - \rho\varepsilon - Y_M + S_k \tag{4}$$

and turbulent dissipation rate (ε):

$$\nabla.\left(\rho\varepsilon\overline{V}\right) = \nabla.\left[\left(\mu + \frac{\mu_t}{\sigma_\varepsilon}\right)\nabla\varepsilon\right] + \rho C_1 S_\varepsilon - \rho C_2\frac{dy}{dx} \times \frac{\varepsilon^2}{k + \sqrt{\nu\varepsilon}} + C_{1\varepsilon}\frac{\varepsilon}{k}C_{3\varepsilon}G_b + S_\varepsilon \tag{5}$$

where: $C_1 = \max\left[0.43, \frac{\eta}{\eta+5}\right]$, $\eta = S\frac{k}{\varepsilon}$, $S = \sqrt{2S_{ij}S_{ij}}$ and model constants are $C_{1\varepsilon} = 1.44$, $C_2 = 1.9$, $\sigma_\varepsilon = 1.0$, $\sigma_k = 1.2$

Table 1 Boundary conditions and solution methods used in simulations

Boundary condition	Solution methods
Inlet: velocity inlet,	SIMPLE algorithm
Turbulent intensity 10%, and hydraulic diameter $D_h = 70.4$ mm	Second order discretization
Air inlet temperature 300 K	Second order upwind
Outlet: outflow	Convergence criteria of
Bottom wall: heat flux $q_w = 1000$ W/m^2	Flow equation: 10^{-6}
Other walls: adiabatic	Energy equation: 10^{-9}

3.3 Data Reduction

Numerical results in the form of velocity contours were presented at different vertical (xy & yz) and horizontal planes (zx). The vertical xy planes (P^2, P^4) are situated upstream and downstream of divider wall at a distance of $0.5D_h$. At the same time, the vertical (yz) plane (P^3) is located at $3.2D_h$, from the origin (Fig. 1). Further, primary flow in the horizontal (zx) plane is depicted using symmetry plane P^5. The temperature at bottom wall (T_w) and the local bulk air temperature (T_{bulk}) were obtained from the numerical solution. The heat transfer coefficient (h) and Nusselt number (*Nu*) distribution were then evaluated using Eqs. 6 and 7, respectively.

$$h = q_w / (T_w - T_{bulk}) \tag{6}$$

$$Nu = hD_h / K \tag{7}$$

3.4 Validation and Verification

Structured grids are generated using Ansys meshing and optimized using the grid independence test (Fig. 2a)

. The present mesh and turbulence model accurately captures flow physics and heat transfer distribution. The comparison of w/W_{max} (W_{max} is duct centerline velocity) at the plane P^1 with PIV results, turbulent velocity profile (1/7th power-law) and laminar velocity profile (Blasius) reveals that the flow is completely turbulent for $Re = 6500$ as shown in Fig. 1d. Heat transfer results are validated against the experiment using LCT technique (Fig. 2b). The numerical simulation accurately captured the primary recirculation region near the divider wall, as observed from the PIV experiment (Fig. 3a, b). However, the presence of the recirculation zone is still visible near the corner area of the first pass, as shown in Fig. 3a.

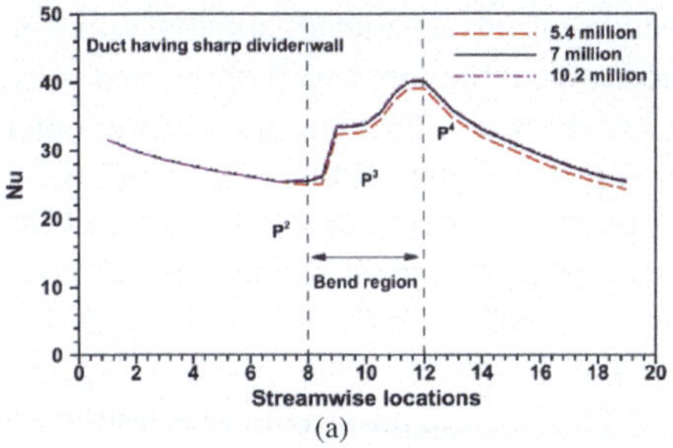

Fig. 2 Spanwise averaged *Nu* at various locations along the streamwise direction **a** at three different mesh sizes **b** for model validation

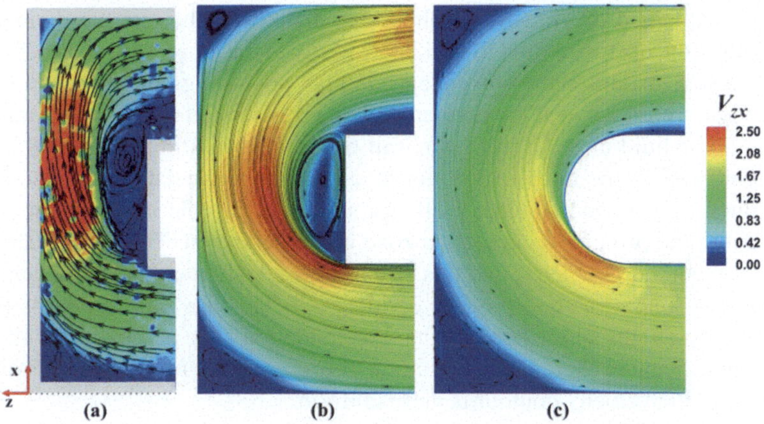

Fig. 3 Streamlined velocity profile of primary flow on horizontal plane P^5 **a** experimental (PIV) result, **b** numerical result with sharp divider wall, and **c** smooth curve divider wall

4 Results and Discussion

The primary and secondary flow progression is explained by velocity contour (V_{mag}) superposed with streamlines at the horizontal plane (P^5) and vertical planes (P^2–P^4) respectively. Finally, the relationship between flow progression and heat transfer has been discussed.

4.1 Flow Progression

The streamlined contour of velocity for the duct with a sharp divider wall is shown in Fig. 3b, and a total of four vortexes can be observed. The incoming flow is uniform and smooth in the first pass, and the flow deviates toward the inner wall just upstream of the bend (P^2) due to the curvature in the bend section. As a result, a radial pressure gradient and a lower pressure zone are formed toward the inner wall of the first pass. Flow accelerates near the inner wall of the first pass and decelerates near the outer wall. The fluid along the outer wall of the first pass impinges on the end-wall of the bend region, forming the first recirculation zone in the duct at the bend corner of the first pass. Moreover, the flow is abruptly redirected by sharp edge of the divider wall. Along with pressure reduction, it causes flow separation and a primary recirculation zone (primary vortex) is formed at the tip of the divider wall, as shown in Fig. 3b. Further downstream, at the outer bend corner in the second pass, a third recirculation zone forms, similar to the first recirculation zone. The primary flow separation zone at the tip of the divider wall extends further downstream in the second pass.

Figure 3b shows the streamlined velocity contour for the duct with a smooth divider wall. The two small recirculation zones at the corners occur for the same reason as explained previously, but their size increases. The circular divider wall shape causes the upstream fluid to stick more easily to the inner wall, so flow remains strongly attached. This causes an enhanced radial pressure gradient in section (P^2), and flow is accelerated near the divider wall and decelerated at the outer wall of the first pass. The fluid close to the outer wall of the first pass impinges farther away on the end-wall of the bend region, forming a larger recirculation zone compared to the duct with a sharp divider wall. As shown in Fig. 3c, the circular divider wall reduces the primary recirculation zone in the bend area. The mainstream fluid follows the curved geometry, and this strong flow attachment leads to a negligible low-pressure zone near the tip of the divider wall. This flow attachment is further reflected downstream of the bend, with no recirculation zone at the inner wall of the second pass. Because there is no separation of primary flow in a duct with a smooth divider wall, the velocity contour is more uniform.

The primary flow observations in the bend region and past literature manifest secondary flow development and its consequences on heat transfer. The primary flow encounters a sharp bend, which causes centrifugal instability [14], resulting in the formation of a pair of counter-rotating vortices at plane P^3 (after 90° rotation of flow) known as the Dean vortex, as seen in Fig. 4. Vortex pair is noted by upper clockwise rotating vortices and lower anticlockwise rotating vortices. The development of vortices enhances the mixing between fluid layers due to the higher velocity gradients in the transverse direction. As shown in Fig. 4a, Dean vortices form away from the sharp divider wall in areas of mainstream flow (where centrifugal instability is induced). Low-velocity magnitude at plane P^3 confirms flow entrapment due to the primary recirculation zone as well as the absence of centrifugal instabilities. Hence, Dean vortices are not induced near the tip of the divider wall. The Dean vortex pair propagates upwards and outwards as the flow advances downstream (180° turn) due to centrifugal forces pushing the main flow outward.

Dean vortices develop near the tip of the smooth divider wall in the region of mainstream flow, as seen in Fig. 4b. Also, the effect of dean vortices extends the entire plane width, reflecting the uniform flow distribution observed in primary flow. Further, the uniform velocity contours are present at vertical planes compared to the duct with a sharp divider wall, and this trend continues downstream the bend in

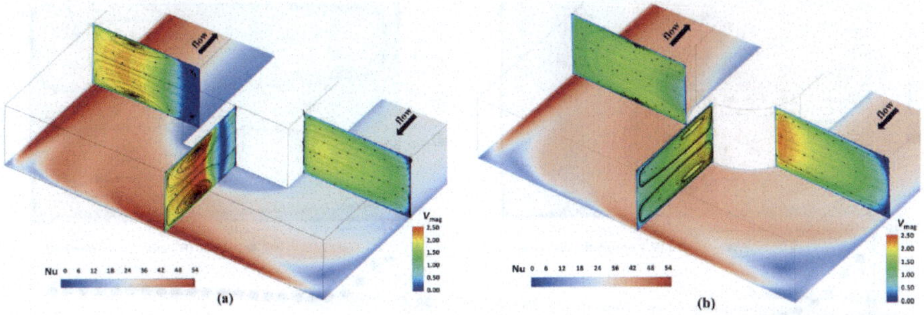

Fig. 4 Flow progression (P^2-P^4) with surface Nusselt number (at bottom wall) distribution across 180° bend having **a** sharp divider wall **b** smooth curve divider wall

the second pass of the duct. Dean vortices develop in mainstream flow, as observed for both geometries, implying that centrifugal instabilities are essential for their development.

4.2 Heat Transfer

The surface heat transfer distribution near the bend region strongly correlates with primary and secondary flows. Upstream of the bend regime with the sharp divider wall, *Nu* values are low due to the boundary layer formation caused by the long entry length of the first pass, which inhibits heat transfer between the surface and the heated air stream. However, there is a significant increase in *Nu* values in the bend regime, particularly in the region of mainstream flow (Fig. 4a). The *Nu* values were observed to be successfully augmented from an order of 20–25 to an order of 50. Although primary flow enhances the heat transfer, the impingement region of dean vortices has slightly higher *Nu* values. The secondary flow induces huge turbulence with high-velocity gradients locally, resulting in higher fluid mixing and momentum transfer between fluid layers. The mixing/momentum transfer mechanism helps in intense augmentation of heat transfer that is reflected by the higher values of local *Nu* nearby the dean vortex. In contrast, heat transfer is lower on the inner side of the bend near the sharp divider wall. It is possibly due to flow entrapment resulting from flow separation by this sharp edge divider. The comparatively uniform primary flow and stretched secondary flow pattern in duct with a smooth curved divider wall results in uniform *Nu* distribution across the bend regime as seen in Fig. 4b.

Now the surface *Nu* at plane P^3 with flow structure is closely presented in Fig. 5. Interestingly for the duct with a sharp divider wall (Fig. 5a), *Nu* is very low in the regime of separated flow (recirculation flow) where flow entrapment takes place. But in the regime of Dean vortices, very intense enhancement in surface *Nu* is observed. *Nu* value on the extent of the separated region is found to be an order of 20, which is intensely raised to approx. 35 (i.e., 1.5 times) on the extent, where Dean vortices

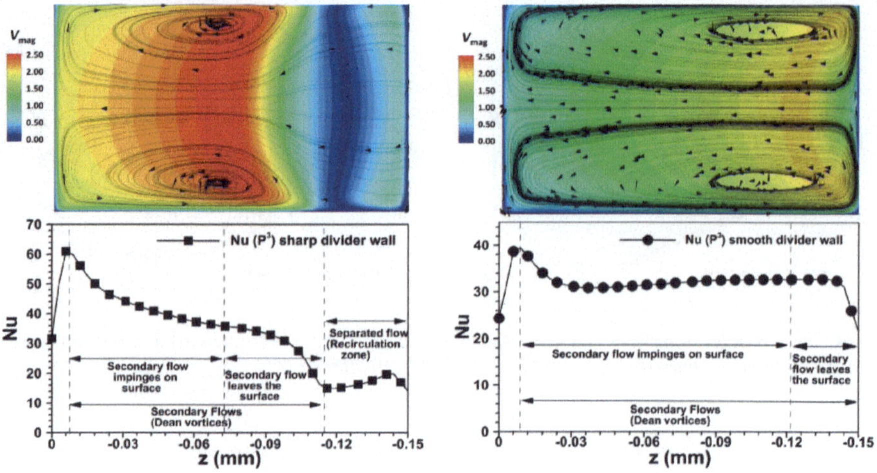

(a) (b)

Fig. 5 Effect of secondary flow structure on surface Nusselt number distribution at vertical plane P^3 for: **a** duct with sharp divider wall and **b** duct with smooth curve divider wall

leave the surface. Subsequently, a further increment is seen in *Nu,* approx. up to 60 at the extent where Dean vortices impinge the surface. Therefore, the surface impingement effect is more pronounced in heat transfer enhancement as compared with the surface leaving effect of Dean vortices. The possible reason is that surface impingement is more significant in breaking the boundary layer than the surface leaving effect. As shown in Fig. 5b, the *Nu* distribution is uniform for the duct with a curved divider wall. The only exception to a higher *Nu* value is the region where the local impinging jet effect of dean vortex is present.

5 Conclusion

The investigation presents detailed flow dynamics and heat transfer behavior across a bend regime in a two-pass rectangular duct with two different divider wall shapes. The divider wall configurations studied are, i.e., the sharp divider wall and the smooth divider wall for an inlet turbulent flow regime. The shape of a divider wall completely changes the flow behavior in the bend region. For the duct with a sharp divider wall, a primary separation bubble is formed near the divider wall, whereas flow remains attached to the curved divider wall. The primary flow significantly affects secondary flow development, modifying the flow pattern and propagation. The secondary flow structure, in the form of counter-rotating Dean vortices, is observed on the transverse plane as the flow turns 90° and propagates in the direction of flow advancement. As observed for both geometries, Dean vortices develop in mainstream flow at the bend regime where centrifugal instabilities exist, implying that centrifugal instabilities are critical for their development. A strong correlation is found between these vortex structures and heat transfer behavior at the bend. An enhancement in localized heat transfer is observed in the regime of Dean vortices due to impinging jet-like effect. The surface averaged Nusselt number is found to be more uniform in the duct with a smooth divider wall. It can be confirmed that the divider wall shape affects the mainstream (primary) flow, which in turn affects the secondary flow magnitude in transverse direction (vertical planes).

References

1. Azzola J, Humphrey JAC, Iacovides H, Launder BE (1986) J Fluids Eng Trans ASME 108:214
2. Wang TS, Chyu MK (1994) J Thermophys Heat Transf 8:595
3. Schabacker J, Bolcs A, Johnson BV (1998) Am Soc Mech Eng 1
4. Liou TM, Tzeng YY, Chen CC (1998) Proc ASME Turbo Expo 4
5. Liou TM, Chen CC (1997) Proc. ASME Turbo Expo 3:167
6. Son SY, Kihm KD, Han JC (2002) Int J Heat Mass Transf 45:4809
7. Saha K, Acharya S (2012) J Turbomach 135
8. Erelli R, Saha AK, Panigrahi PK (2015) Int J Heat Mass Transf 89:667
9. Liu R, Li H, You R, Tao Z (2020) AIP Adv 10
10. Yan H, Luo L, Du W, Wang S, Sunden B, Huang D (2021) Int J Therm Sci 166:106969
11. Reddy DS, Khan MK, Awasthi K (2020) Phys Fluids 32
12. Wang H, Ni MJ, Zhang NM (2022) Phys Fluids 34
13. ANSYS, Inc. (2021) ANSYS Fluent User's Guide, Release 21R1
14. Ali N, Tariq A (2023) Phys Fluids 35:015110

Spot Conductance Measurement Using Infrared Thermography

Khursheed Anwar Khan, T. Altaf Hasan, and Andallib Tariq

Abstract The estimation of thermal contact conductance (TCC) across any contacting surfaces is required for the efficient design and safety of mechanical systems. The accurate estimation of heat flux and temperature drop across the interface is crucial for the assessment of TCC. In the present experimental investigation, stainless steel (SS304) is selected as it is frequently used in various industrial applications. A high-temperature experimental facility has been built to study interfacial heat transfer and TCC estimation. The contact spot conductance has been estimated using steady-state measurements in vacuum utilizing surface temperature information from infrared thermography. The primary objective is to determine the effect of contact pressure (1.7, 4.7, 7.7 MPa) and surface roughness (0.49 and 1.64 μm) on TCC at interface temperatures of 300 °C. Infrared thermography technique is applied for better thermal visualization and accurate prediction of interfacial temperature drop. The result depicts that at lower surface roughness, TCC increases more rapidly with contact pressure. Moreover, experimental results have been suitably normalized and compared to the available theoretical models.

1 Introduction

Understanding and predicting heat transfer through surfaces in contact is an essential aspect of thermal network designs with applications including gas turbines, electronic packaging, internal combustion engine, nuclear reactors, aircraft structural joints, and heat exchangers. The contacting surfaces are not perfectly smooth but exhibit roughness on the microscopic scale. The actual contacts occur only at a few discrete spots, which is less than 2% of the nominal contact area [1]. Recently, Siddappa

K. A. Khan (✉) · A. Tariq
AVTAR (Aerodynamics Visualization and Thermal Analysis Research) Lab, Mechanical and Industrial Engineering Department, Indian Institute of Technology Roorkee, Roorkee 247667, India
e-mail: kanwarkhan@me.iitr.ac.in

T. Altaf Hasan
Department of Mechanical Engineering, Galgotias University, Greater Noida 203201, India

M. A. Siddiqui et al. (eds.), *Advances in Heat Transfer and Fluid Dynamics*, Lecture Notes in Mechanical Engineering, https://doi.org/10.1007/978-981-99-7213-5_13

and Tariq [2] concluded that the variation in actual contact area across contacting surfaces is less than a percent under nominal loading conditions. The contacting spots have a high thermal conductivity than that of the fluid filled in separated gaps. When heat flows from one surface to another, it is constricted to pass through the discrete contact spots, resulting in heat flow resistance and a localized temperature drop. The constriction of heat flow lines through the contacting spots is shown in Fig. 1a. The reciprocal of constriction resistance is termed as thermal contact conductance (TCC) and is expressed as Eq. (1).

$$h = \frac{q}{\Delta T} \tag{1}$$

where, q and ΔT represents interfacial heat flux and temperature drop respectively. TCC basically measures how well heat is transferred between the interface of contacting surfaces. The heat transfer mainly depends on the actual contact area, interface material, and thermo-mechanical properties of the contacting surfaces. Moreover, actual contact area itself depends upon surface topographic parameters and contact pressure. TCC is critical for prior/post-thermal analysis and temperature distribution prediction at high temperatures for structure design and safety evaluation.

Several computational, experimental, and theoretical investigations have been carried out to estimate TCC across the contacting surfaces. Wahid et al. [3] conducted experiments in vacuum at 0.43 MPa contact pressure to estimate TCC and investigated the impact of mean interfacial temperature (40–100 °C) and surface roughness (2.88, 5.4, 14.3, and 21.2 μm) across stainless steel pairs. The outcome shows that surface roughness to asperity slope ratio is a key factor for contact conductance calculations and mean interface temperature has a negligible impact on TCC. Xu

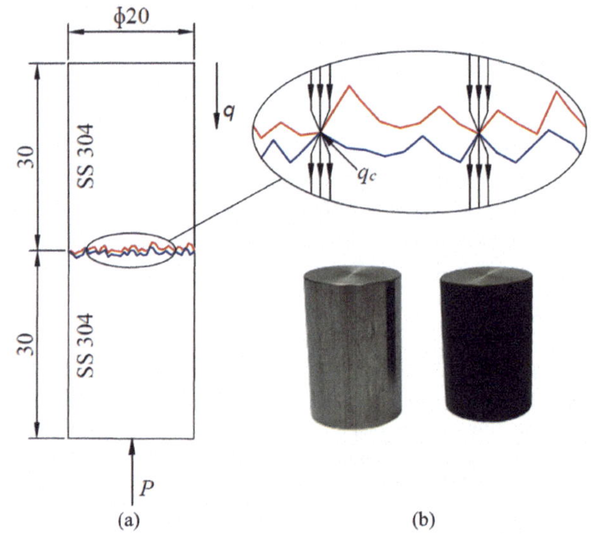

Fig. 1 **a** Diagram of heat flow lines constricted through the contact spots between two contacting surfaces and **b** actual specimen along with the specimen coated with thermographic paint

and Xu [4] estimated TCC at low temperature across stainless steel 304 to investigate the impact of surface characteristics and interfacial temperature (125–210 K) and concluded that both root mean square (rms) roughness and absolute slope are needed for evaluating contact conductance. Tariq and Asif [5] examined steady-state spot conductance using a variety of experimental parameters and compared the results to existing theoretical models. Most models had difficulties in predicting TCC within the boundaries of the experimental constraints. Xing et al. [6] investigated specimen pairs consisting of stainless steel 302, super alloy GH4169, titanium alloy TC11, and die steel 5CrMnMo with contact pressures ranging from 2.96 to 15.68 MPa and interface temperatures ranging from 320 to 550 °C. The author analyzes the influence of various experimental conditions on contact heat transfer using the steady-state heat flux method. The results indicate that TCC typically rises with interface temperature and contact pressure. This variation is more evident at higher temperatures for different contact pressures. Dou et al. [7] studied the impact of surface roughness (0.25 to 2.00 μm), interface temperature (360 to 640 °C) and contact pressure (2.39 to 15.17 MPa) on TCC in an ambient environment across stainless steel interface. The experimental results signify that at higher surface roughness, TCC increases more rapidly with temperature.

In the literature mentioned earlier, thermocouples are widely used to measure temperature inside the conducting body. Chand et al. [8] performed steady-state experiments to evaluate spot and gap conductance across stainless steel contact using thermocouple and infrared thermography. The result reveals that infrared thermography provides better thermal visualization and gives a slightly lower value of TCC than thermocouple. Fieberg and Kneer [9] develop an algorithm based on inverse method to determine TCC using transient infrared temperature data. Steel and aluminum alloy were selected as the testing materials with temperatures and contact pressure ranging from 60 °C to 280 °C and 7.5 to 85 MPa respectively. The results show a linear relationship between contact pressure and TCC, and temperature has almost no impact within the selected temperature range. The author came to the additional conclusion that thermocouples give inaccurate temperature as they alter the thermal characteristics of conducting bodies. Chen et al. [10] developed a data processing technique for estimating thermal contact resistance at high temperatures. An infrared camera is used to precisely measure the temperature drop across thin disk samples. Experiments were carried out on Inconel 718 alloy and ZrB_2/SiC/C to investigate the effect of interface temperature (630–1100 K) and contact pressure (0.1–0.6 MPa). The results show that with increase in interface temperature, thermal contact resistance (TCR) decreases due to high radiative heat transfer at the interface. At high temperatures, the effects of contact pressure are much less pronounced. Helmig et al. [11] utilized infrared thermography technique at nonconforming interfaces and presented a method for inversely quantifying multiscale TCR. The temperature field is distorted when thermocouples are inserted into a heat-conducting body, significantly prejudicing the temperature measurement [12]. Zhang et al. [13] suggested steady-state method as the standard for estimating TCC over transient approach as it is more accurate and reliable. Evidently, better estimation of temperature closest to the interface leads to minimum uncertainty in estimated TCC.

Hence, there is a need to look for the other temperature measurement techniques (infrared thermography and liquid crystal thermography) in the field of TCC estimation. The present work provides a reliable experimental database for TCC across stainless steel pairs using infrared thermography technique. Hot roll bonding joints are widely used in the manufacturing of pressure vessels, nuclear power equipment, heat exchangers, automobiles, etc. The vacuum hot roll bonding prevents the formation of oxides on the surface of materials. The bonding of stainless steel sheets takes place at high contact pressure and temperature of more than 500 °C. In order to initialize the study, steady-state experiments with varying contact pressure (1.7, 4.7 and 7.7 MPa) and surface roughness (0.49 and 1.64 μm) have been performed in vacuum at an interfacial temperature of approximately 300 °C.

2 Experimental Facility

A unique experimental test facility has been designed and fabricated to scrutinize the heat transfer and TCC across contacting specimen. Figure 2a depicts the experimental facility with all the major attachments. The test facility is like the experimental setup reported in earlier publications [5, 14] with certain modifications. The required vacuum of 0.001 Pa can be attained by using dry vacuum pump (Agilent Triscroll 300) along with turbo-molecular pump (Agilent Turbo-V 750). The complete test column assembly is housed inside a vacuum chamber with specimens positioned vertically between the controlled heating and cooling systems as illustrated in Fig. 2b. A stable PID controller is used to control the heating system for producing the requisite interface temperature, which consists of a tubular furnace with a core that is kept in contact with the top specimen. The furnace is covered by ceramic wool to minimize heat loss. The cooling system consists of a copper cylinder block and PID-controlled chiller, and maintains thermal contact with the bottom specimen. A hydraulic loading arrangement in conjunction with a compression type miniature load cell (Honeywell India) has been used to exert axial loading. Rapid response and highly accurate 0.1 mm wire diameter K-type ungrounded thermocouples (Tempsens instrument) along with specialized data acquisition (National Instrument) system have been used to record the temperature distribution closest to the interface at a regulated rate. A high-speed infrared camera of 36 mm fixed focal length (Flir T1030sc) with Zinc-Selenide optical viewport of 90% transmission capability in the infrared range of 2–15 μm is used to acquire and analyze thermal images for surface temperature distribution. The thermal camera has thermal sensitivity of <20 mK in the range of 0 to 2000 °C and can capture images with a resolution of 1024 × 768 HD at a rate of 30 frames per second.

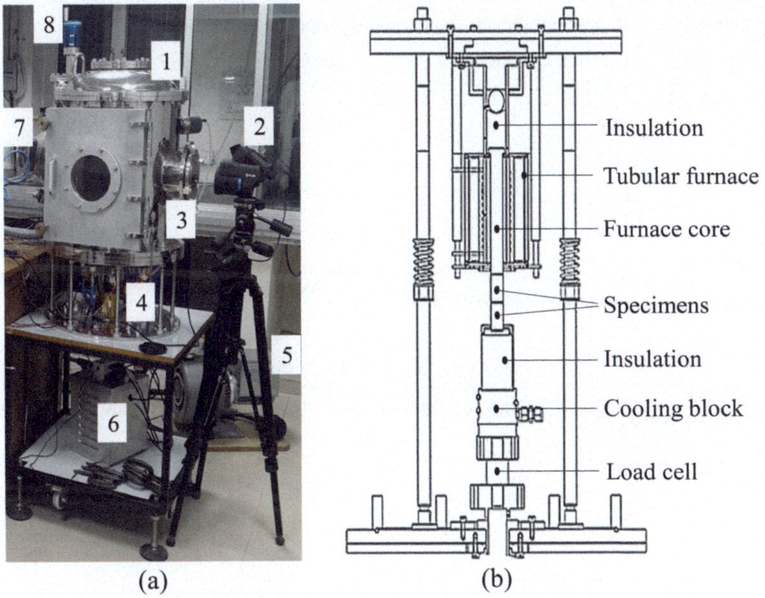

1	Vacuum chamber	2	Infrared camera
3	Zinc Selenide viewport	4	Hydraulic cylinder
5	Dry scroll pump	6	Transformer
7	Turbo molecular pump	8	Gate valve

Fig. 2 **a** Experimental test facility and **b** schematic diagram of specimen column assembly

2.1 Specimen Preparation

Stainless steel (SS304) is selected due to its wide range of thermophysical properties and industrial applications. Table 1 shows the temperature-dependent thermo-mechanical properties of SS304 [15]. The Vickers hardness (H) of SS304 at 25 °C is 3095.76 MPa [5]. The specimens are cylinder-shaped, measuring 30 mm in length and 20 mm in diameter. The desired interface temperature is obtained by averaging the temperature data from thermocouples inserted in the 0.6 mm diameter hole drilled radially up to 10 mm closest to the interface. The desired roughness on the contacting surfaces has been prepared using 100 and 320 grade emery paper. Stylus profilometer (Mitutoyo SJ-412) is used to measure the surface roughness parameters of the prepared rough surfaces. The effective surface roughness parameters are presented in Table 2. The cylindrical surface of the specimens is uniformly painted with black thermographic spray paint (LabIR paint) to raise its emissivity roughly to 0.9 as the highly reflective and low emissivity surface causes inaccurate temperature readings. Figure 1b shows the actual specimen and the black coated specimen. The thermal conductivity of contacting specimen is around 16 W/mK, while that for thermographic paint, it is around 0.5 W/mK. Therefore, significant heat conduction

Table 1 Temperature-dependent thermo-mechanical properties of SS304 [15]

Temperature (°C)	Thermal conductivity (W/m°C)	Young's modulus (GPa)	Poisson's ratio
0	14.6	198	0.29
100	15.1	193	0.30
200	16.1	185	0.30
300	17.9	176	0.31
400	18	167	0.32
600	20.8	159	0.33

Table 2 Surface characteristics of contacting specimens

Grade of emery paper	Surface roughness parameter	
	$R_q\,(\mu m)$	m
100	1.64	0.15
320	0.49	0.07

would take place across the least resistance path, i.e., along the metallic specimen. Thermographic paint allows accurate representation of specimen temperature with minimal temperature difference across the coating thickness. Thus, precise temperature profile across the interface can be obtained using specialized thermographic paint and infrared camera.

3 Experimental Procedure and Data Reduction

TCC across the interface has been estimated using a well-validated steady-state methodology. In this method, interfacial heat transfer and temperature drop have been determined using steady-state temperature data of both specimens. Heat transfer across the interface of conforming rough surfaces is mainly due to contacting nodes conduction, interstitial fluid conduction/convection and radiation [1]. The effect of radiation heat transfer across the interface becomes negligible for temperatures below 300 °C [1, 16] or even 600 °C [17]. The heat transfer through interstitial fluid is negligible in case of vacuum. Now, only contact node conduction is responsible for the transfer of heat, and TCC is referred to as contact spot conductance.

An axial force is applied to the specimen in contact and thermocouples are inserted in the drilled holes. An axial heating and cooling system above and below the contacting specimens initiates 1-D heat transfer across the contact. Furnace and chiller remained operational till the contacting specimen reached a stable condition. Figure 3 shows the transient temperature data using thermocouple up to the steady state. The criterion for fulfilling steady-state conditions is that the variation in temperature measured using thermocouple does not exceed 1 °C in 15 min. In addition, a

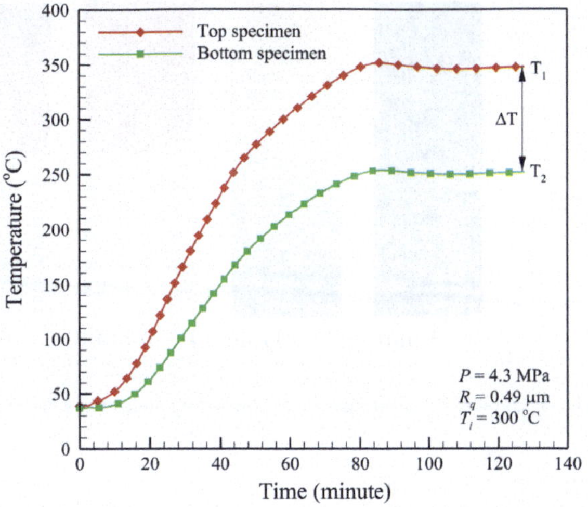

Fig. 3 Transient temperature data using thermocouple up to steady state for a typical run

steady-state thermal image is recorded simultaneously. Once steady state is reached, the applied load is raised to the subsequent level and the process is repeated. For three different contact pressures (1.7, 4.7, and 7.7 MPa) at 300 °C interface temperature, the procedure is repeated with two sets of surface finishes (0.49 and 1.64 m). Figure 4 shows typical thermal images from the experiments up until steady-state conditions have been reached. The region with the highest temperature is represented by red, followed by yellow, green, and blue in decreasing order of temperature. The color transition in the specimens from top to bottom represents the direction of heat transfer. The sharp color change at the interface of contacting specimen is clearly visible, indicating temperature drop at the interface. Thus, accurate estimation of interfacial temperature drop (ΔT) is possible using infrared thermography as it provides temperature closest to the interface. The temperature data for each test condition has been calculated using a frame size of 325 $\times$ 11 pixels which represents roughly 58 mm along the direction of heat flow and 1 mm along the interface. Subsequently, the centerline temperature data is obtained by averaging 11 pixels across the span in order to remove any potential noise effect. Pixels reading on the surface of the specimen is converted into length scale with spatial resolution of 0.173 mm/pixel, and temperature profile is plotted.

Furthermore, a typical steady-state temperature data has been plotted against the axial length of specimen as shown in Fig. 5. The axial temperature profile of each specimen has been accurately fitted with a linear curve using the least squares method from the non-contacting end to the location of the sudden change in temperature. Heat flux across each specimen is obtained by applying Fourier's law within the bulk region of specimen and the interfacial heat flux is obtained by averaging the heat flux of both specimens, i.e., (($q_1 + q_2$)/2). The interfacial temperature drop (ΔT) can be estimated by extending the fitted profile up to the interface for each specimen. The

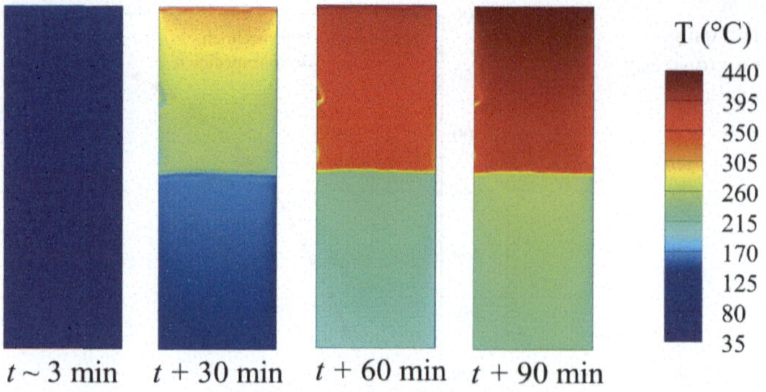

Fig. 4 Typical infrared images of contacting specimens captured at different time intervals

Fig. 5 A typical axial temperature profile of contacting specimens at steady state

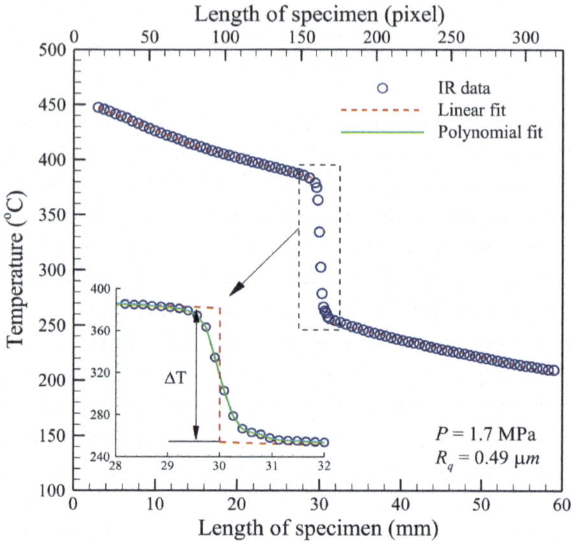

contacting region along with the data points and the temperature drop is presented on enlarged scale in Fig. 5. Finally, the value of TCC can be obtained using Eq. (1).

4 Results and Discussions

TCC has been estimated between SS304 pairs with varying surface roughness (0.49 and 1.64 μm) for various loading conditions (1.7, 4.7, and 7.7 MPa) at interface temperature of 300 °C under vacuum. The impact of contact pressure and surface roughness on TCC are covered in the following sections.

Fig. 6 Variation of TCC with contact pressure for different surface roughness

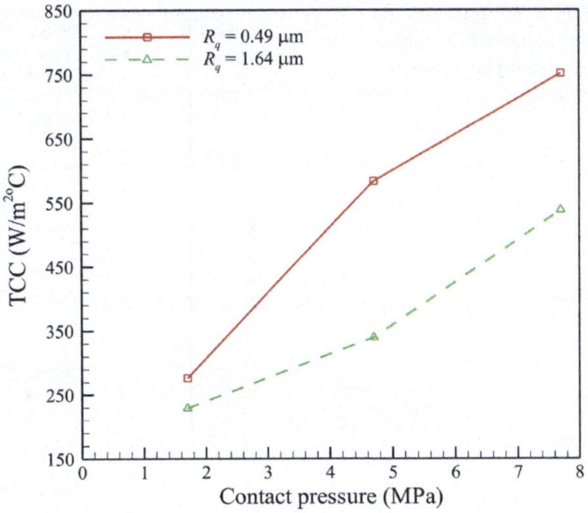

4.1 *Effect of Contact Pressure*

TCC results have been plotted against contact pressure (Fig. 6) for different surface roughness. With an increase in contact pressure, there is a general trend toward an increase in TCC. As contact pressure increases, deformation of contacting asperities causes new asperities to come into contact, increasing the actual contact area. The heat transmission between the surfaces is improved, resulting in a higher heat flux density and a lesser temperature drop across the interface, thereby TCC increases. Furthermore, the estimated TCC ranges from 228.92 to 751.98 W/m^2 °C for present range of surface roughness and contact pressure. The percentage improvement in TCC with increase in contact pressure from 1.7 to 7.7 MPa is 64% for $R_q = 0.49$ µm and 57% for $R_q = 1.64$ µm. Similar types of results have been described in the literature [5–7] under different experimental conditions.

4.2 *Effect of Surface Roughness*

Figure 6 also shows the variation of TCC with surface roughness. At the same contact pressure, it can be seen that contacting specimens with lower surface roughness achieve higher TCC. The distribution and size of micro asperities change with decrease in surface roughness, leading to a large number of contact spots, which raises the actual contact area and, consequently, TCC. It is clear that for contact pressures ranging from 1.7 to 7.7 MPa, the percentage increase in TCC with a 70% reduction in surface roughness varies from 20 to 72%. The percentage increase in TCC is more at higher contact pressure, indicating large deformation of contacting asperities.

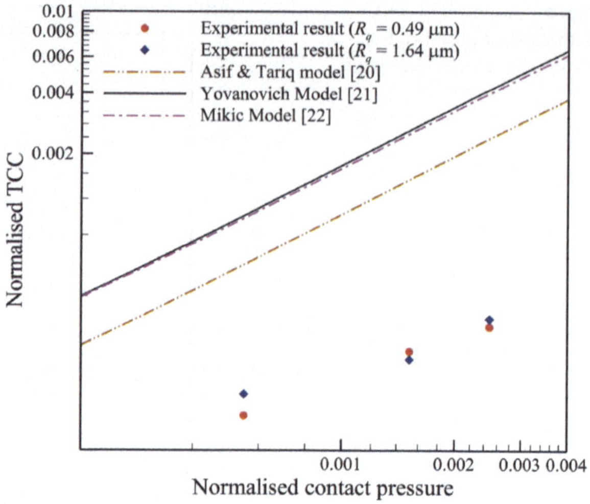

Fig. 7 Comparison of normalized TCC with available plastic contact models

4.3 *Comparison with the Available Theoretical Models*

According to Mikic plasticity index criteria [20], the experimental contacting surfaces undergo plastic deformation. Therefore, the experimental results of TCC and contact pressure have been suitably normalized in accordance with the literature [18–20], for comparison with the available theoretical plastic models. TCC and contact pressure are normalized as $h^* = hR_q/mk$ and $P^* = P/H$ respectively. In the mentioned equations, R_q represents the effective root mean square roughness, m is the effective asperity slope, k denotes the harmonic mean of thermal conductivity, and H is the micro-harness of the softer material. Figure 7 shows the comparison of the normalized experimental results with the available theoretical model. The results do not show greater agreement with the theoretical models, but rather follow the same trend with pressure. The available models are confined to specific situations and if extended over a wider range of thermo-mechanical parameters, tend to underpredict or overpredict experimental TCC values.

5 Conclusion

Steady-state TCC across stainless steel pairs under vacuum has been estimated at high interface temperature using infrared thermography. With the implementation of infrared thermography, better thermal visualization and accurate temperature data estimation closest to the interface are possible, which is difficult to obtain with thermocouples. Results for different loading conditions and surface roughness are shown at a constant interface temperature. TCC increases with an increase in contact pressure and decreases with increase in surface roughness. The available theoretical model overpredicts the experimental TCC at high interface temperature. The developed facility will be used in future to conduct experiments at much higher interfacial temperature range under various surface roughness and contact pressure conditions. Further, the study should be expanded to include non-conforming contacts and transient TCC estimation.

References

1. Madhusudana CV (1996) Thermal contact conductance. Springer, New York, New York, NY
2. Siddappa PG, Tariq A (2020) Tribol Int 148:106358
3. Wahid SMS, Madhusudana CV, Leonardi E (2004) Exp Therm Fluid Sci 28:489
4. Xu R, Xu L (2005) Cryogenics (Guildf). 45:694
5. Tariq A, Asif M (2016) Heat Mass Transf Und Stoffuebertragung 52:291
6. Xing L, Zhang LW, Zhang XZ, Yue CX (2010) Exp Heat Transf 23:107
7. Dou R, Ge T, Liu X, Wen Z (2016) Int J Heat Mass Transf 94:156
8. Chand A, Khan KA, Tariq A (2019) In Proceeding Proceedings 25th National 3rd international ISHMT-ASTFE heat mass transf. conference (Begellhouse, Connecticut, 2019), pp 1097–1102
9. Fieberg C, Kneer R (2008) Int J Heat Mass Transf 51:1017
10. Chen M, Li Q, Zhang P (2020) Int J Heat Mass Transf 157:119749
11. Helmig T, Göttlich T, Kneer R (2022) Int J Heat Mass Transf 186:122399
12. Woolley JW, Woodbury KA (2011) Heat Transf Eng 32:811
13. Zhang P, Xuan YM, Li Q (2014) Exp Therm Fluid Sci 54:204
14. Kumar S, Tariq A (2017) Int J Therm Sci 118:53
15. Deng D, Murakawa H (2006) Comput Mater Sci 37:269
16. Wahid SMS, Madhusudana CV (2000) Int J Heat Mass Transf 43:4483
17. Marotta EE, Fletcher LS (1998) 7th AIAA/ASME Jt. Thermophys. Heat Transf Confe 12
18. Asif M, Tariq A (2016) Exp Heat Transf 29:456
19. Yovanovich MM (1981) AIAA 16th Thermophys Conf
20. Mikić BB (1974) Int J Heat Mass Transf 17:205

Thermal Conductivity Measurement of Novel Waste Tire Rubber-Polypropylene Composite

Khan Zahir Ahmed, Mohammad Faizan, and Mohammad Asif

Abstract As the manufacturing and use of vehicles are increasing globally, the disposal and recycling of waste tire have become a major environmental issue when they are out of service life. One of the potential recycling processes is turning the waste tire into a useful composite as insulating material. The present work produced a blended composite based on waste tire rubber (425 μm) and virgin polypropylene granules in a 4:1 ratio through a single screw filament extruder under controlled conditions. The thermal conductivity of the newly developed insulating material was measured using an in-house fabricated experimental setup built in accordance with ASTM D5470. Since waste tire rubber has good insulating property, it can be combined with thermoplastic materials and converted into insulating material for various thermal applications.

Keywords Waste tire rubber · WTR · Thermal conductivity · Composite · Insulation

Nomenclature

k	Thermal Conductivity (W/mK)
Q	Heat Transfer (W)
$\frac{\Delta T}{\Delta x}$	Temperature gradient (K/m)
HFM	Heat Flow Meter
UHFM	Upper Heat Flow Meter
LHFM	Lower Heat Flow Meter
$\dot{q}$	Heat Flux Density
$\dot{q}_1$	Heat Flux flowing through UHFM

K. Z. Ahmed (✉) · M. Faizan · M. Asif
Department of Mechanical Engineering, ZHCET Aligarh Muslim University, Aligarh, India
e-mail: kzahir@myamu.ac.in

M. Asif
e-mail: masif@zhcet.ac.in

M. A. Siddiqui et al. (eds.), *Advances in Heat Transfer and Fluid Dynamics*, Lecture Notes in Mechanical Engineering, https://doi.org/10.1007/978-981-99-7213-5_14

$\dot{q}_2$	Heat Flux flowing through LHFM
WTR	Waste tire rubber
PP	Polypropylene

1 Introduction

Tire rubber is a complicated mixture of natural and synthetic rubbers, other chemicals, and building materials. A modern automotive tire is made up of approximately twenty different components and ten different rubber compounds. Tire rubber is often composed of a mixture of butadiene rubber, styrene-butadiene rubber polymer, natural rubber, and additional additives like zinc oxide, carbon black, and so on [1]. Since tires are heavy and non-biodegradable, they take up valuable landfill space, and the disposal of used tires has become a major environmental concern worldwide. Many abandoned tires end up in open landfills without being treated. The growing number of garbage disposal lands endangers human health and increases the risk of fires. There are several methods for recycling leftover tire rubber [2]. One technique of recycling waste tire rubber is to generate composite blends of WTR with other waste materials, such as polypropylene, using an extruder machine.

WTR-PP composites are a rubber-thermoplastic blend. It is a high molecular weight polymer that forms a continuous matrix around rubber components. These are typically low-cost typical thermoplastics that can flow under particular conditions and be sculpted into various items. Thermoplastics can be an excellent barrier to keep moisture from entering the composites because they compress and swell but only absorb a tiny amount of water. Polymer composites have exceptional durability, reasonably high stiffness, strength, and a reasonable price compared to other equivalent materials. They are weather- and water-resistant and can be used for various outdoor applications. The recycled composite and WTR industries have risen dramatically in recent years, owing mostly to the automation industries contributing more waste tires and expanding construction which have a significant potential to recycle waste tire rubber in various applications. The three thermoplastics most commonly used for WTR composite are high-density polyethylene (HDPE), polypropylene, and polyvinyl chloride [3–5].

Yang et al. [6] studied the viability of employing rice straw-reinforced composites made from waste tire rubber as construction materials. Researchers produced insulation boards for use in the building using WTR particle composite boards. The panels can prevent impact damage, are adaptable and affordable, and can also be used as insulating boards and other flexural materials in construction. Abu et al. [7] investigated the thermal and physical properties of polyester and waste tire rubber composites. The results showed that the rubber particles were a good filler that can be used with unsaturated polyester to create a composite that acts as insulation. The addition of rubber particles to the polymer matrix reduces the thermal conductivity and the density of composites. The low thermal conductivity (0.144–0.113

W/m K) and very low water retention (2.0%) of rubber-polyester composite show promise for use as a thermal insulator in construction. Carlos et al. [8] studied WTR powder-based composite materials. They created WTR-based tiles with a range of concentrations that can be applied to numerous applications where impact loads are frequent, such as pavement or gym floor tiles. Jeong et al. [9] examined the mechanical characteristics of WTR particles containing mortar composites. Six different concentrations of samples (0%, 5%, 10%, 15%, 20%, and 25%) were tested for flexural and compressive strength. According to experimental findings, waste tire rubber particles replaced 5% of the sand to produce the highest compressive and flexural strength. Nadir et al. [10] examined the successful utilization of WTR in particleboard manufacturing, where manufacturing parameters were specific gravity and water content. The results demonstrated that tire rubber's hydrophobic property enhanced the particleboard's ability to withstand water. Xu et al. [11] studied WTR powder as fillers for wood fibre composites and wood rubber composites. Analysis of the morphological and physical–mechanical characteristics of wood rubber composite panels revealed that WTR might be utilized to modify wood fibre composites' strength and viscoelastic characteristics. WTR addition can also make panels more flexible and hygroscopically stable. Ashori et al. [12] examined the utilization of WTR in hybrid plywood composite panels and studied physical, mechanical, and acoustic characteristics. Overall findings suggest that an increase in WTR content is inversely correlated with the mechanical properties of the manufactured panel and directly correlated with the improvement in physical properties.

Many researchers investigated the WTR's mechanical and morphological properties in cement, pavement material, and polymer blends. Attention should also be directed toward thermal studies of the various combinations of WTR polymers for insulating applications. This research aims to find a way to recycle waste tire rubber (WTR) and turn it into a new composite material by blending it with thermoplastic material where WTR can be used as a filler. Mechanical, physical, and thermal characterizations are carried out to study the workability of WTR with thermoplastic to create new composite materials.

2 Materials and Sample Preparation

2.1 Materials

This research work used waste tire rubber material and virgin polypropylene. The raw waste tire rubber material was bought from a local waste tire rubber supplier as shavings of used car and truck tires with unknown material composition and non-uniform grain sizes. WTR powder (425 μm) was prepared by mechanical shredding, and grading was determined using sieve analysis ASTM C136 for fine and coarse aggregate [13], whereas commercial virgin polypropylene was procured from Reliance Industries Limited (RIL) (Fig. 1).

Fig. 1 Materials **a** WTR powder (425 μm), **b** Virgin polypropylene

Table 1 Characteristics properties of the materials

Properties	WTR (425 μm)	Polypropylene
Color	Black	White
Density (@ 25 °C)	*0.793 gm/cm^3	0.95 gm/cm^3
Melting point	–	170 °C
Water absorption (%)	*0.955	0.03
Thermal conductivity (W/mk)	0.32	0.2

* Calculated values

The moisture content and density of WTR powder (425 m) were measured using the oven-dry method and the density bottle experiment. The properties of PP and WTR are shown in Table 1.

2.2 *Blend and Sample Preparation*

WTR and PP were melt blended in a 4:1 ratio (80% WTR, 20% PP). Melt blending was achieved using a single screw filament extruder with an L/D ratio of 12 with a 5 mm circular die to extract the extrudate as shown in Fig. 2. As the melting point of PP is 170 °C, the extruder machine was run at 50 rpm with a throughput rate of 2.5 kg/hr and a temperature profile of 170 °C (metering zone). A homogenous and continuous extrudate free of air bubbles was obtained. A heated mold at the die end was used to store the molten matrix to give it a cylindrical shape as per the experimental setup sample requirement, as shown in Fig. 3.

Fig. 2 Single screw filament extruder

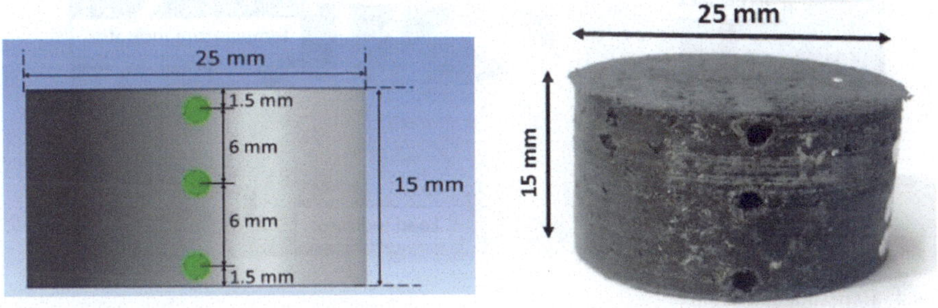

Fig. 3 WTR-PP blend sample (4:1 ratio)

3 Experimental Setup

The thermal conductivity of WTR-PP composite was evaluated using an in-house fabricated experimental setup as shown in Fig. 4. The testing procedure and major components of the setup are described as follows.

3.1 Major Components

3.1.1 Cooling System

A cooling system is used to keep the temperature low by pumping chilled water. The block is drilled with two holes that serve as the input and outflow for the cooling water circulation. For optimal working of the cooling system, inlet temperature of the fluid should be maintained at a constant value.

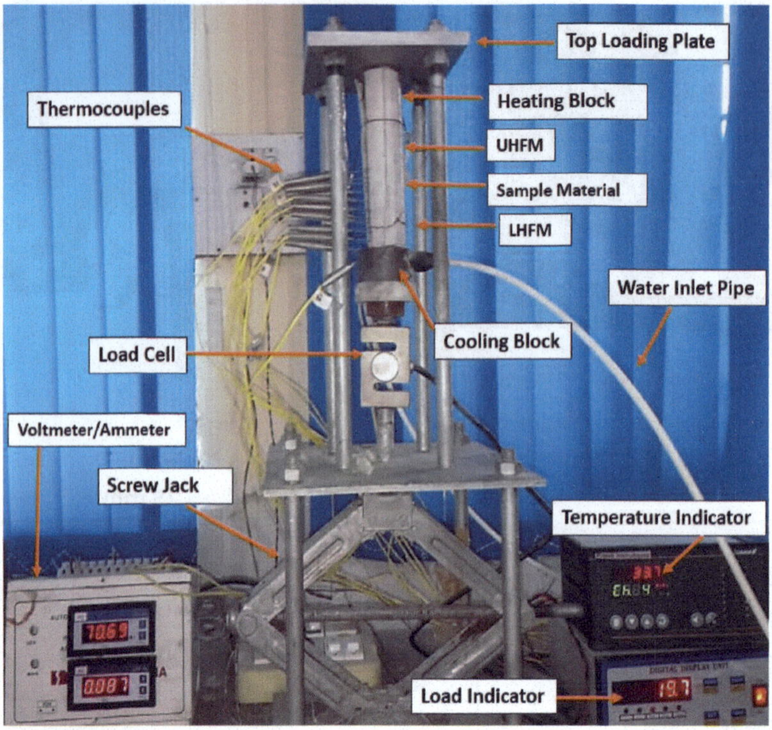

Fig. 4 Experimental setup

3.1.2 Heating System

To maintain a constant heat flux, two high-performance cartridge heater rods of 150 W each were inserted in a cylindrical-shaped copper block, resulting in high-temperature boundary conditions. The top surface serves as an insulating wall, while the bottom surface serves as a conduction route. As the temperature profile inside the heater block is uniform. To provide a steady and monitored heat source, heating rods are connected to a dimmerstat through a voltmeter and an ammeter to provide a constant and measured heat supply.

3.1.3 Temperature Measurement System

T-type thermocouples were installed in holes drilled perpendicular to the sample's symmetry axis as shown in Fig. 5. Each hole was large enough to hold a thermocouple and small thermocouples were used to minimize conduction through the thermocouple leads. To record the temperatures, all the thermocouples were connected to a digital temperature indicator.

Fig. 5 Arrangement of Sample and HFMs

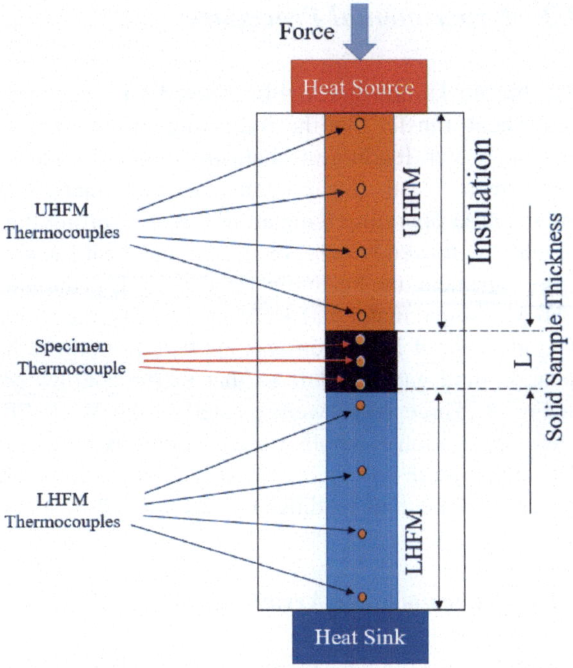

3.1.4 Heat Flow Meter (HFM)

Two cylindrical blocks of known (k) thermal conductivity are used as heat flow meters at the top and bottom of the sample, i.e., UHFM and LHFM. Heat flow meters were made of stainless steel 304, whose thermal conductivity is known with temperature as: $k(T) = 10.67 + 0.0159$ T, where 'T' is in Kelvin [14].

3.1.5 Glass Wool Insulation

To prevent axial heat loss, all the blocks were insulated with glass wool ($k = 0.023$ W/mK) with a critical insulation thickness of 10 mm to minimize the radial heat loss.

3.1.6 Loading System

A one-ton capacity screw jack was used to supply load at the test column to adjust contact pressure according to the requirements. A one-ton universal load cell was utilized to measure the load applied to the test column with the help of a screw jack.

3.2 Experimental Procedure

All the blocks are thoroughly cleaned with acetone before use. It's essential for better heat transfer that the contacting surfaces are as conductive as possible after that the heater, loading mechanism, and cooler have been switched on to start the heat flow across the test column and apply loads. The applied voltage is set as per the required operating temperature. After starting the heater and coolers, the setup should be allowed to achieve steady-state heat transfer for around 120–180 min. In this experiment, the heater operated at 60 °C is mounted at the very top, followed by UHFM, sample material, LHFM, and finally, the cooling block maintained at 20 °C. The glass wool insulation was applied around the blocks. To record temperature data, a small gap was left so that thermocouples could be inserted as presented in Fig. 5. Three holes were created in the WTR-PP sample as shown in Fig. 3. Load application is applied using screw jack measured by the load display unit. The temperature was recorded every 20 min till the change in temperature between current and previous readings was around 0.2 °C and the steady-state was achieved.

3.2.1 Thermocouple Arrangement

Eleven thermocouples were employed in HFMs and testing samples. Arrangements of HFMs, sample and thermocouple are demonstrated in Fig. 5. Each hole was approximately 12.5 mm deep with thermocouple of 1 mm diameter inserted in the hole to record the axial temperature data in the sample and HFMs till the steady-state was achieved.

4 Data Analysis

Thermo-physical experimental data is collected for WTR-PP composite. Heat flux through the WTR-PP sample is calculated using the Fourier's law and thermal conductivity of the WTR-PP composite is determined.

4.1 Heat Flux Calculation

The temperature data at various locations of WTR-PP sample, UHFM, and LHFM is represented in a typical graph as shown in Fig. 6.

The temperature plot of the WTR-PP, UHFM, and LHFM with respect to the thermocouple location is a linear function. Hence the temperature profile in the HFMs and sample along its length may be linearly expressed as $y = mx + c$.

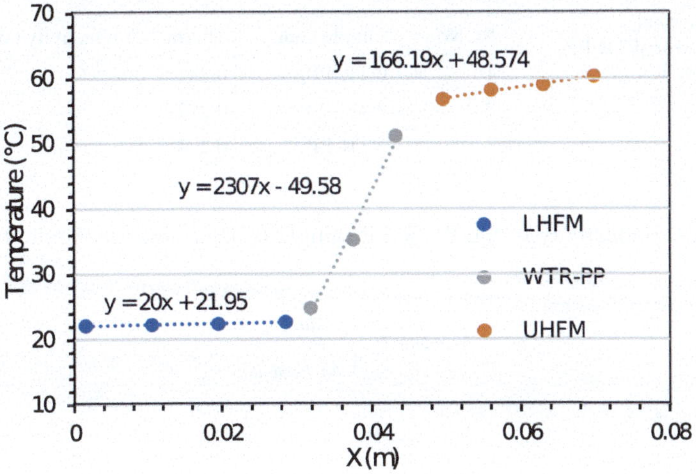

Fig. 6 A Typical plot of steady-state temperatures with locations for LHFM, UHFM, and WTR-PP

where y is the temperature, x is the distance along the line of sample and HFMs and m $= \frac{\Delta T}{\Delta x}$

Heat flux can be evaluated by using the Fourier's law as

$$q = k \times \frac{\Delta T}{\Delta x} \tag{1}$$

4.2 *Estimation of Thermal Conductivity*

The WTR-PP sample, whose thermal conductivity is unknown, is kept between two HFMs made of stainless steel, whose thermal conductivity is known. The axial flow steady heat conduction technique is used to evaluate the thermal conductivity of the sample. Thermocouples are employed to capture temperature profiles of HFMs and samples at various locations (4 thermocouples on HFMs, 3 thermocouples on WTR-PP Sample). Δ T/Δ x is calculated from the temperature profile for HFMs and WTR-PP samples. The thermal conductivity data and temperature profile of two HFMs were used to calculate the heat flux of HFMs ($\dot{q}_1$ and $\dot{q}_2$). The average heat flux ($\dot{q}_{avg}$) for UHFM and LHFM was used for the calculation of the thermal conductivity of sample.

$$\dot{q}_{avg} = \frac{\dot{q}_1 + \dot{q}_2}{2} \tag{2}$$

Table 2 Thermal conductivity for WTR-PP sample codes

Sr. No	Sample code	Thermal Conductivity (W/mK)
1	WTR-PP_1	0.281
2	WTR-PP_2	0.292
3	WTR-PP_3	0.274

Thermal conductivity of the WTR-PP sample is determined using the following formulae

$$k_{sample} = \frac{\dot{q}_{avg}}{\left(\frac{\Delta T}{\Delta x}\right)_{sample}} \tag{3}$$

5 Result and Discussion

Three WTR-PP samples were tested on in-house fabricated setup for thermal conductivity. Operating temperature was controlled by varying input heat flux to maintain the heat source approximately at 80 °C. A screw jack loading system was used to apply the load. The applied load had been kept constant as 10 kg to hold the HFMs and sample in the series array and to create proper contact between HFMs and the WTR-PP. Graphene paste was used between interfaces of sample and HFM to minimize the thermal contact resistance. Thermal conductivity of the sample was calculated using Eq. (3) and the results are reported in Table 2. The thermal conductivity measurements obtained for the WTR-PP samples are satisfactory and in accordance with the published literature for similar materials [15–18].

Percentage deviation in the thermal conductivity values for three samples of WTR-PP was found to be 6.16% and it is mainly due to error in thermocouple location, heat flux measurement, temperature measurement, and minor compositional changes brought on by the manual loading of WTR powder and PP granules into the screw extruder machine during mixing.

6 Conclusion

In the present experimental work, waste tire rubber (425 μm) was blended with virgin polypropylene granules in the 4:1 ratio (80% WTR, 20% PP). Thermal conductivity of the resulting material was measured using an in-house fabricated system. The thermal conductivity of the newly developed material was found to be 0.282 W/m K. As the thermal conductivity of polypropylene alone is around 0.2 W/m K, whereas the thermal conductivity of waste tire material is about 0.32 W/m K, the thermal conductivity of WTR-PP blend seems reasonably accurate as it lies in between the values of the constituent materials. Based on the measured value of thermal conductivity of the novel material, it can be characterized as good insulating material. This blend can be used in various applications, such as roof insulation and water camper layer filling. Successful manufacturing of such composite will not only provide a way of recycling waste tire rubber but also offer a potential replacement for traditional thermal insulators.

References

1. Gupta VK, Gupta B, Rastogi A, Agarwal S, Nayak A (2011) A comparative investigation on adsorption performances of mesoporous activated carbon prepared from waste rubber tire and activated carbon for a hazardous azo dye–Acid Blue 113. J Hazard Mater 186:891–901
2. Hejna A et al (2020) Waste tire rubber as low-cost and environmentally-friendly modifier in thermoset polymers—a review. Waste Manag 108:106–118
3. Azevedo F, Pacheco-Torgal F, Jesus C, Barroso De Aguiar JL, Camões AF (2012) Properties and durability of HPC with tyre rubber wastes. Constr Build Mater 34:186–191
4. Formela K (2022) Waste tire rubber-based materials: Processing, performance properties and development strategies. Adv Ind Eng Polym Res.https://doi.org/10.1016/J.AIEPR.2022.06.003
5. Abdulkadir I, Mohammed BS (2020) effect of waste tire rubber on the durability behavior of cement composites: a review. J Infrastruct Facil Asset Manag 2
6. Yang HS et al (2004) Possibility of using waste tire composites reinforced with rice straw as construction materials. Bioresour Technol 95:61–65
7. Abu-Jdayil B, Mourad AH, Hussain A (2016) Thermal and physical characteristics of polyester-scrap tire composites. Constr Build Mater 105:472–479
8. Revelo CF, Correa M, Aguilar C, Colorado HA (2019) Waste tire rubber powders based composite materials. Miner Met Mater Ser, 437–445. https://doi.org/10.1007/978-3-030-10386-6_50/COVER
9. Jeong D, Cho J, Choo S, Park C-W (2016) Mechanical properties of tire-rubber particles contained mortar composites, 119–121. https://doi.org/10.2991/AMSEE-16.2016.33
10. Ayrilmis N, Buyuksari U, Avci E (2009) Utilization of waste tire rubber in the manufacturing of particleboard, 688–692.https://doi.org/10.1080/1042691090276937624
11. Xu X, Tian F, Li X (2020) Regenerated waste tire powders as fillers for wood fiber composites. BioResources 15:3029–3040
12. Ashori A, Ghofrani M, Rezvani MH, Khojasteh Khosro S (2015) Utilization of waste tire rubber in hybrid plywood composite panel. Polym Adv Technol 26:1034–1040
13. Standard Test Method for Sieve Analysis of Fine and Coarse Aggregates. https://www.astm.org/c0136_c0136m-19.html
14. McGee GR, Schankula MH, Yovanovich MM (1985) Thermal resistance of cylinder-flat contacts: theoretical analysis and experimental verification of a line-contact model. Nucl Eng Des 86:369–381
15. Patti A, Acierno D, Patti A, Acierno D (2019) Thermal conductivity of polypropylene-based materials. Polypropyl Polym Charact Mech Therm Prop. https://doi.org/10.5772/INTECHOPEN.84477
16. Xie X et al (2016) Thermal conductivity, heat capacity, and elastic constants of water-soluble polymers and polymer blends. Macromolecules 49:972–978
17. Marić I (2017) Thermal conductivity of hybrid recycled aggregate—Rubberized concrete. Constr Build Mater 133:516–524
18. Shao J, Zarling J (1995) Thermal conductivity of recycled tire rubber to be used as insulating fill beneath roadways

Thermal Fluid Analysis of Different Combinations of Jet Channel and Air Foil Pillars with CuO–Water Nano Fluid

Deepak Kumar, Mohammad Zunaid, and Samsher Gautam

Abstract Small-scale thermal devices are very effective and interesting to use for the applications which involve large amount of heat flux. The use of Nano particles in the base fluid like water also augments the various properties which help in the heat dissipation from the components subjected to high temperatures. In the current study, the various combinations of jet-channel and pillars have been investigated comparatively. CuO–water-based Nano fluid is used along with the air foil pillars in a heat sink to augment the heat transfer. Comparative analysis has been executed for the different parameters like temperature, heat exchange coefficient, Nusselt number and thermal resistance. The study has been carried out for a constant value of heat flux at the bottom surface. Minimum temperature is found in case of jet-pillar combination. The channel flow diminished the heat exchange coefficient.

Keywords Jet · Nano fluid · Thermal · Fluid · Nusselt number · Thermal resistance

Nomenclature

Re	Reynolds Number
Nu	Nusselt Number
TOHS	Temperature of heat sink
DCOJCP	Different combinations of jet-channel and pillar
DIP	Drop in pressure
COHE	Coefficient of heat exchange
THR	Thermal Resistance
PPR	Pumping power

D. Kumar (✉) · M. Zunaid · S. Gautam
Delhi Technological University Delhi, New Delhi 110042, India
e-mail: deepak209476@gmail.com

M. A. Siddiqui et al. (eds.), *Advances in Heat Transfer and Fluid Dynamics*, Lecture Notes in Mechanical Engineering, https://doi.org/10.1007/978-981-99-7213-5_15

1 Introduction

Lifetime as well as effectiveness of a device can be significantly increased by an improvement in the cooling mechanism. So in the today's scenario requirement of an efficient cooling technology is of great importance. Abo-Zahhad et al. [1] compared the case of jet impingement with the same case with microchannel. For the analysis zone of impingement and width of channel had been taken as the parameters. It was estimated from the study that in case of microchannel effectiveness was more as compared with impingement only. Also temperature uniformity was detected in microchannel with impingement. Friction losses were negligible and the overall efficiency of the system was increased. For low values of rate of flow increase was found in terms of effectiveness. Sorour et al. [2] conducted experimental study to investigate the Nusselt number as a function of Reynolds number, volume fraction of water-dependent Nano fluid. It was predicted from the results that with an increase in Re, Nu also gets increased. Nu also increases with an increase in the volume fraction of Nano particles. The effect of influence ratio was found to be insignificant. A new correlation was also proposed for the Nusselt number. As compared to the water, positive effect of Nano fluid was observed in the investigation. Lelea and Laza [3] performed numerical investigation based on the finite volume method using Nano fluid in case of a micro heat sink. The results of Nano fluid were compared with the water. The geometry of the heat sink was considered for more than one inlet with tangential flow. Low velocity conditions were assumed for the examination of the simulation. The range of the Reynolds number was in between 15 and 100. Constant value of heat flux was taken at the base. With the increase in the flow rate of mass, decreasing trend was noticed in temperature of substrate. Pressure fall was more in case of Nano fluids for a constant value of pumping power.

Alirezaie et al. [4] used two Nano fluids based on ethylene glycol for the study and comparative analysis had been made for the cost of preparation of Nano fluid with the thermal proportion. Results had been validated with the modeling of neural network. So the analysis presents idea about the economic concept of Nano fluid. A new relation was also proposed for the thermal conductivity of Nano fluids. More focus is required on the environmental effects of Nano particles and their use by engineers due to the toxic nature of some of the particles. Anitha et al. [5] investigated the performance of heat exchanger with the help of two Nano fluids. Carbon nanotube—$Al_2 O_3$ and $Fe_3 O_4$ were used as Nano particles in the base fluid. Reynolds number and fraction of Nano fluids were assumed to be design parameters and Nusselt number, rate of heat transfer, etc. were taken as response parameters. Nusselt number and rate of heat exchange were detected to be increased with Reynolds number and rate of mass flow. Among the two Nano fluids, Carbon nanotube—$Fe_3 O_4$ / water was noticed to be more effective. With this Nano fluid effectiveness was increased by fifty percent. Mohammad pour and Lee [6] presented the review of impinging jets both experimental and numerical. Swirling impinging jets were examined to be more effective as compared to the conventional. When the Nano particles were added in the base fluid, increase in critical heat flux was detected. Due to the higher values of viscosity of the fluid sedimentation type problems were observed. Ambreen et al. [7] interpreted the heat sink with aluminum foam along with the application of Nano fluid for hydrothermal attributes. The attributes were studied in the form of Nusselt number, average heat exchange coefficient, and power of pumping. Also the distribution of temperature was discussed in the form of contours. Dynamics had been explained in the form of velocity vectors and streamlines. It was noticed that the porosity of the substrate base plays an important role in the heat exchange property of Nano fluids. Yang et al. [8] devoted their investigation for the case of two heat sinks with the application of Ag Nano fluids. The effect of fraction of the Nano fluid and Reynolds number had been analyzed with the temperature of CPU and heat exchange coefficient. With the augmentation in Reynolds number, temperature of CPU was observed to be decreased. Also the temperature uniformity was more in case of Nano fluid as compared to the water. However, increase in pumping power was also seen due to the frictional effects. The investigation was performed from the 1st and 2nd law view of thermodynamics.

Zheng et al. [9] examined the two shapes of the microchannel from the thermal and hydro point of view. In the first case circular microchannel was taken and in the second case annular microchannel thermal sink was analyzed. Reynolds number, size of dimples, and arrangement were considered to be design parameters. Outer dimples showed the better heat transfer attributes as compared to the inner attributes in case of annular thermal sink. Negligible effect was detected for the arrangement case. Circular case was noticed to be effective from the point of view of energy savings. Bahiraei et al. [10] assessed the 2nd law of thermodynamics in a microchannel thermal sink with water-based hybrid Nano fluids. Microchannel was fitted with ribs of rectangular shape. At the interface of solid and fluid, the generation of entropy was found to be more as compared to the dead area. This generation was due to the friction and exchange of heat. Entropy generation had been observed to be more in the corners in case of solid area. The lowest value of entropy generation due to friction was detected at the center of sink. Entropy generation declined with the augmentation in Reynolds number and weight fraction of Nano particles. Hempijid and Kittichaikarn [11] simulated the two types of thermal sink using computational fluid dynamics as a tool. Direction of flow at inlet and outlet of thermal sink was assumed to main parameter to analyze the pressure and thermos characteristics. The V-shaped thermal sink was noted to be better in terms of thermal performance in differentiation to the other at seventy-five-degree angle. Also the vortices developed at the inlet portion were noted to be affecting the performance of the system. Offsetting the inlet and outlet was found to be affecting the effectiveness of the thermal sink. Mahdavi et al. [12] computed the heat exchange attributes in a rotating disk which was circular and hot using the application of Nano fluid. Ansys fluent was employed to investigate the multiphase flow of air and Nano fluid. The end result of rotation and volume fraction of Nano particles were studied. It was predicted that at higher volume fraction of Nano particles large values of torque is required to enhance the heat exchange. The reason for the same was augmentation in the viscosity of Nano fluid. The value of heat flux was assumed to be constant.

Duan et al. [13] performed numerical simulations in a thermal sink of plate fin type which was cooled with the help of an impinging jet of rectangular shape. Numerical simulations were detected in conformity with experimental results. It was simulated that with augmentation in width at inlet of the jet, fall in pressure dwindles for a constant value of flow rate. Also as the height of fin was increased, pressure drop increased. The sink with and without elliptic bottom nature was discussed. Without elliptic profile, drop in pressure was more as compared to the other. Ramesh, et al. [14] provided an overview of heat exchange in case of heat sink with different heat exchanging techniques both experimentally as well as numerically. The various heat exchange enhancing techniques with jet impingement, microchannel, Nano fluids, and magnet hydrodynamics with the various parameters have been discussed in detail. It was suggested that there is a need for developing low-cost microchannel devices as well as more work is required in case of magnet hydrodynamics. Also use of artificial neural network in these type studies is less. So more attention is required. Datta et al. [15] modeled the stream function and contours of temperature using copper–water Nano fluid in a thermal sink with jet impingement. The plate considered for the study was moving and hot. At different velocities of the plate local and average heat exchange coefficient was analyzed. Also the effect of Reynolds number in the turbulent region was studied on the rate of heat exchange. Relations for the Nusselt number have been represented. The study was carried out using numerical simulations. Re and Nu were found in positive relationship. Chen and Cheng [16] analyzed the concave surface for the purpose of heat exchange using SiO_2 water as Nano fluid and the same was differentiated with the case with water only. For the case of Nano fluid 5.85% rise was observed in the heat exchange coefficient as compared to the water. It was noted that with augmentation in the curvature of the surface, Nusselt number increased. The study was carried out for turbulent flow range. M.S and Venkata Subbaiah [17] studied the impingement of jet with different Nano fluids using Fortran dependent solver. Governing equations were solved using FDM. Reynolds number, concentration of particles, etc. were taken as design parameters. Twenty-five percent augmentation was seen in case of Al_2O_3–water Nano fluid. Also in case of Cu–water Nano fluid sixty percent improvement was detected in the heat exchange effectiveness. Results were validated with the available experimental results.

Pratap et al. [18] studied the relation between Nusselt number and dimensionless spacing with and without Nano fluid. For the investigation constant value of heat flux is assumed. In the case of water as well as Nano fluid, positive relationship was observed between Nu and dimensionless spacing. But the same is more in case of Nano fluid. Circular jet was taken for the analysis and Re was kept constant. Ghanbari and Javaherdeh [19] exploited the use of non-Newtonian Nano fluid to enhance the heat exchange coefficient. The Nano particles were used in different concentrations. Different parameters like friction factor, Nu were determined for the analysis. For 0.2% concentration of graphene, almost 39% increment was detected in heat exchange coefficient. Also 12% increment was seen in thermal conductivity. The experimental work was performed for baffles of circular shape and perforated. Freegah et al. [20] introduced a new design of thermal sink of plate fin type. Pins

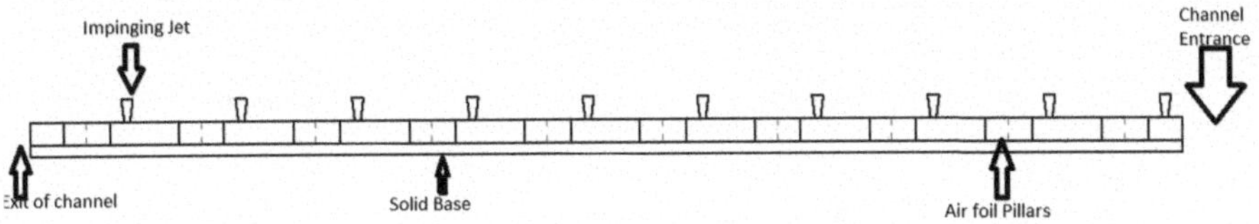

Fig. 1 Combination of Jet-channel-pillar

of semi-round shape are attached to the thermal sink to augment the efficiency. Husain et al. [21] introduced the novel concept of jet impingement heat sink with microchannel and pillar arrangement.

Based on the available literature, novel analysis has been proposed for the different combinations of jet-channel-air foil pillars with the application of CuO–water-based Nano fluid.

2 Geometrical Model

A model of heat sink with air foil pillars using Nano fluid has been proposed. The model is used with combinations like Jet flow, Channel flow, Jet–channel flow, channel-pillar flow, and jet- channel-air foil pillars flow. Air foil pillars are placed in between the jets of impingement. The model is represented with the help of Fig. 1.

3 Results and Discussion

Analysis of heat sink for different combinations has been studied in the form of temperature, pressure drop, coefficient of heat exchange, Nusselt number, thermal resistance, and pumping power for fixed value of mass flow rate. Cuo–water-based Nano fluid is used for the study. The study has been carried out using Computational Fluid Dynamics simulations. The model is validated with the available literature [21].

Figure 2 depicts the variations in the temperature of heat sink with different combinations of Jet, channel, and pillars with the use of CuO–water Nano fluid. TOHS has been found minimal in case of jet and pillar coalition. The presence of pillars leads to division of the boundary layer. Also the jet flow caused high impingement resulting in the lower temperature. Max value is established for the channel flow. The reason for the same is due to advancement of boundary layer.

Figure 3 outlines the variance in pressure drop for separate composition of jet, pillar, and channel flow. For the amalgamation of flow in a channel with jet impingement and pillars, higher drop in pressure is noticed. Due to the increased velocity in this combination drop in pressure augmented.

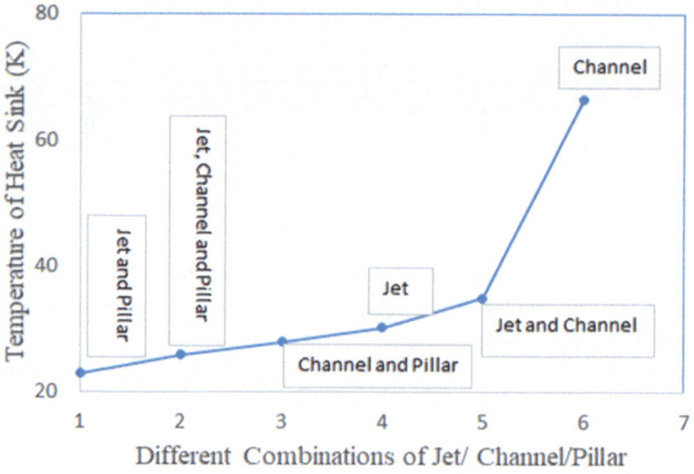

Fig. 2 TOHS versus DCOJCP

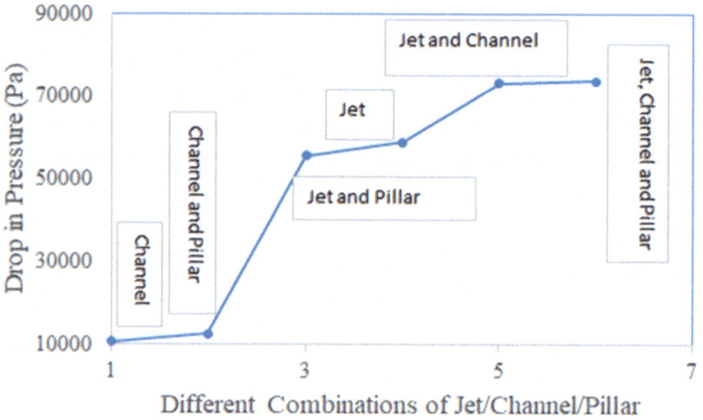

Fig. 3 DIP versus DCOJCP

The variation of coefficient of heat exchange for the distinct combinations with the application of CuO–water Nano fluid is depicted in Fig. 4. Jet and pillars of air foil shape combination resulted into increased value of coefficient of heat exchange while the lower value is found in case of flow in channel. The use of air foil pillars with jet impingement in microchannel flow resulted in high heat exchange coefficient. In contrast to the jet flow, channel flow exhibited lower values. Due to the presence of channel flow vortices diminished which resulted into lower heat exchange coefficient.

The variation in Nu with DCOJCP has been outlined in Fig. 5. The combination of CuO–water Nano fluid with different combinations augmented the Nu. Jet and pillar showed more Nu as compared to the other arrangements. The combinations

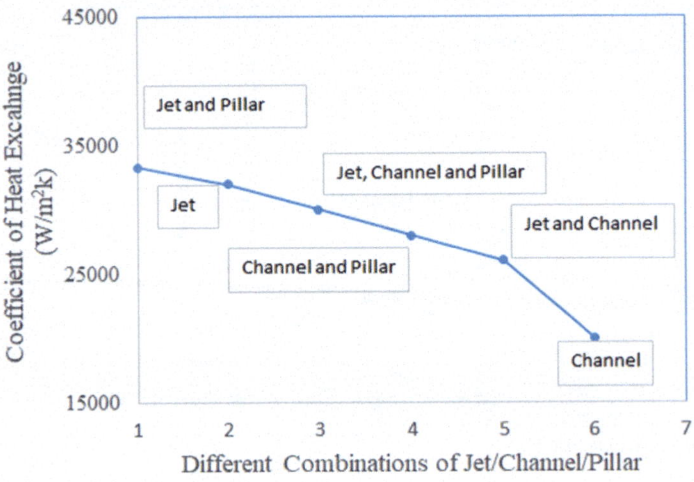

Fig. 4 COHE versus DCOJCP

like jet and channel, channel and pillar, jet, channel and pillars, and jet flow lie in between channel and jet-pillar.

The relation of thermal resistance for several arrangements of jet-channel-pillar is explained with the help of Fig. 6. THR is higher for channel flow and minimum for channel-pillar arrangement. Due to the augmentation in velocity, friction losses are increased resulting into higher THR.

For the distinct arrangements of jet-channel and pillars, change in the pumping power (PPR) has been depicted in Fig. 7. This power is on the higher side for the case of channel-pillar and on the lower side for channel flow. The other combinations lie in the intermediate positions.

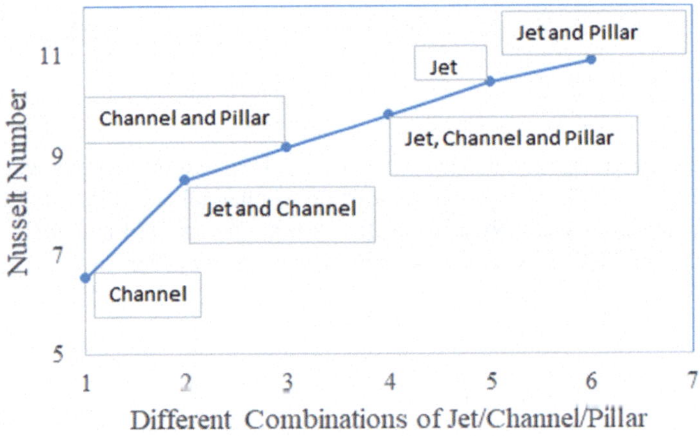

Fig. 5 Nu versus DCOJCP

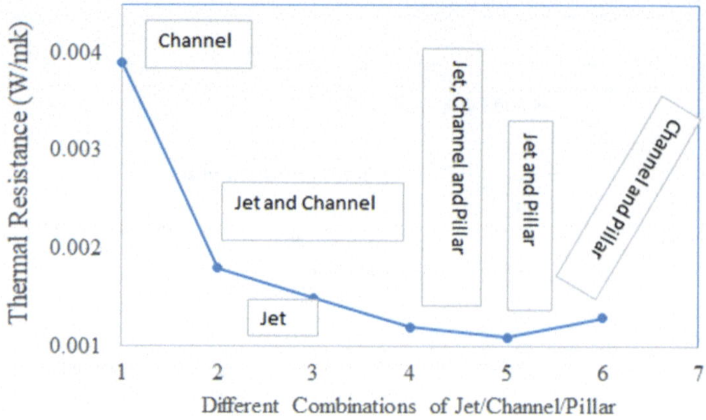

Fig. 6 THR versus DCOJCP

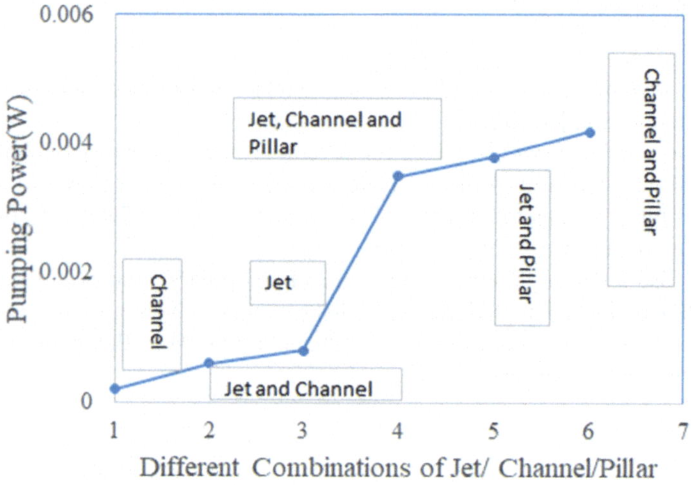

Fig. 7 PPR versus DCOJCP

The association between thermal resistance (THR) and pumping power (PPR) is shown in Fig. 8 for different combinations of jet-channel and pillar along with the application of CuO–water Nano fluid. It has been recognized that for the for the highest value of THR, PPR is lowest and vice-versa. Jet-pillar flow corresponds to higher PPR and low THR.

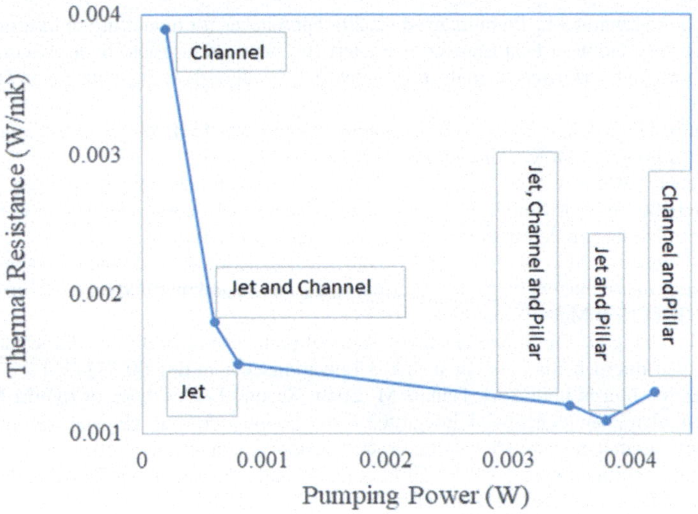

Fig. 8 Relation between THR and PPR

4 Conclusions

The current investigation suggests a sincere idea for the optimum use of different combinations of jet-channel-pillars of air foil shape with the application of CuO–water-based Nano fluids. Power required to drive the fluid flow and the thermal resistance are two important parameters for the optimum combinations for use. At the higher side of the former, lower values are found for the other one. So the combination having minimum value of both can be chosen for further study also. Coefficient of heat exchange and Nusselt number is more for jet-pillar and jet-channel-air foil pillar arrangement. So jet-pillar arrangement with Nano fluid can be a better option for the heat dissipation.

References

1. Abo-Zahhad EM, Ookawara S, Radwan A, El-Shazly AH, Elkady MF (2019) Numerical analyses of hybrid jet impingement/microchannel cooling device for thermal management of high concentrator triple-junction solar cell. Appl Energy 253:113538
2. Sorour MM, El-Maghlany WM, Alnakeeb MA, Abbass AM (2019) Experimental study of free single jet impingement utilizing high concentration SiO nanoparticles water base nanofluid. Appl Thermal Eng, 114019
3. Lelea D, Laza I (2014) The water based Al_2O_3 Nano fluid flow and heat transfer in tangential microtube heat sink with multiple inlets. Int J Heat Mass Transf 69:264–275
4. Alirezaie A, Hajmohammad MH, Hassani Ahangar MR, Hemmat Esfe M (2018) Price-performance evaluation of thermal conductivity enhancement of Nano fluids with different particle sizes. Appl Therm Eng 128:373–380

5. Anitha S, Loganathan K, Pichumani M (2020) Approaches for modelling of industrial energy systems: correlation of heat transfer characteristics between magneto hydrodynamics hybrid Nano fluids and performance analysis of industrial length-scale heat exchanger. J Therm Anal Calorim
6. Mohammad Pour J, Lee A (2020) Investigation of nanoparticle effects on jet impingement heat transfer: a review. J Molecular Liquids, 113819
7. Ambreen T, Saleem A, Woo Park C (2020) Analysis of hydro-thermal and entropy generation characteristics of Nano fluid in an aluminum foam heat sink by employing Darcy-Forchheimer-Brinkman model coupled with multiphase Eulerian model. Appl Thermal Eng, 115231
8. Yang L, Huang J, Mao M, Ji W (2020) Numerical assessment of Ag-water Nano-fluid flow in two new microchannel heatsinks: thermal performance and thermodynamic considerations. Int Commun Heat Mass Transfer 110:104415
9. Zheng L, Zhang D, Xie Y, Xie G (2016) Thermal performance of dimpled/protruded circular and annular microchannel tube heat sink. J Taiwan Inst Chem Eng 60:342–351
10. Bahiraei M, Jamshidmofid M, Dahari M (2020) Second law analysis of hybrid Nano fluid flow in a microchannel heat sink integrated with ribs and secondary channels for utilization in miniature thermal devices. Chem Eng Process Process Intensification, 107963
11. Hempijid T, Kittichaikarn C (2019) Effect of heat sink inlet and outlet flow direction on heat transfer performance. Appl Thermal Eng, 114375
12. Mahdavi M, Sharifpur M, Meyer JP, Chen L (2020) Thermal analysis of a Nano fluid free jet impingement on a rotating disk using volume of fluid in combination with discrete modelling. Int J Therm Sci 158:106532
13. Duan Z, Lv X, Ma H, Su L, Zhang M (2019) Analysis of flow characteristics and pressure drop for an impinging plate fin heat sink with elliptic bottom profiles. Appl Sci 10(1):225
14. Ramesh KN, Sharma TK, Rao GAP (2020) Latest advancements in heat transfer enhancement in the micro-channel heat sinks: a review. Arch Comput Methods Eng
15. Datta A, Kumar S, Halder P (2019) Heat transfer and thermal characteristics effects on moving plate impinging from cu-water nano fluid jet. J Thermal Sci
16. Chen W, Cheng J (2020) A numerical analysis on the heat transfer of jet impingement with Nano fluid on a concave surface covered with metal porous block. Heat and Mass Transfer. Heat Mass Transfer 56:3071–3083
17. MS A, Venkatasubbaiah K (2020) Numerical investigation of jet impingement flows with different Nano fluids in a mini channel using Eulerian-Eulerian two-phase method. Thermal Sci Eng Progress, 100585
18. Pratap A, Baghel YK, Patel VK (2020) Effect of impingement height on the enhancement of heat transfer with circular confined jet impingement using Nano fluids. Mater Today Proc
19. Ghanbari S, Javaherdeh K (2019) Thermal performance enhancement in perforated baffled annuli by nanoporous graphene non-newtonian nano fluid. Appl Thermal Eng, 114719
20. Freegan B, Hussain AA, Falih AH, Towsyfyan H (2019) CFD analysis of heat transfer enhancement in plate-fin heat sinks with fillet profile: investigation of new designs. Thermal Sci Eng Progress, 100458
21. Husain A, Ariz M, Al-Rawahi NZH, Ansari MZ (2016) Thermal performance analysis of a hybrid micro-channel, -pillar and -jet impingement heat sink. Appl Therm Eng 102:989–1000

CFD Study of Winged Pin–Fin Heat Sink

Krishna Veera Kumar and Md. Shaukat Ali

Abstract The present study numerically investigates the forced convection from the pin–fin type of the heat sink with an inline and a staggered arrangement of fins for varying Reynolds numbers. It further investigates the effect of attaching wings along the length of the fins on flow and heat transfer characteristics. It has been observed that the Nusselt numbers increase and friction factor decreases with increase in Reynolds numbers irrespective of the shape of the fins as well as the way the fins are arranged. The staggered arrangement of pin fins provided better thermo-hydraulic performance than the inline arrangement and the performance is further improved when the wings are embedded on the downstream side of the fins.

1 Introduction

The transfer of energy from one object to another object or within the same object due to temperature difference is referred to as heat transfer. Heat transfer analysis is essential for designing the components of the system in order to prevent the impairment due to overheating of electronic and mechanical devices and to improve the efficacy of the thermal systems. The pin–fin heat sink (PFHS) is a component which helps in dispersing the heat generated in an electronic or a mechanical device to a fluid source in order to keep the temperature at a desirable level in the devices. Pin fins are extended surfaces which are utilized to disperse the heat by any method of heat transfer like conduction, convection or both. The main factors that affect the performance of a heat sink are the velocity of the cooling medium, the material and the geometry of the fins. A short review of the previous investigation carried out on this topic is given below.

Yang et al. [1] experimentally studied the overall performance of the pin–fin type of the heat sink that has distinct cross-sections like circular, elliptical and square shaped. The elliptical pin fin caused lowest pressure drop for both inline as well as staggered

K. V. Kumar · Md. Shaukat Ali (✉)
Department of Mechanical Engineering, Sreenidhi Institute of Science and Technology, Hyderabad 501301, Telangana, India
e-mail: shaukatali@sreenidhi.edu.in

M. A. Siddiqui et al. (eds.), *Advances in Heat Transfer and Fluid Dynamics*, Lecture Notes in Mechanical Engineering, https://doi.org/10.1007/978-981-99-7213-5_16

arrangement. The staggered arrangement of elliptic pin fin provided lowest thermal resistance for fixed surface area and pumping power. Sahin and Demir [2] investigated the thermal performance and design parameters that are most beneficial to the heat sink with circular perforated fins. Test results have proven that the most important parameters influencing performance are Reynolds (Re) numbers, fin height and fin spacing. Ismail et al. [3] conducted numerical and experimental studies on cooling performance of four microprocessor heat sinks, namely, Pentium-III, Pentium-IV, AMD-Athlon and AMD-Duron. They obtained close agreement between the CFD and experimental results and found that the Pentium-IV heat sink provides the best performance than the other types of heat sinks.

Zhou and Catton [4] quantitatively evaluated the thermo-hydraulic performance of the plate type pin–fin heat sink (PPFHS) with varying pin width and wind velocity. They observed that overall the elliptical and NACA 0050 PPFHS provides best performance. Yuan et al. [5] have simulated PPFHS to test various parameters influencing the thermal hydraulic performance and found significant effect of pin height velocity of air.

Li et al. [6] have investigated Y-shaped bifurcation plates in a water cooled straight micro-channel heat sinks and found that they provide better thermal performance than corresponding straight channels. Moreover, they suggested that proper tuning of the angle between the Y-arms could further improve overall thermal performance. Razavi et al. [7] performed a computational analysis of hydrothermal energy in a PPFHS with splitters attached to the pin fins. The provision of a splitter was found to be advantageous in terms of pressure drop as well as thermal resistance, with further improvement in performance at higher heat flux and air velocity. Al-Damook et al. [8] investigated the thermal and airflow behaviour of a PFHS with multiple perforations experimentally and numerically. The results showed that the Nusselt number increases and the pressure drop decreases with an increase in the number of perforations inside the pin fin along the flow direction. The increase in the Nusselt number was attributed to an increase in surface area and jet-like flow through the perforations.

Saraireh [9] performed CFD simulation of plate-fin type of heat sink and circular pin–fin type of heat sink by varying the inlet velocity between 2 and 12 m/s and found better performance in case of the latter. Al-Damook et al. [10] computationally explored the advantages of the perforated pin fins implementing a conjugate heat transfer approach. They reported that the rate of heat transfer improves and pressure drop diminishes with the increase in the number of perforations in the pins, with an added advantage of savings in material cost and weight.

Maji et al. [11] have carried out a numerical study of the heat transfer enhancement characteristics of the pin–fin heat type of the sink with perforations like circular, diamond and elliptical in the pin fins of circular, square and elliptical shapes arranged in inline and staggered manner. They found higher heat dissipation and lower pressure penalty in the case of perforated fins than solid fins up to certain number and size of the perforations, particularly for staggered fin arrangement. Elliptical fins with circular perforations provide higher exergy efficiency.

Tijani and Jaffri [12] experimentally and numerically investigated the thermo-hydraulic characteristics of solid and perforated pin–fin and flat plate type of the heat sink. It is reported that perforated pin–fin and flat plate types of the heat sinks provided improved thermal efficiency compared to solid ones. Using Infrared thermography, Caliskan et al. [13] have investigated the heat transfer from cylindrical pin fins and hexagonal pin fins in a rectangular duct and concluded that the hexagonal pin fins provided better thermal performance.

Using CFD and optimization techniques, Maji et al. [14] investigated the influence of knurling and perforations on the effectiveness of a pin–fin heat sink. They observed faster heat transfer up to certain number and size of the perforations, which was further enhanced with knurling on the pin surface but at the cost of pressure penalty. Choudhary et al. [15] experimentally investigated the heat dissipation and fluid-flow characteristics of pin–fin embedded with wings by varying the Reynolds number, pitch between fin and length of the wing and finally they reported that the heat transfer and the pressure drop for both inline as well as staggered arrangements increases with decrease in fin pitch. The addition of wings causes heat transfer intensification with a moderate pressure penalty.

Hosseinirad et al. [16] have numerically studied the efficacy of a square pin–fin heat sink with a forward and backward arrangement of arched and wavy splitters. It has been concluded that pin fins with forward-arched splitters provide maximum heat transfer coefficient and performance.

Khattak and Ali [17] reviewed different types of heat sinks and their effectiveness, and found that perforated fin heat sinks are rendering better heat transfer performance and staggered type fin arrangement has better result compared to inline arrangement. Further, they observed that there is a negligible effect of fin orientation in the case of heat sinks filled with PCM.

The present objective is to perform CFD investigation of cylindrical aluminium pin–fin heat sink in which the fins are arranged in an inline and staggered manner with and without wings longitudinally attached behind the pin fins. The forced convection heat transfer and flow field study was carried out in a duct of size (L × W × H) 2300 × 180 × 80 mm for five different Reynolds numbers (Re) of 7583, 9858, 12,133, 14,408 and 16,683. A conjugate heat transfer model was adopted for the heat transfer study. The heat sink was placed on the wider bottom side of the duct. The fin diameter, fin pitch and wing height are fixed as d = 10 mm, 2d and 0.2d for all arrangements. Length of the fin is the same as the duct height of 80 mm. Results are presented in terms of augmentation Nusselt number (Nu/Nu_o), normalized friction factor (f/f_o) and thermo-hydraulic performance (THP) parameter $(Nu/Nu_o)/(f/f_o)^{1/3}$, where Nu_o and f_o are, correspondingly, the Nusselt number and the friction factor without fins.

2 Materials and Methods

Figure 1 shows the flow domain which consists of a 1500 mm long inlet section, 275 mm long test section and 525 mm long outlet section. The different configurations of the PFHS along with fin and wing dimensions are illustrated in Fig. 2. The conjugate heat transfer methodology is used in this study which involves simultaneous calculation of conduction heat transfer within the fins and convective heat transfer to the air flowing across the fins.

The standard k-ε turbulence model is implemented to define the turbulent flow of fluid through the duct using SIMPLE algorithm. Convergence is assumed to be obtained if the normalized residual errors of the continuity, the energy, the momentum are not more than 10^{-5}, 10^{-8} and 10^{-3}.

2.1 Boundary Conditions and Data Reduction

The boundary conditions are assumed to be velocity inlet and pressure outlet with free stream temperature of 300 K. A constant heat flux of 768 W/m^2 is supplied to the base of the PFHS and an adiabatic wall conditions on rest of the walls. The boundary conditions are also illustrated in Fig. 3.

The different thermo-physical quantity required for the study is determined using the following formulae shown in Eqs. 1–7.

The Nusselt number (Nu) is calculated as follows

$$Nu = \frac{hD_h}{k} \tag{1}$$

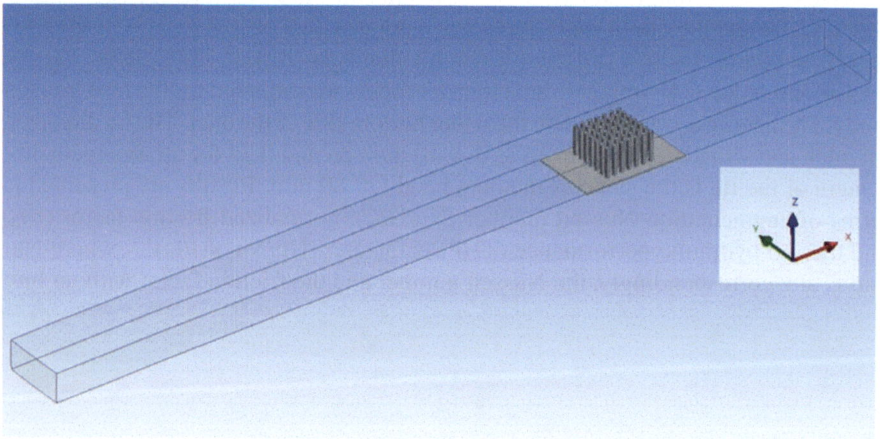

Fig. 1 Computational domain

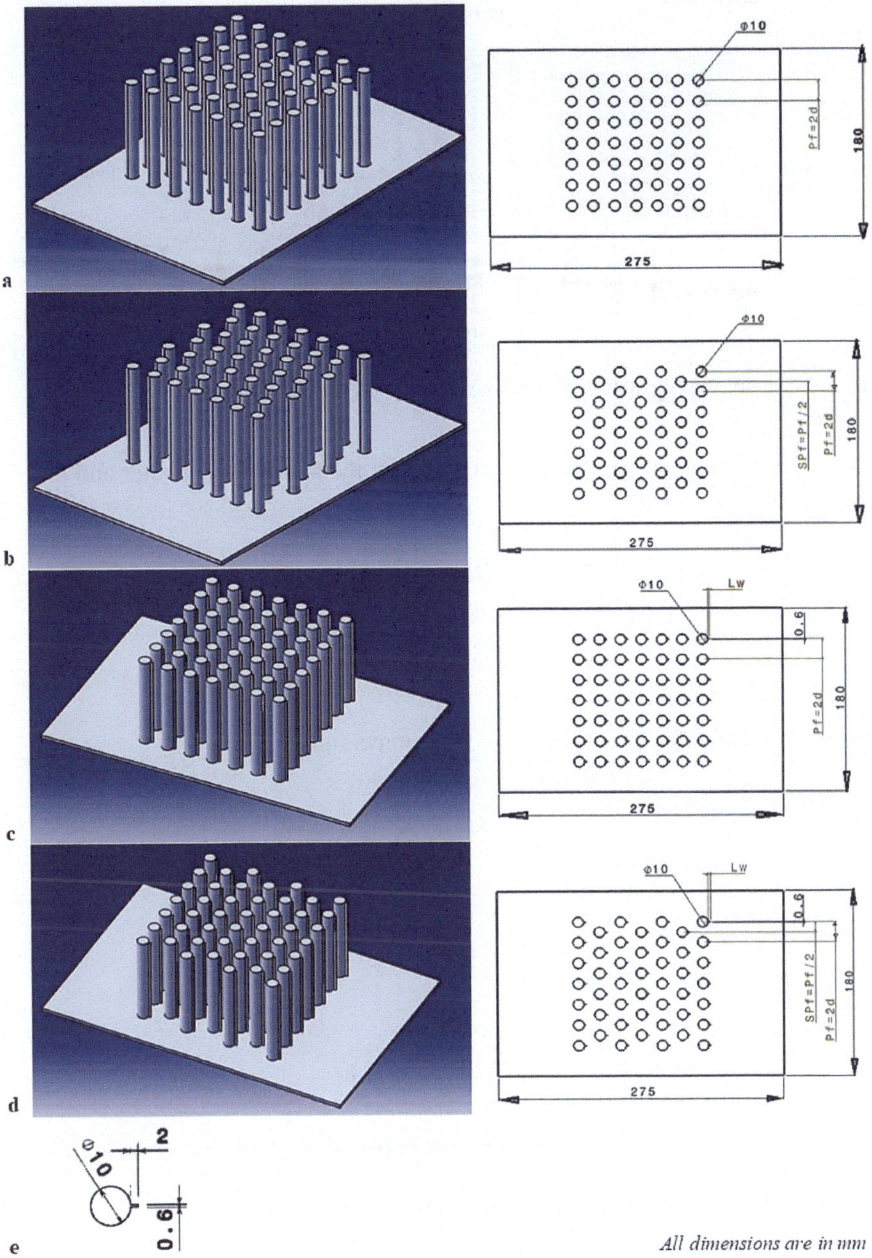

Fig. 2 Schematics of pin–fin heat sink investigated in the present study **a** Inline without wing, **b** Staggered without wing, **c** Inline with wing, **d** Staggered with wing, **e** Dimensioned view of pin fin with wing

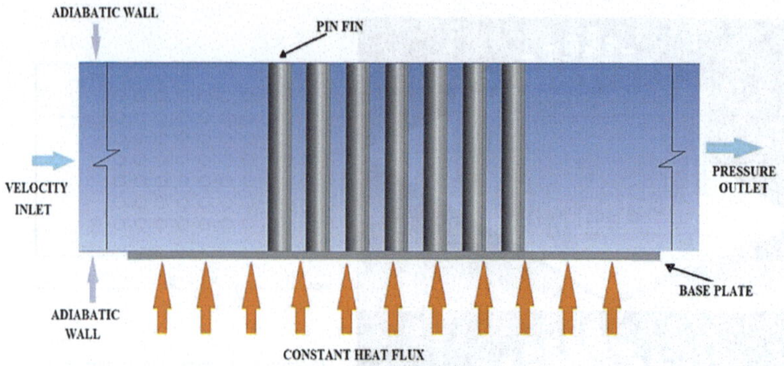

Fig. 3 Schematic representation of boundary conditions

where k is the thermal conductivity of air, D_h is the hydraulic diameter and h is the heat transfer coefficient

$$D_h = \frac{4A_d}{p} \tag{2}$$

$$h = \frac{Q''}{T_s - T_f} \tag{3}$$

where Q″ is Heat flux and T_f is mean film temperature

$$Q'' = \frac{Q}{A} \tag{4}$$

$$T_f = \frac{T_i + T_o}{2} \tag{5}$$

Friction factor

$$f = \frac{\Delta p}{\left(\frac{L_t}{D_h}\right)\left(\rho \frac{v^2}{2}\right)} \tag{6}$$

where v is the air velocity, L_t is the distance between two pressure tapings and Δp is the pressure drop.

Thermal Hydraulic Performance is calculated as follows.

$$THP = \frac{Nu/Nu_o}{(f/f_o)^{1/3}} \tag{7}$$

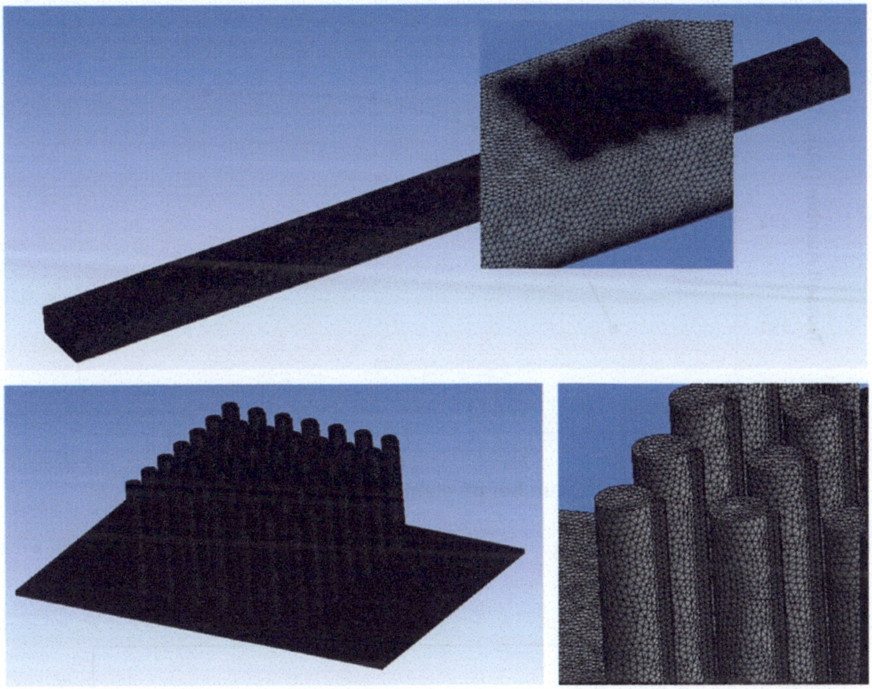

Fig. 4 Typical mesh generated for the inline arrangement of pin–fin heat sink with wing

where Nu_o and f_o are Nusselt number and friction factor for un-finned plate. Figure 4 shows a typical discretization scheme within the solid fin geometry and the flow domain.

3 Results and Discussion

If the number of elements is higher, the result obtained will have a more accurate result, but it may take a much longer simulation time. Therefore, a grid independence test is performed by varying the number of grid elements for the case shown in Fig. 5. It was observed that variation in Nusselt number is negligibly small after 2.63 Millions of elements and therefore, a comparable number of elements were considered for all the other fin configurations.

The accuracy of the present work is tested as shown in Fig. 6 wherein the value of Nu for base plate without pin fins as a function of Re is plotted along with the published results reported in Sahin and Demir [2] and Choudhary et al. [15]. The maximum deviation in average Nusselt number is found to be approximately 12.9% and 9% respectively.

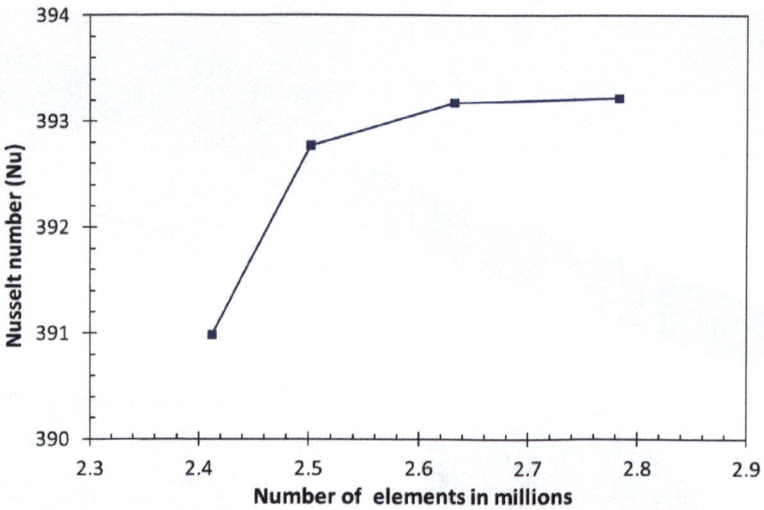

Fig. 5 Grid independence test result for the inline arrangement of pin–fin heat sink without wing at Re = 7583

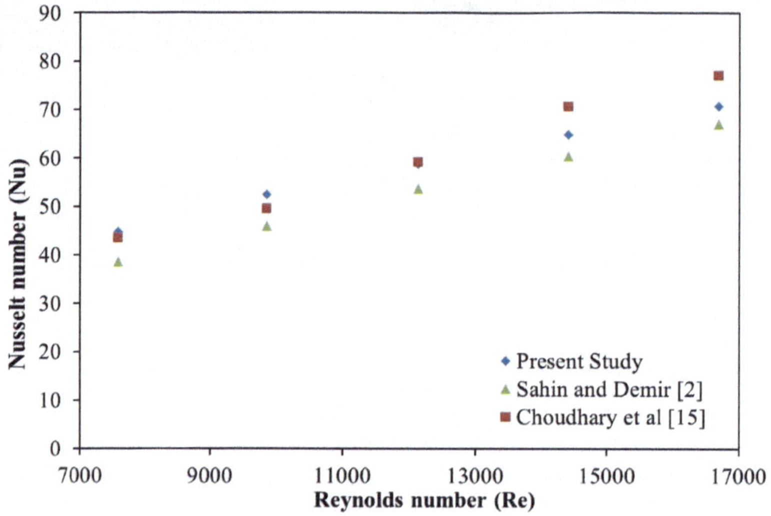

Fig. 6 Validation of Nu versus Re for smooth plate

3.1 Effect on Heat Transfer Rate

Figures 7 and 8 show the effect of Re, fin arrangement, and fin shape on h-factor and Nu respectively. It can be observed from Fig. 7 that with an increase in Reynolds number, h-factor increases irrespective of the fin configuration. This may be attributed

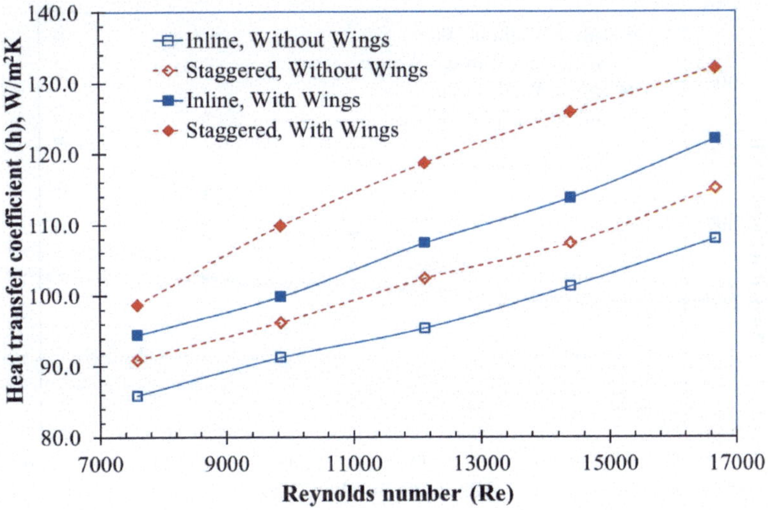

Fig. 7 Variation in h with Re for inline and staggered arrangements of fins with and without wing

to faster heat dissipation at higher air velocity and the resulting perturbation. Staggered fin arrangements provide a better heat transfer coefficient compared to their inline counterparts. For inline arrangement without the wing, the maximum h value is about 108 W/m^2K and 122.1 W/m^2K that with the wing, whereas, for a staggered arrangement without the wing, the maximum h value is 115.1W/m^2K, that with the wing it is 132.2 W/m^2K. This shows that with an increase in air inlet velocity, also the heat transfer coefficient increases.

Figure 8 shows that the Nu increases with an increase in the Re for all the fin geometry as well as the way they are arranged. However, the staggered arrangement of fins provides higher Nu compared to inline one due to higher perturbation in case of the former. Further, it can be observed that the fins with wings have higher Nu compared to those without wings.

3.2 *Effect on Friction Factor*

Figures 9 and 10 show how the Reynolds number, fin arrangement and fin shape affect friction factor (f) and the normalized friction factor (f/f_o). The value of f as well as f/f_o decreases with an increase in Re for all fin arrangements and shapes. Staggered arrangement gives higher f and f/f_o compared to inline arrangement. This may be due to more frequent head-on obstructions to the high-speed flow emanating from the upstream rows of fins (Fig. 14b). The inline as well as staggered winged fin arrangements provide still higher friction, which may be attributed to higher flow restrictions due to the presence of the wings.

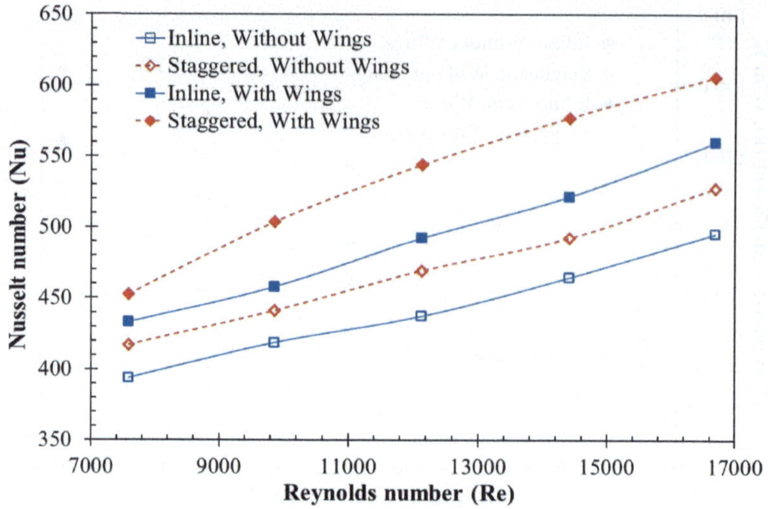

Fig. 8 Variation in Nu with Re for inline and staggered arrangements of fins with and without wing

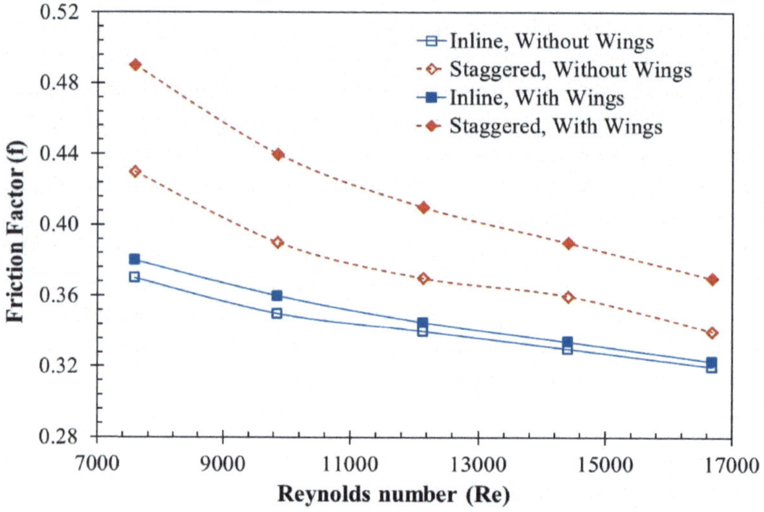

Fig. 9 Variation in f with Re for inline and staggered arrangements of fins with and without wing

3.3 Heat Transfer Augmentation

Figure 11 shows the variation in Nu/Nu_o with Re for inline, and staggered arrangements of fins with and without wings. The value of Nu/Nu_o decreases with the increase in the Reynolds number irrespective of the fin shape and how they are arranged. However, staggered arrangement provided higher Nu/Nu_o as compared

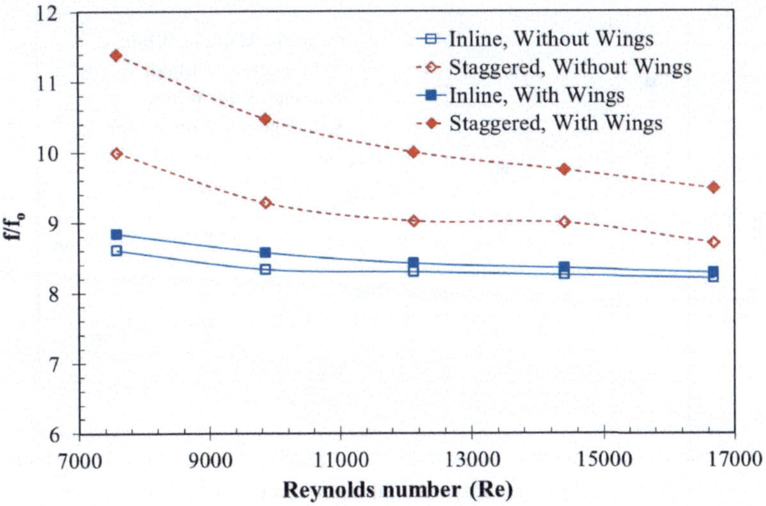

Fig. 10 Normalized f versus Re for inline and staggered arrangements of fins with and without wing

to inline one. The winged pin fins deliver higher heat transfer enhancement for all the Reynolds numbers and fin arrangement. The maximum Nu/Nu_0 is found to be 10.1 and 9.3 for staggered and inline arrangement of fins with wings respectively, corresponding to Reynolds number of 7583.

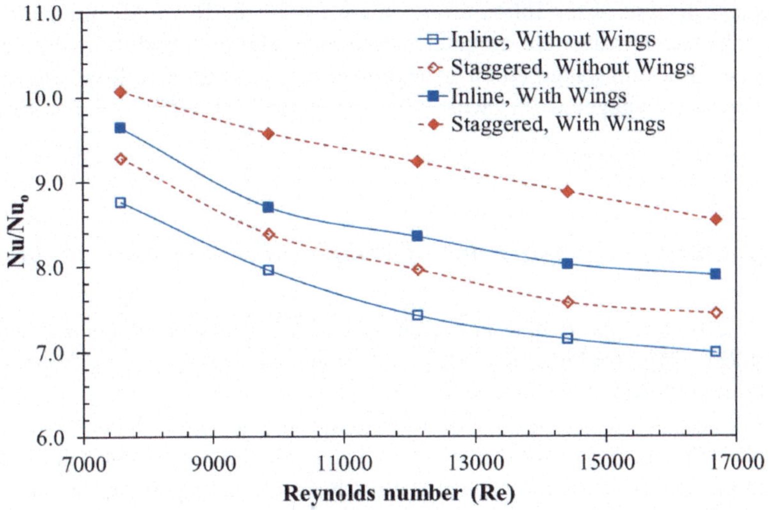

Fig. 11 Variation of Nu/Nu_0 with Re for Inline and Staggered arrangements of fins with and without wings

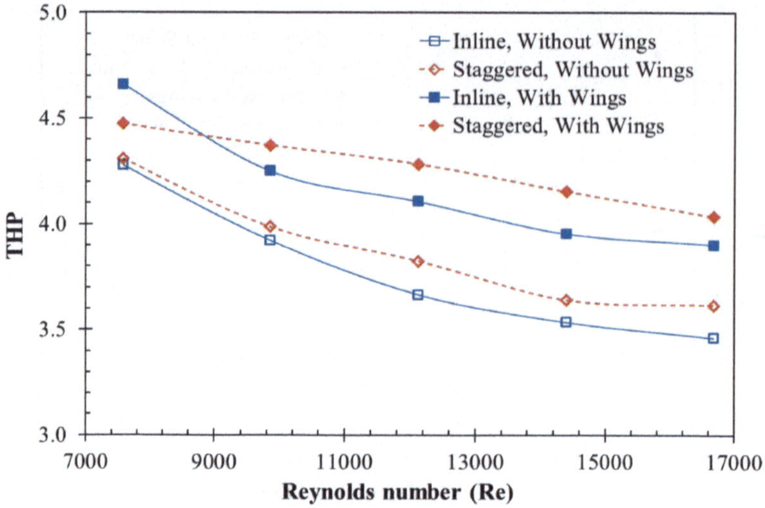

Fig. 12 THP versus Re for inline and staggered arrangements of fins with and without wing

3.4 Thermo-Hydraulic Performance

The THP parameter, $(Nu/Nu_0)/(f/f_0)^{1/3}$ is observed as shown in Fig. 12 to optimize the heat sink parameters for maximum heat transfer augmentation at the cost of pressure penalty. The figure shows that THP value decreases as Re increases for all the fin geometry and fin arrangement. However, the staggered one shows higher THP as compared to that of the inline arrangement for all the Reynolds numbers. THP further enhances if wings are mounted. Maximum THP is found to be 4.7 and 4.5 for winged inline and staggered fin arrangements corresponding to Re=7583 respectively. However, staggered fins with wings provides best THP at higher Reynold numbers.

3.5 Qualitative Comparisons of Temperature and Flow Field

Figure 13a, b show the temperature contours of a PFHS with wings and without wings at Re = 7583 for inline and staggered arrangements respectively. The temperature of fins are lower at the inlet zone than outlet zone of the test section. This is due to the air is getting heated up as it passes through the heat sink. It can be observed from figures that the portion of the base plate where pin fins are mounted shows lower temperature, which clearly indicates that effectiveness of using pin fins in dissipating heat from the base plate at a faster rate as compared to un-finned plate. Compared to the inline arrangement, the staggered one shows lower base plate temperature indicating faster heat dissipation resulting in higher heat transfer rate and Nusselt number which further enhance with the introduction of wings (Figs. 7 and 8).

Figure 14a, b represent the velocity vector plots for both the inline and the staggered configuration of PFHS without and with wings corresponding to Reynolds number of 7583. The velocity at the inlet is the minimum, after entering into the pin–fin heat sink the velocity of the fluid increases because the fluid passes through the lesser area between the pin fins. In case of staggered arrangement, the fluid strikes the succeeding pin–fin straight after coming out of the gaps of the previous row of

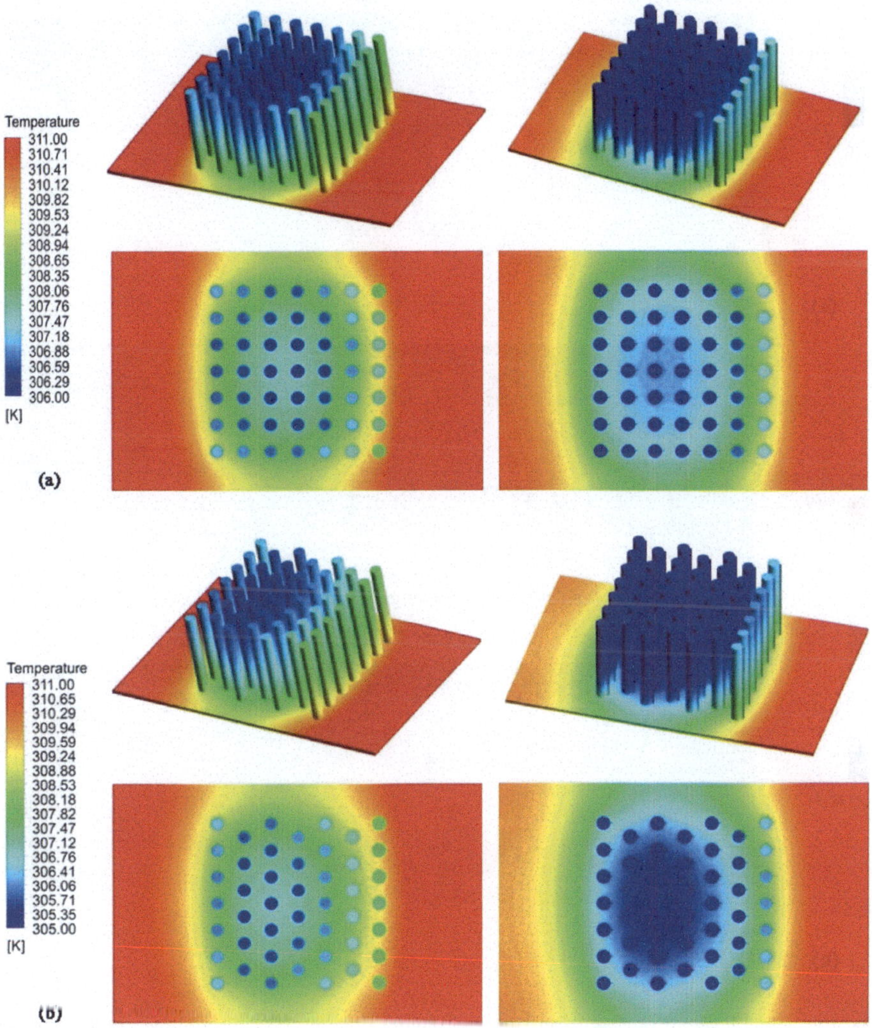

Fig. 13 Temperature contours (with and without wings) at Re = 7583 for **a** inline and **b** staggered arrangements

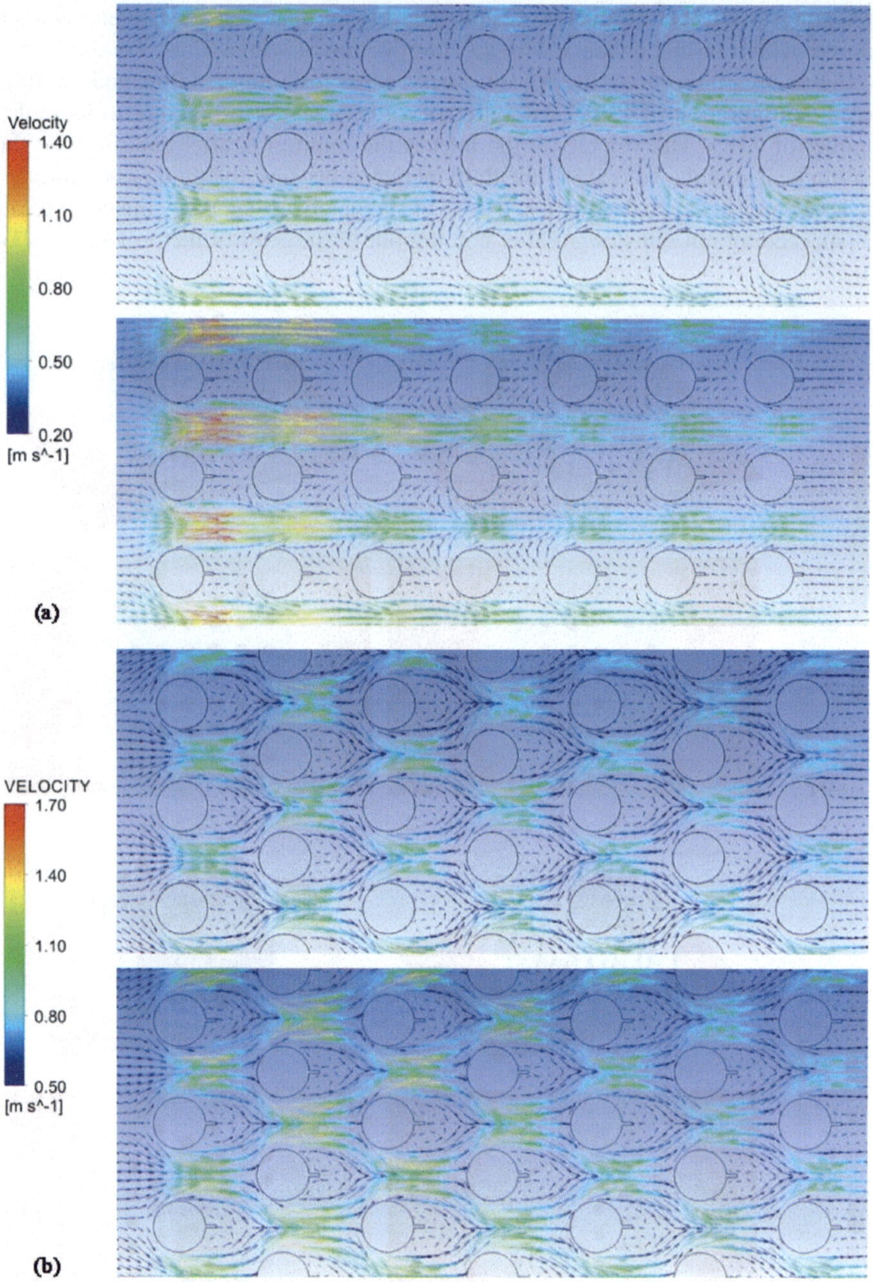

Fig. 14 Velocity vector plot (with and without wings) at Re = 7583 for **a** inline and **b** staggered arrangements

fins. This causes greater perturbation in flow resulting in higher friction factor and heat transfer enhancement as compared to inline arrangement which is evident from Figs. 10, 11, 13a, b. Moreover, the flow is perturbed to a greater extent in case of fins with wings resulting in higher heat dissipation as well as friction.

4 Conclusion

The present study uses conjugate heat transfer model to analyse circular PFHS with the wings as well as without the wings, arranged in inline, and staggered manner, has been performed with the aid of ANSYS Fluent 18.0. A constant heat flux boundary condition is used at the back of the base plate of the PFHS with air as working fluid. A pitch of $P_f = 2d$ for inline arrangement and semi-pitch of $P_f/2$ for staggered arrangement is used for the fluid flow and the heat transfer studies for Re of 7583, 9858, 12,133, 14,408 and 16,683. The outcomes are as follows:

- Nu increases with increase in Re for all the fin configurations (inline or staggered fin arrangement, with or without wing).
- Friction factor gradually decreases as Re increases for all the fin configurations.
- The staggered arrangement of pin fins provides higher Nusselt number than inline arrangement due to increase in turbulence.
- By attaching wings to the circular pin fins at the downstream side, the better turbulence has been created in flow which has resulted in better heat transfer augmentation but at the cost of pressure penalty.
- Staggered fin arrangement provides better THP as compared to inline one and THP further improves with the introduction of the wings.

References

1. Kai-Shing Y, Wei-Hsin C, Ing-Yong C et al (2007) A comparative study of the airside performance of heat sinks having pin fin configurations. Int J Heat Mass Transf 50:4661–4667
2. Bayram S, Alparslan D (2008) Thermal performance analysis and optimum design parameters of heat exchanger having perforated pin fins. Energy Conv Manag 49:1684–1695
3. Ismail MA, Abdullah MZ, Mujeebu MA (2008) A CFD-based experimental analysis on the effect of free stream cooling on the performance of microprocessor heat sinks. Int Commun Heat Mass Transf 35:771–778
4. Feng Z, Ivan C (2011) Numerical evaluation of flow and heat transfer in plate-pin fin heat sinks with various pin cross-sections. Numer Heat Transf A Appl 60(2):107–128
5. Wuhan Y, Jiyun Z, Tso CP et al (2012) Numerical simulation of the thermal hydraulic performance of a plate pin fin heat sink. Appl Therm Eng 48:81–88
6. Li, Yanlong, Fengli Z, Dengt S et al (2014) Laminar thermal performance of microchannel heat sinks with constructal vertical Y-shaped bifurcation plates. Appl Therm Eng 73:183–193
7. Razavi SE, Osanloo B, Sajedi R (2015) Application of splitter plate on the modification of hydro-thermal behavior of PPFHS. Appl Therm Eng 80:97–108

8. Amer A-D, Kapur N, Summers JL et al (2015) An experimental and computational investigation of thermal air flows through perforated pin heat sinks. Appl Therm Eng 89:365–376
9. Mohammad S (2016) Computational fluid dynamics simulation of plate fin and circular pin fin heat sINKS. Jordan J Mech Ind Eng 10(2):99–104
10. Amer A-D, Kapur N, Summers JL et al (2016) Computational design and optimisation of pin fin heat sinks with rectangular perforations. Appl Therm Eng 105:691–703
11. Ambarish M, Dipankar B, Kumar PP (2017) Numerical investigation on heat transfer enhancement of heat sink using perforated pin fins with inline and staggered arrangement. Appl Therm Eng 125:596–616
12. Salami TA, Binti JN (2018) Thermal analysis of perforated pin-fins heat sink under forced convection condition. Procedia Manuf 24:290–298
13. Sinan C, Ayla D, Isak K (2018) Experimental investigation of heat transfer from different pin fin in a rectangular channel. Exp Heat Transf 32:376–392
14. Ambarish M, Dipankar B, Kumar PP (2018) Effect of knurled fin surface on thermal performance of perforated fin heat Sink. J Thermophys Heat Trans
15. Vishwjeet C, Manoj K, Kumar PA (2019) Experimental investigation of enhanced performance of pin fin heat sink with wings. Appl Therm Eng 155:546–562
16. Hosseinirad E, Khoshvaght-Aliabadi M, Hormozi F (2019) Effects of splitter shape on thermal-hydraulic characteristics of plate-pin-fin heat sink. Int J Heat Mass Transf 143:118586
17. Zulfiqar K, Muhammad AH (2019) Air cooled heat sink geometries subjected to forced flow: a critical review. Int J Heat Mass Transf 130:141–161

Development, Characterization and Performance Evaluation of Graphene Oxide Nanofluid in EG:Water-Based PV/Thermal System

Ajiv Alam Khan and Syed Mohd Yahya

Abstract The advanced nanofluid with good stability is demand of industry as well as solar thermal systems. In industrial application, Graphene Oxide (GO) nanofluid formulated with Ethylene Glycol (EG): water (W) is usually known well for good stability along with high thermal conductivity. In this research, GO nanofluid is characterized for exploring thermal, optical, and suspension stability under certain conditions and then utilized as working fluid in photovoltaic thermal (PV/T) system. SEM, UV–vis, FTIR and TGA analysis are performed for morphology, optical and thermal stability characterization respectively. Thermal conductivity measurements of the GO/EG:W nanofluid revealed an enhancement of 9.5% at 40 °C compared to water. It also displayed good long-term stability, with a zeta potential of 30.3 mV. The long-term stability with time as well as stability at high temperature of GO nanofluid give a good room for future researchers to use it as coolant in solar PV panel. Because large increment in thermal conductivity of EG:W nanofluid of GO may act very amazing media for heat transport which resulted in enhancement of performance of PV solar panels. Therefore the electrical as well as thermal efficiencies of the PV/Thermal system improved.

Keywords GO · EG:W · Nanofluid · Zeta potential

Nomenclature

GO	Graphene oxide
EG	Ethylene glycol
W	Water

A. A. Khan (✉) · S. M. Yahya
Department of Mechanical Engineering, Sustainable Development & Acoustic Research Laboratory, ZHCET Aligarh Muslim University, Aligarh 202002, India
e-mail: azim7alam@gmail.com

S. M. Yahya
e-mail: smyahya@zhcet.ac.in

M. A. Siddiqui et al. (eds.), *Advances in Heat Transfer and Fluid Dynamics*, Lecture Notes in Mechanical Engineering, https://doi.org/10.1007/978-981-99-7213-5_17

UV-vis	Ultraviolet visual spectroscopy
TEM	Transmission electron microscopy
FESEM	Field emission scanning electron microscopy
PVT	Photo-voltaic thermal

1 Introduction

The eighteenth century is the century of industrial revolution, so energy became one of the greatest challenges among all the challenges faced by mankind. Due to rapid growth of the world population, demand of energy has increased massively in last few decades. Hence to meet the demand of energy the rapid depletion in conventional sources of energy takes place. So the fossil fuel-based fuels such as petroleum, coal and natural gas have been depleting, which is very serious issue for the future development of the world. However, fossil fuel resources are rapidly depleting due to exploration to meet the energy demand of the countries. If these conventional sources of energy are exhausted, it is impossible to replenish them.

Photovoltaic-thermal collectors [1], a cogeneration technology, which combine the components of photovoltaic cell and solar thermal absorber, have been shown to be a viable solution because they maximize the use of netted heat inside the photovoltaic cell while also enhancing PV efficiency. Chow et al. [2] have investigated that every 10° rise in the solar cell temperature results in to decrease in conversion efficiency of Photo Voltaic cells by 5%. So if we combine the both, the photovoltaic panel with the thermal system so that both help each other and the combination is known as Photovoltaic-Thermal System.

The efficiencies of these energy harnessing technologies are observed usually in the range of approximately 50–70% [3] and 15–20% [4] respectively. At present, the hybrid photovoltaic thermal system (PV/T) developed by integrating both PV and solar thermal systems receives considerable attentions by the researchers as it improves the energy generation efficiency [4, 5]. The working fluid used in PV/T system plays key role in absorbing heat from solar module and transferring it to the collector. The thermal conductivity of commonly used base fluids such as water, oil or ethylene glycol is usually increased by colloidal mixing of nanoparticles (1–100 nm). The suspended nanoparticles in base fluid help improve its thermal properties and thus enhance the overall performance of the system [6]. The reported literature studies [3, 5] reveal that the commonly used metal-based nanoparticles in preparing nanofluids are either metals (Zn, Fe, Cu, Al) or metal oxides (Al_2O_3, ZnO, Cu_2O, TiO_2, etc.). Many researchers use carbon-based nanomaterials such as fullerenes (carbon molecule Cn where $n > 20$) [7], carbon nanotubes (carbon allotropes with cylindrical nanostructures) [6] and graphene (caron with two-dimensional allotropic form) [8]. Concerning PV/T system's overall efficiency, thermal and electrical efficiency, surface temperature, entropy generation and energy loss, different researchers investigated various input parameters with nanofluids including types, size, shape and concentration of nanoparticles, types of base fluid, stability, viscosity and flow rate experimentally/numerically [9, 10].

Recently, the hybrid nano-fluids, composed of two distinct nanoparticles or amalgamated nano-powder suspended in a base fluid, increase capabilities in all domains of heat transfer and most notably in solar thermal collectors, including concentrating kinds and flat plate [11]. It has also been suggested that synthesis nanofluids are more effective as compared to mono nanofluids [12–14]. When combination of two nanoparticles is used in solar collectors the solar thermal efficiency of solar collectors improves significantly [15, 16]. The researchers have also reported the enhancement of effectiveness of parabolic solar collectors [17]. Due to the synergistic effect of nanoparticles, the good characteristics of traditional heat transfer fluids (like ethylene glycol, water, and oil) and the nano-fluids incorporating single nanoparticles can indeed be enhanced [18, 19]. The combined nano-fluids' increased absorption capacities are enticing leading to a significant increase in solar thermal collector efficiency. Khan et al. evaluate the performance of water-based $Fe3O4/SiO2$ hybrid nano-fluid in PV thermal system [20].

2 Materials and Methods

2.1 Preparation of Aqueous Ethylene Glycol Nanofluids

To prepare the nanofluids, the graphene nano-platelets were procured from SRL lab. These graphene nano-platelets dispersed with the water-ethylene glycol in a volumetric ratio of 60:40 and mass fractions were kept in range of 0.1–0.3%. The required mass of the graphene nano-platelets was weighed with the help of calibrated digital balance. One kilogram of water-ethylene glycol in the ratio of 60:40 by volume was prepared. To achieve maximum stability of nanofluid a surfactant (NPE 400) in volumetric ratio of 0.1% of the base fluid is added. A speed stirrer, which was running about 300 rpm, was used to disperse the graphene nano-platelets in water ethylene glycol. The stirrer run for about 20 min to achieve high level of stability of nanofluid. The nanofluid prepared by stirrer may have agglomerations of nano-particles. Hence to break these agglomerations, an ultrasonic homogenizer was used. There may be chances of damage of nano-platelets, so to minimize this damage, the sonication of nano-fluid was performed for 10 min at 20 kHz and 150 W.

2.2 UV–Vis Analysis of the Aqueous Ethylene Glycol Nanofluids

An Ultraviolet–visible spectroscopy (UV–Vis) of model: Perkin Elmer Lambda 750 was used to record the optical absorbance measurements. The absorption data was recorded at room temperature for the wavelength within 800 to 200 nm range. The adjusted scan speed was 266.75 nm/min with the 860 nm monochromatic.

2.3 Viscosity Measurement of Aqueous Ethylene Glycol Nanofluids

The measurement of viscosity was carried out by Rheometer Anton Paar model MCR92. T-Ramp measurement i.e. the investigation of variation of viscosity with temperature was carried out in the temperature range of 20–80 °C for all samples (60 mL). T-Ramp measurements for the pure aqueous ethylene glycol, aqueous ethylene glycol nanofluids including 4 different types of nanoparticles and in different concentrations were performed in identical condition to assure the uniformity of the measurements.

2.4 Microstructure and Morphological Analysis of Graphene

The morphology of the synthesized Graphene nanoflakes was monitored with a scanning electronic microscopy (SEM) of model; TESCAN, VEGA3 and energy-dispersive x-ray spectroscopy (EDX) model; OXFORD INSTRUMENT). A digital ion coater (COXEM Co, SPT-20) was utilized for Pt coating on the samples in the fixed current of 3 mA for 300 s for taking SEM images.

2.5 Thermal Conductivity Measurements

The thermal conductivity of synthesized aqueous ethylene glycol nano-fluids with different concentrations was measured with the help of thermal properties analyzer (Tempos) from Meter Group (USA) with improved proprietary algorithm. The sensor of the analyzer was KS-3 with the length of 60 mm and dimension of 1.3 mm. The selected sensor is able to measure the thermal conductivity in the range of 0.02–2 W/m K in the accuracy range of ±10%. Thermal conductivity measurements for all prepared samples were executed in 4 different temperatures including 25, 40, 55 and 70 °C. Firstly, the prepared samples were poured into a vial followed by locating the KS-3 sensor inside the sample. Then the vial was placed inside a water bath (MEMMERT, WNB22) for stabilizing the temperature to achieve accurate results. Thermal conductivity measurements for all samples in certain temperatures were repeated for 5 times to assure the accuracy of the obtained data.

3 Results and Discussion

Figure 1a illustrates the FESEM image of Graphene Oxide nanoflakes, and FESEM and TEM images of GO-based aqueous nanofluids. Upon formulation of GO nanofluid with E:W, the GO particles show some surface wrinkling and well dispersed distribution in fluid.

Figure 1a presents a typical micrometer scale flake of GO, which displays some surface wrinkling. Figure 1c presents a representative TEM sample showing well dispersed and transparent sheets of GO [22]. Typically, suspensions with absolute zeta potential values greater than 30 mV are physically stable, while suspensions below 20 mV have limited stability, and suspensions below 5 mV experience rapid aggregation. Thus, the GO-based aqueous nanofluid is physically stable and it was only after 6 months its zeta potential dropped down to 30.3 mV.

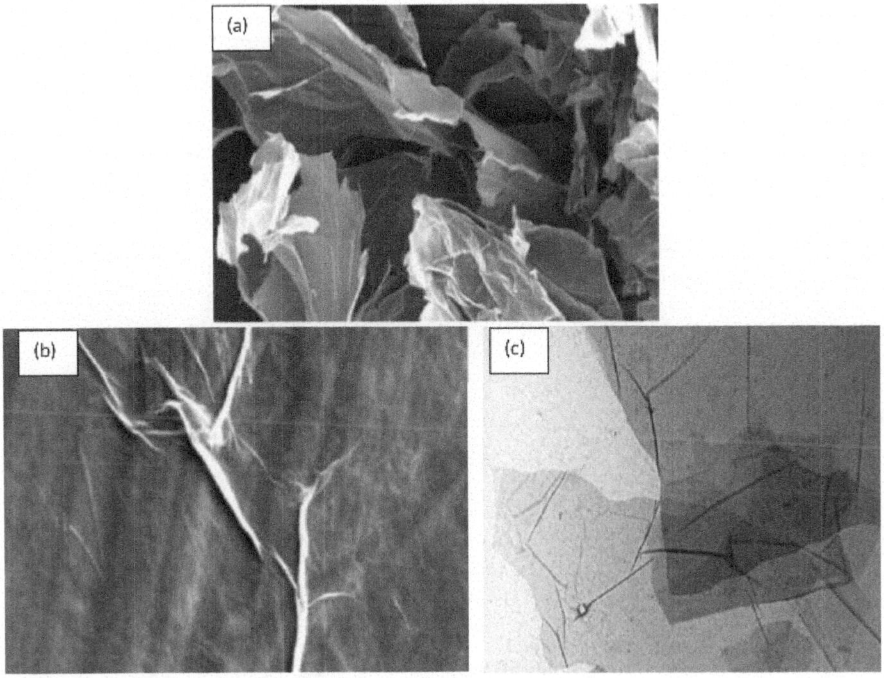

Fig. 1 **a** FESEM image of Graphene Oxide nanoflakes, **b** FESEM and **c** TEM images of Graphene Oxide -based aqueous nano-fluids (taken from Ref. [21])

3.1 Thermal Conductivity Analysis

The thermal conductivities of different nanofluids such as Al_2O_3, TiO_2, hybrid (Al_2O_3-TiO_2) and Graphene in the aqueous EG:W at 0.1% concentrations and aqueous EG:W are presented in Fig. 4. Among others, the GO nanofluid shows higher thermal conductivity (Figs. 2 and 3).

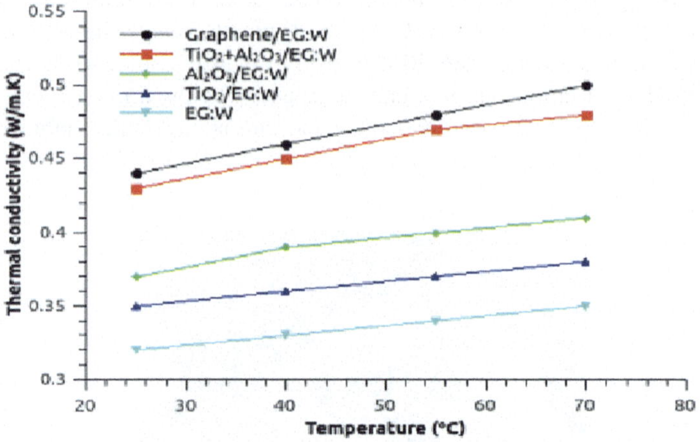

Fig. 2 Comparative thermal conductivity analysis of different nanofluids with increasing temperature for 0.1wt.% concentration

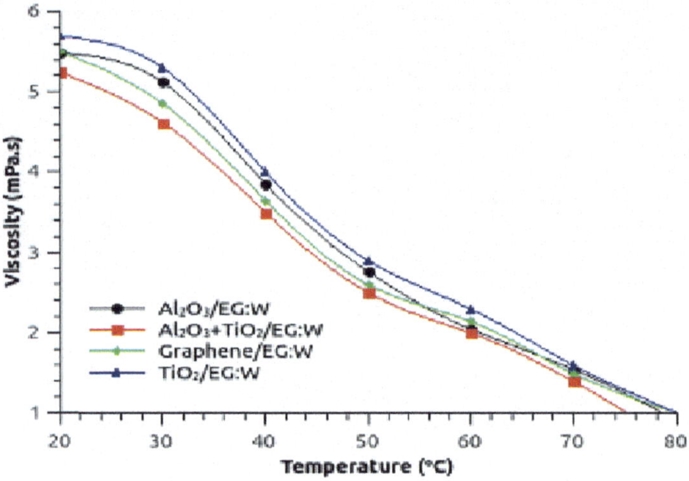

Fig. 3 Viscosity variation of different evaluated nanofluids with increasing temperature at a concentration of 0.1wt.%

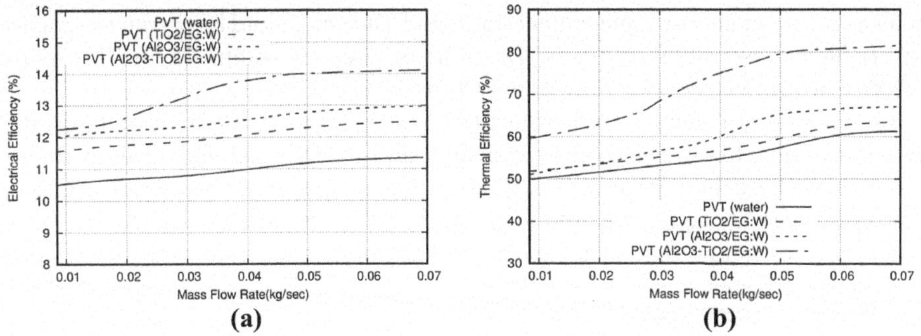

Fig. 4 **a** The variation of electrical efficiency of PVT system and **b** Thermal efficiency of PVT system with mass flow rate with different kinds of coolant at irradiance of 1000 W/m^2 and 0.1% wt concentration

The order of thermal conductivity as shown in Fig. 2 is GO/EG:W > TiO_2 + Al_2O_3/EG:W > Al_2O_3/EG:W > TiO_2/EG:W > EG:W. The thermal conductivity of EG:W is found to be increased from 0.32 to 0.44 W/m.K at 25 °C due to addition of graphene oxide. It is also seen from Fig. 4 that the thermal conductivity rises linearly with increasing the temperature. The thermal conductivity of GO nanofluid is found to be raised from 0.44 to 0.5 when temperature is increased from 25 to 70 °C. The temperature grows at the same rate of each of temperature step. The cause of this phenomenon can be related to the Brownian motion of the particles in the nanofluid [23–26]. When the temperature increases, the kinetic energy of the nanoparticle inside the nanofluid also increases leading to faster collision among the nanoparticles. Hence the thermal conductivity is enhanced as also reported by Chiam and Azmi [27], Esfe and Karimipour [28] and Sati and Shende [29]. These authors have also reported the linear enhancement in thermal conductivity with the increase of temperature.

3.2 Viscosity Analysis

In Fig. 3, the viscosity of the nanofluids with respect to temperature and wt.% concentration of the nanoparticles is depicted. The temperature is varied from 20 to 80 °C in 50 steps. The study of viscosity variation is very important as the nanofluids are under constant flow conditions. During such applications, the rheological properties must remain intact and should be in favor of pumping power required. From the figures, it is evident that the viscosity of the nanofluids is marginally affected by the concentration of the nanoparticles while the temperature has shown a promising effect. In Fig. 3, a comparative analysis of all four nanofluids with lowest viscosity in comparison to base fluid is depicted. The hybrid nanofluid has provided the lowest of all viscosity at almost same temperatures and TiO_2 nanofluid has given a comparatively higher viscosity than others. The order of viscosity within 30–50 °C is TiO_2/EG:W > Al_2O_3/EG:W > GO/EG:W > $TiO_2 + Al_2O_3$/EG:W.

From an overall analysis of thermal and viscosity property of the nanofluids, the hybrid nanofluid has shown more promising results than the individual nanoparticles suspended in base fluid. It can also be inferred that irrespective of nanoparticle usage, at temperatures above the room temperature (here 50 °C) the increase/decrease in viscosity is found to seize and at the near boiling point of water the viscosity is same at all concentrations.

3.3 Performance of Nanofluid in PVT System

In this study, the temperature of PV panel is maintained in permissible limit by different coolants in the back of PV panel. Figures 4a and b present the electrical and thermal efficiencies of PV/T with different nanofluids having a nanoparticle concentration of 0.1% in the base fluid. The electrical efficiencies observed for the PV/T with TiO2/EG:W, Al2O3/EG:W, Al2O3-TiO2/EG:W and GO/EG:W are 12.2, 12.3, 12.5 and 13.5% respectively at the maximum flow rate of 0.07 kg/s. Similarly, at the same flow rate, the thermal efficiencies observed for those systems are 62, 64, 68 and 76% respectively. Both electrical and thermal efficiencies for GO/EG:W nanofluid are observed to be better than other tested fluids. The enhancement in thermal and electrical efficiency is due to the better heat removal capacity of the nanofluid as compared to other coolants. This also indicates the improvement in thermal conductivity of the nanofluid. Because almost same trend for the improvement of thermal conductivity of nanofluids is shown in above graph of thermal conductivity in Fig. 4.

4 Conclusions

4.1 The Following Conclusions Are Drawn from the Study:

(i) Prior to processing, the FESEM image of GO nanoflakes shows an agglomeration of large micrometer scale sheets displaying surface wrinkling and folding. Upon formulation of GO nanofluid, the GO particles display some surface wrinkling and well dispersed distribution in fluid. It was physically stable as the zeta potential even after 6 months was around 30.3 mV which was greater than the required value (30 mV) assigned for stability.

(ii) The order of thermal conductivity for the tested fluids is GO/EG:W > $TiO_2 + Al_2O_3$/EG:W > Al_2O_3/EG:W > TiO_2/EG:W > EG:W. The thermal conductivity of EG:W is found to be increased from 0.32 to 0.44 W/m.K at 25 °C due to addition of graphene oxide. The thermal conductivity of GO nanofluid is further found to be raised from 0.44 to 0.5 when temperature is increased from 25 to 70 °C.

(iii) The thermal degradation for almost all the nanofluids is seen to begin in the range at nearly 55 to 70 °C. Approximately 3 to 10% of mass is lost in this temperature range. The viscosity is decreased with the increase of temperature and the order of viscosity within 30–50 °C is found be TiO_2/EG:W > Al_2O_3/EG:W > GO/EG:W > $TiO_2 + Al_2O_3$/EG:W.

(iv) The electrical efficiencies observed for the PV/T with TiO_2/EG:W, Al_2O_3/EG:W, Al_2O_3-TiO_2/EG:W and GO/EG:W at 0.1% concentration level are observed by 12.2, 12.3, 12.5 and 13.5% respectively at the maximum flow rate of 0.07 kg/s. Similarly, at the same flow rate, the thermal efficiencies observed for those systems are 62, 64, 68 and 76% respectively. This demonstrates that both electrical and thermal efficiencies for GO/EG:W nanofluid are better than other tested fluids. Among others, the PV/T system with GO/EG:W system shows the highest electrical and thermal efficiencies. Both electrical and thermal efficiencies are found to be further increased when the concentration level is increased.

(v) Both electrical and thermal efficiencies increase with the increase of mass flow rate. The electrical efficiency for Al_2O_3/EG:W, TiO_2/EG:W, Al_2O_3-TiO_2/EG:W increased from 11.65 to 12.5%, from 12 to 13% and from 12.25 to 14.10% respectively for the increase of mass flow 0.01 to 0.07 kg/s. Similarly, the thermal efficiencies of these working fluids are increased from 62 to 64%, from 64 to 68%, and from 68 to 81.5% respectively when the mass flow rate is increased from 0.01 to 0.07 kg/s. All the analyzed properties and performances of different nanofluids indicate GO/EG:W as better working fluid for PV/T compared to others.

5 Future Scope

The combination of two different nanoparticles in a base fluid improves system performance when compared to the mono-nanofluid. The performance of the system also improved when graphene is used as compared to hybrid nanofluid. Furthermore, hybrid nanofluid shows the better stability than mono-nanofluid. Future researchers may investigate the effect of cooling for high weight concentration of graphene and compare it with the others. Also, there is a need to evaluate the effect of nanoparticles on the cell temperature of PV panel. More comprehensive study can be started to acknowledge the overall performance of the system.

References

1. Ahmed F, Al Amin AQ, Hasanuzzaman M, Saidur R (2013) Alternative energy resources in Bangladesh and future prospect. Renew Sustain Energy Rev 25:698e707
2. Chow TT, Hand J, Strachan P (2003) Building-integrated photovoltaic and thermal applications in a subtropical hotel building. Appl Therm Eng 23:2035–2049. https://doi.org/10.1016/S1359-4311(03)00183-2
3. Choudhary S, Sachdeva A, Kumar P (2020) Influence of stable zinc oxide nanofluid on thermal characteristics of flat plate solar collector. Renewable Energy 152:1160–1170. https://doi.org/10.1016/j.renene.2020.01.142
4. Bhakre S, Sawarkar P, Kalamkar V (2021) Performance evaluation of PV panel surfaces exposed to hydraulic cooling—a review. Sol Energy 224:1193–1209. https://doi.org/10.1016/j.solener.2021.06.083
5. Hossain F, Karim M, Bhuiyan A (2022) A review on recent advancements of the usage of nano fluid in hybrid photovoltaic/thermal (PV/T) solar systems. Renew Energy. https://doi.org/10.1016/j.renene.2022.01.116
6. Tembhare S, Barai D, Bhanvase B (2022) Performance evaluation of nanofluids in solar thermal and solar photovoltaic systems: a comprehensive review. Renew Sustain Energy Rev 153:111738. https://doi.org/10.1016/j.rser.2021.111738
7. Zaytseva O, Neumann G (2016) Carbon nanomaterials: production, impact on plant development, agricultural and environmental applications. Chem Biol Technol Agric 3:17
8. Wahab A, Khan M, Hassan A (2020) Impact of graphene nanofluid and phase change material on hybrid photovoltaic thermal system: exergy analysis. J Clean Prod 277:123370. https://doi.org/10.1016/j.jclepro.2020.123370
9. Qeays I, Yahya S, Asjad M, Khan Z (2020) Multi-performance optimization of nanofluid cooled hybrid photovoltaic thermal system using fuzzy integrated methodology. J Clean Prod 256:120451. https://doi.org/10.1016/j.jclepro.2020.120451
10. Parsa S, Yazdani A, Aberoumand H, Farhadi Y, Ansari A, Aberoumand S, Karimi N, Afrand M, Cheraghian G, Ali H (2022) A critical analysis on the energy and exergy performance of photovoltaic/thermal (PV/T) system: the role of nanofluids stability and synthesizing method. Sustain Energy Technol Assess 51:101887. https://doi.org/10.1016/j.seta.2021.101887
11. Sundara LS, Sharmab KV, Singha MK, Sousaa ACM (2017) Hybrid nanofluids preparation, thermal properties, heat transfer and friction factor—a review. Renew Sustain Energy Rev 68:185–198
12. Yahya SM, Anwer SF, Sanghi S (2013) Enhance heat transfer and fluid flow in a channel behind photovoltaic panel in hybrid PV/T system. Indus Eng Chem Res 52:18413–18420
13. Gupta M, Singh V, Kumar S, Kumar S, Dilbaghi N, Said Z (2018) Up to date review on the synthesis and thermophysical properties of hybrid nanofluids. J Clean Prod 190(169):92
14. Ranga Babu JA, Kiran Kumar K, Srinivasa RS (2017) State-of-art review on hybrid nanofluids. Renew Sustain Energy Rev 77(551–56):5
15. Bellos E, Tzivanidis C (2018) A review of concentrating solar thermal collectors with and without nanofluids. J Therm Anal Calorim 1–24
16. Bellos E, Said Z, Tzivanidis C (2018) The use of nanofluids in solar concentrating technologies: a comprehensive review. J Clean Prod 196:84–99
17. Bellos E, Said Z, Tzivanidis C (2018) Thermal analysis of parabolic trough collector operating with mono and hybrid nanofluids. Renew Sustain Energy Rev 26:105–115
18. Minea AA, El-Maghlany WM (2018) Influence of hybrid nanofluids on the performance of parabolic trough collectors in solar thermal systems: recent findings and numerical comparison. Renew Energy 120:350–364
19. Azwadi CSN, Adamu IM, Jamil MM (2016) Preparation methods and thermal performance of hybrid nanofluids. J Adv Rev Sci Res. 24(1):13–23
20. Khan AA, Danish M, Rubaiee S, Yahya SM (2022) Insight into the investigation of Fe3O4/SiO2 nanoparticles suspended aqueous nanofluids in hybrid photovoltaic/thermal system. Clean Eng Technol 100572

21. Yu X et al (2018) Investigation on synthesis, stability, and thermal conductivity properties of water-based SnO2/reduced graphene oxide nanofluids. Materials 11(1):38
22. Chamsaard W (2019) Synthesis, Characterisation and thermo-physical properties of highly stable graphene oxide-based aqueous nanofluids for low-temperature direct absorption solar collectors and solar still desalination. Thesis of Doctor of Philosophy, Murdoch University
23. Chiam H et al (2017) Thermal conductivity and viscosity of Al2O3 nanofluids for different based ratio of water and ethylene glycol mixture. Exp Thermal Fluid Sci 81:420–429
24. Esfe MH et al (2017) Optimization, modeling and accurate prediction of thermal conductivity and dynamic viscosity of stabilized ethylene glycol and water mixture Al2O3 nanofluids by NSGA-II using ANN. Int Commun Heat Mass Transfer 82:154–160
25. Manimaran R et al (2014) Preparation and characterization of copper oxide nanofluid for heat transfer applications. Appl Nanosci 4(2):163–167
26. Aslfattahi N, et al (2019) Experimental investigation of thermal stability and enthalpy of eutectic alkali metal solar salt dispersed with MgO nanoparticles. 10(6):1112–1119
27. Esfe MH et al (2015) Experimental study on thermal conductivity of ethylene glycol based nanofluids containing Al2O3 nanoparticles. Int J Heat Mass Transf 88:728–734
28. Sati P, Shende RC, Ramaprabhu S (2018) An experimental study on thermal conductivity enhancement of DI water-EG based ZnO (CuO)/graphene wrapped carbon nanotubes nanofluids. Thermochim Acta 666:75–81
29. Sunil J et al (2019) The thermal properties of CaO-Nanofluids. Vacuum 161:383–388

Experimental Studies on Pool Boiling of Water-Based Al_2O_3, CuO and Hybrid Nanofluids

Mohammad Saalim, M. Altamush Siddiqui, and Suhail Ahmad Khan

Abstract Experiments on pool boiling of distilled water and nanofluids: Al_2O_3-water, CuO-water and Hybrid (Al_2O_3-water and CuO-water, mixed in equal volume) have been performed on a heated nichrome wire that is submerged in distilled water and the nanofluids. This study aims to find increase in the critical heat flux (CHF) due to hybrid nanofluids mixed in equal volumes of Al_2O_3-water and CuO-water nanoparticles. The volume concentrations of nanoparticles were varied from 0.01 to 0.05%. The heat flux to the test heater was impressed and the corresponding current and voltage supply were recorded using a data logger until the wire started becoming red hot and then broke which is referred to as the critical heat flux. Experiments show that there is an enhancement of critical heat flux (CHF) on varying the vol. %age concentration. It is found that the CHF is quite high in case of the hybrid nanofluids as compared to the base fluid that is distilled water and single particle nanofluids. Enhancement of CHF to around 91.06% in the hybrid nanofluids has been estimated when compared with the distilled water. The heat flux versus excess temperature has been plotted for different concentrations of the nanoparticles in the Al_2O_3-water, CuO-water and hybrid nanofluids (Al_2O_3 + CuO/water). The heat transfer coefficient (HTC) has also been shown for different concentrations of the nanoparticles in the three nanofluids under study with heat flux for the nucleate boiling region.

Keywords Pool boiling · Al_2O_3-water · CuO-water · Hybrid-water · CHF

1 Introduction

Boiling is a phenomenon in which liquid is heated up to its boiling point to convert into the vapour. It can be either pool boiling or flow boiling. The liquid in a pool boiling is still, and its movement close to the surface is mainly caused by natural convection and mixing caused due to the bubble growth and detachment. While in the flow boiling, fluid circulates inside a heated pipe or channel.

M. Saalim (✉) · M. A. Siddiqui · S. A. Khan
Department of Mechanical Engineering, ZHCET Aligarh Muslim University, Aligarh, India
e-mail: mohammadsaalim684@gmail.com

M. A. Siddiqui et al. (eds.), *Advances in Heat Transfer and Fluid Dynamics*, Lecture Notes in Mechanical Engineering, https://doi.org/10.1007/978-981-99-7213-5_18

Scientists and engineers are interested in investigating pool boiling because of the rising need for high energy dissipation in electrical devices and computing devices. Boiling heat transfer coefficient (HTC) and critical heat flux (CHF) are the two parameters that need critical examination on boiling heat transfer [1]. The heated surface when covered in the vapour in the film boiling region just away from the CHF delays the evacuation of heat produced and degrades heat transfer. CHF is an unwanted phenomenon because it causes the heat exchanger surface to burn as a result of the surface overheating. The performance of heat transfer systems will be poor due to the low thermal characteristics of common fluids like water, engineering oil, ethylene glycol (EG) and others. Numerous methods to improve the CHF have been researched [2]. Nanofluids have improved thermo-physical properties and heat transfer capability, making them viable heat transfer fluids that can be used in numerous devices to improve heat transfer performance. The solid nanoparticles (such as gold, silver and platinum) and metal oxides (such as alumina, zirconia, silica and titania) are dispersed into the traditional liquids like distilled water and ethylene glycol (EG) [14,15,16]. Addition of the nanoparticles increases the heat transfer rate particularly the critical heat flux (CHF) in the pool boiling case.

In the last few decades, the researchers have studied the effect of various nanoparticle materials, including silicon, aluminium and titanium oxides on CHF. You et al. [3–8] found the enhancement in CHF up to 200%, even at very low concentration (less than 1% by volume) of nanoparticles. They carried out experimental study on pool boiling of water-based alumina nanofluid over a square Cu-flat heater. They revealed that the CHF rises considerably (by around 200%) at low concentrations (less than 0.01 vol. % age) in comparison to pure water. Vassallo et al. [4] conducted experiments using the NiCr wire that was suspended horizontally in nanosolutions of silica. The maximum heat flux (CHF) was found to be nearly twice as much with the silica micro-solution as it was with pure water. Bang and Chang [9] applied Al_2O_3-water nanofluids in varying concentrations (0.4–0.5 vol %) to a horizontal flat, smooth surface. They ascertained that CHF increases about 32% for horizontal pool boiling and 13% for vertical pool boiling. Kim et al. [10] utilized the water-based nanofluids consisting of Al_2O_3, ZrO_2 and SiO_2 nanoparticles and stated that the nanoparticles' concentration and the wettability of the heating surface are the major factors that improve heat transfer. Further, Kim et al. [11] conducted pool boiling experiments and discovered an increase in CHF from 0.001 to 0.01%. Additionally, the CHF value did not increase when the concentration was increased from 0.01 to 0.1%. Shariq et al. [12] conducted pool boiling experiments and found the enhancement in CHF, that was around 38.5% at 0.03 vol% of silica particle concentration.

Fig. 1 XRD of alumina nanoparticles

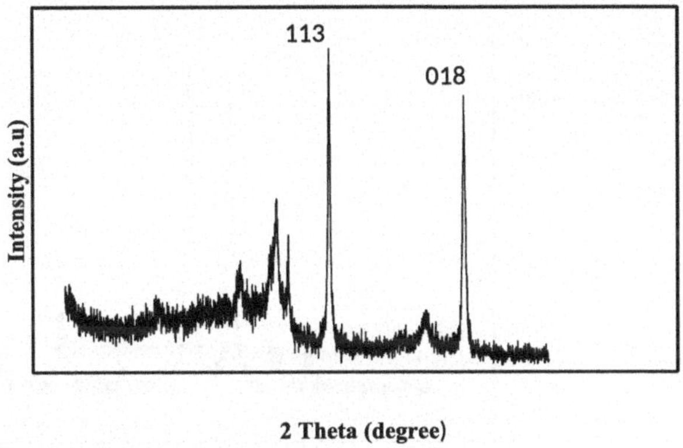

2 Characterization of Nanoparticles

2.1 X-Ray Diffraction

A very successful non-destructive method is the use of X-ray diffraction to characterize materials. It describes structures, phases, preferred crystal orientations (texture) and other structural characteristics like crystallinity, strain and crystal defects. Monochromatic X-ray beams are incident on the sample which reflects back depending upon topography structure of the materials. The XRD peaks are produced by particular angles from each set of lattice planes in a sample. As a result, the XRD pattern can be used to locate periodic atomic structures in a substance. The X-ray diffraction graph of the Al_2O_3 and CuO powders used in the experiment were recorded at room temperature with the help of BRUKER AXS D8 ADVANCE (Germany) using an X-ray beam with Cu Kα radiation. The principle of X-ray diffraction is the constructive interference of monochromatic X-rays with a material. The XRD graph, Intensity versus 2theta of Al_2O_3 and CuO are shown in Figs. 1 and 2. Peak values are observed at 2theta equal to 45.8 degree and 66.8 degree in case of Al_2O_3, with exquisitely index plane (113) and (018) respectively while in case of CuO they appear at around 35.6 and 38.8 degrees with exquisitely index plane (111) and (022) respectively which are completely in accordance with the International Centre for Diffraction Data (ICDD).

2.2 Scanning Electron Microscope (SEM)

A scanning electron microscope (SEM) is a class of electron microscope that takes pictures by moving a focused beam of electrons over the surface of a sample. When the electrons interact with the atoms in the sample, they produce different signals that tell us about the sample's topography and composition. SEM images are used to analyse

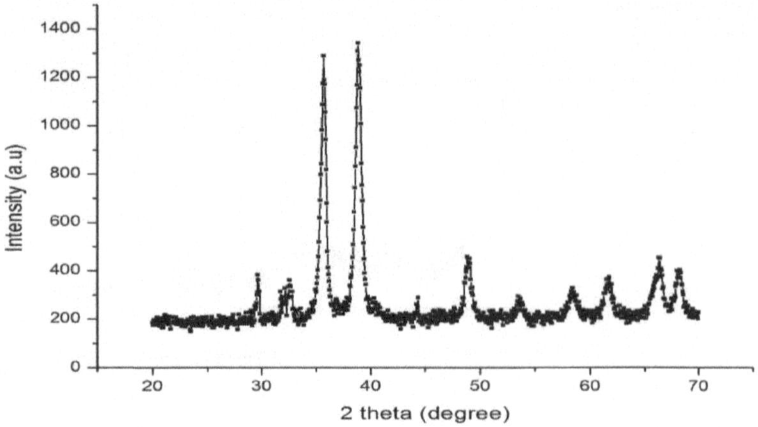

Fig. 2 XRD of CuO nanoparticles

the surface topography of the composite's worn-out surfaces as well as powder particles. This is also carried out at USIF AMU Aligarh. SEM was used to examine the nanostructure of aluminium oxide and copper oxide powders. Figures 3 and 4 depict SEM images of these powders. Figure 3 shows aluminium oxide nanopowder having spherical and some asymmetrical shapes with a broad size distribution. Figure 4 shows CuO nanopowder having mostly dendritic and some asymmetrical shapes with vast size distribution morphology.

3 Preparation of Nanofluids

In the present work, the regular nanofluids (Al_2O_3-water, CuO-water) and hybrid nanofluids (Al_2O_3 + CuO/water) are synthesized by using a two-step method. The known weight of dry particles of Al_2O_3, CuO is dispersed into distilled water. The weight of the nanoparticles for the desired volume concentrations (0.01–0.05%) are calculated using Eq. 1. To ensure the stability of the nanofluids for longer period of time, the sodium dodecyl benzene sulphonate (SDBS) and Polyvinyl chloride (PVC) surfactants (weight equal of 1/10th of weight of nanoparticles) are added to synthesize the Al_2O_3-water nanofluid and CuO-water nanofluid respectively. The solutions are then allowed to stir for around 30 min on a magnetic stirrer to ensure the proper mixing of solution. The solution is further agitated inside the ultrasonic bath for 60 min to break the agglomerates of particles to make the stable nanofluids.

Two-step procedures are also used to prepare samples of hybrid nanofluids [13]. The regular nanofluids (Al_2O_3-water, CuO-water) were prepared in equal volume separately for each concentration and then mixed. The mixed solution is then sonicated for 60 min to synthesize the water-based hybrid nanofluids to obtain desired

Fig. 3 SEM images of alumina nanoparticles

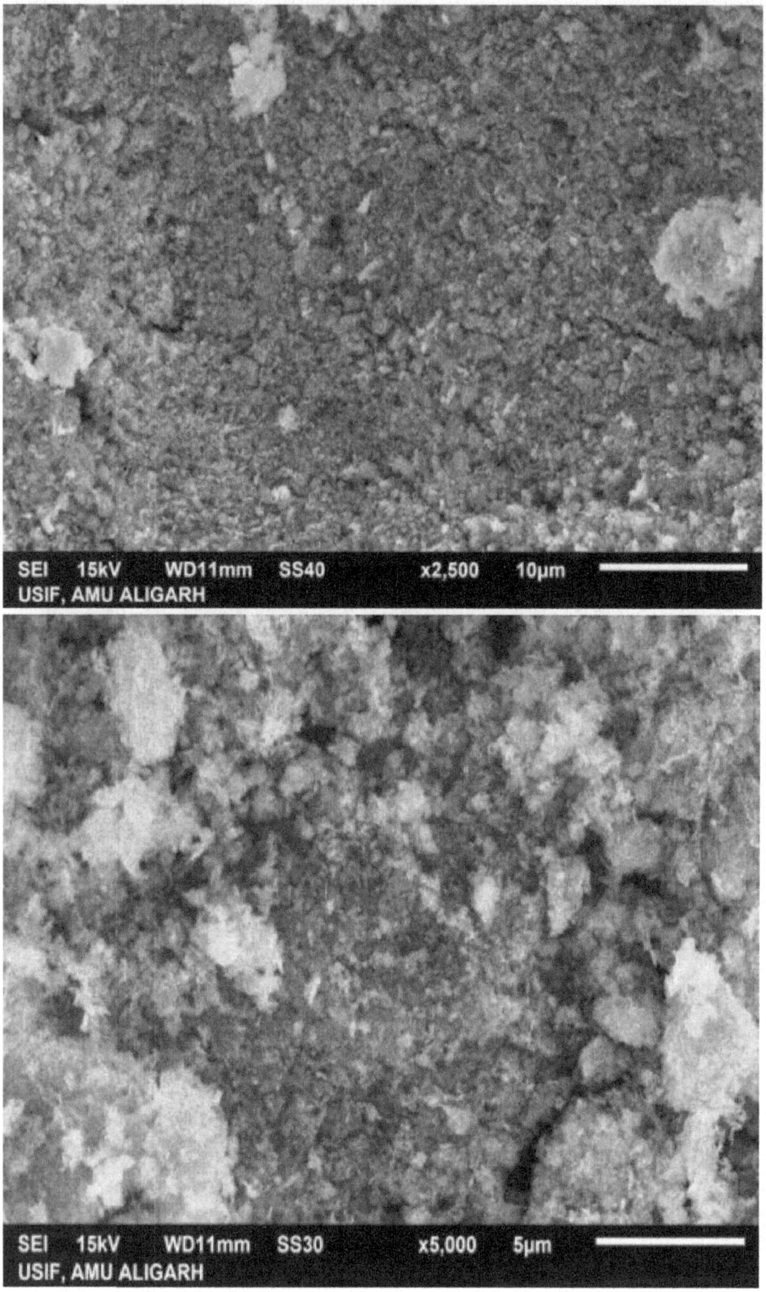

Fig. 4 SEM images of CuO nanoparticles

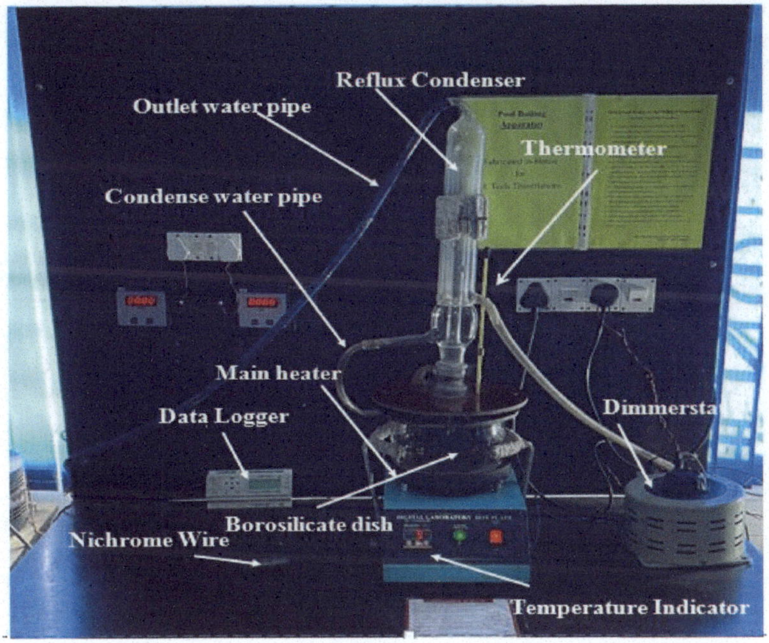

Fig. 5 Experimental setup

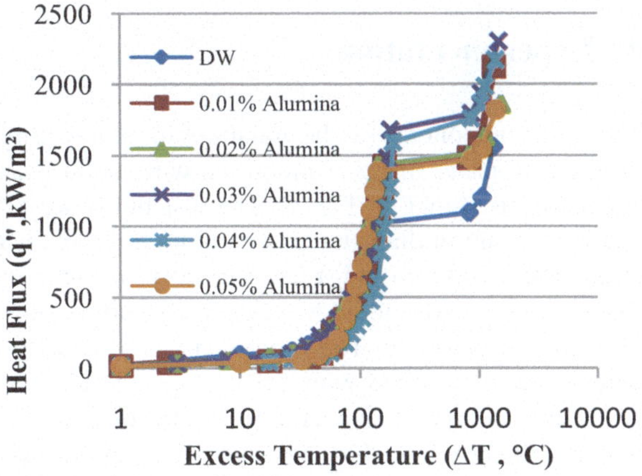

Fig. 6 Heat flux versus temperature plot for Alumina-DW

volume concentration of nanoparticles (0.01–0.05% volume concentration) as shown in Table 1.

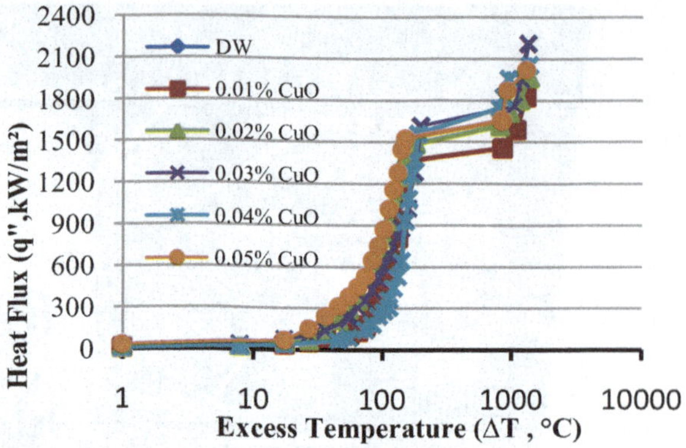

Fig. 7 Heat flux versus temperature plot for CuO-DW

Table 1 Volume fractions of the nanoparticles used to form hybrid nanofluid

(Ø Al_2O_3) % in distilled water	(Ø CuO) % in distilled water	Ø Hybrid = (ØAl_2O_3 + Ø CuO), in distilled water %
0.005	0.005	0.01
0.01	0.01	0.02
0.015	0.015	0.03
0.02	0.02	0.04
0.025	0.025	0.05

4 Experimentation

The main components for the apparatus of pool boiling consist of a borosilicate glass dish, a test heater made of nichrome wire, main heater, a data logger and a reflux condenser as shown in Fig. 5. The test fluids are filled into the borosilicate dish that is 180 mm in diameter and 110 mm high. A nichrome wire heater (0.112 mm in diameter, 100 mm long) fixed properly is submerged in the fluids under study. The nichrome wire should be completely submerged in distilled water or nanofluid of fixed concentration. The setup consists of an external plate heater on which the borosilicate dish, filled with 1.5 L of nanofluid, was placed. The fluid is initially heated to raise its bulk temperature. For the purpose of heating the liquid, the external heater is directly linked to the mains. Through firmly fastened copper lead studs, a nichrome wire serving as a test heater is horizontally submerged in the borosilicate dish. The heater measures 0.112 mm in diameter and 100 mm in length. A dimmerstat is used to control the voltage and current of nichrome wire heater. The melting point of nichrome wire is 1400 °C. Experiments are carried out by controlling the AC current through a dimmerstat. With the help of a data logger, the electrical energy input to the test heater is measured. The setup was created to examine the pool boiling up to the critical heat flux point. By gradually raising the applied voltage across the test heater, it is possible to regularly increase the heat flow from the wire. Both the bulk temperature of the water and the temperature of nanofluids were measured using calibrated thermometers. For maintaining the bulk temperature, the external heater can be turned on and off independently as needed. For the purpose of maintaining a consistent concentration, atmospheric pressure and liquid pool height, the experimental system is connected to a reflux condenser. Two concentric columns make up the reflux condenser; the outer column has cold water running through it. The jacket of water flowing outside cools and condenses the vapour as it rises through the inner column of the reflux condenser from the boiling liquid. After that, it returns to the borosilicate dish.

The experiment is repeated with the nichrome wire heater immersed in the borosilicate dish which was filled with 1.5 L of either distilled water, Al_2O_3-water, CuO-water or hybrid nanofluids (Al_2O_3-water and CuO-water, mixed in equal volume). The fluid inside the dish should every time completely cover the nichrome wire heater. The bulk heater is turned on until the liquid in dish reaches its saturation point then the bulk heater is switched off and test heater (nichrome wire) is switched on. By gradually increasing the power supply, the voltage and current readings are captured after 30 s at each point until the wire failure occurs during the saturated pool boiling experiments. The experiments are repeated for varying nanoparticle concentrations for different nanofluids.

5 Data Analysis

To determine the necessary volume concentration, the weight of nanoparticles is calculated using the following equation.

$$Volume\ Concentration\ (\phi),\ \% = \frac{\frac{W_{np}}{\rho_{np}}}{\frac{W_{np}}{\rho_{np}} + \frac{W_{bf}}{\rho_{bf}}} \tag{1}$$

The following relation is used to compute the heat flux (q″) through the wire.

$$Heat\ Flux\left(q''\right) = \frac{V \cdot I}{\pi \cdot D \cdot L} \tag{2}$$

Determine the heat transfer coefficient (HTC) by using following relation:

$$Heat\ Transfer\ Coefficient\ (h) = \frac{q''}{Tw - Tsat} \tag{3}$$

6 Results and Discussion

6.1 Heat Flux Versus Excess Temperature

The pool boiling experiment is done with Al_2O_3, CuO, and hybrid nanofluids at volume concentrations of 0.01, 0.02, 0.03, 0.04, and 0.05%. Heat flux for pool boiling is calculated using Eq. (2). Heat flux increases with wall temperatures in all the cases shown in Figs. 6, 7 and 8. The CHF increases when nanoparticles are added to distilled water. The CHF of hybrid nanofluids is higher than Al_2O_3 and CuO nanofluids. Compared to distilled water, the critical heat flux of hybrid nanofluids is 91.06% higher at 0.03% volume concentrations. The increased critical heat flux is due to the presence of two different nanoparticles in base fluids resulting in hybrid nanofluids having a higher thermal diffusivity than single nanofluid, which improves wettability. The increased critical heat flux is also caused by the random movement of nanoparticles in the fluid and more surface area. The CHF of hybrid nanofluids goes up as the volume concentration increases. Increasing volume concentrations up to 0.04% and 0.05% are used to estimate the CHF at such concentrations.

In pool boiling, the bubble formation rate and its departure from the test surface have a significant effect on heat transmission. The variation between the boiling curves of distilled water and nanofluids at various concentrations is clearly shown in Figs. 6, 7 and 8. Bubbles form over the heated surface during nucleate boiling heat transfer to the fluid. On the surface, vapour bubbles start to form, develop, and finally detach. When heat fluxes are higher for water, small bubbles form and cover the heating surface. When nanoparticles are added to the fluid, nanoparticles accumulate on the test surface. The capillary action of nanofluids on the heater surface is increased by the nanoparticle deposition, which has an influence on bubble dynamics. Figures 9, 10 and 11 show log–log plot of q" versus ΔT for different concentrations of the nanoparticles. They represent linear relation between ln(q") and ln(ΔT) up to the nucleate boiling region. Due to the capillarity action in the nanoporous

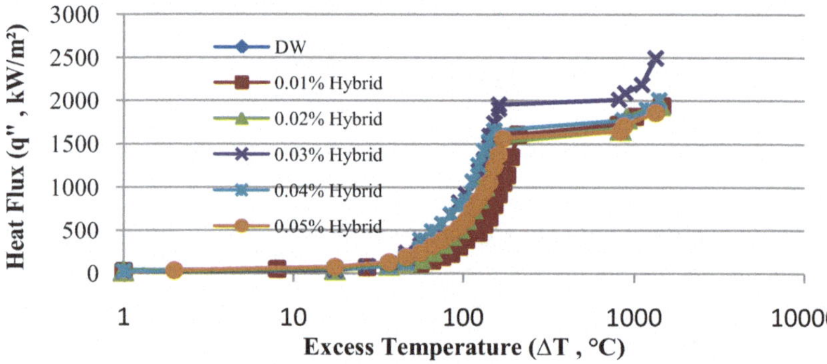

Fig. 8 Heat flux versus temperature plot for Hybrid (Al_2O_3 + CuO/DW) nanofluids

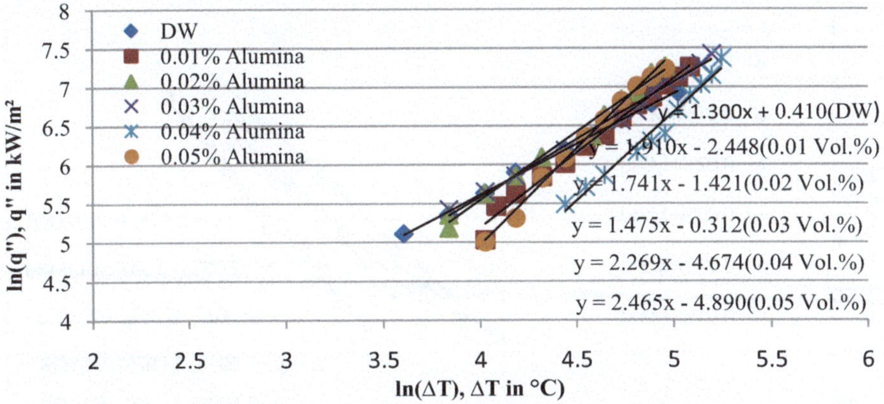

Fig. 9 Heat flux versus excess temperature (log–log plot) for the nucleate boiling region (Alumina-DW)

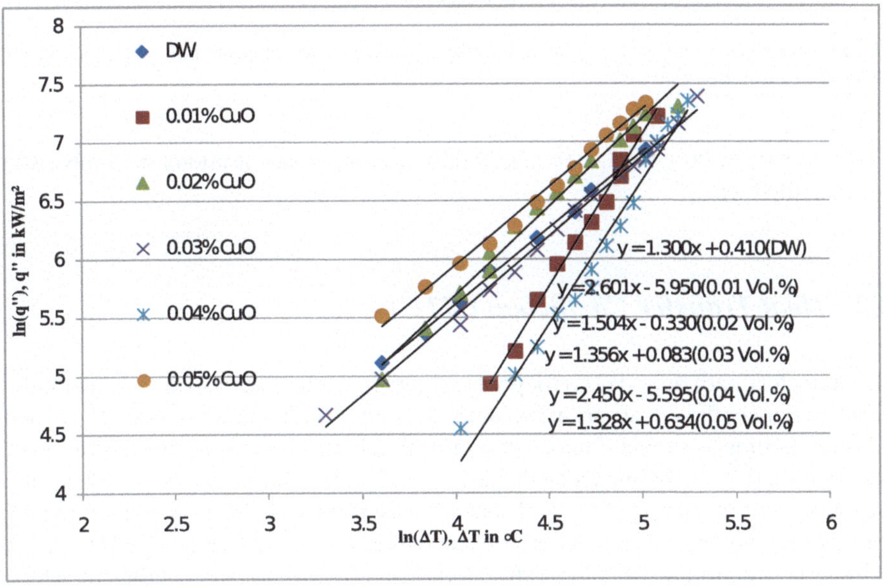

Fig. 10 Heat flux versus excess temperature (log–log plot) for the nucleate boiling region (CuO-DW)

layer of the test surface, the heat flux decreases as volume concentration increases. Nanofluids' boiling curves approach the water curve as volume concentration rises. When volume% concentration increases, CHF is delayed by the surface rewetting effect. The maximum CHF is acquired with a concentration of 0.03 volume percent. The enhancement in CHF of hybrid nanofluids is 57.45, 49.36, 91.06, 61.87 and

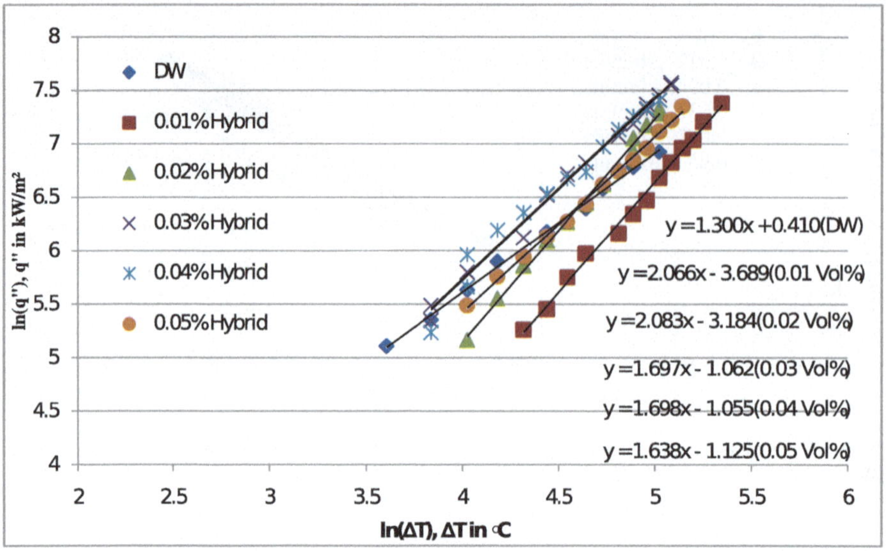

Fig. 11 Heat flux versus excess temperature (log–log plot) for the nucleate boiling region (Al_2O_3 +CuO/DW) nanofluids

54.47% at 0.01, 002, 0.03, 0.04 and 0.05% volume concentrations in comparison with distilled water.

6.2 *Heat Transfer Coefficient (HTC)*

Heat transfer coefficient (HTC) for pool boiling is calculated using Eq. 3. Nanofluids' heat transfer coefficient at various volume concentrations is shown in Figs. 12, 13 and 14. Pool boiling with nanofluids has significant challenges due to the deposition of nanoparticles. It has an impact on the testing surface roughness and bubble departure frequency. When nanoparticles settle on the test surface, they form nanolayers, which increases the number of nucleation sites and bubble formation frequency, HTC is high at lower concentrations, and the wall superheats. The heat transfer coefficient, which is influenced by the fluid's properties, fluid dynamics, and heating surface geometry, governs convective heat transfer. In comparison to single phase heat transfer, boiling involves a higher heat transfer coefficient. The variation in the heat transfer coefficient during the pool boiling in both pure water and nanofluid is depicted in Figs. 12, 13 and 14. As the boiling proceeds, the heat transfer coefficient reduces quickly as the boiling process progresses in the free convection zone and then increases gradually in the nucleate boiling region up to the CHF (peak value), then dramatically drops to a very low level in the film boiling region. In comparison to distilled water, the nanofluid has a higher heat transfer coefficient. Similar scenarios apply to nanofluids having

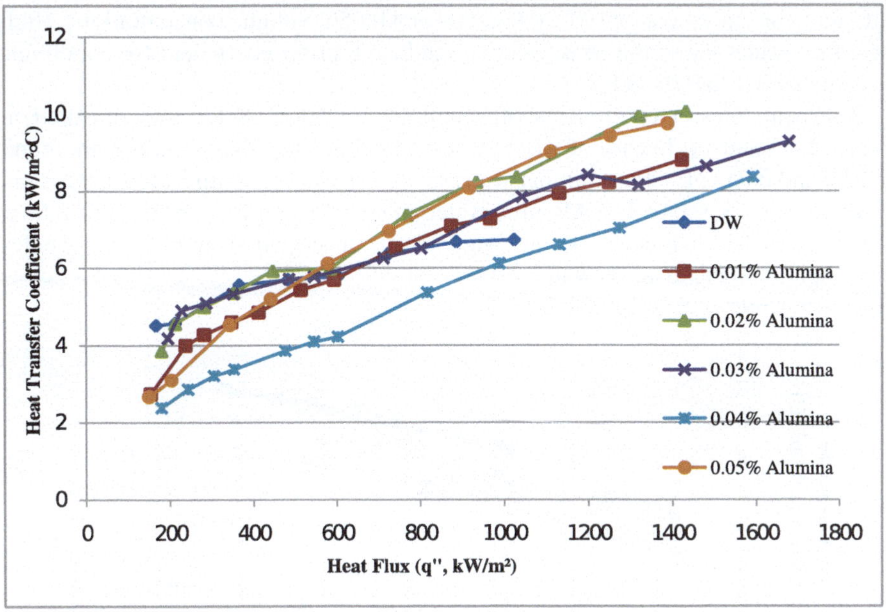

Fig. 12 Heat transfer coefficient versus heat flux for the nucleate boiling region (Alumina+DW)

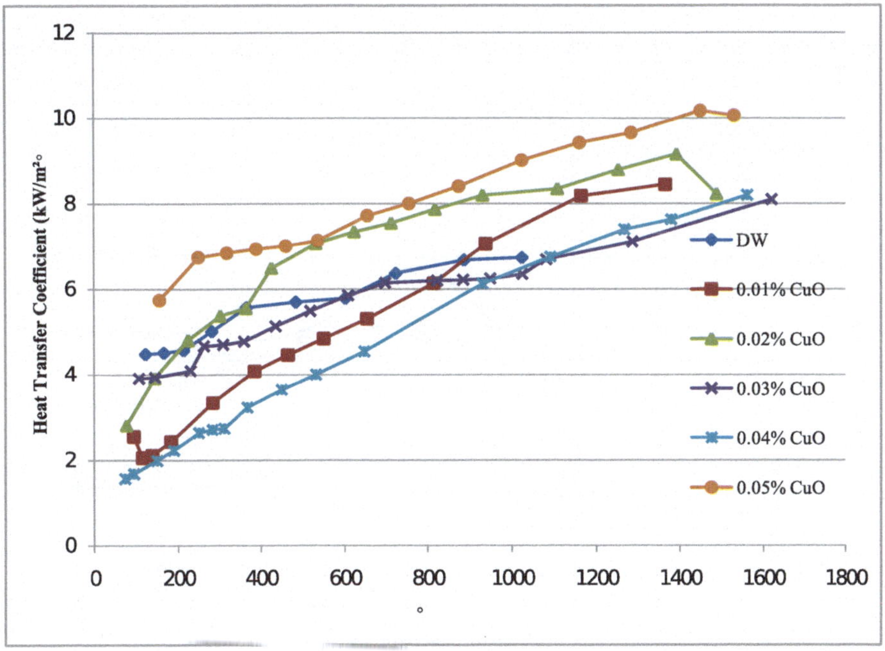

Fig. 13 Heat transfer coefficient versus heat flux for the nucleate boiling region (CuO+DW)

different concentrations of 0.02, 0.03, 0.04 and 0.05% volume concentrations. High heat flux values cause the peak value of the heat transfer coefficient for nanofluids to contain 0.01 to 0.05 vol %.

The heat transfer coefficients of nanofluids and pure water are compared in Figs. 12, 13 and 14. In case of Al_2O_3-water nanofluids the highest values are found at 0.02 and 0.05 vol.% for Alumina-water as well as CuO nanofluids. It is lower than distilled water at 0.04vol.% Alumina-water and CuO nanofluids. At high heat flux, the heat transfer coefficients become high for all concentrations of the nanoparticles as compared to the distilled water.

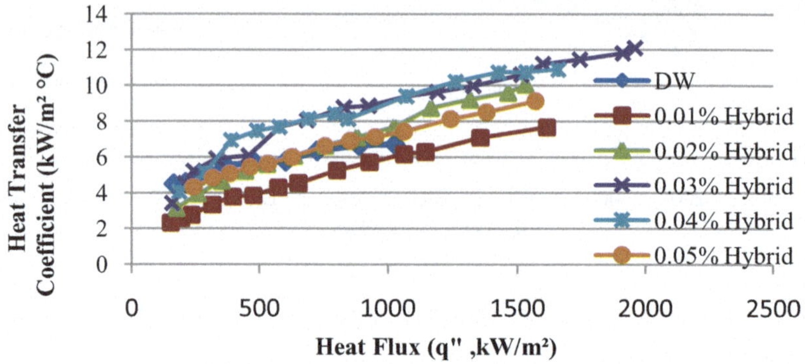

Fig. 14 Heat transfer coefficient versus heat flux for the nucleate boiling region (Al_2O_3+CuO/DW) nanofluids

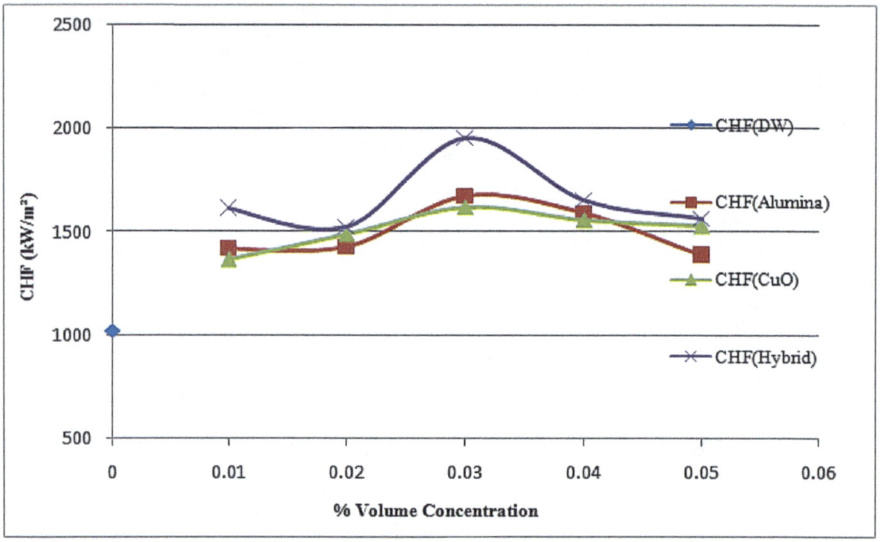

Fig. 15 Critical heat flux versus %volume concentration of nanofluids

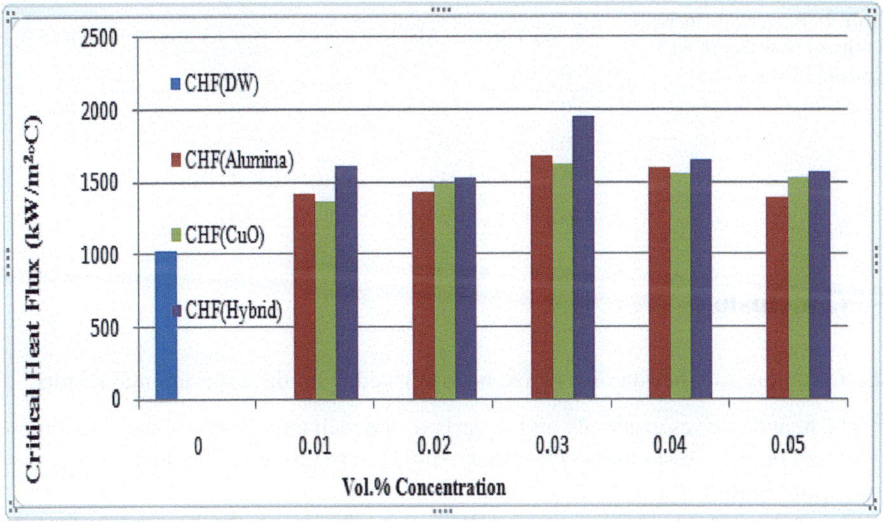

Fig. 16 Critical heat flux versus vol.% of nanofluids

For the case of Hybrid (Al_2O_3-water + CuO-water) nanofluids, the highest value of heat transfer coefficient is found for 0.03 and 0.04 Vol. % concentrations, while it is lowest at 0.01 vol. % concentration. For 0.02 and 0.05 Vol. % concentrations, the HTC is near to the distilled water at low heat flux, increasing subsequently at high heat flux; lowest for 0.01 Vol. % concentrations.

6.3 Critical Heat Flux Versus Concentration

The critical heat flux obtained for different nanofluids at various concentrations is shown in Fig. 15. They increase with concentration, reach to a maximum at around 0.03 vol. % concentration and then decrease. The hybrid nanofluid shows very high increase in the CHF, while the Al_2O_3 and CuO have very close values. The same result has been presented in the form of bars as shown in Fig. 16, which are also showing increase in CHF at 0.03 Vol. % concentration. The percentage increase in CHF of the nanofluid at a given concentration from that of the distilled water is tabulated in Table 2. The hybrid nanofluid shows considerable rise in CHF with around 91% at 0.03 vol.% concentration.

Table 2 Percentage increase in critical heat flux (CHF) from distilled water

% Vol. conc.	(Al_2O_3-water) %	(CuO-water) %	Hybrid %
0.01	38.96	33.36	57.45
0.02	39.90	45.33	49.36
0.03	63.85	58.22	91.06
0.04	55.52	52.51	61.87
0.05	35.49	49.47	54.47

7 Conclusion

The following salient outcomes have been derived from the experimental results:

- The heat flux rises gradually and linearly as the wall temperature rises. The superheated liquid is responsible for transferring heat in this area of natural convection or quiet boiling.
- In comparison to distilled water (DW), the bubbles on the heating surface of nanofluids have a lower contact angle, which results in increased wetting of the heating surface.
- The low thermal conductivity of the vapour limits heat transfer between the wire and the liquid, raising the wire's temperature. The critical heat flux is the heat flux at this point.
- The wire turns red hot when the critical heat flux is reached. As the heat flux increases, the red hot length increases. The wire's red hot conditions signify the radiation heat transmission, which rises with temperature.
- In comparison to distilled water, nano-fluids are found to have greater heat transfer coefficients at the critical heat flux. In the case of nanofluids, the greater heat transfer coefficients occur at high heat flux values.
- In pool boiling, CHF is higher in hybrid nanofluids than in single-type nanofluids. When compared to distilled water, the boiling curves of hybrid nanofluids shifted to the left. As concentrations increased, the curves shifted to the right because of the particle deposition, resulting in increasing capillarity action.
- The enhancement of CHF in case of hybrid nanofluids is 57.45, 49.36, 91.06, 61.87 and 54.47% at volume concentrations of 0.01, 0.02, 0.03, 0.04 and 0.05% compared with distilled water.
- The enhancement in CHF of nanofluid may be due to the nanoparticles deposition over the heated surface, which increases the nucleation site, surface roughness, and consequently wettability, which delays burn out of the wire.
- Due to their high Brownian motion, hybrid nanofluids have superior thermal conductivity than single type nanofluids. Due to their particular surface area, smaller nanoparticles (Al_2O_3, CuO) have stronger thermal conduction. These two types of different nanoparticles, when present in base fluids, facilitate the transfer of thermal energy from heated source to the surrounding fluids. So, it is possible to think of using various nanoparticles in base fluids to have superior thermo physical properties and heat transfer rates.

Nomenclature

D	Diameter of nichrome wire (mm)
L	Length of nichrome wire (mm)
V	Voltage (V)
I	Current (A)
h	Heat transfer coefficient ($kW/m^2 \circ C$)
K	Thermal conductivity (W/mK)
q"	Heat flux (kW/m^2)
W_{np}	Weight of nanoparticles (gram)
W_{bf}	Weight of base fluid
ρ_{np}	Density of nanoparticles (kg/m^3)
ρ_{bf}	Density of base fluids (kg/m^3)
T_w	Wire temperature (∘C)
T_{sat}	Saturation temperature of base fluid
ΔT ($T_w - T_{sat}$)	Excess temperature (∘C)
CHF	Critical heat flux (kW/m^2)
DW	Distilled water
HTC	Heat transfer coefficient ($kW/m^2 {}^\circ C$)
XRD	X-ray diffraction
SEM	Scanning electron microscope

References

1. Fang X, Chen Y, Zhang H, ChenW, Dong A,Wang R (2016) Heat transfer and critical heat flux of nanofluid boiling: a comprehensive review. Renew Sustain Energy Rev
2. Li W, Dai R, Zeng M, Wang Q (2020) Review of two types of surface modification on pool boiling enhancement: passive and active. Renew Sustain Energy Rev
3. You SM, Kim J, Kim KH (2003) Effect of nanoparticles on critical heat flux of water in pool boiling heat transfer. Appl Phys Lett 83(16):3374–3376
4. Vassallo P, Kumar R, D'Amico S (2004) Pool boiling heat transfer experiments in silica-water nano-fluids. Int J Heat Mass Transfer 47:407–411
5. Bang IC, Chang SH (2005) Boiling heat transfer performance and phenomena of Al_2O_3-water nano-fluids from a plain surface in a pool. Int J Heat Mass Transfer 48:2407–2419
6. Milanova D, Kumar R (2005) Role of ions in pool boiling heat transfer of pure and silica Nanofluids. Appl Phys Lett 87:233107
7. Kim H, Kim J, Kim M (2006) Experimental study on CHF characteristics of water-TiO_2 nano-fluids. Nucl Eng Technol 38(1)
8. Kim SJ, Truong B, Buongiorno J, Hu LW, Bang IC (2006) Study oftwo-phase heat transfer in nanofluids for nuclear applications. In: Proc.ICAPP'06, USA, paper6005
9. Kim H, Kim J, Kim MH (2006) Effect of nanoparticles on CHF enhancement in pool boiling of nano-fluids. Int J HeatMass Transf.
10. Kim SJ, Bang IC, Buongiorno J, Hu LW (2007) Study of pool boiling and critical heat flux enhancement in nanofluids. Bull Polish Acad Sci Techn Sci

11. Zafar S, Adil M, Azhar M et al (2021) Experimental and numerical study of pool boiling and critical heat flux enhancement using water based silica nanofluids. Heat Mass Transfer 57:1593–1607
12. Sarafraz MM, Hormozi F, Silakhori M, Peyghambarzadeh SM (2016) On the fouling formation of functionalized and non-functionalized carbon nano tube nano-fluids under pool boiling condition. Appl Therm Eng
13. Ahmad Khan S, Altamush Siddiqui M (2020) Numerical studies on heat and fluid flow of nanofluid in a partially heated vertical annulus. Heat Transfer 49(3):1458–1490
14. Husain S, Khan SA, Siddiqui MA (2021) Wall boiling of Al2O3-water nanofluid: Effect of nanoparticle concentration. Prog Nucl Energy 133:103614
15. Usmani R, et al (2021) Numerical investigation on natural convection of hybrid nanofluid Al2O3–MWCNT/water inside a vertical annulus. In: IOP conference series: materials science and engineering, vol. 1146(1). IOP Publishing
16. Khan SA, et al (2022) Numerical investigation and implementation of the Taguchi based entropy-ROV method for optimization of the operating and geometrical parameters during natural convection of hybrid nanofluid in annuli. Int J Therm Sci 172:107317

Numerical Analysis of an HVAC System Incorporated with Latent Heat Energy Storage System

Rahul Kumar Sharma, Ashish Kumar, and Dibakar Rakshit

Abstract Owing to India's climatic conditions, India's energy outlook has predicted a three-fold increase in ownership and growth in energy consumed for cooling to 650 TWh by 2040. Further, due to airborne viruses like COVID-19, the health and air-conditioning societies modified guidelines and suggested incorporating air from the ambient to dilute the indoor air and reduce contaminant levels. However, mixing ambient air with recirculating air results in an additional cooling load in the heating, ventilation, and air-conditioning system (HVAC). The study aims to design a shell and tube heat exchanger to reduce the HVAC load by cooling the fresh air from ambient temperature and subsequently mixing it with the recirculated air. The fresh air at 315 K flows through the heat exchanger with calcium chloride hexahydrate as a phase change material (PCM). Further, two configurations of heat exchangers are examined. In one, air flows through the inner tube, and PCM is in the annulus, while in the second, the air flows in the annulus, and the PCM is in six radially equidistant tubes. The designed heat exchanger with PCM in annulus and PCM in tubes led to an average energy saving of 3 and 2.7%, respectively, for 3600 s of operation.

1 Introduction

The building sector has drawn special attention worldwide as it is responsible for up to 40% of the total energy consumption and contributes to global greenhouse gas emissions [1]. In addition, due to climate change, the energy required to cool indoor

R. K. Sharma · A. Kumar · D. Rakshit (✉)
Indian Institute of Technology Delhi, Hauz Khas, New Delhi, India
e-mail: dibakar@iitd.ac.in

R. K. Sharma
e-mail: esz198528@iitd.ac.in

A. Kumar
e-mail: ashish.k@galgotiasuniversity.edu.in

A. Kumar
Department of Mechanical Engineering, Galgotias University, Greater Noida, India

M. A. Siddiqui et al. (eds.), *Advances in Heat Transfer and Fluid Dynamics*, Lecture Notes in Mechanical Engineering, https://doi.org/10.1007/978-981-99-7213-5_19

spaces is further estimated to increase by three times by 2040 [2]. Furthermore, for a developing nation like India, with an increase in purchasing power and adverse summers, the number of air-conditioning (AC) units is predicted to rise to 1.6 per household from the present level of 0.8 units per household. The increase in units will increase the electrical energy consumption to 650 TWh by 2040 [3]. Hence, enhanced energy-efficient air conditioners, more efficient cooling systems, thermally efficient buildings, and cool roofs [4] are required to mitigate the increased energy demand.

The phase change material (PCM) absorbs energy at nearly constant temperature and has a high latent energy storage capacity [5]. PCM absorbs thermal energy when the operating temperature is above its melting temperature and releases energy below the melting temperature. These properties make them suitable for solar thermal applications [6], the agricultural industry [7], photovoltaic thermal systems [8], and enhancing energy performance and thermal comfort in buildings [9]. Since PCM increases the thermal mass of a building envelope, it can be used to insulate the building envelope and manage a building's temperature. The application of PCM to ceiling panels effectively reduces energy consumption through active cooling systems [10]. PCM used in brick walls reduced 40–60% of heat transfer for the PCM wall under summer conditions [11, 12]. Window glazing with PCM has effectively reduced the cooling load [13]. PCM-air heat exchangers (PAHX) can provide an efficient and effective solution to the problem of ambient air at a higher temperature. Yanbing et al. [14] combined the PAHX system with a night ventilation strategy for the moderate climate of Beijing and achieved a considerable enhancement in indoor thermal performance. Osterman et al. [15] studied system thermal performance for free cooling and heating in Slovenia and observed its heating and cooling potential for different weather conditions.

The COVID-19 outbreak had an impact on both human life and in order to reduce the contamination levels in enclosed spaces, the Indian Society of Heating, Refrigerating and Air-Conditioning Engineers (ISHRAE), American Society of Heating, Refrigerating and Air-Conditioning Engineers (ASHRAE), and health societies have suggested the incorporation of fresh air in indoor spaces [16, 17]. During the summer, the ambient temperature in Delhi may reach 315 K, as per the observation from the weather station at the Indian Institute of Technology Delhi. In conventional air-conditioning systems, the air from the room is recirculated in the air-conditioning system. However, as seen in Fig. 1, when fresh air is added, some of the total air is replaced by the ambient air at ambient conditions. As a result, the energy use of the AC unit and, consequently, the energy used by the building will increase.

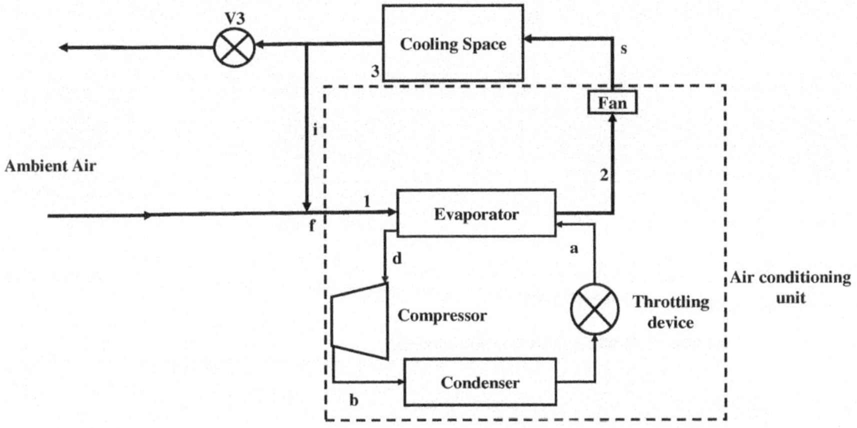

Fig. 1 Schematic layout of a conventional air-conditioning system with fresh air incorporation

The existing studies have suggested the PCM capabilities to enhance buildings' energy efficiency via different designs and configurations. However, its application with the HVAC system in buildings is not being explored. The reports by various agencies indicate a steep rise in HVAC systems in the coming decades. The present numerical study uses a PCM-based shell and tube heat exchanger to make an AC system working on vapour compression refrigeration cycle (VCRS) with fresh air incorporation more energy efficient. The heat exchanger is used to cool the fraction of ambient air before mixing it with the recirculation air, reducing the AC unit's cooling load and making it more energy efficient. Further, the effect of incorporating PCM in different configurations is analyzed in the study.

2 Physical Domain

In the present work, a tube in tube and shell and tube type latent heat energy storage system (LHESS) is numerically studied for cooling ambient air supplied to an air-conditioning system. The length of the LHESS is 1500 mm, while two heat exchanger configurations i.e., tube in tube and shell and tube, with different positioning of PCM are considered for a comparative analysis. In the tube in tube arrangement, air flows through the inner pipe of diameter 200 mm, and the PCM is placed in the annulus of thickness 20 mm, as shown in Fig. 2a. In the shell and tube arrangement, the air flows through the annulus with an inner diameter of 200 mm, while the PCM is placed in six radially equidistant internal tubes of an inner radius of 38 mm, as shown in Fig. 2b. The PCM considered for the study is calcium chloride hexahydrate, having a melting temperature of 303 K.

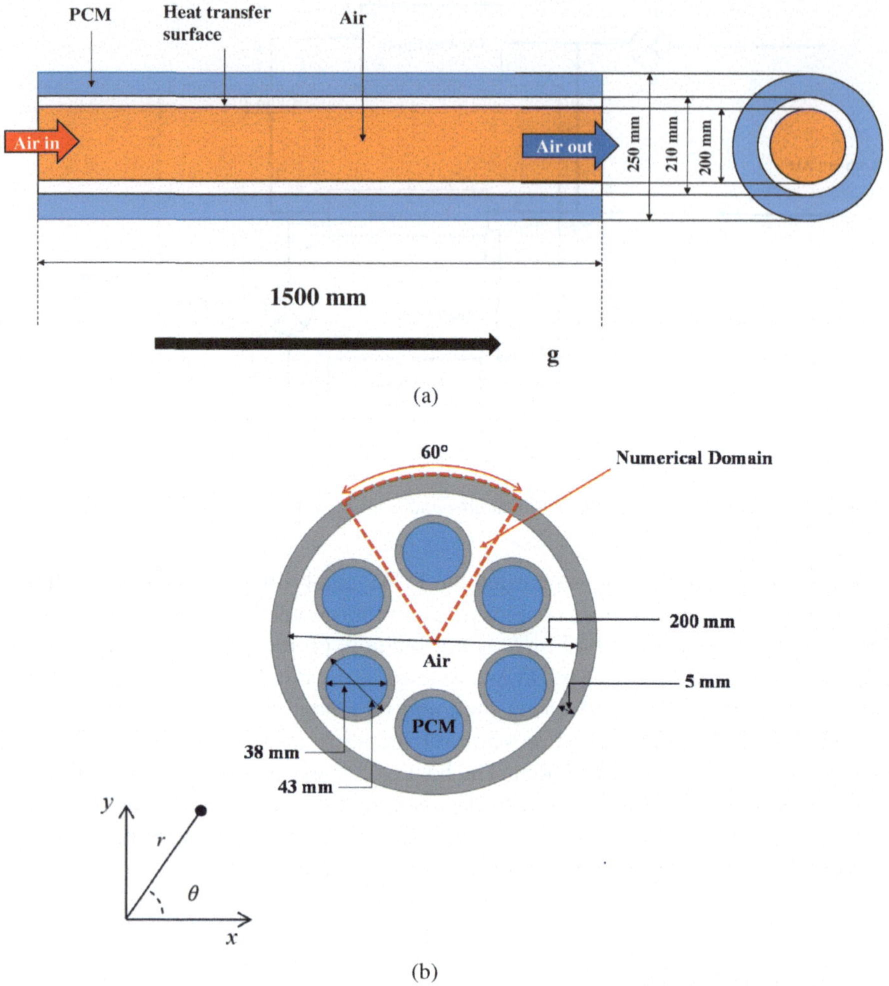

Fig. 2 **a** Schematic of PCM-based tube in tube heat exchanger with PCM in the annulus **b** Physical domain of PCM-based shell and tube heat exchanger with PCM in the inner tube

To reduce the computing time and effort, the one-sixth axisymmetric three-dimensional domain of the LHESS is considered for analysis by imposing symmetrical boundary conditions, as seen in Fig. 2b. The axisymmetric conditions considered will imitate the heat transfer and flow phenomena of the whole heat exchanger. The thermophysical characteristics of calcium chloride hexahydrate, air, and aluminium are provided in Table 1. After the air passes through the exchanger, it is mixed with a component of recirculated air and sent to the air-conditioning unit. The air-conditioning system uses a vapour compression refrigeration cycle with a compression ratio of 5 to keep the working environment at 27 °C.

Table 1 Thermophysical properties of PCM, air and aluminium [18]

Properties	$CaCl_2.6H_2O$ [19]	Air	Aluminium
C_p (kJ/kg.K)	2.230	1.006	8.710
ρ (kg/m^3)	1538	1.225	2719
k (W/m/K)	0.546	0.242	202.4
μ (kg/m.s)	0.1	1.789×10^{-5}	–
Latent heat (kJ/kg)	170	–	–
β (K^{-1})	0.0005	–	–
T_m (K)	303	–	–

3 Mathematical Modelling

In the LHESS based heat exchanger, the transient heat transfer and fluid flow are solved with a three-dimensional axisymmetric numerical model. PCM's phase change behaviour is shown using an enthalpy porosity method based on a fixed grid. For the development of the numerical model, a few assumptions are made: (a) the phase change of the PCM takes place at a constant temperature; (b) the impacts of expansion and contraction during the solidification and melting of PCM are not taken into account; and (c) the thermophysical properties of PCM, air and pipe material are isotropic and temperature-invariant, besides the density of PCM which varies in the direction of gravity (z-direction). The continuity, momentum and energy conservation equations in the vector form are as follows [5, 18]:

Continuity equation:

$$\frac{\partial(\rho_i)}{\partial t} + \nabla.(\rho_i V_i) = 0 \tag{1}$$

Momentum equation:

$$\frac{\partial(\rho_i V_i)}{\partial t} + V_i \nabla.(\rho_i V_i) = -\nabla P + \nabla^2[\mu V_i] + A V_i + S_z \tag{2}$$

Energy equation:

$$\frac{\partial(\rho_i c_{p,i} V_i)}{\partial t} + V_i \nabla.\left(\rho_i c_{p,i} V_i\right) = \nabla^2[k_i T_i] + S_{TP} \tag{3}$$

The source term in the momentum equation for PCM represents the Boussinesq approximation in the z-direction and considers no change in thermophysical properties other than the fluid's density and can be expressed as.

$$S = -A\frac{(1-\beta)^2}{\delta^3 + b} \tag{4}$$

where β is the thermal expansion coefficient, and T_m is the melting temperature of PCM, while parameter A is the mushy zone parameter. The source term in Eq. 3 is known as nodal latent heat revolution with time and is a function of temperature [5]. The source term can be represented as [20]:

$$S_{TP} = -\frac{\partial(\rho\Delta E)}{dt} \tag{5}$$

where ΔE represents the nodal heat value content, and its value varies from zero to latent heat of fusion.

Initial and boundary conditions

In the present study, the initial temperature condition is considered slightly less than the melting point of the PCM to minimize the sensible heating of PCM before melting. The initial temperature condition mathematically can be represented as:

$$T(r, \theta, z, 0) = 315K \tag{6}$$

The boundary conditions to solve the mass, momentum and energy conservation equation for air and PCM zone are.

(a) For the HTF(air) domain:

(i) No slip condition at the wall of the tube. i.e. u = 0.
(ii) At inlet tube, z = 0, $u_z = u_{in}$ and $T_z = 315$ K.
(iii) At the outlet tube, p = 0.

(b) For the PCM domain:

(i) At the wall of the shell tube. i.e. u = 0
(ii) $\left.\frac{\partial T}{\partial n}\right|_{shellsidewall} = 0$
(iii) At the inner wall of the shell, $\left.\frac{\partial T}{\partial r}\right|_{r=\frac{D_i}{2}} = 0$
(iv) At θ = -−30° and θ = 30°, symmetric boundary condition is imposed. i.e., $\left.\frac{\partial \phi}{\partial n}\right|_{sym} = 0$ and $\phi.\vec{n} = 0$

Φ represents u or T, and n is the unit vector normal to the surface.

4 Solution Procedure

The governing Eqs. (1), (2), and (3) are iteratively solved using the Finite Volume Method (FVM), and the SIMPLE scheme is used for pressure–velocity coupling. The spatial discretization of momentum and energy equations is carried out using power law. Ansys Fluent V2021 is a commercially available software to solve governing equations with suitable initial and boundary conditions. The convergence criteria for continuity, momentum and energy equation are set to 10^{-3} and 10^{-6}, respectively.

The energy in the LHESS is stored in the form of latent heat and sensible heat, and their estimation is based on the average PCM temperature and the liquid fraction of PCM. Equations 8 and 9 can be used to calculate the total sensible heat and latent heat during the energy storage process [21]:

$$Q_{sensible} = \sum\nolimits_{t=0}^{t} (T^{t}{}_{avg,p} - T^{t-1}{}_{avg,p}) \tag{7}$$

$$Q_{latent} = \sum\nolimits_{t=0}^{t} m L_{latent} (\epsilon^{t} - \epsilon^{t-1}) \tag{8}$$

where m, t, t_{avg}, p and ϵ denote the mass of PCM in the heat exchanger, the total time of operation, the average PCM temperature and the liquid fraction of PCM, respectively.

5 Energy Consumption by Air-Conditioning System

The conventional AC system with fresh air incorporation is illustrated in Fig. 1. The AC operates on a vapour compression and refrigeration (VCR) system. The temperature of the air inside the office space is managed as per the ISHRAE standards, i.e., 27 °C DBT and relative humidity of 50% [16, 22, 23]. The fresh air and the recirculated air mix adiabatically before entering the AC unit [23].

The total sensible heat extracted from the space (Qs) is given as:

$$Q_s = \dot{m}_a (1.005 + 1.88\omega)(T_{mix} - T_s) \tag{9}$$

Here $\dot{m}_a$ is the total mass of air and calculated as:

$$\dot{m}_a = \dot{m}_{ai} + \dot{m}_{ao} \tag{10}$$

The present study supplies a 0.42 kg/s mass to cool the office space.
The enthalpy (h_{mix}) and temperature (T_{mix}) of the mixture of air is obtained by:

$$h_{mix} = \frac{\dot{m}_{ai} h_i + \dot{m}_{ao} h_{out}}{\dot{m}_{ai} + \dot{m}_{ao}} \tag{11}$$

$$T_{mix} = \frac{\dot{m}_{ai} T_i + \dot{m}_{ao} T_o}{\dot{m}_{ai} + \dot{m}_{ao}} \tag{12}$$

The load on the chiller ($Q_{chiller}$) is dependent on the temperature of mixed air, and the supply air temperature is estimated as:

$$Q_{chiller} = \dot{m}_a (h_1 - h_s) \tag{13}$$

$Q_{chiller}$ then determines the mass of the refrigerant required to achieve the desired refrigeration effect.

$$\dot{m}_{ref} = \frac{Q_{chiller}}{(h_a - h_d)} \tag{14}$$

Since in a VCR system, the compressor is the primary energy-consuming unit, so for the study, we have assumed the total energy consumed by the AC system corresponds to the energy consumed by the compressor in the system. The work required by the compressor is given by:

$$W_{comp} = \dot{m}_{ref} (h_{bs} - h_d) \tag{15}$$

The efficiency of a compressor depends on the operating pressures [8], hence calculating the work required by the AC system:

$$\eta_{comp} = \frac{h_{bs} - h_d}{h_b - h_d} = 0.85 - 0.046667 \frac{p_2}{p_1} \tag{16}$$

$$W_{ac} = W_{comp} / \eta_{comp} \tag{17}$$

6 Model Validation

In order to prove that the developed model provides correct results, an initial run was performed and compared with experimental data from Agyenim et al. [24] for a horizontal concentric tube heat exchanger with erythritol, which has a melting point of 390 K. The average temperature profile of the PCM versus the time between in PCM was compared with the numerical results from the model and plotted in Fig. 3, representing that the current predictions and experimental results are in good agreement.

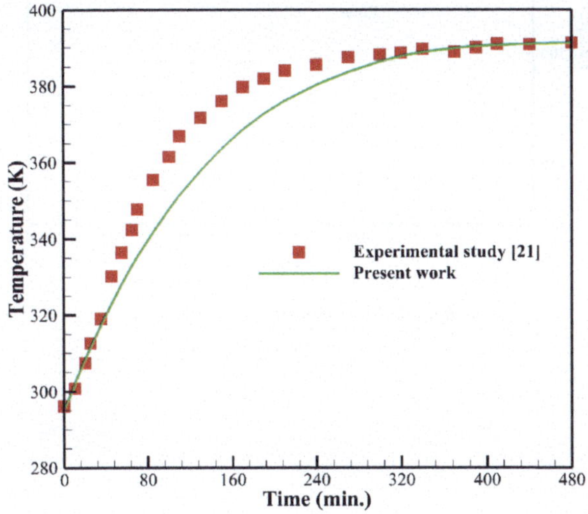

Fig. 3 Validation of the present model for the temperature of PCM with an experimental study by Agyenim et al. [22]

7 Grid Independence Study

A grid independence analysis is performed to determine the ideal grid size for LHESS. Three grid sizes are explored, with the number of cells increasing by a factor of two. The coarse grid has 354,806 cells, whereas the fine and finer grids have 685,350 and 1,211,560 cells, respectively. In the computational domain, the output air temperature variation with time is compared for different mesh sizes and represented in Fig. 4. The highest deviation of output air temperature is observed to be 0.16 °C when the number of cells is further increased from 354,806 (coarse grid) to 685,350 (fine grid) and reduces to 0.014 °C when the number of cells increases from 685,350 to 1,211,560 (finer). As a result of this analysis, the fine grid can be considered for further study. The time step is selected to 0.1 s since the difference in temperature distribution in LHESS diminishes as the time step is decreased.

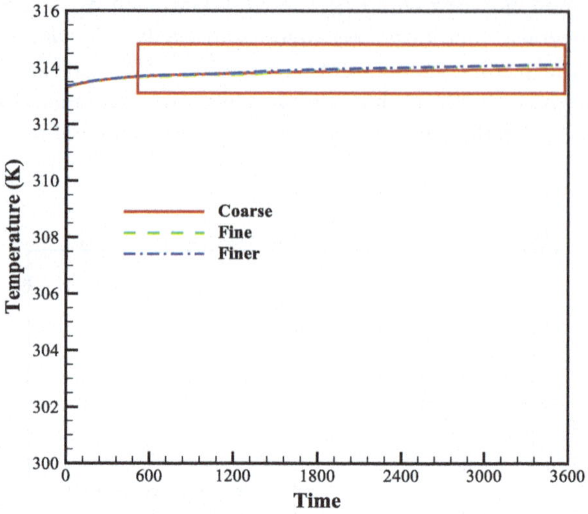

Fig. 4 Grid independence study

8 Results

The heat transfer characteristics of the LHESS are discussed, followed by the energy savings obtained by incorporating such a system into an air-conditioning unit. For the LHESS, two configurations are considered; in one configuration, the PCM is filled in the annulus around the pipe carrying air, while in the second configuration, PCM is filled in six internal tubes.

9 Melting Fraction

The temporal variation of the melt fraction of the PCM, for the PCM stored in the inner tube and in the annulus, is shown in Fig. 5. It can be observed that for PCM in inner tubes, the melting starts at the time instant of 100 s. Further, the latent heat storage in the PCM increases and complete melting is achieved at 1100 s. Subsequently, the energy is stored only in the form of sensible heat after complete melting. For the case where PCM is in the annulus, the melting starts at 200 s of operation, and after 3600 s of operation, 93.5% of the PCM melts. Compared to the arrangement with PCM in the annulus, the augment in the heat transfer area between air and PCM is 20% for the same mass of PCM. Hence, less melting time is observed in the latter configuration of shell and tube LHESS.

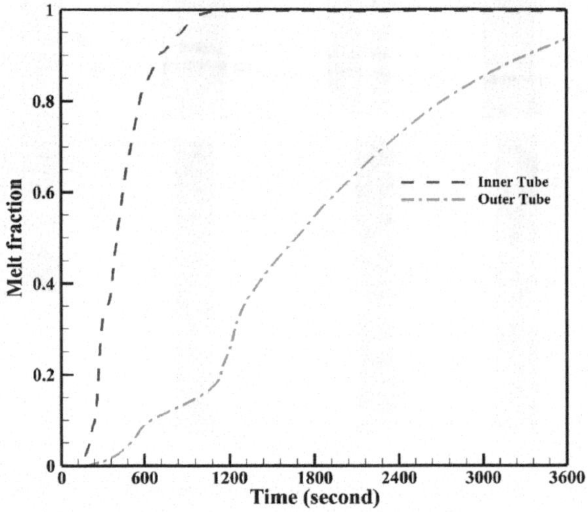

Fig. 5 Temporal variation of melt fraction for different configurations of heat exchanger

The contour of the melt fraction for PCM in the LHESS for both configurations is represented in Fig. 6. It can be seen from Fig. 6a that at time instants of 600 s and 1800s, the melted PCM at the top of the heat exchanger is greater than that in the bottom area for the LHESS with PCM in the annulus. In the initial stage of operation, conduction predominates over convection. However, once the melting of PCM begins, the effect of convection heat transfer is evident and dominates the conduction heat transmission over time. The melted PCM floats upwards because of the buoyancy effect and temperature difference, and the heat transfer happens due to natural convection. It is found that the PCM melt percentage increases from 54.5% after 1800s to 93.3% after 3600 s.

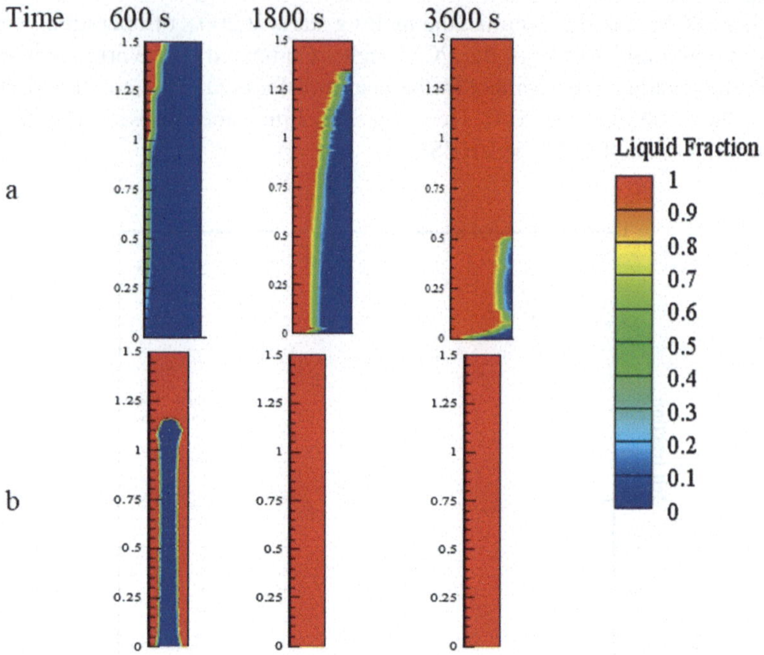

Fig. 6 Melt fraction of PCM for **a** PCM in the annulus and **b** PCM in the inner tube

Further, Fig. 6b represents the contour of melted PCM for a heat exchanger with PCM in the inner tube. After 600 s of operation, the melt front of PCM represents the heat transfer through the curved surface of the PCM tube, and almost 50% of the PCM melts. The unmelted portion can be seen in the lower middle portion of the tube, attributed to natural convection heat transfer. As the PCM melts completely after 1100 s, the heat transfer to PCM in latent heat ceases, and the energy stored afterwards is in sensible heat.

10 Air Outlet Temperature

Figure 7 illustrates the variation of air temperature at the outlet of the heat exchanger with time. The air at the inlet is considered at a temperature of 315 K, imparts its thermal energy to the PCM and then comes out at a lower temperature from the other end. The heat exchanger with PCM in the annulus shows lesser fluctuation in the air outlet temperature as the energy is mainly stored in latent heat. For the heat exchanger with PCM in the inner tube, a sudden rise is observed after 1100 s denoting the complete melting of PCM, and hence the storage in the form of latent heat ceases. Further, the temperature at the outlet after 1500 s is greater for the heat exchanger with PCM in the inner tube than the heat exchanger with PCM in the annulus.

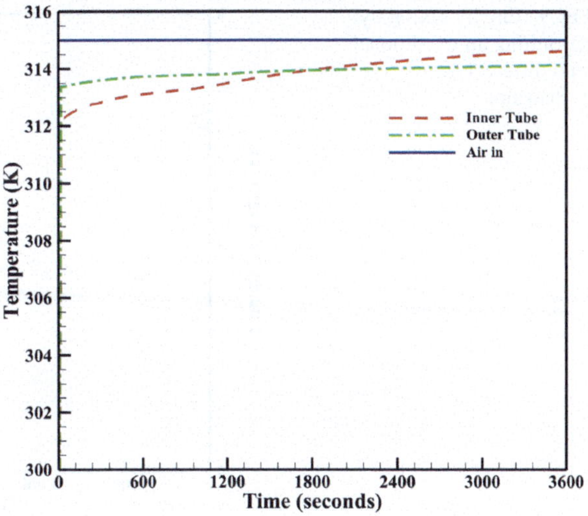

Fig. 7 Temporal variation of outlet air temperature for different configurations of heat exchanger

Figure 8 shows the air temperature distribution at a cross-section located at the exit of the heat exchanger. The air temperature at the surface near the PCM shows a higher drop, and it can go up to 5 K, as represented by the contours in Fig. 8. The average air temperature at the outlet is 1.63 K, 1.31 K and 1.13 K after 600 s, 1800s and 3600 s of operation, respectively. When the same mass of PCM is deployed in six equidistant inner tubes, the average air temperature drops to 2.41 K, 1.79 K and 1.21 K after 600 s, 1800 s and 3600 s of operation, respectively.

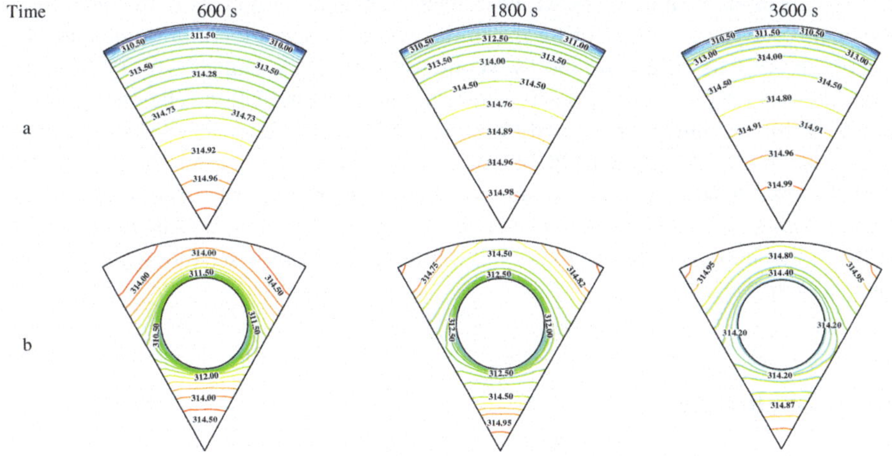

Fig. 8 Air temperature (K) at the outlet for the heat exchanger with **a** PCM in the annulus and **b** PCM in the inner tube

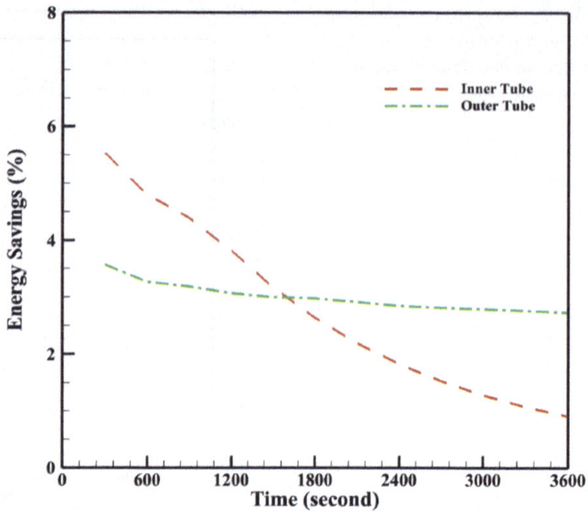

Fig. 9 Energy saving by retrofitting air conditioner with different heat exchangers

11 Energy Saving

In the present study, the cooled air from the heat exchanger is fed to an HVAC unit working on the VCRS cycle. In a VCRS system, the compressor consumes a significant proportion of energy consumed by the unit; hence, for simplification, the energy consumed by the compressor is considered the total energy consumed by the HVAC unit. The variation of energy saving with time is represented in Fig. 9. For the initial 100 s, energy savings are not calculated due to marginal heat transfer in the form of the latent heat to the PCM. For a heat exchanger with PCM in the inner tube, the energy saving is 5.5% during the initial phase of operation and then decreases steeply to 0.9% to the end at 3600 s. While for the heat exchanger with PCM in the annulus, the energy saving is nearly constant with less fluctuation. The energy saved is 3.57% for the initial phase and drops to 2.57% at the end of 3600 s. The peak energy saving in LHESS with PCM stored in the inner tube is more than that in the LHESS with PCM placed in the annulus due to increased heat transfer area in the former configuration. However, the average energy saving in the heat exchanger with PCM in the annulus is found to be 3%, while for PCM in inner tubes, the average energy saving is 2.7%.

12 Conclusions

An approach to lower the energy consumption of an air-conditioning unit using fifty percent fresh air per circulation is investigated. The AC system is based on the VCR cycle and operates for 3600 s at 315 K ambient air temperature. A PCM-based shell-and-tube heat exchanger is considered to reduce the temperature of air supplied at ambient conditions. PCM absorbs thermal energy as latent heat when fresh hot air flows through LHESS. The heat transfer performance of the heat exchanger is analysed, and its effect on energy consumed is investigated. Comparative analysis is carried out by placing the same mass of PCM in the annulus and subsequently in the inner tube of the shell and tube type LHESS. It is found that for the PCM filled in the annulus, the outlet air temperature is reduced by an average of 1.13 K for 3600 s of operation. Integrating this heat exchanger into a VCR-based AC system results in maximum and average energy savings of 3.57 percent and 3 percent, respectively, for 3600 s. While in the second configuration, the outlet air temperature is reduced by an average of 1.21 K for 3600 s of operation. The average energy saving with PCM in the internal tube is 2.7%. Hence for designing such a system, PCM in annulus provides efficient and enhanced performance. The proposed technology will give an efficient solution to the AC system's rising energy consumption. The research will act as a guide for the design of PCM heat exchangers with VCR systems.

Acknowledgements The authors would like to thank IIT Delhi HPC facility for computational resources, Department of Science and Technology, Government of India "Different Energy Vector Integration for Storage of Energy"—Grant number—TMD/CERI/MICALL19/2020/03(G), Department of Science and Technology, Government of India "India-Trento Programme for Advanced Research"—Grant number—INT/Italy/ITPAR-IV/RE/2018/G, and Yardi Systems for providing the necessary resources to carry out the present study under the research grants titled "Design and Demonstration of Sustainable Building Infrastructure" —Grant number—IITD/IRD/RP03743/111718.

Nomenclature

List of Acronyms

AC	Air conditioning
DBT	Dry bulb temperature
HTF	Heat Transfer Fluid
PCM	Phase Change Material
HVAC	Heating Ventilation and Air Conditioning
LHESS	Latent Heat Energy Storage System
VCR	Vapour Compression Refrigeration

List of Symbols

A	Contact Area (m^2)
Q	Heat Flow Rate (W)
T	Temperature (K)

W	Work Done (W)
h	Enthalpy per kg (kJ/kg)
k	Thermal Conductivity (K/W)
$\dot{m}$	Mass Flow Rate (kg/s)
P	Pressure (Pa)

List of Subscripts

a	Ambient
avg	Average
comp	Compressor
mix	Mixed Air
ref	Refrigerant
s	Supply Air

Greek Letters

η	Efficiency
ρ	Density (m3/kg)
μ	Viscosity (kg/m.s)
β	Thermal expansion coefficient (K^{-1})

References

1. Commercial Buildings Energy Consumption Survey (CBECS) (2009) User's Guide to the 2012 CBECS Public Use Microdata File
2. Global Status Report for Buildings and Construction Sector (2019) UNEP–UN Environment Programme. https://www.unep.org/resources/publication/2019-global-status-report-buildings-and-construction-sector. Accessed 3 Nov 2022
3. Energy Agency I. India Energy Outlook 2021 World Energy Outlook Special Report
4. Verma R, Kumar S, Rakshit D, Premachandran B (2023) Design and optimization of energy consumption for a low-rise building with seasonal variations under composite climate of India. J Sol Energy Eng 145. https://doi.org/10.1115/1.4054831
5. Kumar A, Saha SK (2018) Latent heat thermal storage with variable porosity metal matrix: a numerical study. Renew Energy 125:962–973. https://doi.org/10.1016/J.RENENE.2018.03.030
6. Kumar A, Shahi P, Saha SK (2018) Experimental study of latent heat thermal energy storage system for medium temperature solar applications. https://doi.org/10.11159/htff18.152.
7. Shalaby SM, Bek MA, El-Sebaii AA (2014) Solar dryers with PCM as energy storage medium: a review. Renew Sustain Energy Rev 33:110–116. https://doi.org/10.1016/J.RSER.2014.01.073

8. Barthwal M, Dhar A, Powar S (2021) The techno-economic and environmental analysis of genetic algorithm (GA) optimized cold thermal energy storage (CTES) for air-conditioning applications. Appl Energy 283:116253. https://doi.org/10.1016/J.APENERGY.2020.116253
9. Kumar Sharma R, Yagnamurthy S, Rakshit D (2022) Energy analysis of a phase change material embedded heat exchanger for air conditioning load reduction in different Indian climatic zones. Sustain Energy Technol Assess 53:102776. https://doi.org/10.1016/J.SETA.2022.102776
10. Yahaya NA, Ahmad H (2011) Numerical investigation of indoor air temperature with the application of PCM gypsum board as ceiling panels in buildings. Procedia Eng 20:238–248. https://doi.org/10.1016/J.PROENG.2011.11.161
11. Saxena R, Rakshit D, Kaushik SC (2020) Experimental assessment of Phase Change Material (PCM) embedded bricks for passive conditioning in buildings. Renew Energy 149:587–599. https://doi.org/10.1016/J.RENENE.2019.12.081
12. Saxena R, Rakshit D, Kaushik SC (2019) Phase change material (PCM) incorporated bricks for energy conservation in composite climate: a sustainable building solution. Sol Energy 183:276–284. https://doi.org/10.1016/J.SOLENER.2019.03.035
13. Fasi MA, Budaiwi IM (2015) Energy performance of windows in office buildings considering daylight integration and visual comfort in hot climates. Energy Build 108:307–316. https://doi.org/10.1016/J.ENBUILD.2015.09.024
14. Yanbing K, Yi J, Yinping Z (2003) Modeling and experimental study on an innovative passive cooling system—NVP system. Energy Build 35:417–425. https://doi.org/10.1016/S0378-7788(02)00141-X
15. Osterman E, Butala V, Stritih U (2015) PCM thermal storage system for 'free' heating and cooling of buildings. Energy Build 106:125–133. https://doi.org/10.1016/J.ENBUILD.2015.04.012
16. Kapur V, Agarwal B, Baliga G, Bhambure JM, Garg V, Mathur J, et al (2020) ISHRAE COVID-19 guidance document for air conditioning and ventilation. Indian Soc Heat Refrigerat Air Cond Eng 15
17. Washington State Department of Health (2020) Ventilation and air quality for reducing transmission of COVID-19. 19:1–4
18. A phase change material (PCM) based novel retrofitting approach in the air conditioning system to reduce building energy demand
19. Das D, Sharma RK, Saikia P, Rakshit D (2021) An integrated entropy-based multi-attribute decision-making model for phase change material selection and passive thermal management. Decision Anal J 1:100011. https://doi.org/10.1016/J.DAJOUR.2021.100011
20. Kumar A, Saha SK (2016) Energy and exergy analyses of medium temperature latent heat thermal storage with high porosity metal matrix. Appl Therm Eng 109:911–923. https://doi.org/10.1016/J.APPLTHERMALENG.2016.04.161
21. Kumar A, Saha SK (2021) Performance study of a novel funnel shaped shell and tube latent heat thermal energy storage system. Renewable Energy 165:731–747. https://doi.org/10.1016/J.RENENE.2020.11.023
22. Indoor I, Quality E, Second S (2019) ISHRAE_IEQ_Feb_26_2019_public_draft
23. Refrigeration and Air Conditioning https://www.expresslibrary.mheducation.com/pdfreader/refrigeration-air-conditioning50163303. Accessed 18 May 2021
24. Agyenim F, Eames P, Smyth M (2009) A comparison of heat transfer enhancement in a medium temperature thermal energy storage heat exchanger using fins. Sol Energy 83:1509–1520. https://doi.org/10.1016/J.SOLENER.2009.04.007

Optimisation for Optimum Nozzle Design of the Hybrid Jet-Pillar-Microchannel Heat Sink

Jyoti Pandey and Mohd. Zahid Ansari

Abstract The present study conducted an optimisation study for geometric variables of the jet nozzle of the multi-jet impingement and pillar-integrated microchannel heat sink for best performance. Nozzle diameter, nozzle angle, and nozzle diameter decreasing factor are the three design variables and minimisation of the thermal resistance is considered as the objective function. A regression model is developed using Minitab software and optimised using response surface methodology and further contribution of each variable is observed by performing analysis of variance. Nozzle angle exhibited the maximum contribution in varying the objective function that is 47.84% and thus thermal resistance was reduced by 15.8% by changing the nozzle angle from optimum conditions.

1 Introduction

Drastic escalation of temperature in the electronic component is the most concerning consequence of miniaturisation of the devices and adding up large number of functionalities in the small space with the time as it would shorten the lifespan and working efficiency of the device. Discovery of microchannel heat sink gained popularity for thermal management of high-power density micro-scale devices owing to its compact size and higher surface area to volume ratio. Tuckerman and Pease [1] corroborated that liquid-based cooling employed with microchannel can solve the future high heat dissipation problem in compact electronic devices. Numerous passive and active techniques were investigated and tested for superior hydrothermal performance of the heat sink. Techniques were focused on reducing boundary layer thickness, increasing surface area to volume ratio, improving flow mixing, inducing chaotic advection, etc. in the modified heat sink design. Jet impingement is one of the active techniques, capable of reducing boundary layer thickness, enhancing flow mixing and inducing chaos by disrupting channel flow as fluid is injected normal to

J. Pandey (✉) · Mohd. Z. Ansari
MEMS and Microfluidics Lab, Mechanical Engineering Department, PDPM Indian Institute of Information, Technology, Design, and Manufacturing, Jabalpur, MP 482005, India
e-mail: 1813602@iiitdmj.ac.in

M. A. Siddiqui et al. (eds.), *Advances in Heat Transfer and Fluid Dynamics*, Lecture Notes in Mechanical Engineering, https://doi.org/10.1007/978-981-99-7213-5_20

the surface. Higher heat transfer coefficient was achieved and hot spot was effectively treated using jet impingement in the heat sink.

Trapezoid-shaped channel surpassed circular and rectangular-shaped channel in terms of thermal performance of the microchannel heat sink with single slot jet impingement while circular channel provided highest pressure drop [2]. Lowering jet width and microchannel height contributed in decreasing surface temperature [3]. Temperature uniformity, thermal resistance and thermodynamic performance index of the channel can further be improved significantly by using multiple nozzles instead of single. Tran et al. [4] conducted study for multi-nozzle microchannel heat sink and found near to 62 and 47.3% reduction in pumping power with respect to single layer and double layer microchannel heat sink respectively. Moreover, with reducing the channel length from 10 to 1 mm, almost 62%, 10 times and 12 times improvement was obtained in thermal resistance, temperature uniformity and pressure drop respectively. Also, in case of multi-nozzle microchannel heat sink, thermal performance index and thermal resistance were improved by reducing the rib width and enlarging the channel aspect ratio.

Naphon et al. [5] investigated that approximately 18.56% heat transfer coefficient was enhanced in the microchannel heat sink by impinging jet of TiO_2 nanofluid with 0.015% concentration. Heat transfer coefficient was amplified and pressure drop was reduced even by lowering nozzle level height and increasing diameter. Mixing of spent fluid and jet of fresh fluid in case of multi-nozzle microchannel heat sink limits the cooling efficiency of the jet impingement. Also, fluid coming from the upstream nozzles obstructs the jet of fluid in the downstream region to strike directly on the bottom wall of the channel. Effusion holes were created between nozzles to limit the fluid cross-flow between neighbouring jets [6, 7].

Several hybrid multi-jet impingement microchannel heat sink flow and thermal characteristics were studied for further improvement of the microchannel heat sink performance [8, 9]. Gan et al. [10] conducted study for microchannel heat sink having jet impingement and dimples with side outlets and found improved cooling performance as well as smaller pumping power. Huang et al. [11] investigated microchannel heat sink performance for the combination of jet impingement on dimple structures of concave, convex and mixed shape. Convex dimples provided the superior overall performance as they contributed in disrupting the boundary layer whereas concave-shaped dimples provided worst performance as it created a dead zone in the flow path that blocked the fluid flow. Moreover, larger dimple radius and height provided better heat transfer performance [12, 13]. Husain et al. [13] performed numerical study for hybrid microchannel heat sink including jet impingement and pillar and found higher heat transfer coefficient for low jet pitch to jet diameter ratio and high standoff to jet diameter ratio, whereas, lower pressure drop was obtained for high jet pitch to jet diameter ratio and low standoff to jet diameter ratio.

Jet impingement heat transfer characteristic performance was significantly affected by the nozzle geometric design. Heat transfer coefficient was increased and bottom wall temperature was lowered by decreasing the size of the jet nozzle diameter along the fluid flow direction in the microchannel. However, highest temperature uniformity was obtained by using the equal size jet nozzle diameter along the length of the microchannel heat sink. In addition, with increasing the jet nozzle diameter along the channel flow produced complex flow pattern as spent fluid flow was blocked and thus increased the temperature gradient at the bottom wall [14, 15]. Maximum wall temperature as well as thermal entropy generation rate was lowered by decreasing the jet nozzle angle against the flow direction whereas pressure drop and frictional entropy generation rate were hiked [15]. Prandtl number, Reynolds number and area ratio (ratio of total jet area and heater area) were found the influencing parameters for the heat transfer performance of the micro-jet array [16]. Strong impingement effect was obtained by reducing the jet nozzle diameter, thus bottom wall temperature was lowered while weak impingement effect was obtained with increasing the number of jet nozzles [17]. Zhang et al. [2] proposed that by optimising the single slot jet including trapezoid shaped microchannel height, bottom width and corner angle, bottom wall temperature can be reduced by 12.22, 14.85 and 7.15% respectively.

This study aimed on optimising the hybrid jet impingement and pillar included microchannel heat sink design numerically from the view of optimum nozzle geometry. Jet nozzle angle, nozzle diameter and nozzle diameter reduction factor are the three design variables chosen to be optimised. The optimisation is performed considering minimum thermal resistance as the objective function.

2 Geometric Model and Numerical Scheme

Microchannel heat sink with array of nozzle jets for impingement and pillars for flow obstruction in Fig. 1 was designed in the view of electronic cooling application. Channels of 400 μm (H_C) depth were etched on the copper substrate of 600 μm (H) height and 24 mm (L) length. Substrate total width (W) was sectioned into multiple channels of 700 μm width pitch (W_P), which is the summation of channel width (W_C) of 500 μm and substrate fin width (W_F) of 100 μm kept on each side of the channel. Circular-shaped nozzles of diameter, $D_N = 0.2$ mm, 0.25 mm and 0.3 mm were made on the glass substrate of 400 μm height (H_N) on equal distance of 2.4 mm (S). Jet of water as coolant was injected from the nozzles normal to the channel bottom surface, supplied with 100 W/cm^2 constant heat flux. After striking the bottom surface, fluid was allowed to flow parallel to the channel length in both the direction and ejected from the outlet. Circular pillars of 0.2 mm diameter (D_P) were located in between the two nozzles to act as an obstructer. To save the cost and time consumed in the numerical simulation, only half length of the single channel was considered for the computational domain shown in Fig. 2. Jet nozzle inclination angles (θ) considered in the study include $\theta = 30°, 45°, 60°$, and 90° whereas nozzle diameter was decreased along the channel length by a decreasing factor (d) from the base diameter of the nozzle (D_N) that includes $d = 0.00$ mm, 0.01 mm, 0.02 mm, and 0.03 mm.

Hybrid microchannel heat sink system with boundary conditions was simulated on the computational fluid dynamics (CFD) based commercial software Ansys FLUENT [18]. Mesh was generated with 1.3 million nodes, satisfying independency test.

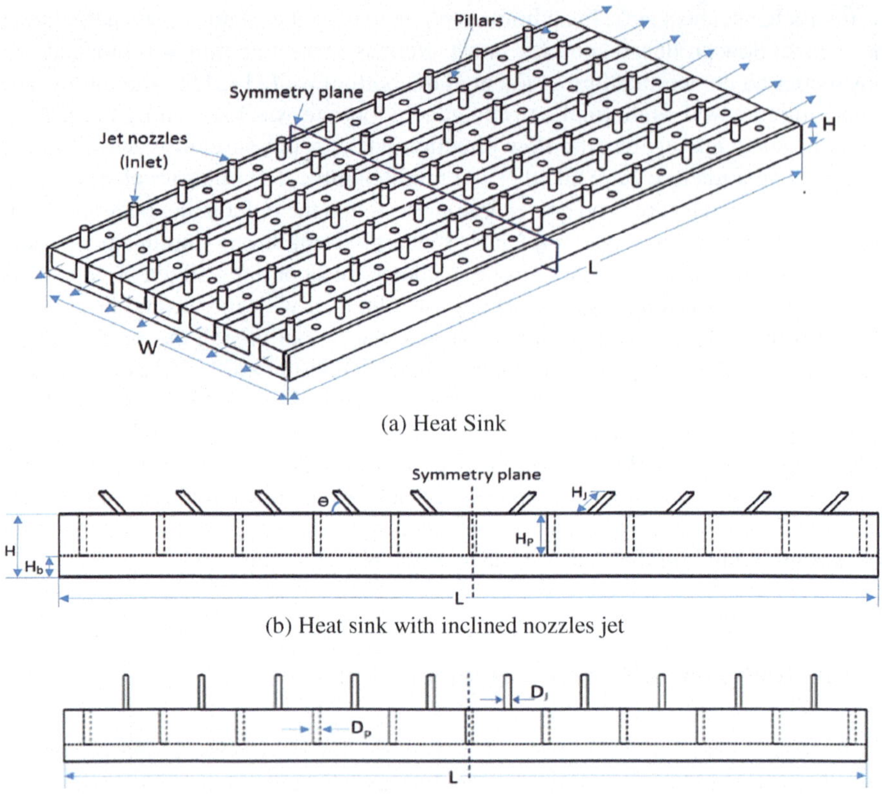

(a) Heat Sink

(b) Heat sink with inclined nozzles jet

(c) Heat sink with decreasing nozzle diameter

Fig. 1 Schematic of the multiple jet impingement and pillar integrated microchannel heat sink

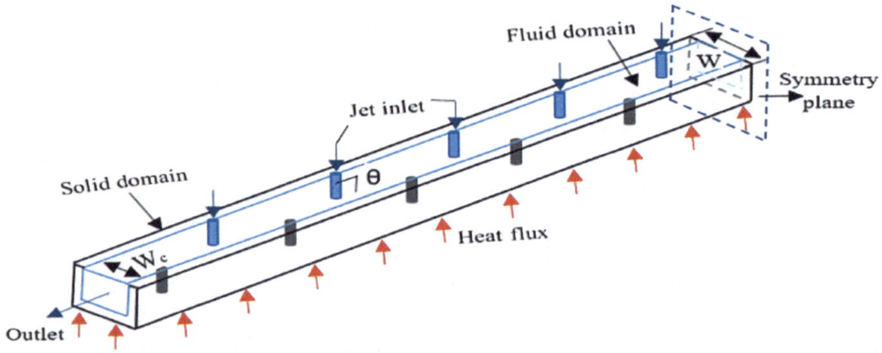

Fig. 2 Computational domain with boundary conditions of the multiple jet impingement and pillar integrated microchannel heat sink

Table 1 Properties of the solid substrate and fluid

Property	DI Water	Copper
Thermal conductivity, k (W/m^{2}K)	0.6	387.6
Specific heat capacity, C_p (J/kgK)	4182	381
Density, ρ (kg/m^3)	998.2	8978
Viscosity, μ (kg/ms)	0.001	–

Simplified fundamental governing equations such as continuity, momentum and energy were solved numerically to compute the characteristic parameters that can define the heat sink hydrothermal performance. Governing equations were simplified assuming steady state, laminar and incompressible fluid flow. In addition, fluid and substrate properties, mentioned in Table 1, were assumed independent of temperature; and radiation, viscous dissipation, and gravity effects were neglected.

Fluid domain:

Continuity equation,

$$\nabla.(\rho V) = 0 \tag{1}$$

Momentum equation,

$$V.\nabla(\rho V) = -\nabla P + \nabla.(\mu \nabla V)) \tag{2}$$

Energy equation,

$$V.\nabla(\rho C_P T) = \nabla.(k \nabla T) \tag{3}$$

Solid domain:

Energy equation,

$$\nabla.(k_{subs} \nabla T_{subs}) = 0 \tag{4}$$

Fundamental equations were discretised using second-order upwind algorithm and pressure-based solver was used for fluid flow behaviour and conjugate heat transfer analysis in the microchannel heat sink. Pressure and velocity-related variables were coupled using SIMPLE (Semi-Implicit Method for Pressure Linked Equation) scheme and residual minimisation criteria was adopted to cease the number of iterations in the numerical solution. Constant mass flow rate of fluid at a fixed temperature and uniform heat flux boundary conditions were applied at the channel inlet and bottom wall surface respectively. Channel outlets were considered to be at atmospheric boundary condition and remaining wall surfaces were considered adiabatic. No slip fluid flow condition was assumed at all the channel walls. All the boundary conditions are expressed mathematically below in Table 2.

Table 2 Boundary conditions

Boundary condition	Expression
Jet inlet	$u = u_{in}$, $v = 0$, $w = 0$, $T = T_{in}$
Outlet	$P_t = P_{atm}$, $\frac{\partial T_f}{\partial n} = 0$
Bottom surface	$-k_s \frac{\partial T_s}{\partial x} = q''$
Other walls	$\frac{\partial T_s}{\partial n} = 0$ (adiabatic), $u = v = w = 0$ (No slip)
Symmetry faces	$\frac{\partial T}{\partial n} = 0$ (adiabatic), $\frac{\partial V}{\partial n} = 0$
Solid–fluid interfaces	$k_s \frac{\partial T_s}{\partial n} = k_f \frac{\partial T_f}{\partial n}$, $T_s = T_f$, $u = v = w = 0$

3 Optimisation Methodology

Three design variables (α, δ, ϕ) were considered for the optimisation among various hybrid heat sink geometric variables from the view of designing optimum nozzle geometry. Design variables were based on changing the diameter of the nozzle, D_N; decreasing the nozzle diameter along the channel length by factor, d; changing the angle of the nozzle against the flow direction, θ. Design variables were normalised and defined while keeping the channel height (H_C) and pitch width ($W_P = W_C + W_F$; summation of channel width, W_C and fin width, W_F) constant as: α ($= D_N/W_C$) represents ratio of nozzle diameter and channel width, δ ($= d/W_C$) represents ratio of factor by which nozzle diameter was reduced along the channel length and channel width and ϕ ($= (\theta_o - \theta)/\theta_o$) represents ratio of nozzle inclination angle to the normal angle ($\theta_o = 90°$). Range of design variables chosen was $0.5 \leq \alpha \leq 0.75, 0 \leq \delta \leq 0.075, 0.33 \leq \phi \leq 1$. Optimisation is performed considering minimisation of the thermal resistance as objective function using polynomial-based response surface method. A second-order polynomial function is fitted between response function and design variables in the response surface approximation and applied sequential quadratic programming to search the optimal point. The mean absolute error (MAE) associated with the regression and numerical model is calculated as

$$MAE = \frac{1}{N}\sum_{i=1}^{N}\frac{\left|R_{th\_pred} - R_{th\_actual}\right|}{R_{th\_actual}} \times 100\% \tag{5}$$

The performance of the model was analysed on the basis of the minimum error and maximum correlation coefficient (R) expressed as:

$$R^2 = \frac{\sum_{m=1}^{N}\left(R_{th\_pred} - \overline{R}_{th\_pred}\right)\left(R_{th\_actual} - \overline{R}_{th\_acual}\right)}{\sqrt{\sum_{m=1}^{N}\left(R_{th\_pred} - \overline{R}_{th\_pred}\right)\sum_{m=1}^{N}\left(R_{th\_actual} - \overline{R}_{th\_actual}\right)}} \tag{6}$$

4 Results and Discussion

The numerical model is validated by the experimental work of Tuckerman and Pease [1] for fluid flow in a microchannel and Wang et al. [19] for jet impingement to ascertain the accuracy of the numerical method adopted in this study as shown in Fig. 3 and found numerical results in close proximity to the experimental results.

On-linear regression analysis is performed to develop the response model of the objective function, R_{th} using minitab [20] in terms of normalised nozzle geometric variables that is α, δ and ϕ as shown below:

$$R_{th} = 1.232 + 6.691 \times \alpha + 0.754 \times \phi + 7.34 \times \delta - 5.341 \times \alpha \times \phi \\ - 43.3 \times \alpha \times \delta + 15.10 \times \phi \times \delta \quad (7)$$

The correlations formulated are applicable for water as coolant in the jet impingement pillar-integrated microchannel heat sink with normalised nozzle geometric variables within a specified range of $0.5 \leq \alpha \leq 0.75$, $0 \leq \delta \leq 0.075$, $0 \leq \phi \leq 0.6667$. Empirical relations predicted values of Rth are compared with the actual numerical data as shown in Fig. 4. Predicted Rth data values at the most deviate from the actual value by ± 6.5%, R^2 and MSE error associated with the R_{th} is 94.30% and 2.898% respectively.

Analysis of Variance (ANOVA) was performed as presented in Table 3 to find the relative contribution of each geometric design variable. It is found that ϕ exhibits the maximum contribution that is 47.84% on affecting the objective function followed by α having 26.97%, and least contributed by δ that is 14.50%. Figure 5 and Fig. 6 showed the effect of each design variable and their interaction on the objective function for the better cooling performance of the MJI-P-MCHS. It is found from Fig. 5 that the mean of the objective function value is continuously increasing with increasing the α that means thermal resistance will be higher for larger diameter of

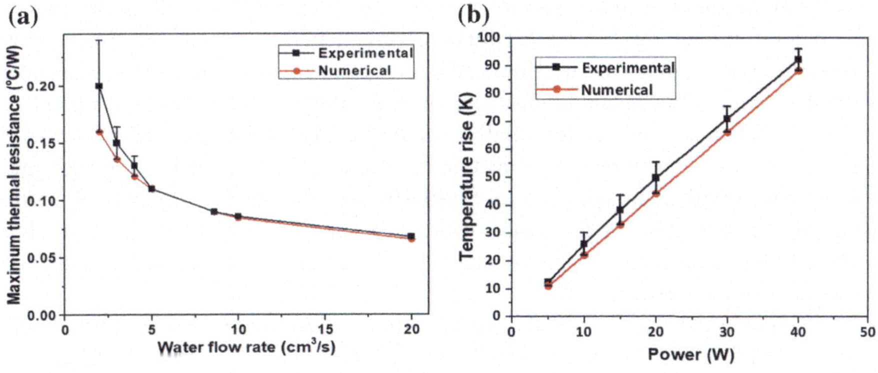

Fig. 3 The numerical and experimental results are compared to validate the numerical model for **a** microchannel heat sinks [1] and **b** jet impingement techniques [19]

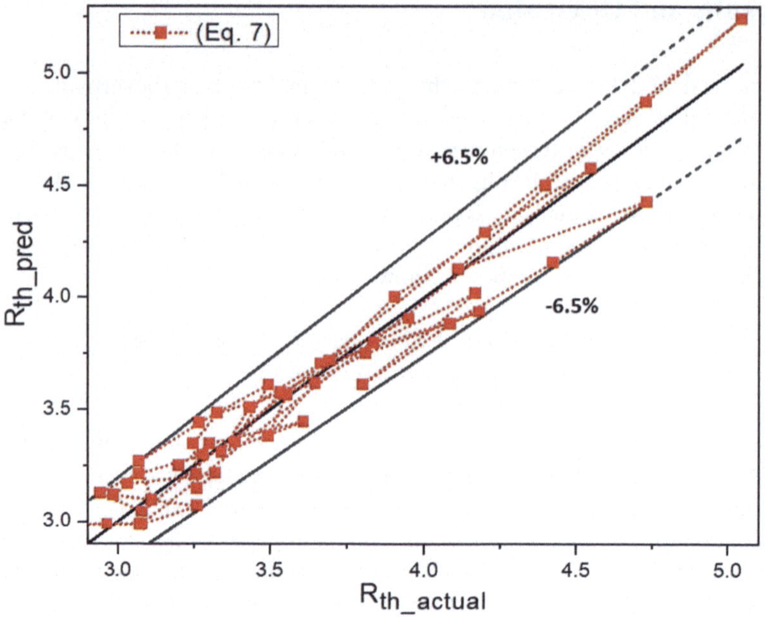

Fig. 4 Comparison of the actual and predicted value of the thermal resistance

the jet which is attributed to the smaller velocity of the jet impinged from the larger size nozzle. Similarly, mean of the objective function value continuously increased as the ϕ increases that means thermal resistance will be higher for larger angle of the jet which is attributed to the smaller path length of the jet in the fluid domain. In addition, mean of the objective function value decreased fast as the δ increased but later on slow down after 0.04 that means thermal resistance will be higher for equal diameter of the jet along the length while it is reduced with deceasing the jet diameter along the flow which is attributed to the enhanced flow mixing by higher velocity jet impinged by the smaller diameter nozzle in the downstream. Interaction plots in Fig. 6 presented that for constant δ, objective function mean value is lower for smaller value of the α and larger value of ϕ. In addition, for constant ϕ, objective function mean value is lower for smaller value of the α but larger value of the δ for higher α whereas for smaller α, objective function mean value first decreased and then increased with increasing the δ. Similarly, for constant α, objective function mean value is lower for larger value of the ϕ but larger value of the δ for higher ϕ whereas for smaller ϕ, objective function mean value first decreased and then increased with increasing the δ.

Regression equation is optimised for the minimum R_{th} and found optimum value as $R_{th} = 2.965 \times 10^{-3}$ W/K at $\alpha = 0.4$, $\phi = 0.667$, and $\delta = 0.04$. Furthermore, variation in R_{th} is observed by changing the design variables from optimum condition as presented in Fig. 7. R_{th} was reduced by 8.3, 15.8 and 3.5–3.7% by changing the α from 0.4 to 0.6, ϕ from 0.667 to 0 and δ from 0.04 to 0 or 0.06 respectively.

Table 3 ANOVA of each design variable

Design variable	Seq SS	Contribution (%)
α	3.690	26.97
ϕ	6.545	47.84
δ	1.984	14.50
Error	1.463	10.69
Total	13.681	100.0%

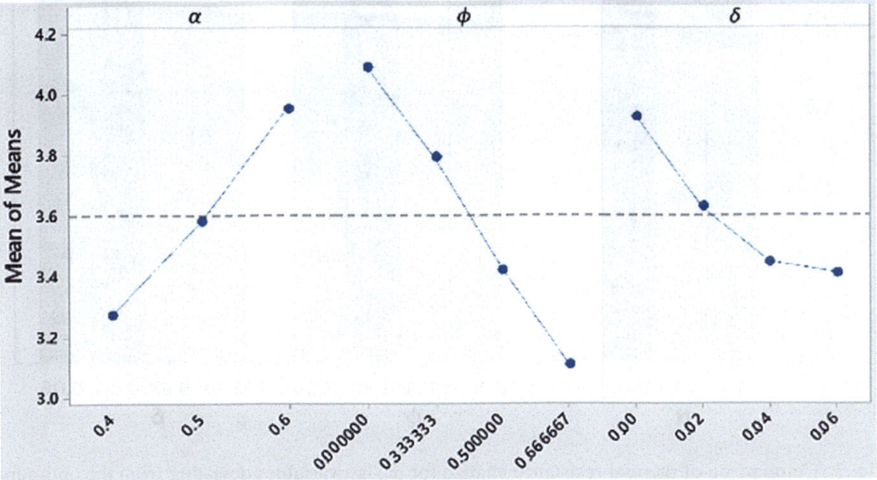

Fig. 5 Main effect plots for the means of the objective function for each design variable

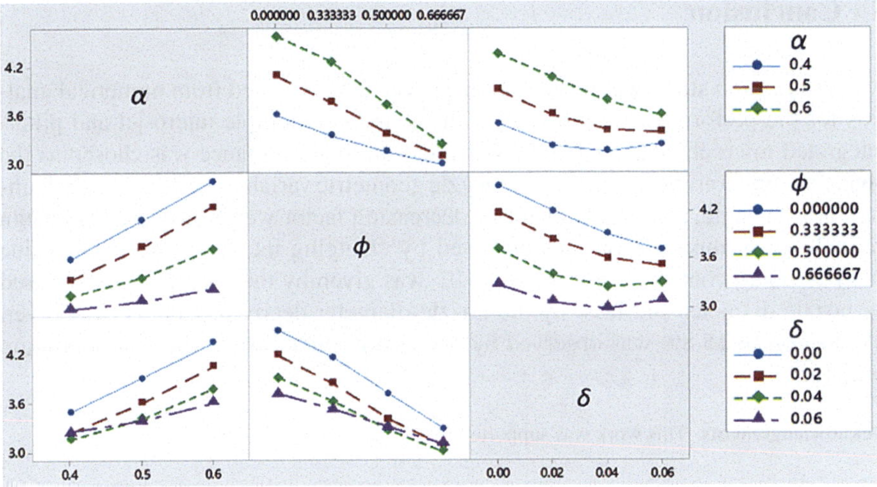

Fig. 6 Interaction plots for the means of the objective function for each design variable

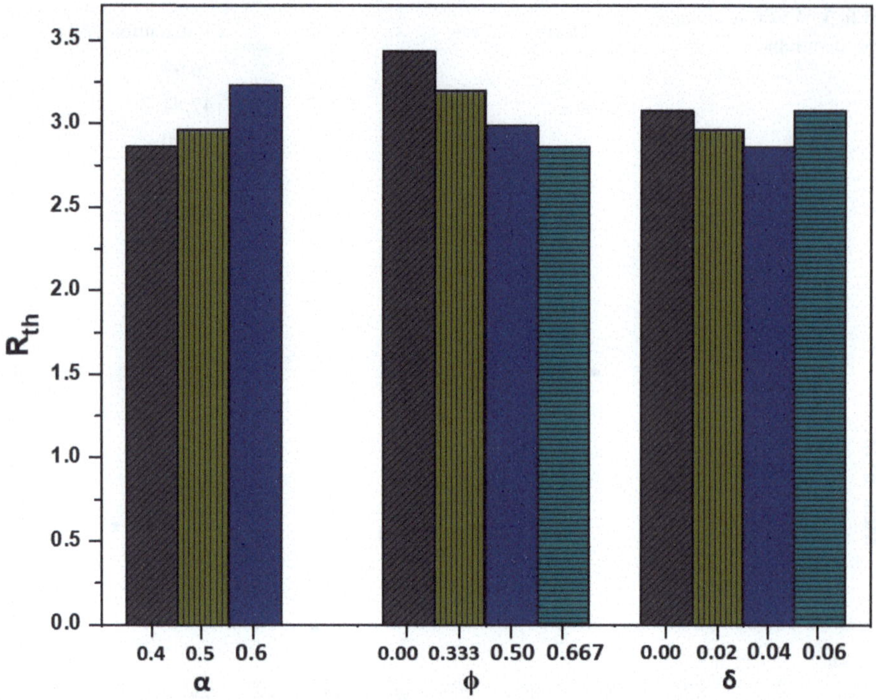

Fig. 7 Comparison of thermal resistance change for design variables deviating from the optimum value

5 Conclusion

An optimisation study was performed using the data obtained from numerical analysis for the performance improvement of the hybrid multiple micro-jet and pillar-integrated microchannel heat sink. Minimum thermal resistance was chosen as the objective function and normalised jet nozzle geometric variables such as nozzle diameter, nozzle angle, and nozzle diameter decreasing factor were considered as design variables. An improvement was observed by changing the design variables value and maximum contribution that is 47.84% was given by the nozzle angle followed by nozzle diameter and least by the nozzle diameter decreasing factor. Moreover, a reduction of 15.8% was observed by decreasing the nozzle angle from optimum condition.

Acknowledgements This work was supported by IIITDM Jabalpur, India.

References

1. Tuckerman DB, Pease RFW (1981) IEEE Electron Device Lett. EDL-2(5):126–129
2. Zhang Y, Wang S, Ding P (2017) Int J Heat Mass Transf 113:295–309
3. Sung MK, Mudawar I (2006) Int J Heat Mass Transf 49:682–694
4. Tran N, Chang Y-J, Teng J-T, Dang T, Greif R (2016) Int J Heat Mass Transf 101:656–666
5. Naphon P, Nakharintr L, Wiriyasart S (2018) Int J Heat Mass Transf 126:924–932
6. Husain A, Ariz M (2017) Int J Engg 30(10):1599–1608
7. Husain A, Al-Azri NA, Al-Rawahi NZH (2015) J Thermophys Heat Transf 30(2):1–7
8. Pandey J, Husain A, Ansari MZ (2022) Proc IMechE Part C J Mech Eng Sci 236 (17):9814–9827
9. Pandey J, Ansari MZ, Husain A (2021) Int J Num Methods Heat Fluid Flow 32(8):2659–2681
10. Gan T, Ming T, Fang W, Liu Y, Miao L, Ren K, Ahmadi MH (2020) J Thermal Anal Calorimetry 141:45–56
11. Huang X, Yang W, Ming T, Shen W, Yu X (2017) Int J Heat Mass Transf 112:113–124
12. Tingzhen M, Cunjin C, Wei Y, Wenqing S, Wei F, Nan Z (2018) J Therm Sci 27(3):321–330
13. Husain A, Ariz M, Al-Rawahi NZH, Ansari MZ (2016) Appl Therm Engg 102:989–1000
14. Sung MK, Mudawar I (2008) Int J Heat Mass Transf 51:4614–4627
15. Pandey J, Ansari MZ, Husain A (2021) J Mech Sci Tech 35(12):5753–5764
16. Michna GJ, Browne EA, Peles Y, Jensen MK (2011) Int J Heat Mass Transf 54:1782–1790
17. Kim SH, Shin H-C, Kim S-M (2019) J Mech Sci Tech 33(7):3555–3562
18. ANSYS Fluent (2017) version 18.0, Canonsburg, PA, US
19. Wang EN, Zhang L, Jiang L, Koo J-M, Maveety JG, Sanchez A, Goodson KE, Kenny TW (2004) J Microelectromech Syst 13(5):833–842
20. Ryan BF, Ryan TA Jr, Joiner BL (2017) Minitab, version 18. PA, US

Investigation of Laminar and Turbulent Natural Convection Combined with Radiation in a Square Enclosure with a Vertical Partition

Mariyam Ali and Anil Kumar Sharma

Abstract The present research work focuses on numerical simulation of combined natural convection and radiation inside a square enclosure with a vertical partition having different heights. The distance of the partition is varied from the hot wall for establishing the results. The left wall is considered to be the hot, the right wall as the cold and the horizontal top and bottom walls are considered to be adiabatic. The analysis is done in both the laminar and in the turbulent range for the Rayleigh number in range of 10^3 to 10^{10}. The partition is considered to have a fixed thickness which is taken to be L/20. Also, it is assumed to be adiabatic. Air is taken as the working medium. It has been concluded from the results that an increase in radiative Nusselt number is much higher than that of convective Nusselt number for turbulent flow regime. It was found to be 379% as the Rayleigh number increases from 10^8 to 10^{10}. Also, the increase in Nu_T is almost 352% when Ra is increased from 10^8 to 10^{10}. However, as the partition height is increased the value of Nu_T decreases to 15 and 28% for Ra 10^5 and 10^{10} respectively.

1 Introduction

The study of natural convection and radiation inside cavities with different orientations and boundary conditions has been a topic of interest for the past decades by the thermal engineering researchers. The heat transfer and fluid flow are the major factors studied in these researches. However, often in these studies the combined effect of natural convection and radiation is not considered. So, in this research work the combined effect of natural convection and radiation is studied in both the laminar and turbulent range. Bahlaoui et al. [1] in their research work present the results of mixed convection and radiation inside a cavity which is heated from below and consists of a vertical adiabatic partition on this horizontal heated surface. The results are shown with the help of streamline and isotherm contours. Han and Baek [2] in

M. Ali · A. K. Sharma (✉)
Department of Mechanical Engineering, Faculty of Engineering & Technology Jamia Millia Islamia (A Central University), New Delhi 110025, India
e-mail: asharma1@jmi.ac.in

M. A. Siddiqui et al. (eds.), *Advances in Heat Transfer and Fluid Dynamics*, Lecture Notes in Mechanical Engineering, https://doi.org/10.1007/978-981-99-7213-5_21

their paper studied the phenomenon of heat transfer and fluid flow by considering three cases, in the first case only natural convection is considered, in the second case convection along with surface radiation is considered and in the third case convection with surface/gas radiation is considered. Both the partitions were considered to be adiabatic. The results obtained showed that the surface radiation had greater effect than the gas radiation on the heat transfer phenomenon. Again, the result is presented with the help of streamlines and temperature contours along with the calculation of Nusselt number. Zemani et al. [3] carried out a numerical simulation inside a cavity having hot and cold vertical walls with a partition placed parallel to the horizontal walls on the hot wall. The Rayleigh number was considered between 10^6 and 3.7×10^9. Finite volume method was used to solve the governing equations. Time evolutions and temperature contours were presented to demonstrate the ongoing simulation along with the calculation of Nusselt number. Also, the results show that the presence of partition highly abrupt the fluid flow and heat transfer inside the cavity. Nakhi and Chamka [4] in their research work presented the results of laminar natural convection inside a square enclosure consisting of an inclined partition attached to the vertical hot wall while the rest of the three sides of the cavity were cooled from outside. Brown and Solvason [5] performed an experimental study of the phenomenon of natural convection inside rectangular openings with two cases, in the first case considering a vertical partition and in the second case considering a horizontal partition. The Grashof number for the present study was considered between 10^6 and 10^8. The results were presented in the form of Nusselt number calculated from the study. Nagaraja [6] in his research work carried out a numerical simulation of natural convection in a square enclosure with partition. The Rayleigh number for the present study is considered in the laminar range i.e., 10^3 to 10^6. Similar to the other studies, the results in this also are presented in the form of streamline, isotherm contours and the Nusselt number. From the above studies we can see that most of the studies are done either in laminar region or in turbulent region. In the present study, the numerical computation is done both in the laminar region and in the turbulent region. Also, the cavity considered for the analysis consists of a vertical partition.

2 Physical Model and Problem Description

In the present study, a square cavity is considered with vertical walls as hot and cold walls while the horizontal walls are assumed to be adiabatic as shown in Fig. 1. The vertical partition considered in the present research work is assumed to be adiabatic. Air is taken as the working fluid having a Prandtl number of 0.71 and also the Boussinesq approximation is adopted. All other physical parameters are assumed to be constant. The present study is carried out for Rayleigh number 10^3, 10^5, 10^8 and 10^{10} resulting in the analysis in both the laminar and turbulent regions. In order to study the effect of partition height, the length of partition is varied as L, 3L/4, L/2 and L/4 as in Cases 1, 2, 3 and 4 respectively. Likewise, the partition distance is varied as L/4, L/2 and 3L/4 respectively as in Cases a, b and c.

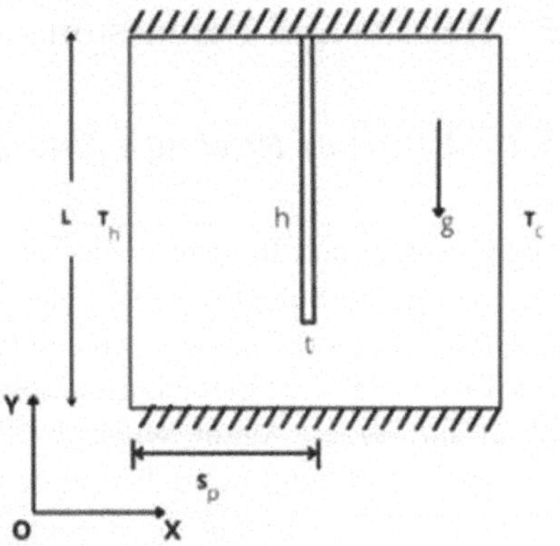

Fig. 1 Computational domain

3 Solution Methodology

The numerical simulation for the present study is done using ANSYS FLUENT 2022R1 which is a finite volume-based solver. The pressure velocity coupling is done using Coupled algorithm. Also, the pressure interpolation is done by Second order scheme. For solving momentum and energy equation Second order Upwind is used. The surface emissivity value for all the general cases is considered to be 1 and Discrete Ordinate is used as the radiation model. SST K-Ω model is adopted as the turbulence model after detailed model sensitivity analysis beyond 80 × 80.

4 Grid Sensitivity Test

The grid sensitivity analysis has been carried out for Ra 10^5 for case 1b. Grid sizes 80 × 80, 100 × 100, 120 × 120 and 140 × 140 have been chosen for the analysis. Table 1 shows that the changes in the Nu values are almost negligible. Therefore, 100 × 100 has been chosen for further analysis for all the cases. Also, the test is done for turbulent flow in the enclosure.

Table 1 Grid sensitivity study for different grid sizes for Rayleigh number 10^5

Grid	Nu (numerical)	% Change with numerical results
80 × 80	5.910	–
100 × 100	5.902	−0.135
120 × 120	5.906	0.067
140 × 140	5.905	−0.016

5 Results and Discussion

5.1 Effect of Rayleigh Number

Heat transfer and fluid flow characteristics in the enclosure are studied in both the laminar and turbulent regimes. The variation of total Nusselt number with Rayleigh number is depicted in Fig. 2. It can be seen that as the Rayleigh number is increased from 10^3 to 10^5 i.e., in the laminar range the total Nusselt number increases to a value of around 202%. Again, when the Rayleigh number is increased from 10^8 to 10^{10} i.e., in the turbulent range the percentage increase in total Nusselt number is 343%. Thus, we can see that as the Rayleigh number is increased either in the laminar or in the turbulent region, the value of Nu_T increases with it in the direct proportion.

5.2 Effect of Partition Height

In the research work in order to study the effect of partition height on the overall heat transfer and fluid flow, different cases of partition height are considered. In case 1 the height of the partition is taken as L, in case 2 it is taken as 3L/4, in case it is considered as L/2 and for the 4th case it is taken as L/4. It is observed that as the partition height is increased from L/4 to L, the value of total Nusselt number decreases to 15% for Rayleigh number 10^5 and 28% for Rayleigh number 10^{10} as shown in Fig. 3. Thus, it leads us to the observation that both the stature of partition and Nu_T are inversely proportional to each other.

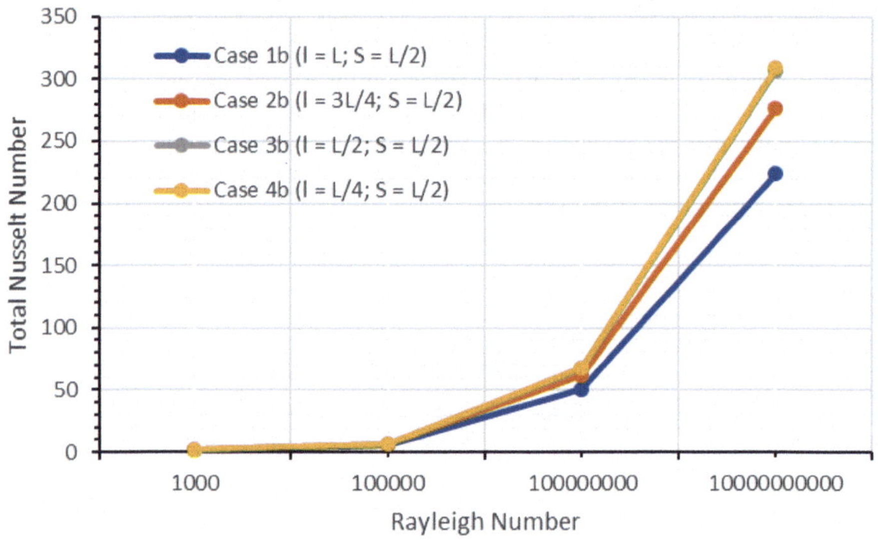

Fig. 2 Variation of total Nusselt number with Rayleigh Numbers along the hot wall for central position of partition wall and different partition height

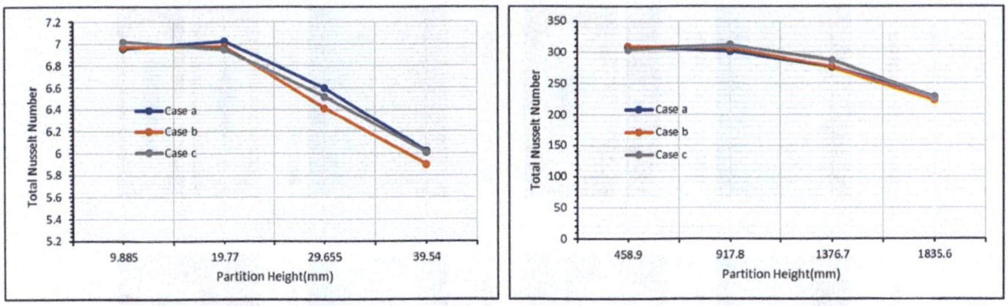

(a)Graph of h versus Nu_T for Ra 10^5 (b) Graph of h versus Nu_T for Ra 10^{10}

Fig. 3 Variation of total Nusselt number with partition length for different partition distance from hot wall for Rayleigh number 10^5 and 10^{10} respectively

5.3 Combined Radiation and Natural Convection

The study of natural convection and radiation is done inside a square cavity with a vertical partition. All these analyses are done both in the laminar and turbulent range for Rayleigh number 10^3 to 10^{10} and as such the results are presented in the form of isotherm and streamline contours as shown in Figs. 4, 5, 6 and 7 respectively. For all the Rayleigh number and for all the considered cases the value of surface emissivity is taken as 1 i.e., $\varepsilon = 1$. In the laminar range i.e., for Ra $= 10^3$ and 10^5 it can be observed from the streamline contours for Case 'a' when the partition is close to the hot wall, that the fluid flow is in the clockwise direction with the fluid rising from the bottom of right side of the partition, rising above than after coming in contact with the top adiabatic wall flows along the left wall in the downward direction thus forming uniform flow of the fluid. When the partition is present at the center, the fluid flows on both the side of the partition as in Case 'b'. While when the partition is present near the cold wall in Case 'c' the fluid flow occurs on the right side of the partition. Also, from the temperature contours for Ra 10^3 we can observe the contours are almost parallel to the vertical walls. However, for Ra 10^5 the isotherms are observed to be disrupted. Similarity is breaking and unsymmetrical flow behavior is seen along the vertical sides.

The streamline and isotherm contours for turbulent range are shown in Figs. 6 and 7 for Rayleigh number 10^8 and 10^{10} respectively. For all the cases of partition height and distance it can be observed that the fluid flows very close to the enclosure walls and the partition wall in the clockwise direction. The isotherm contours for Ra 10^8 are almost parallel to the horizontal walls but as the Ra number is increased to 10^{10} the contours get disturbed.

Figure 8 shows the variation of Nu_C, Nu_R and Nu_T with Rayleigh number. It can be observed that there is a slight variation in Nu_C as the Ra varies from 10^3 to 10^5 however beyond that the change is significant to a value of almost 290% as Ra changes from 10^8 to 10^{10}. For Nu_R the percentage change from 10^8 to 10^{10} is extremely high similar to a value of 379%. Also, it can be observed that the value of Nu_R is greater than the value of Nu_C. The increase in Nu_C and Nu_R results in the increase of Nu_T with the increase in Rayleigh number.

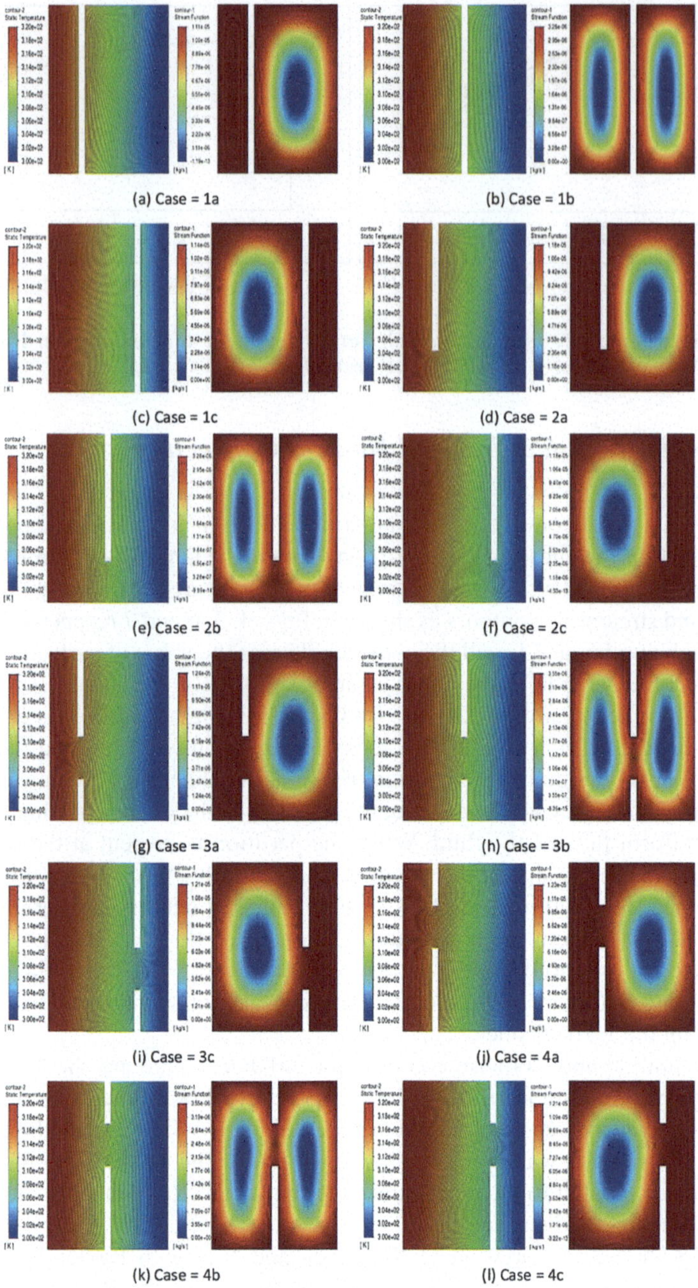

Fig. 4 Temperature and streamline contours for natural convection and radiation for Rayleigh number 10^3 for all cases of partition height and distance from hot wall

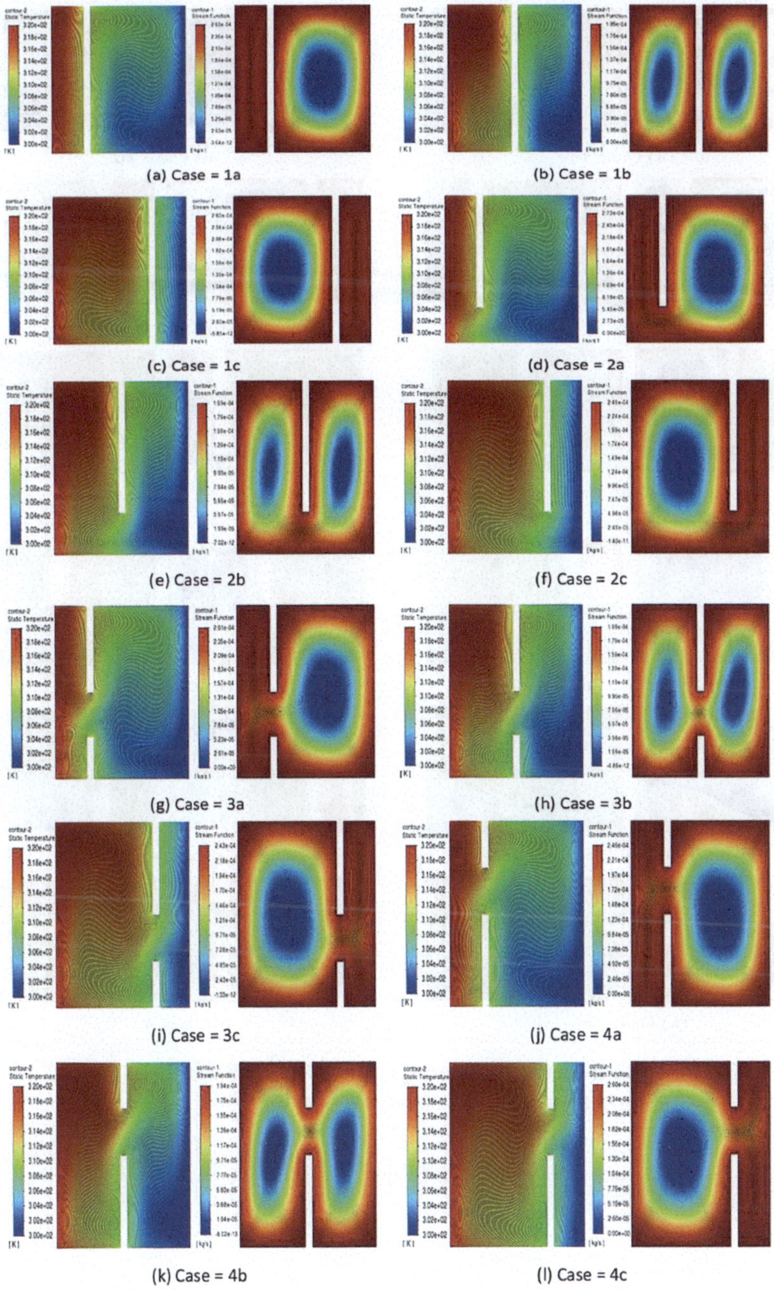

Fig. 5 Temperature and streamline contours for natural convection and radiation for Rayleigh number 10^5 for all cases of partition height and distance from hot wall

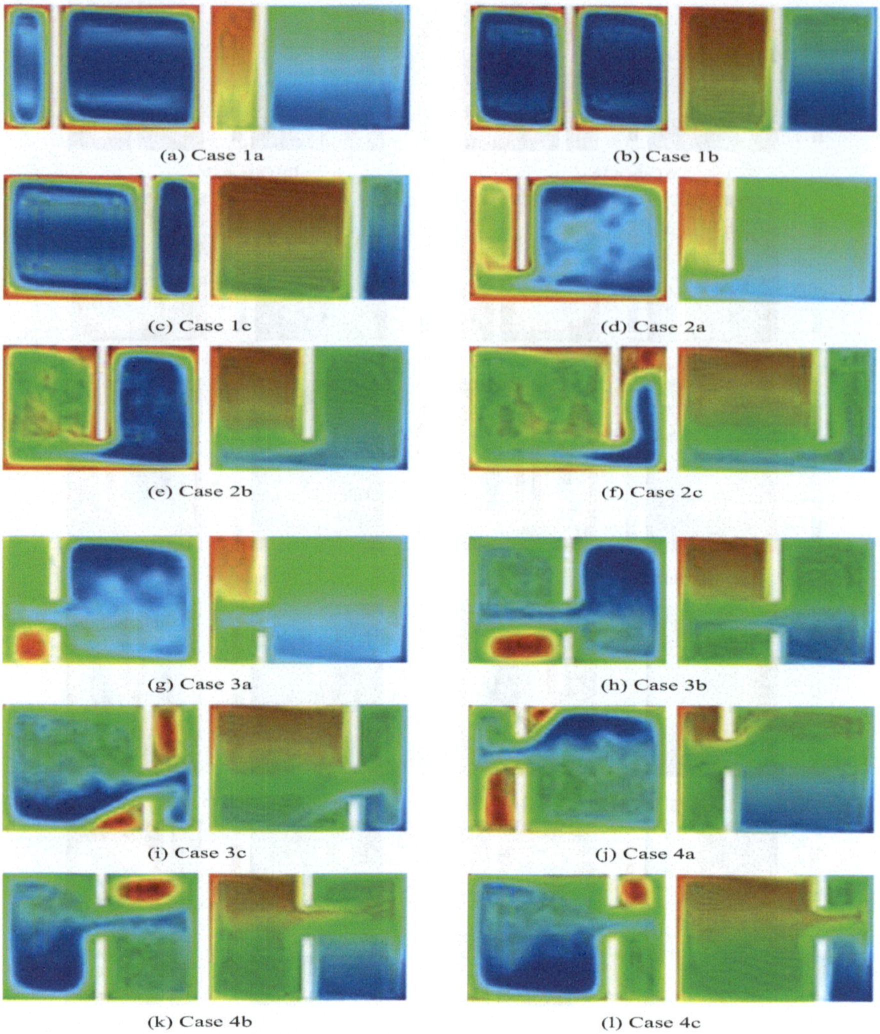

Fig. 6 Temperature and streamline contours for natural convection and radiation for Rayleigh number 10^8 for all cases of partition height and distance from hot wall

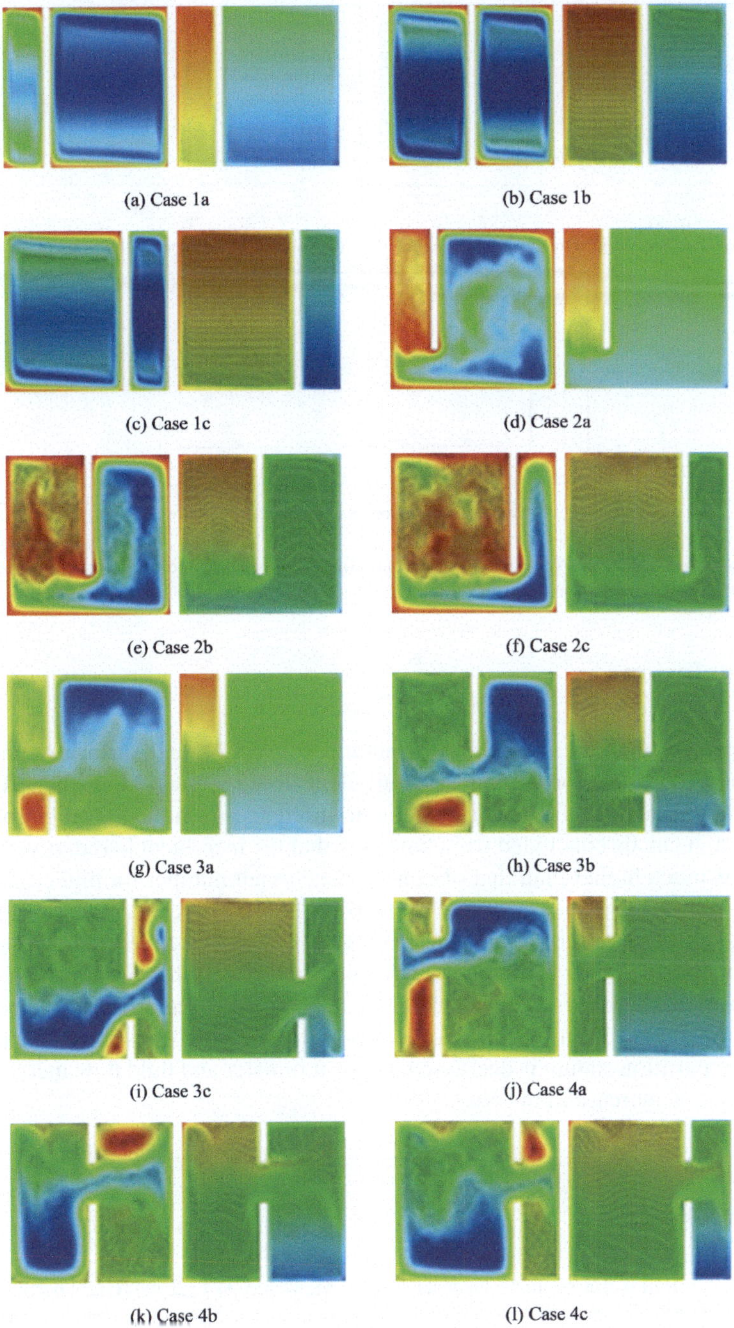

(a) Case 1a
(b) Case 1b
(c) Case 1c
(d) Case 2a
(e) Case 2b
(f) Case 2c
(g) Case 3a
(h) Case 3b
(i) Case 3c
(j) Case 4a
(k) Case 4b
(l) Case 4c

Fig. 7 Temperature and streamline contours for natural convection and radiation for Rayleigh number 10^{10} for all cases of partition height and distance from hot wall

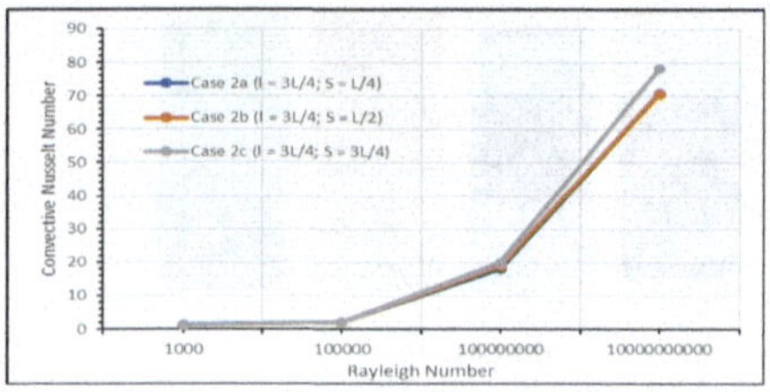

(a) Variation of Nu_C along Ra for Cases 2a, 2b and 2c

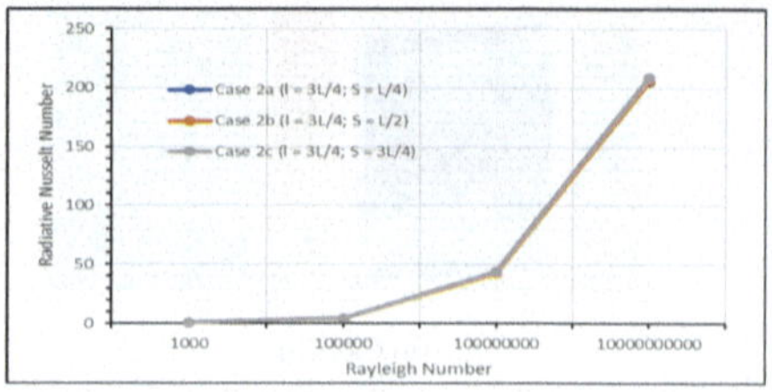

(b) Variation of Nu_R along Ra for Cases 2a, 2b and 2c

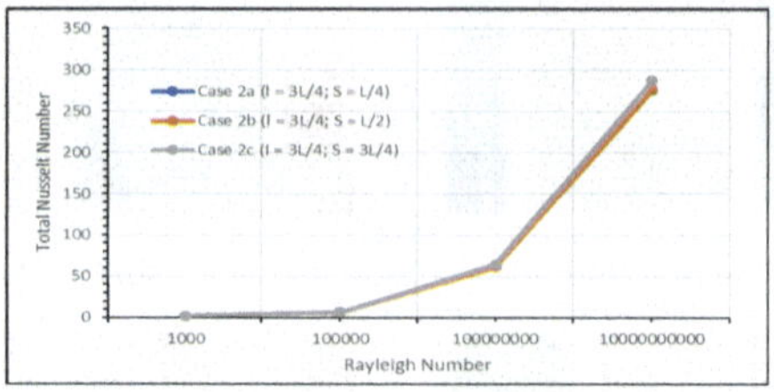

(c) Variation of Nu_T along Ra for Cases 2a, 2b and 2c

Fig. 8 Variation of convective, radiative and total Nusselt number for combined natural convection and radiation for hot wall for $\varepsilon = 1$ in r case 1 for Rayleigh number $10^3 - 10^{10}$

6 Conclusion

This paper presents the interaction of natural convection and surface radiation in partitioned enclosure for laminar and turbulent flow regimes. More than 70 cases were analyzed and the following systematic inference from the investigation were addressed. It can be concluded from the study that the increment in radiative Nusselt number is much higher than that of convective Nusselt number for higher Rayleigh number. When Ra number increases from 10^8 to 10^{10} the percentage increase in Nu_C and Nu_R are observed to be almost 290% and 379% respectively. Thus, radiation plays a major role in net heat transfer from the enclosure. It has been observed from the decrease in the values of about 15% and 28% for Ra 10^5 and 10^{10} as the partition height is increased from L/4 to L respectively. Hence, this leads to the conclusion that as the partition stature is decreased, the heat transfer and fluid flow increase thus resulting in an increase in the Nu_T.

References

1. Bahlaoui A, Raji A, Hasnaoui M, Ouardi C, Naïmi M, Makayssi T (2011) Height partition effect on combined mixed convection and surface radiation in a vented rectangular cavity. J Appl Fluid Mech 4(1):89–96
2. Han CY, Baek SW (2000) The effects of radiation on natural convection in a rectangular enclosure divided by two partitions. Numer Heat Transfer Part A 37:249–270

3. Zemani F, Sabeur-Bendehina A (2018) Transient natural convection in partitioned enclosures. Mech Mech Eng 22(4):1015–1030
4. Ben-Nakhi A, Chamkha AJ (2007) Conjugate natural convection in a square enclosure with inclined thin fin of arbitrary length. Int J Therm Sci 46:467–478
5. Brown WG, Solvason KR (1962) Natural convection through rectangular openings in partitions. Int J Heat Mass Transf 5:859–868
6. Nagaraja R, Aswatha, Vijay Kumar VM, Seetharamu KN (2020) Natural convection in a square enclosure with partitions. Mukt Shabd J 9(6)

Effect of Fuel Spray Geometries on Diesel Combustion Characteristics for Developed Combustion Chamber

Inayat Hussain, Ajit K. Dubey, Sanaur Rehman, Saty Dev, Pushpendra Yadav, and Md. Nazeem Khan

Abstract This study investigates the effect of geometries of fuel spray on diesel combustion characteristics, i.e., ignition lag (IL) for diesel engines. The IL is measured for two nozzle spray designs, i.e., hollow conical spray (HCS) and solid conical spray (SCS). Various cylinder air pressures (CAP), hot surface temperatures (HST), and injection pressures (IP) are considered operating parameters. An experimental setup is fabricated for testing and analysis to reduce IL. The IL for the fuel sprays is recorded through an oscilloscope based on the time lag between the incidents of injection and the presence of flame inside the combustion chamber. The results show that the IL for HCS is (4.9 ms) lesser than SCS (5.2 ms), corresponding to minimum IP, maximum AP and HST. The IL decreases by changing the IP from 100 to 200 bar for HCS and SCS. The increasing HST from 350 to 550 °C and CAP from 15 to 25 bar provide reduced IL. The IL with HCS is 6.12% lower than the SCS. However, the maximum IL of 36.6 ms is obtained by solid conical spray whereas 32.6 ms by HCS corresponds to same parameters i.e. minimum values of HCS, IP, and CAP. The findings show that the hollow conical spray design is more suitable for the diesel engine considering the fuel saving, smooth functioning, and other combustion characteristics.

Keywords CI engine · Nozzle spray designs · IL

I. Hussain · A. K. Dubey
Department of Mechanical Engineering, Indian Institute of Technology Roorkee, Roorkee 247667, Uttarakhand, India

S. Rehman
Department of Mechanical Engineering, Zakir Hussain College of Engineering and Technology AMU Aligarh, Aligarh 202002, UP, India

S. Dev (✉) · P. Yadav
Department of Mechanical Engineering, SRMS College of Engineering and Technology, Bareilly 243202, UP, India
e-mail: satyadevyadav08@gmail.com

Md. N. Khan
Department of Mechanical Engineering, Delhi Technological University, Delhi 110042, India

M. A. Siddiqui et al. (eds.), *Advances in Heat Transfer and Fluid Dynamics*, Lecture Notes in Mechanical Engineering, https://doi.org/10.1007/978-981-99-7213-5_22

Acronyms and Notations

CI	Compression Ignition
HCS	Hollow Conical Spray
SCS	Solid Conical Spray
CAP	Cylinder Air Pressures
IP	Injection Pressures
CFD	Comp. Fluid Dynamics
DSB	Double Swirl Piston Bowl
MSB	Multi-Swirl Piston Bowl
RB	Reentrant Piston Bowl
BB	Base Piston Bowl
CO	Carbon Mono Oxide
°C	Degree (unit of temp)
bar	Unit of Pressure
kg/m^3	Unit of Density
K	Kelvin (unit of temp)
IL	Ignition Lag

1 Introduction

Diesel engines are used in various commercial and consumer applications, such as powering devices like large ships, locomotives, commercial trucks, power generators, and automobiles [1–3]. IL is assigned to the time consumed by both the physical and chemical delays for CI engines. In physical delay, the fuel will inject inside the combustion chamber and it will take some time to mix with air, but in chemical delay, the delay will occur because of time consumption to complete the reaction. Some physical factors affect the IL, such as injection quantity of fuel, drop size, load, engine speed, etc. Some critical factors that affect the IL are shown in Fig. 1. In most CI engines, the IL is shorter than the injection duration [2, 4–6].

Gupta and Khan [2] involved different operating parameters for different geometries of fuel spray and finally concluded that high injection pressure and high temperature inside the combustion chamber plays a significant role in providing a better effect of IL for compression ignition engine with diesel fuel [2].

Rehman and Zaidi [7] investigated the IL of diesel engines corresponding to different parameters such as HST, CAP, and IP with varying values. The findings show that HST is the most significant parameter for controlling the IL in the diesel engine. The reduction in IL was obtained 75–80% by properly controlling hot surface temperature. Another study by Rehman [8] performed similar analyses for the performance of diesel engines. The authors concluded that the processing conditions have a significant role in minimizing IL for smooth combustion in the diesel engine. The

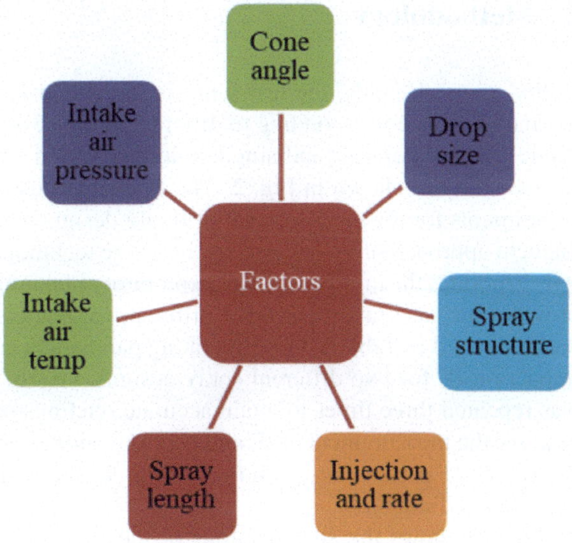

Fig. 1 Factors affecting the IL in CI engine

results also indicate that the increasing hot surface temperature leads to a substantial reduction in IL.

Sener et al. [9] experimentally analyzed the effects of the bowl's shape, spray angle, and fuel injection pressure on ignition phenomena and also obtained various emissions. The significant results of the CFD simulations were found as DSB and MSB improve the air–fuel mixture with CAP for RB, MSB, BB, and DSB corresponding the angles of spray such as 120, 140, 150° at IP of 200 and 1000 bar. They also examined that piston bowl geometry considered in this study reduces the rich-fuel regions. The geometry of the combustion chamber, spray angle, and IP play a significant role in improving the fuel/air mixture, combustion, and minimizing emission. On increasing the IP, indicated thermal efficiency increases while CO and soot particles emissions decrease.

The decrease of IL plays an essential role in smooth combustion in CI engines, better environmental conditions, fuel saving, engine life, good human health, and economy. The IL is affected by many physical and chemical parameters. Many previous studies investigated the effects of operating parameters, such as injection quantity or load, drop size, cylinder pressure, hot surface temperature, etc., on IL in CI engines. The former studies [2, 4–6, 9] analyzed the impact of such parameters on existing and conventional systems. The present study investigated the effect of different spray designs and processing conditions, i.e. IP, HST, and CAP on IL with the help of its own developed setup. The results show significant improvement in the performance of the engine.

2 Methodology

In this research study, the test setup is fabricated in the lab to measure the IL of marine engines corresponding to different spray designs and processing conditions. Pintle (hollow conical) and single hole type (solid conical) nozzles are considered for the study as shown in Fig. 2. The fabricated test setup is shown in Fig. 3. The experiments for the measurement of IL are designed using the Taguchi method. The Taguchi approach offers precise and effective technique for design and optimization. It incorporates the improvement of exploratory planning of experiments. This method provides the easy investigation of different parameters with orthogonal arrays [10]. As mentioned in Table 1, three operating parameters are considered to measure their impact on IL for two different spray designs, i.e. HCS and SCS. Each experiment was repeated three times to attain accurate readings of responses. The oscilloscope used for the measurement of IL and pressure gauges and temperature controller fixes the values of parameters as shown in Fig. 4. The major specifications of setup are provided in Table 2.

Table 3 shows the specifications of the fuel injection pump, oscilloscope, and pressure gauge. The specifications of fuel considered in this study are provided in Table 4.

The major steps in measuring the IL are summarized in Fig. 5. The process starts with selecting designs for nozzle spray i.e., HCS and HCS. Different CAP, HST, and injection pressures (IP) are considered significant factors.

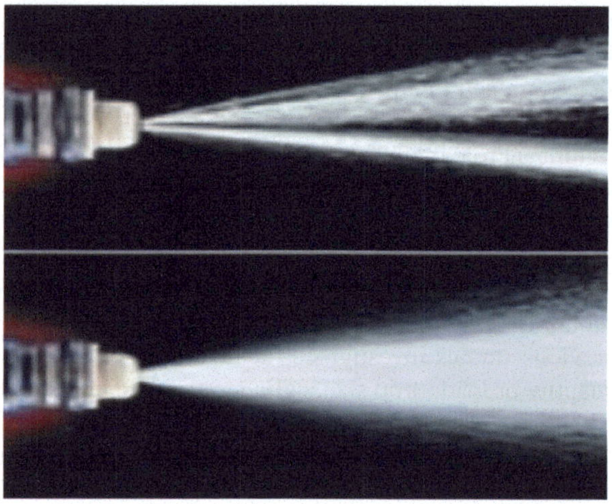

Fig. 2 Hollow conical spray and solid conical spray. *Source* [11]

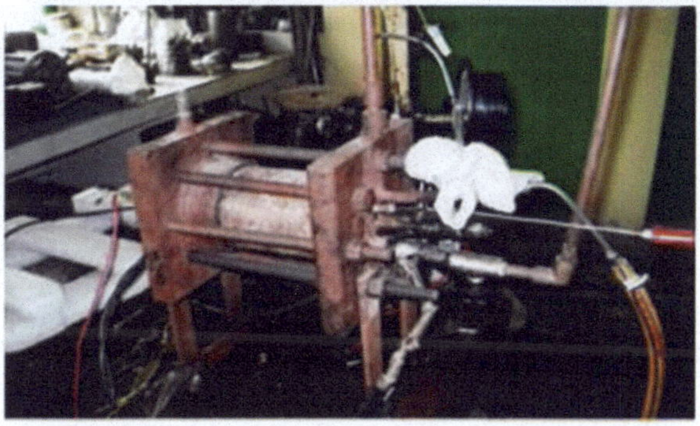

Fig. 3 Pictorial view of the experimental setup

Table 1 Operating parameters and their levels

S. No.	Parameters	Levels		
		Low (1)	Medium (2)	High (3)
1	IP (bar)	100	150	200
2	HST (°C)	350	450	550
3	CAP (bar)	15	20	25

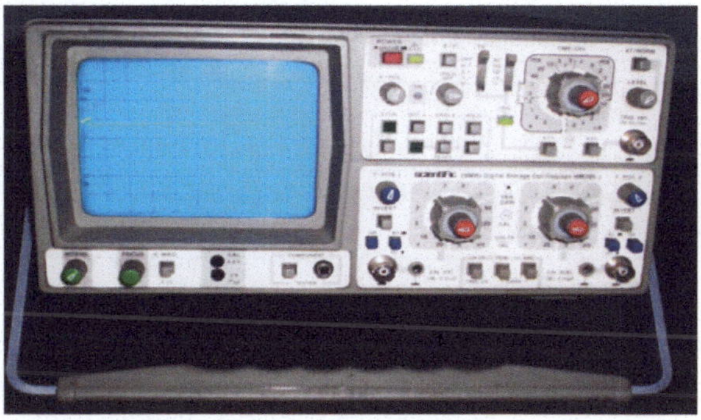

Fig. 4 Oscilloscope used for IL measurement

Table 2 Specifications of setup

S. No.	Particular	Value/type	Unit
1	Length	14	cm
2	Diameter	90	cm
3	Material	Stainless steel	–
4	Nozzle types	Pintle and single hole	–
5	Temp. controller	UTC	–
6	Injection start	Piezo-electric sensor	–
7	Combustion start	Photo sensor	–
8	Max. temperature	800	°C
9	Max. pressure	500	bar

Table 3 Details of setup components

Components	Specifications	Value	Unit
Fuel injection pump (Bosch)	Stroke length	25.4	mm
	Hole size (bore)	9.5	mm
	Limit of fuel delivery	10–140	mm^3
Oscilloscope (Scientech)	Type	Dual Trace Digital Storage	–
	Model	201, 20 MHz–2 mV	–
Pressure gauge (Ki)	Pressure	0–700, 0–140	kg/cm^2

Table 4 Properties of fuel/diesel

Property name	Value	Unit
API gravity	38.98	°C
Density	830	kg/m^3
Aniline point	74	°C
Cetane number	55.36	–
Critical temperature	725.9	K
Critical pressure	20.89	bar

Source [12]

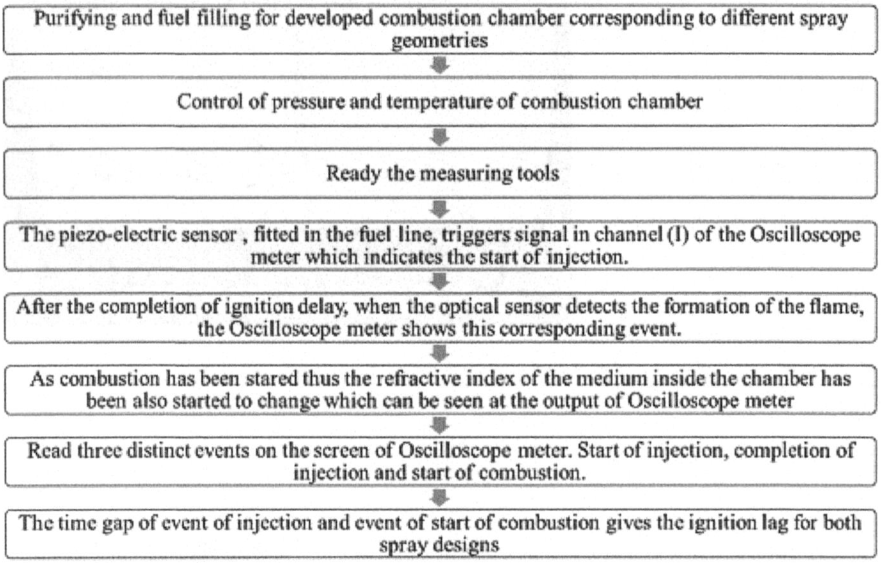

Fig. 5 Procedure of IL measurement

3 Results and Discussions

The values of IL concerning the designed experiments are shown in Table 5 for selected designs. The results show that the ILs for HCS are (4.9) lesser than SCS (5.2), corresponding to minimum IP, maximum AP and HST. The IL with HCS is 6.12% lower than the SCS. It shows that the HCS design is more suitable to the diesel engine considering the fuel saving, smooth functioning and the combustion characteristics.

Table 5 Responses corresponding to the designed experiments

Operating parameters			Responses	
IP (bar)	HST (°C)	CAP (bar)	IL [HCS] (ms)	IL [SCS] (ms)
1	1	1	32.5	36.5
1	1	1	32.4	36.4
1	1	1	32.6	36.6
1	2	2	16.25	10.6
1	2	2	16.30	10.5
1	2	2	16.50	10.7
1	3	3	5.00	5.40
1	3	3	4.90	5.30
1	3	3	5.10	5.20
2	1	2	19.6	32.33
2	1	2	19.5	32.20
2	1	2	19.7	32.4
2	2	3	9.20	7.80
2	2	3	9.10	7.90
2	2	3	9.30	7.70
2	3	1	9.60	7.20
2	3	1	9.70	7.30
2	3	1	9.50	7.10
3	1	3	11.2	35.5
3	1	3	11.1	35.3
3	1	3	11.3	35.4
3	2	1	11.6	13.4
3	2	1	11.7	13.5
3	2	1	11.8	13.3
3	3	2	8.00	6.80
3	3	2	8.10	7.00
3	3	2	7.90	6.90

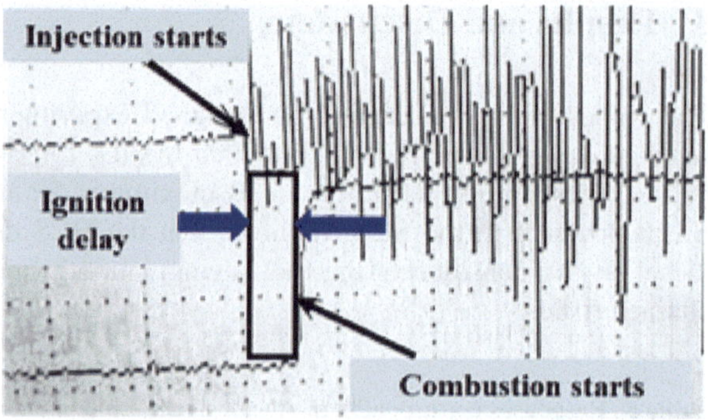

Fig. 6 Regions of combustion

The oscilloscope is used to measure the events of IL. Different regions of fuel combustion are represented in Fig. 6. It shows the lag as the difference between injection and ignition starting.

Figure 7a shows the effect of IP at 100, 150, and 200 bar corresponding to CAP 15 bar and HST 350 °C. Here IL decreases from 100 to 200 bar IP for HCS also it decreases from 100 to 200 bar and slightly increases from 150 to 200 bar for SCS. Figure 7b shows the effect of IP at 100, 150, and 200 bar corresponding to CAP 15 bar and HST 450 °C. Here IL decreases from 100 to 200 bar IP for HCS, whereas it decreases from 100 to 150 bar and slightly increases from 150 to 200 bar for SCS. Figure 7c shows the effect of IP at 100, 150, and 200 bar corresponding to CAP 15 bar and HST 550 °C. Here IL is almost constant from 100 to 150 bar IP but slightly decreases from 150 to 200 bar for HCS. In case of SCS IL decreases from 100 to 150 bar IP, whereas it is somewhat increasing from 150 to 200 bar.

Figure 8a shows the effect of IP at 100, 150, and 200 bar corresponding to CAP 20 bar and HST 350 °C. Here IL decreases from 100 to 200 bar IP for HCS and it is constant from 100 to 150 bar and slightly increasing from 150 to 200 bar for SCS. Figure 8b shows the effect of IP at 100, 150, and 200 bar corresponding to CAP 20 bar and HST 450 °C. Here IL decreases from 100 to 200 bar IP for HCS, whereas it decreases from 100 to 150 bar and slightly increases from 150 to 200 bar for SCS. Figure 8c shows the effect of IP at 100, 150, and 200 bar corresponding to

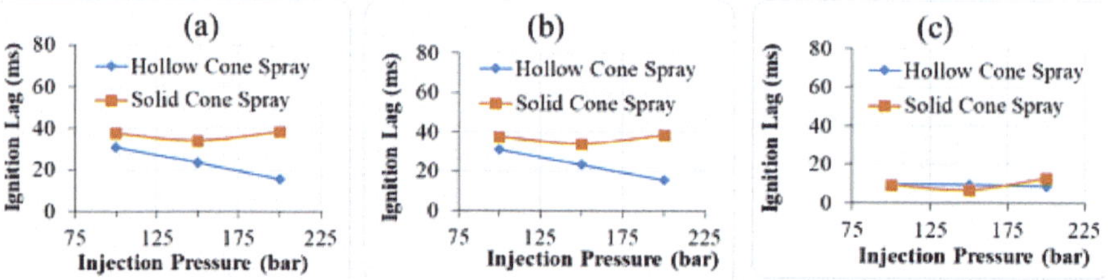

Fig. 7 Effect of IP on IL for both fuel sprays for IP 100–200 bar and CAP 15 bar **a** HST 350 °C **b** HST 400 °C **c** HST 450 °C

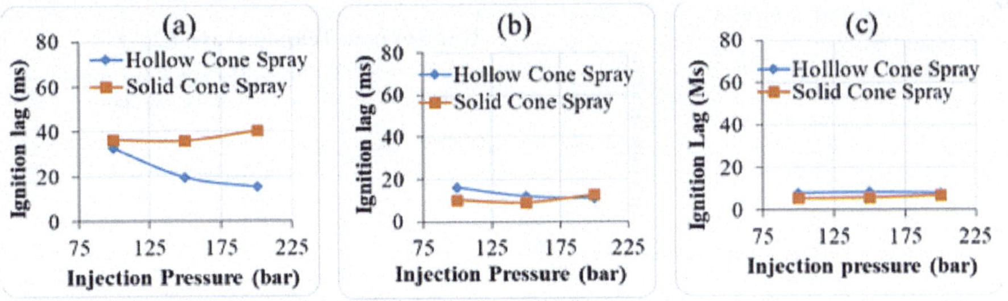

Fig. 8 Effect of IP on IL for both fuel sprays for IP 100–200 bar and CAP 20 bar **a** HST 350 °C **b** HST 400 °C **c** HST 450 °C

CAP 20 bar and HST 550 °C. Here IL is almost constant from 100 to 200 bar for both the fuel sprays.

Figure 9a shows the effect of IP at 100, 150, and 200 bar corresponding to CAP 25 bar and HST 350 °C. Here IL decreases from 100 to 200 bar IP for HCS also decreases from 100 to 150 bar and increases from 150 to 200 bar for SCS. Figure 9b shows the effect of IP at 100, 150, and 200 bar corresponding to CAP 25 bar and HST 450 °C. Here IL is almost constant from 100 to 200 bar IP for both the fuel sprays. Figure 9c shows the effect of IP at 100, 150, and 200 bar corresponding to CAP 25 bar and HST 550 °C. Here IL is almost constant from 100 to 200 bar for both fuel sprays.

Figure 10 shows the effect of CAP at 15, 20, and 25 bar corresponding to HST 550 °C and IP 100 bar. Here IL decreases from 10 to 25 bar CAP for both HCS and SCS but is lesser for HCS than SCS.

Increasing the physical land at high injection pressure reduces the physical lag for diesel engines [13]. The high compression ratio increases air temperature, cylinder air pressure, and self-ignition temperature which helps to reduce the ignition temperature. High intake temperature reduces the IL in the CI engine. Increasing the cylinder air temperature helps fast chemical reactions because of higher vaporization [3]. Different designs of fuel sprays also play a significant role in reducing IL, which is

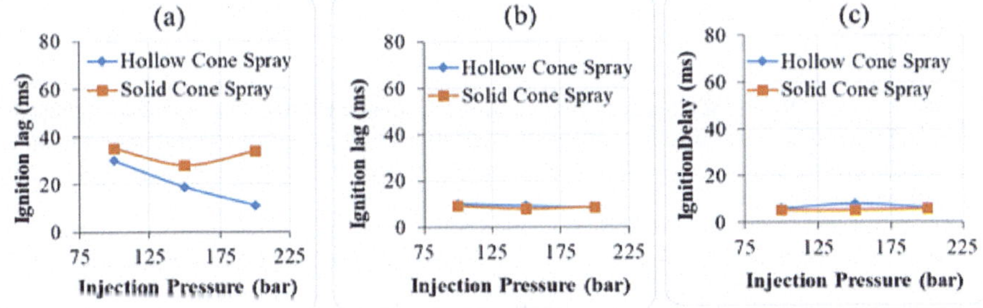

Fig. 9 Effect of IP on IL for both fuel sprays for IP 100–200 bar and CAP 25 bar **a** HST 350 °C **b** HST 400 °C **c** HST 450 °C

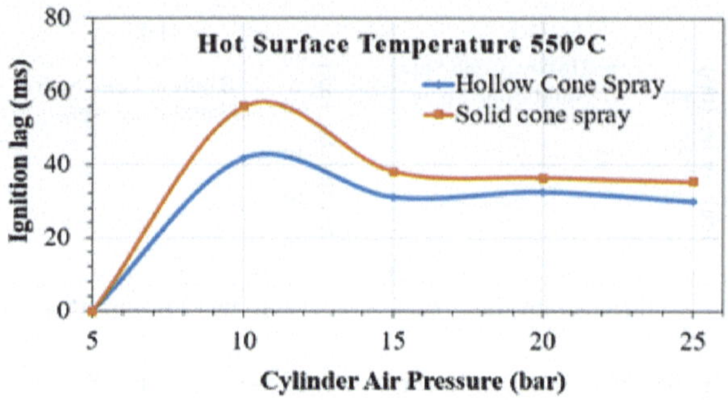

Fig. 10 Comparative results of HCS and SCS

the key to better combustion characteristics for a healthy environment by controlling emissions [14].

4 Conclusions

The present study investigated the effect of geometries of fuel spray on diesel combustion characteristics, i.e., IL for diesel engines. The testing is performed on its own developed setup. The results show that the ILs for HCS are (4.9) lesser than the SCS (5.2), corresponding to minimum IP, maximum CAP, and HST. The IL with HCS is 6.12% lower than the SCS. The IL decreases by changing the IP from 100 to 200 bar for HCS and SCS. The increasing HST from 350 to 550 °C and CAP from 15 to 25 bar provide reduced IL. The IL with HCS is 6.12% lower than the SCS. However, the maximum IL of 36.6 ms is obtained by solid conical spray whereas 32.6 ms by HCS corresponds to same parameters i.e. minimum values of HCS, IP, and CAP. It shows that the HCS design is more suitable for the diesel engine considering the fuel saving, smooth functioning, and combustion characteristics. The study findings will help the researchers and reactionaries of diesel engine applications.

References

1. Bhuiya MMK, Rasul MG, Khan MMK, Ashwath N, Azad AK (2016) Prospects of 2nd generation biodiesel as a sustainable fuel—part: 1 selection of feedstocks, oil extraction techniques and conversion technologies. Renew Sustain Energy Rev 55:1109–1128. https://doi.org/10.1016/j.rser.2015.04.163
2. Gupta T, Khan ME (2019) Variation in ignition delay with changing the environment inside the combustion chamber for different sprays. Int J Innov Technol Explor Eng 8(10):3346–3353. https://doi.org/10.35940/ijitee.J1213.0881019
3. Salman M, Kim SC (2019) Effect of cylinder air pressure and hot surface temperature on ignition delay of diesel spray in a constant volume combustion chamber. Energies 12(13). https://doi.org/10.3390/en12132565

4. Lata DB (2018) Effect of injection timing on the ignition delay of ethanol blended diesel engine. 5(5):995–1002. www.jetir.org995
5. Miron L, Chiriac R, Brabec M, Bădescu V (2021) Ignition delay and its influence on the performance of a diesel engine operating with different diesel—biodiesel fuels. Energy Rep 7:5483–5494. https://doi.org/10.1016/j.egyr.2021.08.123
6. Sanaur Rehman AAK, Nafis MS, Waheed MA (2022) Experimental investigation on combustion and performance characteristics of a direct injection diesel engine. Int J Energy Clean Environ 23(6):1–17
7. Rehman S, Zaidi K (2012) Study of ignition delay of certain blended petroleum fuel sprays in a cylindrical combustion chamber. In: International conference on emerging trends in Engineering and Technology College of Engineering, Teerthanker Mahaveer University [2012] study, vol 2, no 4, pp 35–46
8. Rehman S (2014) Experimental study of hot surface temperature effect on diesel and blended diesel exhaust emissions. In: International conference on energy and environment, pp 15–17
9. Sener R, Yangaz MU, Gul MZ (2020) Effects of injection strategy and combustion chamber modification on a single-cylinder diesel engine. Fuel 266(January):117122. https://doi.org/10.1016/j.fuel.2020.117122
10. Vishwas M, Basavaraj CK, Vinyas M (2018) Experimental investigation using Taguchi method to optimize process parameters of fused deposition modeling for ABS and Nylon materials. Mater Today Proc 5(2):7106–7114. https://doi.org/10.1016/j.matpr.2017.11.375
11. Gupta T, Khan ME (2019) Effect of varying hot surface temperature on ignition delay characteristics for different fuel sprays. Int J Recent Technol Eng 8(4):2906–2910. https://doi.org/10.35940/ijrte.d6599.118419
12. Rehman S (2018) Hot surface ignition and combustion characteristics of sprays in constant volume combustion chamber using various sensors. Cogent Eng 5(1). https://doi.org/10.1080/23311916.2018.1464879
13. Shi Z et al (2020) Effect of injection pressure on the impinging spray and ignition characteristics of the heavy-duty diesel engine under low-temperature conditions. Appl Energy 262(January):114552. https://doi.org/10.1016/j.apenergy.2020.114552
14. Bergthorson JM, Thomson MJ (2015) A review of the combustion and emissions properties of advanced transportation biofuels and their impact on existing and future engines. Renew Sustain Energy Rev 42:1393–1417. https://doi.org/10.1016/j.rser.2014.10.034

An Estimation of Various Performance Parameters of CI Engine Fueled with Diesel and Biodiesel Blends

Brahma Nand Agrawal, Charan Singh, and Shailendra Sinha

Abstract A series of test experimental analyses have been performed to analyze the performance parameters of CI engine fueled with diesel and biodiesel blends. The various biodiesel blends have been prepared as K0L20D80, K5L15D80, K10L10D80, K15L5D80, and K20L0D80 with the combination of Karanja and Linseed biodiesel along with diesel. The results indicate that the performance parameters of blends have been given adjacent results to diesel. A higher concentration of Karanja oil as compared to linseed oil in diesel blends has given better results for performance parameters. The results demonstrated that K15L5D80 gives better results compared to all other blends and is close to diesel for the performance parameters of CI engine.

Keywords CI engine · Performance parameters · Karanja oil · Linseed oil · Diesel

1 Introduction

Extensive research was started to find substitute sources of conventional fuel. Biodiesel came as one of the alternatives for fossil fuels. It is formed mainly by conversion of edible oils [1]. In an experimental study, the longer ignition delay, and pressure rise have been achieved at propanol diesel blends [2]. A study done on ignition delay was studied using blends of n-pentane and DEE with regular diesel. A type of optical method is used for the measurement of ignition delay for this experimental setup. During the experiment, various different pressures and temperatures

B. N. Agrawal (✉)
Department of Mechanical Engineering, Manav Rachna International Institute of Research and Studies, Faridabad, India
e-mail: brahmaagrawal.set@mriu.edu.in

C. Singh
Department of Mechanical Engineering, Galgotias University, Greater Noida, India

S. Sinha
Department of Mechanical Engineering, Institute of Engineering and Technology, Lucknow, India

M. A. Siddiqui et al. (eds.), *Advances in Heat Transfer and Fluid Dynamics*, Lecture Notes in Mechanical Engineering, https://doi.org/10.1007/978-981-99-7213-5_23

of air coming inside the combustion chamber have been studied. The investigation was done on a diesel engine combustion system fueled with blends of n-Pentane and DEE with pure/regular diesel. The ignition delay characteristics were studied. The result reveals that by increasing the temperature and pressure of hot air, the ignition delay of diesel engine fuel spray decreases. Results also show that at low ignition temperatures, effect of methyl group is more dominant whereas at higher temperature the alkyl group is more effective. The main factors for ignition delay are temperature and the pressure of hot air inside the engine combustion chamber. The results say that the DEE reduces the ignition delay [3]. An experimental test investigation has been done under varying load conditions (0, 2, 4, 6, 8, and 10) kW, with diesel engine having blends of Karanja oil to be tested, and SVO, with diesel. Results showed that the BSFC decreases as rise in blend percentage of regular diesel while brake thermal efficiency (BTE) increases. Carbon dioxide, carbon monoxide, and some amount of hydrocarbon emission decrease but Nitric oxide shows slight increasing trend in comparison to FD. The results reveal that the emission characteristics of KB20 were discovered useful [4]. The engine can simultaneously achieve high efficiency and low emissions so that it has attracted much interest. Use of blend fuel Dimethyl ether (DME)/Natural Gas (NG) is used to investigate the effect by (SOC) timing. The experimental results are focused on to parameter variations of SOC timing, and heat flux, and revealed that the quantity of DME/NG supplied increased in blend a 2-stage heat release rate would be seen than that of quantity supplied. It was clearly noticed that changes in DME can advance SOC while having to reduce inlet temperature. Supplying 5% DME would make a change of 6.7-CAD in SOC [5]. An (HCCI) combustion is demonstrated to see the ability of lower engine emission by characteristics of even mixture and lower temperature of combustion at part loads and low speed with mode variation of CI-HCCI and vice versa in real application. Great variations in IMEP value and combustion occur during mode switch causing unfavorable performance of diesel engine. The test research deals with diesel engine HCC Reliance by early injection of fuel with negative valve overlap (NOV) have been investigated. To increase the smooth working of IMEP during switching of modes based on fuel injection strategy did with significant improvement in EGR smoothness of CA50 [6]. The results reveal that due to turbulent nature of the fuel stream I system, breakup occurs at an early stage at higher air flow rate [7]. Various blend of pine oil and additives addition was used to improve characteristics like performance of engine, emissions coming out, and combustion without any modification of a DI engine. Biofuel significantly degrades the DI engine performance values. To overcome this challenge pine oil with an additive (1,4-dioxane) was selected for study and two variant concentrations of Pine oil were utilized (P20, P40) at additive proportions of (5 ml, 10 ml). A comparison of the mixture and diesel value was conducted and an increase of 5.77 and 3.88% in fuel consumption and brake thermal efficiency was noticed respectively. A contemplative of emissions and decreases in pollutant is noted (22.41% CO, 29.19% HC, 22.81% smoke) leaving CO_2 and NO_x at peak loads is described. Change in NO_x and CO_2 value increase about 7.24 and 25.42%. Parameters like heat release rate and peak pressures of pine oil mixture gave inferior values with respect to regular diesel. P20 is the best blend when mixed with

10 ml additive and gives a close value than that of regular diesel [8]. The estimated performance analysis for engine is conducted. The result shows that highly reactive fuel changes the operation condition and optimum combustion significantly. A great influence of unburned hydrocarbon, nitrogen oxide and carbon monoxide is seen on addition of n-heptane. Simulation shows pure biogas and n-heptane operation generates greater results with practical HCCI engine solutions [9]. A diesel engine powered by fish oils and waste palm cooking oil is conducted to see the characteristics change in performance and emission with other parameters. During experiment modified engine with different parameters is used with advanced compression ratio of (17.9:1), injection timing (23.9° bTDC), injections pressure (231 bar), modified Pistons head geometry and speed at 1500 rpm. In experiment PMBD20% shows close resemblance to regular diesel fuel [10]. A review on combustion, emission with performance of CI, CRDI, and HCCI engine operating on different fuels is discussed. The use of mechanical governor fuel injection system to be changed to and Micro-controller controlled system to control injections of fuel at any injection pressure and current angle with improved BTE of Compressed ignition engine with change in emissions levels is reviewed [11]. The earliest start of combustion covers more area in the CP-time diagram while using regular diesel–biodiesels–DEE blends [12]. The ignition sources of dual-fuel combustion were biodiesel derived from soybeans oil [13]. Effects of blends on the cylinder pressure, heat release rates, and emissions were measured. For all the blends of SMEs, NO_x emissions were increased. The experimental test results are validated with Diesel-RK software [14]. Characteristics of performance and emission were studied on the effects of thermal barriers coated pistons fueled with mahua-biodiesel on regular diesel engine and compared with those results of neat diesel fuel. Coating of 0.25 mm thickness of Al_2O_3 material is used on pistons, cylinder walls and the valves of the engine with same compression ratio of the engine. A series of experiments were performed while engine was fueled with regular diesel and B20 and results were compared with and without coating [15].

Many performance parameters should be added to attention. In this study, major base parameters were investigated at CR 18, when it is fueled with 0% Karanja oil, 20% linseed oil, and 80% regular diesel by volume and mixed properly, named as K0L20D80, 5% Karanja oil, 15% linseed oil, and 80% regular diesel, named as K5L15D80, 10% Karanja oil, 10% linseed(flax) oil and 80% regular diesel, named as K10L10D80, and 15% Karanja oil, 5% linseed oil and 80% regular diesel, named as K15L5D80, 20% Karanja oil, 0% linseed oil and 80% regular diesel, named as K20L0D80. This test study has been done to analyze and find optimum blends for performance parameters of the CI engine.

2 Materials and Methods

There are two types of biodiesels used in this study which are obtained from Karanja and Linseed oil.

Table 1 Physical properties of fuel

S. No.	Name of property	Unit	Karanja oil	Linseed oil	Pure diesel
1	Viscosity	cS	27.9	29.3	2.79
2	Density	kg/m^3	911.5	925.7	848.7
3	Calorific value	MJ/kg	34.1	30.7	43.8
4	Cloud point	°C	13	−3.9	−10.2
5	Pour point	°C	4.1	1.01	−14.98
6	Flash point	°C	206	240	76.5

Karanja Oil

Karanja oil is extracted from millettia pinnate tree which is a native of South and Southeast Asia. The seeds on the plant can be extracted from fifth year of plant life and increase each year until it stabilizes in tenth year. The harvest season is spring usually and each seed weighs around 1.101–1.79 g with an oil content of 27.1–38.99% according to season. A single tree can supply around 10–50 kg of seeds on different conditions. The method of oil extraction may change according to the person extracting the oil by generally expeller pressing, solvent extraction, and cold pressing are preferred.

Linseed Oil

Linseed oil or flax seed oil is used widely for many things from paint binder, making of varnish for wood finishing, edible oil for dietary supplement, and various chemical manufacturing. The oil is obtained from linseed seeds by the process of expeller pressing and solvent extraction. Within eight weeks of seed sowing plant can attain a height of 10–15 cm and around 35–45% oil can be extracted from a kg of dry linseed seeds (Table 1) [16, 17, 18].

3 Experimental Arrangement and Method

3.1 Experimental Setup

The present study covers the investigation of performance characteristics of the CI engine. The diesel and blended with linseed and Karanja biodiesel was used as a fuel for investigation. Figure 1 shows the schematic diagram of experimental setup. The computer is connected to the engine. Engine-soft-LV has been commissioned to analyze various characteristics of engine. Different load conditions were applied to the engine connected with the dynamometer. The engine specification is given in Table 2.

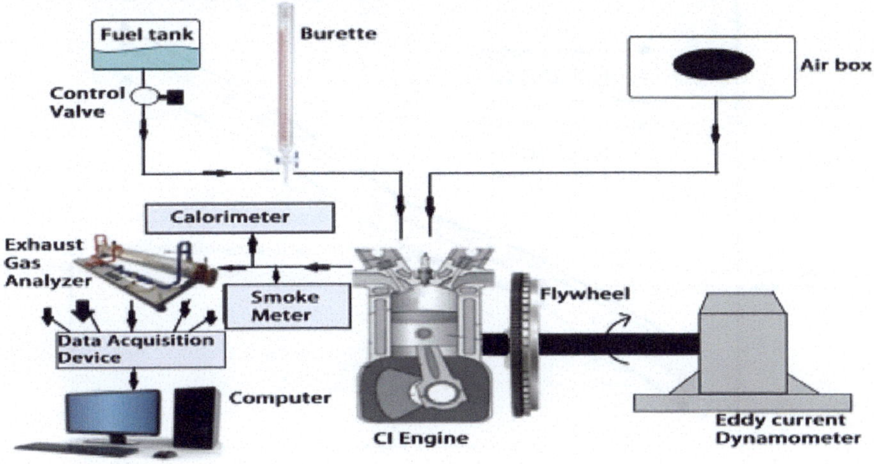

Fig. 1 Experimental setup

Table 2 Engine specifications

Engine specifications	
Engine type	Four stroke
Rated power capacity	3.5 kW @ 1500 rpm
Compression ratio available	18
CR ranging	13–18

3.2 Experimental Procedure

The procedure adopted during test investigation is as follows.

a. First, all the blends of Karanja oil and linseed oil were prepared.
b. To test and investigate of performance parameters of the CI engine, engine needs to run for fifty minutes to become stable in its conditions while fueled with diesel. The data acquisition devices collect all data at constant speed of engine and various loads.
c. All previous steps repeat for blends K0L20D80, K5L15D80, K10L10D80, and K15L5D80, K20L0D80.
d. Evaluated data were analyzed and from the observed values, the various graphs show the performance parameters.

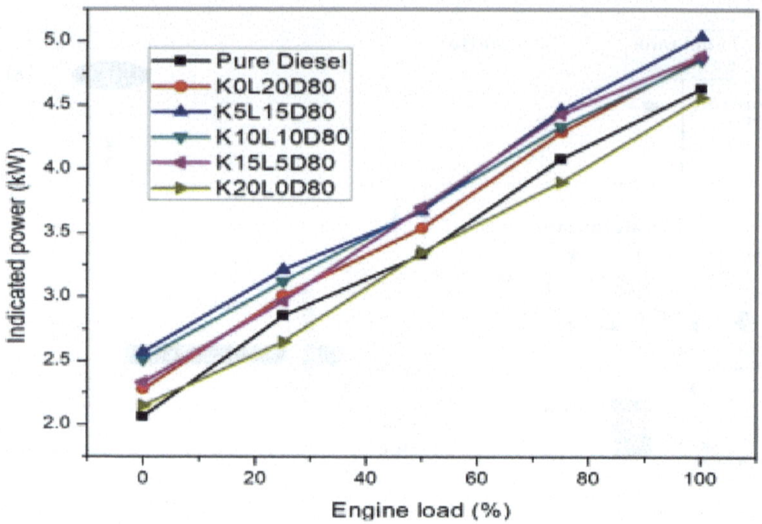

Fig. 2 Indicated power versus engine load

4 Results and Discussion

4.1 Performance Parameters

This section deals with various performance parameters of engine fueled with various biodiesel blends and diesel, and discussed below.

4.1.1 Indicated Power

It is been detected from Fig. 2, that all blends except K20L0D80 give larger indicated power than that of regular diesel for all loads. At full loads, the blend K5L15D80 gives more indicated power by amount of 500 W to regular diesel. At 75% of engine load both blends K5L15D80 and K15L5D80 give nearly same results but greater than pure diesel by amount of 410 W.

4.1.2 Mechanical Efficiency

This study reveals that from Fig. 3, blend K20L0D80 give higher mechanical efficiency while comparing other blends and diesel at entire ranges of engine loads. Nevertheless, at no load conditions, all blends of Karanja and linseed oil give very similar result for mechanical efficiency to that of pure diesel. The blend K20L0D80 gives 3% more mechanical efficiency compared to diesel at full loads.

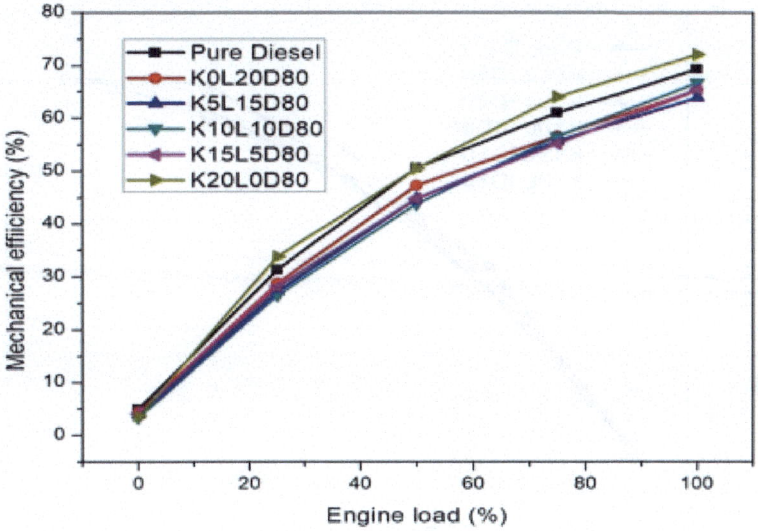

Fig. 3 Mechanical efficiency versus engine load

4.1.3 Brake Thermal Efficiency

It was clear that K15L5D80 has more BTE than that of other blends and pure/regular diesel from 35 to 70% of engine loads as Fig. 4. However, except 35 to 70% of load, pure diesel has greater brake thermal efficiency than any other blend except K15L5D80. At 50% load condition, 6% higher brake thermal efficiency was achieved by blend K15L5D80 than that of pure/regular diesel.

4.1.4 Brake-Specific Fuel Consumption

Brake-specific fuel consumption is a sign of effective fuel combustion. Figure 5 displays the BSFC of diesel, Karanja, and linseed oil mixtures for various loads. From 0 to 22% of load blend K20L0D80 has higher BSFC than other blends along with diesel. Due to improved combustion, the BSFC of blends and pure/regular diesel drops as loads rise.

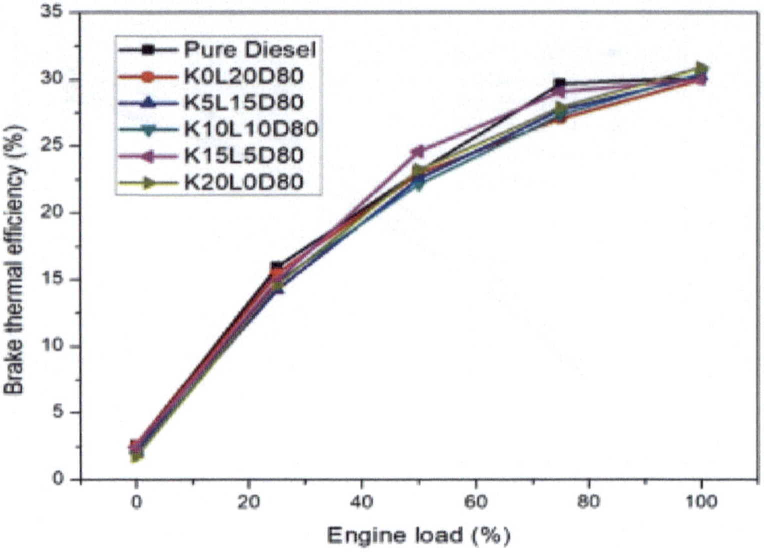

Fig. 4 BTE versus engine load

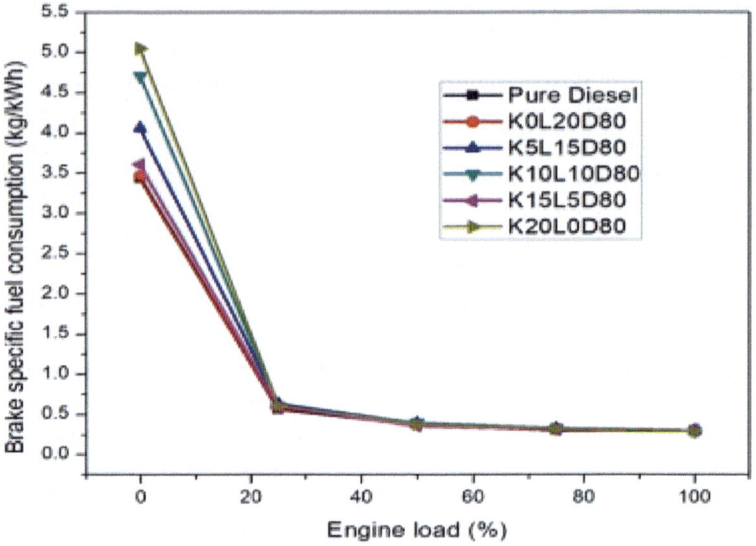

Fig. 5 BSFC versus engine load

5 Conclusion

The concluding remarks can be drawn from the study as follows.

- The blend K15L5D80 gives close result to pure diesel for cylinder pressure and mass fraction burned as comparable to pure diesel.
- The heat releasing rate of blend K20L0D80 is greater than that of all other blends of linseed oil and diesel also.
- The mean gas temperature is also close to pure diesel at blend K15L5D80.
- The rate of pressure rise is comparable to pure/regular diesel at blend K20L0D80.
- It has been shown in the test research that at all blends K5L15D80, noticed highest indicated power and indicated mean effective pressure than pure diesel and all blends except blend K20L0D80 at no load condition to full loads.
- Pure diesel has 3% more mechanical efficiency than that of blend K20L0D80 at full loads.
- At blend K15L5D80 higher the brake thermal efficiency has been achieved from 35 to 70% of engine load than that of diesel. However, on other engine load conditions brake thermal efficiency is comparable to diesel.
- Brake-specific fuel consumption is more at blends K20L0D80 from 0 to 22% of engine loads.
- Its analysis forces us to assume that there should be an optimal ratio of the combination of Karanja oil, linseed oil, and diesel. The present investigation noticed better combustion attributes and performance parameters at blend K15L5D80. This study concludes that higher concentration of Karanja oil can be used as partial substitution for the diesel.

References

1. Yunus TM, Atabani AE, Anjum I, Badarudin A, Khayoon MS, Triwahyono S (2014) Recent scenario and technologies to utilize non-edible oils for biodiesel production. Renew Sustain Energy Rev 37:840–851. https://doi.org/10.1016/j.rser.2014.05.064
2. Muthaiyan P, Gomathinayagam S (2016) Combustion characteristics of a diesel engine using propanol diesel fuel blends. J Inst Eng Ser C 97(3):323–329. https://doi.org/10.1007/s40032-016-0229-x
3. Nand Agrawal B (2017) Combustion characteristics of diesel combustion system using blended diesel: an experimental study. Am J Mech Appl 5(4):34. https://doi.org/10.11648/j.ajma.20170504.12.
4. Nayak SK, Mishra PC, Kumar A, Ranjan G, Nayak B (2017) Environmental effects experimental investigation on property analysis of Karanja oil methyl ester for vehicular usage. Energy Sources Part A Recover Util Environ Eff 39(3):1–7. https://doi.org/10.1080/15567036.2016.1173131
5. Ezoji H, Shafaghat R, Jahanian O (2019) Numerical simulation of dimethyl ether/natural gas blend fuel HCCI combustion to investigate the effects of operational parameters on combustion and emissions. J Therm Anal Calorim 135(3):1775–1785. https://doi.org/10.1007/s10973-018-7271-2

6. Xu M, Gui Y, Yao Deng K (2015) Fuel injection and EGR control strategy on smooth switching of CI/HCCI mode in a diesel engine. J Energy Inst 88(2):157–168. https://doi.org/10.1016/j.joei.2014.06.005
7. Garai A, Pal S, Mondal S, Ghosh S, Sen S, Mukhopadhyay A (2017) Experimental investigation of spray characteristics of kerosene and ethanol-blended kerosene using a gas turbine hybrid atomizer. Sadhana - Acad Proc Eng Sci 42(4):543–555. https://doi.org/10.1007/s12046-017-0624-x
8. Mebin Samuel P, Devaradjane G, Gnanamoorthi V (2020) Performance enhancement and emission reduction by using pine oil blends in a diesel engine influenced by 1, 4-dioxane. Int J Environ Sci Technol 17(3):1783–1794. https://doi.org/10.1007/s13762-019-02455-8
9. Kozarac D, Taritas I, Vuilleumier D, Saxena S, Dibble RW (2016) Experimental and numerical analysis of the performance and exhaust gas emissions of a biogas/n-heptane fueled HCCI engine. Energy 115:180–193. https://doi.org/10.1016/j.energy.2016.08.055
10. Nayak SK, Hoang AT, Nayak B, Mishra PC (2021) Influence of fish oil and waste cooking oil as post mixed binary biodiesel blends on performance improvement and emission reduction in diesel engine. Fuel 289(December 2020):119948. https://doi.org/10.1016/j.fuel.2020.119948
11. Khandal SV, Banapurmath NR, Gaitonde VN, Hiremath SS (2017) Paradigm shift from mechanical direct injection diesel engines to advanced injection strategies of diesel homogeneous charge compression ignition (HCCI) engines—a comprehensive review. Renew Sustain Energy Rev 70(November 2016):369–384. https://doi.org/10.1016/j.rser.2016.11.058
12. Rajendiran G, Subramanian R, Venkatachalam R, Nedunchezhian N (2014) Experimental investigation on combustion analysis of multicylinder direct injected diesel engine using diesel–biodiesel–DEE as alternative fuel. Int J Ambient Energy (October 2014):37–41. https://doi.org/10.1080/01430750.2012.740419
13. Han S, Hyun S, Sik C (2014) Bioethanol and gasoline premixing effect on combustion and emission characteristics in biodiesel dual-fuel combustion engine. Appl Energy 135(x):286–298
14. Al MF, Bhatti SK (2014) Experimental and computational investigations for combustion, performance and emission parameters of a diesel engine fueled with soybean biodiesel-diesel blends. Energy Procedia 52:421–430. https://doi.org/10.1016/j.egypro.2014.07.094
15. Babu RTS, Kannan M, Lawrence P (2016) Performance analysis of low heat rejection diesel engine, using Mahua oil bio fuel. Int J Ambient Energy 0750(September):3–8. https://doi.org/10.1080/01430750.2016.1222963
16. Agrawal BN, Sinha S, Kuzmin AV, Pinchuk VA (2019) Effect of vegetable oil share on combustion characteristics and thermal efficiency of diesel engine fueled with different blends. Therm Sci Eng Prog 14:100404, ISSN 2451-9049. https://doi.org/10.1016/j.tsep.2019.100404
17. Agrawal BN, Sinha S, Singh SK, Dubey KK, Kumar M (2021) An experimental study on performance parameters of diesel engine fueled with blends of diesel fuel and linseed biodiesel. In Sikarwar BS, Sundén B, Wang Q (eds) Advances in fluid and thermal engineering. Lecture notes in mechanical engineering. Springer, Singapore. https://doi.org/10.1007/978-981-16-0159-0_23
18. Sinha S, Agrawal BN, Ujjwal, Rajput S (2023) Comparative study of performance parameters of CI engine fueled with various blends of biodiesel. In Sikarwar BS, Sharma SK, Jain A, Singh KM (eds) Advances in fluid and thermal engineering. FLAME 2022. Lecture notes in mechanical engineering. Springer, Singapore. https://doi.org/10.1007/978-981-99-2382-3_28

Combustion and Performance Analysis of a CI Engine Fuelled with Soybean Oil Biodiesel Blends

Muhammad Aazam Waheed, Mohammad Sufiyan Nafis, Anas Anwar Khan, and Sanaur Rehman

Abstract In the present experimental investigation, comparative analysis of combustion and performance characteristics of a DI diesel engine fuelled with diesel and soybean biodiesel blends was done. Biodiesel was produced through catalytic transesterification method and its blends with diesel were prepared in various concentrations (B10, B20 and B30) by volume. In BX blend, X indicates percentage concentration of the biodiesel in the diesel fuel. Analysis of various combustion and performance parameters such as in-cylinder pressure, rate of pressure rise, net heat release, ignition delay period, brake thermal efficiency and brake specific fuel consumption were performed at different engine loads (20, 40, 60, 80 and 100%) keeping constant the engine speed. Combustion parameters obtained are the averages of five engine cycles. It is observed that with increasing loads, higher values of various combustion parameters are found for all fuels. Results indicated better combustion and performance parameters at full engine load for all fuels. Increase in biodiesel concentrations result in higher in-cylinder pressure, net heat release, rate of pressure rise and lower ignition delay periods at full load. It is found that at full load condition, S30 and S20 blend have shown improved combustion over S10 blend. There has been a marginal improvement in performance parameters for S10 blend at full load.

Keywords Biodiesel · Soybean · Diesel engine · Combustion · Performance

Nomenclature

ATDC	After Top Dead Centre
BSFC	Brake Specific Fuel Consumption
BTE	Brake Thermal Efficiency

M. A. Waheed (✉) · M. S. Nafis · A. A. Khan · S. Rehman
Department of Mechanical Engineering, Aligarh Muslim University, Aligarh, India
e-mail: universitypoly752@gmail.com

S. Rehman
e-mail: sanaur.rehman@zhcet.ac.in

M. A. Siddiqui et al. (eds.), *Advances in Heat Transfer and Fluid Dynamics*, Lecture Notes in Mechanical Engineering, https://doi.org/10.1007/978-981-99-7213-5_24

CA	Crank Angle
CI	Compression Ignition
CR	Compression Ratio
DI	Direct Injection
EOC	End Of Combustion
ICE	Internal Combustion Engines
NHR	Net Heat Release
TDC	Top Dead Centre
ROPR	Rate of Pressure Rise
SOC	Start Of Combustion
SOI	Start Of Injection

1 Introduction

Global energy consumption has risen dramatically since the twentieth century as a result of industrialization, population growth, and a multi-fold increase in energy demand. The work produced by the burning of fossil fuels such as coal and petroleum accounts for over 95% of worldwide transport energy. Even modern transportation and its reliance on traditional sources of energy such as fossil fuels are hastening their depletion in every sector of the world [1]. Furthermore, coal and petroleum combustion contribute to serious environmental issues such as climate change, global warming, and air pollution. The continued use of fossil fuels because of rising global energy demand has prompted scientists and researchers to look for innovative and efficient alternative fuels that are both renewable and non-polluting, with the potential to reduce global warming and pollution [2, 3]. Biodiesel has proved to be the most efficient and promising alternative fuels for diesel engines, since it is oxygenated, nontoxic, biodegradable, and ecologically acceptable, and it is made up of fatty acids and may well be synthesized from vegetable oils, animal fats, using a transesterification method. Furthermore, it is free of sulphur and minerals, and its high oxygen concentration. The biggest benefit of utilising biodiesel is that it can be utilised in current engines without requiring any modifications [4, 5].

When compared to diesel, several prior studies have indicated that biodiesel provides the same or slightly superior thermal efficiency, as well as lower or equal BSFC values for higher loads. Schumacher's [6] investigation on a diesel engine operating on soybean blends at 10, 20, 30, and 40% concentrations showed that engines running on soybean/diesel fuel blends produced lower levels of particulate matter, total hydrocarbons, and carbon monoxide, but higher levels of nitrogen oxides (NO_x). The B20 blend reduced emissions significantly while sacrificing engine performance marginally. Sidharth and Naveen [7] studied performance and emissions of ternary fuel blends of diesel, biodiesel, and octanol, and it was found that the blend had better BTE and BSEC than diesel due to more oxygen content and comparable properties. In comparison to pure diesel, biodiesel and octanol blends have a higher oxygen concentration which aids in better combustion and performance [7]. Aydin et al. [8] analysed three different types of fuel including diesel, B20, BK20 (80% biodiesel and 20% kerosene) and their effects on CO, NO_x, and performance parameters. Kerosene was used as an additive to approach the properties of biodiesel to diesel. The experimental results showed that the exhaust emissions for BK20 were fairly reduced as compared to diesel fuel and B20. Also, performance was enhanced with increased BTE and decreased BSFC [8]. Similarly, Devaraj et al. [9] investigated the effects of two oxygenated additives namely, DMC (Dimethyl-carbonate) and Pentanol (n-P), on ignition patterns of biodiesel/diesel blends in the diesel engine. BSFC was lowered by 0.4 and 0.5 g/kW with 0.3 and 0.6% increase in BTE by blending DMC and Pentanol to base fuel respectively. The peak pressure of the base fuel was increased by 2.3 and 3.1 bar by adding DMC and pentanol, respectively. Abed et al. [10] studied the WCO effects on a single cylinder diesel engine with three blends with concentrations of 10, 20, and 30%. The thermal efficiency of WCO biodiesel blends was recorded to be inferior than that of the diesel fuel, although BSFC was found to be higher. Furthermore, as the concentration of the blend grew, so did the thermal efficiency and BSFC decreased. Hariram and Kumar [11] used a mixture of two biodiesels in equal weight ratio namely cucurbita pepo L (pumpkin) and Tectona grandis (teak) seed oil for the synthesis of a biodiesel with 5-mL diethyl ether as additive. It was observed that in B20 blend with DEE, there was 35.5% increment in BTE and 25.8% reduction in BSFC at maximum load for 200 mL pumpkin and teak biodiesel blend mixed with diesel. Also, at maximum load, the emission of CO reduced by 26.67%, CO_2 reduced by 12.96%, and NO_2 reduced by 26.57%. Vinothkannan and Pushparaj [12] experimentally investigate the use of soybean oil methyl ester and its blends (5, 10, 15 and 20%) in a single cylinder direct injection diesel engine with diesel and with varying loads. The combustion study such as in-cylinder, maximum peak pressure, rate of heat release and ignition delay are analysed and found that SOME 20% blend exhibits maximum cylinder pressure and better heat release characteristics. Zhu et al. [13] observed an opposite trend, with almost comparable maximum pressure values for the biodiesel and diesel fuel at low and medium engine loads and higher cylinder pressure values for the biodiesel at high engine loads.

Annisa et al. [14] tested the performance and emissions of soybean based WCO. The tests were carried out at 4 loads in a four-cylinder turbocharged diesel engine and reveals that the BTE of BD5W10 was up to 8.92% higher than diesel at low load. When compared to diesel, the BSFC of BD5W10 rose by 11.66% at medium load. CO emissions were greater in BD5W10 than in diesel and B10. When compared to diesel at high load, CO emissions in BDSW10 were reduced by 12.8 percent. Gupta et al. [1] studied the performance and exhaust emissions of a single cylinder diesel engine driven by soybean biodiesel blends (B10, B20, and B50) at various engine speeds. According to the data, the BSFC increased but the torque decreased. Biodiesel blends lowered CO and HC emissions considerably. Yehliu et al. [15] tested a production engine using soybean biodiesel (B100), low-sulphur diesel oil, and a synthetic fuel at varied speeds, loads, and fuel injection settings. When compared to diesel oil, B100 produced more SFC but had a poorer thermal efficiency. Pasupathy and Mohan [16] investigated the performance, combustion and emission of a turbocharged engine by varying the compression ratios for the blends and compared with pure diesel. 16:1, 17:1, and 18:1 compression ratios were used for the engines. The turbocharger had a significant impact on performance and emissions, according to the findings. When compared to other compression ratios, the performance of 18:1 was better. Without a turbocharger, increasing the compression ratio increases NO_x and CO_2 while lowering CO and HC emissions. When compared to B20 and diesel, B10 mix has lower NO_x and HC emissions and slightly higher CO emissions without the turbocharger. Overall emissions are reduced by the turbocharger. Kumar and Jaikumar [17] looked at the efficiency and combustion characteristics of a single cylinder diesel engine that ran on WCO. According to the data, the WCO mix improved in-cylinder pressure and delay period. When compared to neat diesel, neat WCO showed reduced peak pressure in all operating conditions. All the studied fuels improved their BTE as engine power increased (BMEP). The maximum BTE was calculated to be 26% at a maximum BMEP of 4.4 bar, but it was 30.5% with clean diesel. Although numerous research studies have been conducted in the past on performance characteristics and a few combustion parameters for biodiesel blends of soybean, the literature lacks an in-depth combustion and performance study of these blends. Because incomplete combustion of the fuel in the cylinder is the primary cause of increased emissions and poor engine performance, combustion analysis is the most effective way to address the problems associated with inefficient combustion. As a result, with an engine speed of 1500 rpm, the influence of load fluctuation as well as blend variation on several combustion parameters such as in-cylinder pressure, NHR, ROPR, and ID period is presented in this article.

2 Experimental Section

The block diagram of the experimental model is shown in Fig. 1. Tables 1 and 2 provide the technical characteristics of the engine test bed and the combustion parameters, respectively. Experiments are conducted on a two-cylinder, four-stroke, water-cooled, DI diesel engine coupled with a hydraulic dynamometer. A diesel engine, orifice meter, air box, pressure transducer (sensor), manometer, hydraulic dynamometer, K-type thermocouple, CA sensor, piezo powering unit and computer make up the experimental setup. The air flow rate entering the engine cylinder through the air box was calculated using an orifice metre, and the density of the upstream air was calculated using the inclined manometer. The air suction temperature of the engine was measured using a temperature sensor and a K-type thermocouple was used to measure the cylinder gas temperature. CA was measured in 1° with the help of a crank encoder and a data collection card was used for data collection. To adjust the load on the engine, a hydraulic dynamometer was used in conjunction. A clear glass tube fuel metre was connected to the fuel tank to measure the mass flow rate of fuel entering the engine cylinder by measuring the time it took for the fuel level to drop. All combustion parameters were measured for 1° CA for a mean cycle of 100. Table 3 shows the accuracy of the several equipment employed.

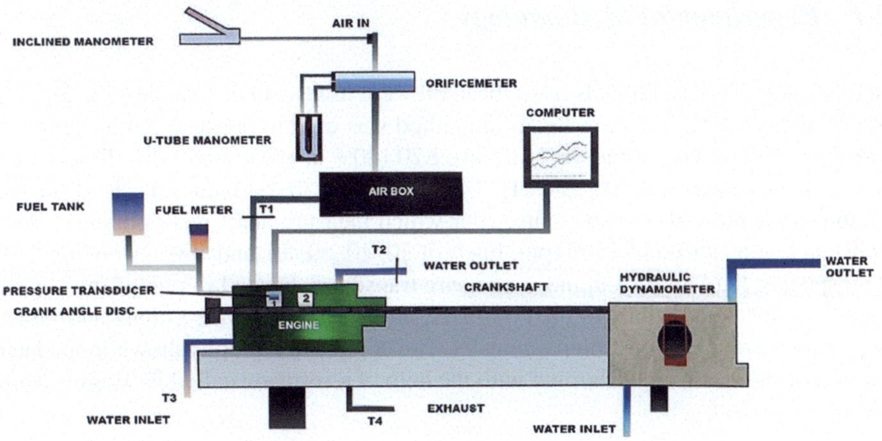

Fig. 1 Block diagram of a DI diesel engine test bed [18]

Table 1 Engine test bed details [18]

Engine type	Kirloskar, TV2, 2-cylinder, 4-stroke, water cooled, diesel engine
Power	10 kW @ 1500 rpm
Cylinder bore	87.50 mm
Stroke length	110.00 mm
Compression ratio	17.50
Swept volume	661.45 cc
Injection timing	27° bTDC
Orifice diameter	25 mm
Fuel pipe diameter	34.50 mm
Ambient temperature	27 °C
Fuel injection pressure	200 bar

Table 2 Combustion parameters [18]

Specific gas constant	1.00 kJ/kg K
Air density	1.17 kg/m^3
Adiabatic index	1.41
Polytrophic index	1.13
Number of cycles	05

Table 3 Accuracy of Instruments Used [18]

Instrument	Accuracy
Pressure sensor (model S111A22)	±0.1 mV/psi
K type thermocouple	±2.2%
Hydraulic dynamometer	±0.1 kg
CA sensor	1 deg (resolution)

2.1 Experimental Methodology

In this study, biodiesel blends of soybean oil were used as fuels in a diesel engine. A traditional catalytic transesterification method was used to produce biodiesel in the lab. S10 (10% soybean and 90% diesel), S20 (20% soybean and 80% diesel), and S30 (30% soybean and 70% diesel). The engine was first driven at no load for ten minutes to achieve the steady state, after which measurements were obtained. At a constant engine speed of 1500 rpm, loads of 20, 40, 60, 80, and 100% were applied. At a specific load on the engine, a pressure transducer located in one of the engine cylinders that sent signals to the computer, allowing to generate graphs and tables depicting various combustion parameters. The combustion graphs shown in the later section of the text were generated with the help of a software called IC Engine Soft.

2.2 Biodiesel Production Process and Fuel Properties

The biodiesel fuel utilised in this test were produced by transesterification of soybean oil with ethanol, and catalysed with sodium hydroxide (NaOH) as the base. The catalyst concentration was determined to be 1% by weight of oil, or 10 g per 1 L of oil. The alcohol to oil ratio employed in the transesterification process was 1:6 by weight of oil. The reaction temperature was 60 °C, and the stirring speed was set to 1000 rpm. To speed up the reaction, both samples were heated in an electric heater to a temperature ranging from 50 to 70 °C. The amount of ethanol measured was combined with sodium hydroxide (NaOH) and agitated until all of the catalyst had interacted with ethanol while the oil was heated. The alcohol-catalyst mixture from a separate beaker was then put into the oil, which was kept at a reaction temperature of 60 °C, and agitated constantly for 1 h till the glycerine layer separated completely. The mixture was further left overnight to separate into two clean layers after which the immiscible solution was then poured into a funnel to separate the glycerine from the ester. Finally, the soybean oil esters were rinsed numerous times with room temperature distilled water until all of the FFA had been drained. Lastly, this synthesized biodiesel was heated at 100 °C for 15 min to eliminate any remaining water in the biodiesel. Various test samples of soybean oil biodiesel were made with diesel at 10, 20, and 30% concentrations, denoted as S10, S20, and S30.

All the test fuels properties such as kinematic viscosity, dynamic viscosity, and density were tested in Combustion and Pollution Control Laboratory, ZHCET, A.M.U using a kinematic viscosity bath, hydrometer. At 40 °C, the kinematic viscosity of all fuels was measured. Up to the specified point, the setup was filled with 25 L of distilled water. Following that, a 25 mL fuel sample was obtained in a clean, dry, and calibrated viscometer. The test fuel was suspended in the viscosity bath for 30 min until it reached the temperature of 40 °C. Later, a suction was applied to the viscometer's thicker limb, and the test fuel was made to flow from mark (C) to mark (G), with the flow time being recorded. The kinematic viscosity of all test fuels

Table 4 Properties of all test fuels

Parameters/test fuels	Diesel	S10	S20	S30
Density (kg/m^3)[a]	820	825	837	845
Kinematic viscosity (cS)[a]	3.65	5.70	6.49	7.24
Dynamic viscosity (mPa s)[a]	2.993	4.70	5.43	6.11
Calorific value (kJ) [19]	42,200	41,689	41,378	41,067
Cetane number [19]	48	48.4	48.8	49.2

[a]Measured properties

was calculated by multiplying flow duration by the kinematic viscosity constant. Similarly, a hydrometer with a range of 800–850 kg/m^3 was chosen to calculate the density. A 500 mL sample of the test fuel was placed in a beaker, and the hydrometer was submerged in it for 2 min before being allowed to settle. The density readings were later noticed. Table 4 lists the properties of all test fuels.

3 Results and Discussion

3.1 Combustion Characteristics

3.1.1 Cylinder Pressure

One of the most significant elements to consider while examining combustion characteristics is cylinder pressure. In a compression ignition engine, the peak cylinder pressure, or pressure developed in the initial stage of combustion, is determined by the burning of fuel proportion during the premixed burning phase. As a result, cylinder pressure shows how well the air/fuel mixture burns in the combustion chamber. Figure 2a depicts the fluctuation in cylinder pressure with load for diesel and (b), (c), and (d) for several biodiesel blends of soybean oil of S10, S20 and S30 concentrations, and (e) shows the several blends comparison at full load for an engine running at 1500 rpm. Figure 2a shows how the cylinder pressure of a diesel engine increases as the engine load increases from 20 to 100%. Diesel has a maximum cylinder pressure of 63.6 bar, which is roughly at 12° aTDC. Increased fuel flow to the engine owing to rising fuel requirement causes pressure to increase with load [20].

Under blend comparison, as shown in Fig. 2b–d cylinder pressure of soybean blends increases with increasing load due to higher energy release. Diesel has a peak cylinder pressure of 63.6 bar at full load, which is greater than all biodiesel blends. 62.6 bar (S10), 62.3 bar (S20), and 61.13 bar (S30) are the peak pressures for biodiesel blends. Despite the fact that S10 blend had pressure values equivalent to diesel, all other tested fuels had identical in-cylinder pressure curves, implying that they had similar combustion processes in the engine, namely premixed combustion followed

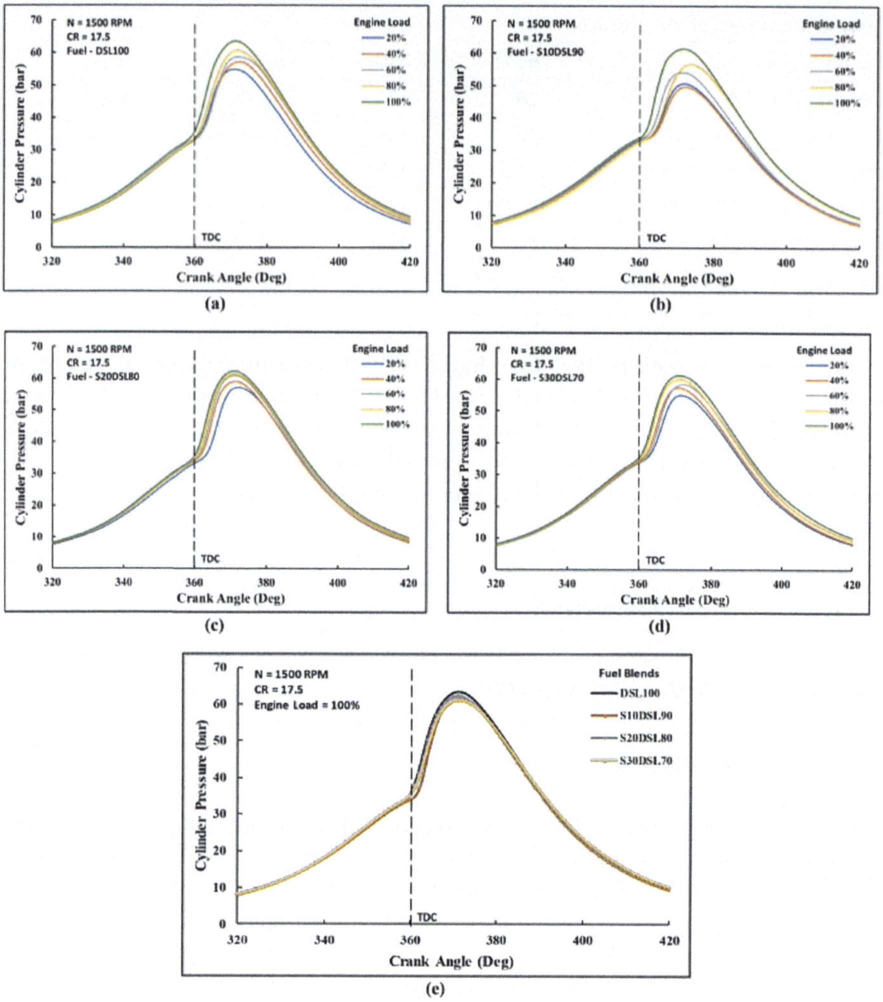

Fig. 2 Variation of cylinder pressure versus crank angle for diesel and soybean oil blends

by diffusion combustion. With the addition of soybean blending, combustion has advanced at all engine loads, with peak pressures for blends reaching near 10 aTDC. Because biodiesel has a larger bulk modulus and density than diesel [21], it requires earlier injection time. Furthermore, a higher cetane number for soybean mixing has resulted in quicker ignition of the blends into the combustion chamber, resulting in higher pressure peaks [22]. The ignition delay duration is another important aspect in the combustion phenomena in an IC engine. It is defined as the time difference between SOI and SOC, and it consists of physical and chemical delay periods, where physical delay refers to the time it takes for the fuel to evaporate and atomize, and chemical delay describes the time taken by fuel to ignite and a flame to appear [23].

Figure 5 shows that the delay period for the rising load is longer for diesel than for most biodiesel blends, with the exception of S10 blend in a few engine load scenarios, owing to the experimental and computational error. Despite this, the general trend for biodiesel blends appears to be declining. The lower cetane number of the diesel fuel results in a longer ignition delay period as compared to biodiesel. Biodiesel has a lower volatility and a shorter ignition delay than the diesel fuel and also has a slightly higher viscosity. As a result, lighter compounds develop, which may ignite more quickly, resulting in a shorter ignition delay.

Furthermore, biodiesel has a larger cetane number, it is easier and faster to ignite, causing the delay period to decrease [22, 24]. As a result, a longer diesel delay period allowed for more air/fuel mixture aggregation in the combustion chamber, resulting in a higher premixed peak for diesel fuel following ignition. Higher cetane numbers and shorter delay times for biodiesel blends, on the other hand, enabled their combustion to advance, resulting in a somewhat shorter premixed pressure peak that recorded a few crank angle degrees before diesel, as confirmed by [25, 26].

3.1.2 Net Heat Release

Net heat release is a significant and helpful measure for studying the combustion process and determining the length and intensity of combustion. NHR is a result of rapid burning of air-fuel mixture amassed during the ignition delay interval in the combustion chamber, according to authors in [27]. Under various engine loads, Fig. 3a shows the variance in net heat release versus crank angle for diesel, and (b), (c), and (d) for soybean oil blends (in 10, 20, and 30% concentrations). For all blends with initially increasing load, it can very well be observed that NHR is also increasing. A higher NHR value results from more heat production from the chemical energy of the fuel during combustion. Thus, rising engine loads result in greater combustion quality because more heat energy is liberated as cylinder pressures rise. Longer ignition delay periods cause more air-fuel mixture to be gathered, resulting in quick ignition of all the charge present and hence higher NHR when the engine load is increased from 20 to 60%. The rate of rise in the NHR is found to become constant at engine loads of 80–100%, which is owing to generally constant delay times with even higher loads, as proven by [28].

Figure 3b–d show a negative heat emission at the start of the compression phase. In addition, the accumulation of fuel promotes vaporisation, which is a phenomenon in which heat is absorbed and then released as negative heat. Diesel and biodiesel mixtures undergo similar combustion phases. The air/fuel combination quickly ignites during the premixed combustion period at the end of the delay period, and heat release becomes positive. Additionally, the quantity and velocity of air/fuel mixing may be regulated at this stage to control the burn rate. For a 100% engine load, Fig. 3e shows the variation of NHR versus crank angle for diesel, and soybean blends at full load. Peak values of heat release and location for diesel, S10, S20, S30, blends for full engine load are 57.12 J/deg, 362 CA; 48 J/deg, 362 CA; 50.85 J/deg, 361 CA; 49.02 J/deg, 362 CA respectively. Diesel has a far greater peak NHR than its blends,

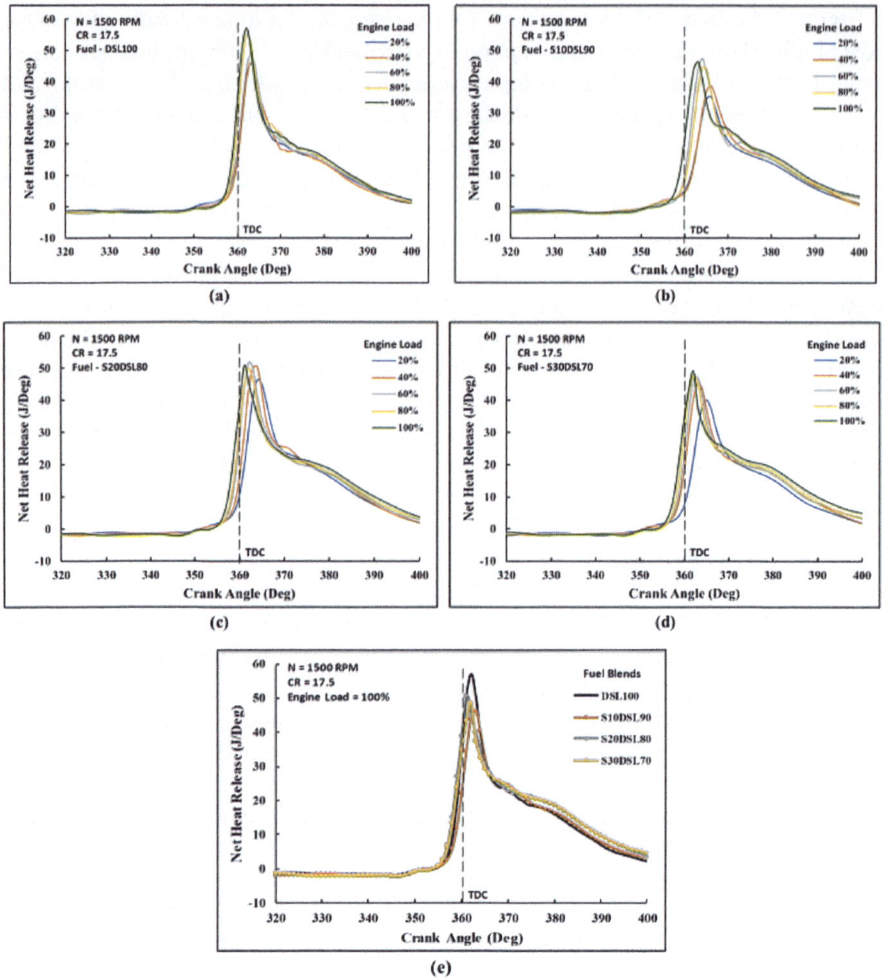

Fig. 3 Variation of net heat release versus crank angle for diesel and soybean oil blends

owing to the increased volatility and better mixing of diesel fuel with air. Second, unlike biodiesel, diesel has a better spray formation and has a lower viscosity, which favours premixed combustion. Another key explanation for the greater NHR peak and its farthest position for diesel, 362 CA, is increased build-up of fuel mixture over the longer delay periods by authors in [29, 30] who also mentioned these observations. With the rise in blending percentage, soybean blends show an accurate trend of advancing and decreasing NHR. Due to shortest delay period, observed from Fig. 5, and highest cetane number of S30 blend, the NHR value recorded for it is least and is found at TDC, i.e., 360 CA.

Furthermore, biodiesel blends demonstrated greater heat release and combustion acceleration, mostly due to the biodiesel's higher oxygen content, which resulted in faster and more complete burning of the biodiesel fuel than diesel [1, 31].

3.1.3 Rate of Pressure Rise

The ROPR during engine operation represents the roughness of the combustion, the peak cylinder pressure, the smoothness of the operation, and the power produced. The rate of combustion of the air/fuel mixture collected in the engine cylinder during the premixed combustion phase determines ROPR. Figure 4 shows how the ROPR varies with crank angle for diesel and soybean oil blends under different loads. The diesel and all blends' statistics indicate that ROPR grows non-linearly with load and tends to become constant at the greatest engine loads of 80,100%. Increased ROPR readings for rising load indicate the burning of a large amount of injected fuel collected during the ignition delay. Longer delay times result in a higher rate of pressure rise, owing to the fast and rapid combustion that occurs following the initial flame kernel. Thus, for diesel and all biodiesel blends, the charges amassed during the initially longer delay periods burn, resulting in a high liberation of heat release, a rise in cylinder pressure, and, eventually, an increase in ROPR. Furthermore, when the delay period decreases for the greatest loads, the quick, intense, and uncontrolled burning shortens and controlled combustion begins, resulting in consistent ROPR [32, 33].

Figure 4a–d show the effect of mixing soybean oil with diesel on ROPR for varied engine loads. When the load is increased for all blends, the ROPR peak shifts towards TDC. This is again attributed to the earlier start of combustion, before TDC, for the blends. Furthermore, the greater the cetane number of the blends, the faster the air/fuel combination burns, advancing the ROPR peak towards TDC [33]. As per Fig. 4e, the value of ROPR and its position for diesel, S10, S20, and S30 under full load conditions are 4.75 $dP/d\theta$, 361 CA; 4.09 $dP/d\theta$, 361 CA; 4.88 $dP/d\theta$, 361 CA; 4.37 $dP/d\theta$, 361 CA respectively. Diesel ROPR for full load is the highest reported. It's because of the effective mixing of air/fuel mixture during the premixed combustion. Furthermore, the premixed combustion NHR is recorded to be greater for diesel, which may address for the faster ROPR [33]. Finally, as the blending percentage increases, the magnitude of ROPR for the blends reduces, which might be attributed to decreased energy content, low flowability, and lower volatility of biodiesel in the blends, which can reduce fuel-air mixing, which might affect blend combustion [27] (Fig. 4).

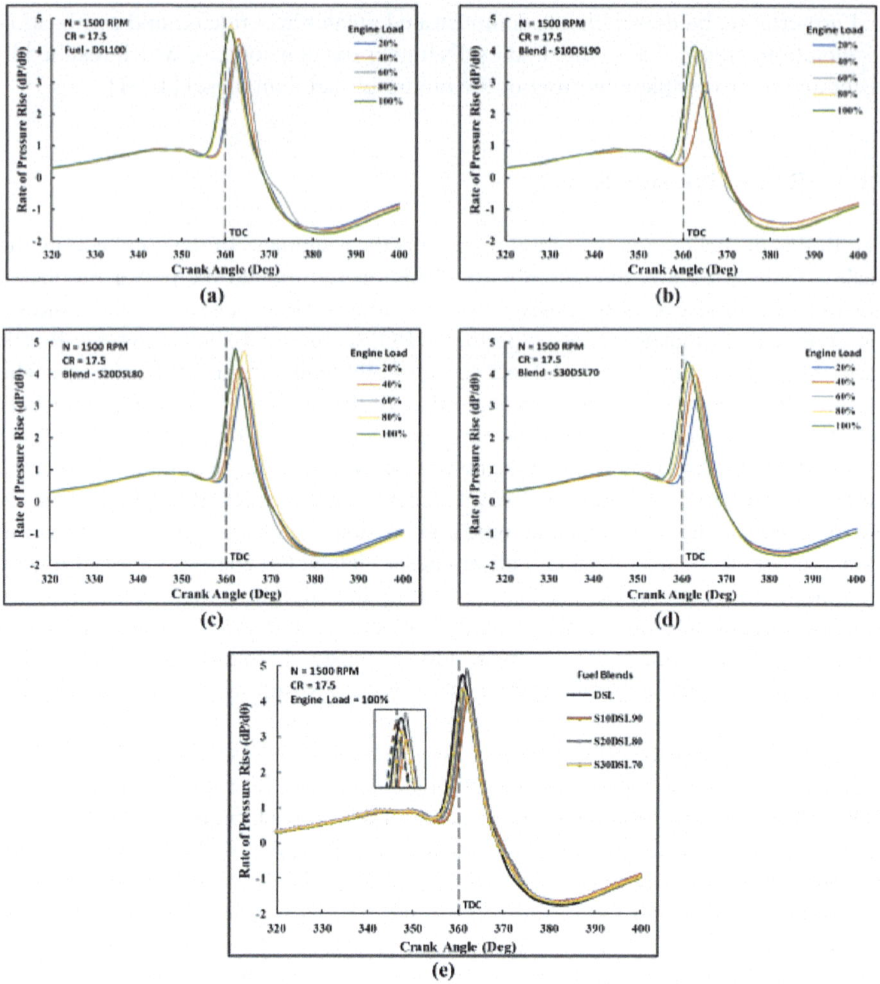

Fig. 4 Variation of rate of pressure rise versus crank angle for diesel, and soybean oil blends

3.2 *Performance Parameters*

3.2.1 Brake Thermal Efficiency

The conversion of fuel energy into useful work obtained at the crankshaft is referred to as BTE efficiency. In other words, it is the product of the flow rate of injected fuel mass and the lower heating value, and it is the ratio between the output power and the energy introduced through fuel injection. Under increasing engine loads, Fig. 6 shows the variation in experimental BTE for diesel and different biodiesel blends in S10, S20, and S30 concentrations. The BTE bar chart shows that as loads have

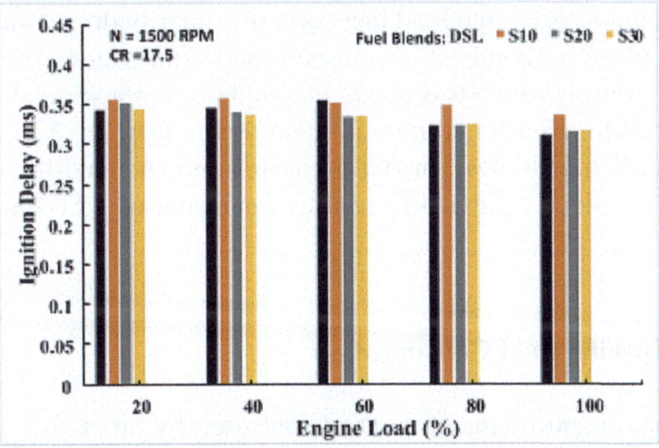

Fig. 5 Variation of ignition delay period for various fuel blends with varying load

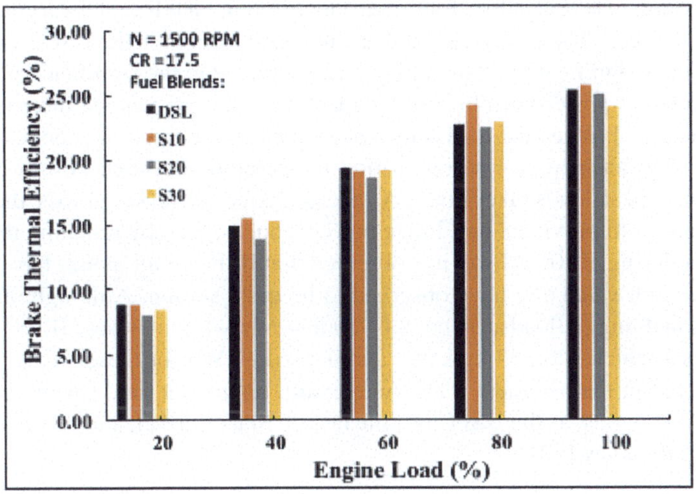

Fig. 6 Brake thermal efficiency for various fuel blends with varying load

increased, the BTE has increased for every blend under any operating state. Improved spray properties, air/fuel mixing, adequate in-cylinder pressure, temperature and heat release, and shorter delay periods observed at higher loads have all resulted in effective combustion for the highest loads. At 20% and 100% load, the BTE for diesel is 8.82% and 25.4%, respectively. When considering the blend variations, the highest efficiencies are achieved at 100% engine load for all test fuels. The BTE of S10 blend at full engine load is higher than diesel. However, as compared to pure diesel, the efficiency of biodiesel blends are marginally reduced at lower loads, except for the cases of S10 blend, which have shown almost similar efficiency as diesel for lower loads while exceeding diesel at higher loads. This improvement in BTE for these biodiesel blends as engine load increases is primarily due to more combustion reaction taking place as combustion pressure and temperature rise. Whereas the decrease in BTE for other blends is due to incomplete combustion observed because of high air/fuel ratio. Similar results are also studied by [1, 22, 34]. Thus, it is concluded that, utilising biodiesel in a CI engine results in better diffusion combustion and less intense premixed combustion, resulting in increased BTE for a few biodiesel blends [22].

3.2.2 Brake Specific Fuel Consumption

BSFC is a measurement of the amount of fuel used by an engine to produce one unit of power per hour. It is a critical performance metric that aids in the analysis of an engine's performance and the calculation of its fuel economy. BSFC is based on fuel characteristics and assesses the engine's efficiency [35, 36]. At constant engine speed, variation in BSFC values for diesel and different blends under varying loads is shown in the Fig. 7. The results reveal that when load increases, the BSFC value of all blends reduces significantly, which may be attributable to the significant proportion increase in brake power with load in comparison to the percentage increase in fuel consumption [37]. When the load is increased, the cylinder pressure and temperature rise, leading to an increase in combustion temperature and, as a result, enhanced combustion at peak loads takes place. As a result, under increasing loads, the amount of fuel required to provide a unit brake power reduces, thereby reducing the BSFC. When considering at 100% load, it is observed that BSFC of all blends have dropped to the minimum value and have observed to be quite similar. Although, among all the blends and engine loads, blends of S10 have shown the closest BSFC value to that of diesel which is 0.335 kg/kWh, and 0.337 kg/kWh for diesel. This have also been reflected in their highest BTE. As a result, on greater load for biodiesels are highly advantageous, as the BSFC is quite near to that of diesel, and there is increase in thermal efficiency [22].

4 Conclusions

In the present experimental investigation, biodiesel blends of soybean oil were tested in a direct injection diesel engine at various engine load conditions. The blends were compared with diesel fuel in terms of various combustion and performance parameters. Conclusions drawn from the study are as follows:

1. Combustion parameters like in-cylinder pressure, NHR and ROPR were recorded to be highest at full load condition for all test fuels. It is observed that with increasing loads, higher values of various combustion parameters are found for all fuels. Furthermore, for a particular blend with increase in engine load, ignition delay continuously reduced for all test fuels.

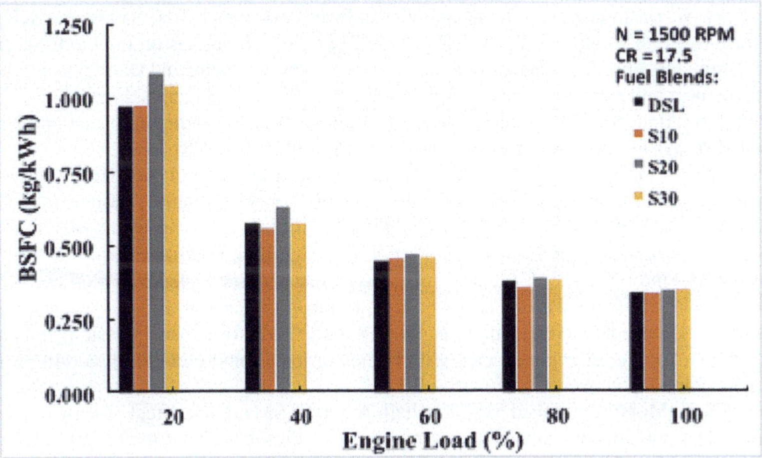

Fig. 7 Brake specific fuel consumption for various fuel blends with varying load

2. At higher loads, performance parameters like BTE and BSFC were also enhanced. Highest BTE was found at 100% load in the range of (23–25%). Similarly, for all test fuels, the lowest BSFC was noted at the maximum load. Thus, best combustion and performance parameters were observed at full engine load for all test fuels.
3. Blending concentration has significant effects on an engine's performance and combustion parameters. Increase in biodiesel concentrations result in higher in-cylinder pressure, net heat release, rate of pressure rise and lower ignition delay periods at full load.
4. It is found that at full load condition, S30 and S20 blend have shown similar combustion characteristics and improved combustion over S10 blend. There has been a marginal improvement in performance parameters for S10 blend at full load.

References

1. Gupta R, Bhalla J, Mourya S (2016) Performance analysis of a diesel engine using the soybean oil based biodiesel. Indian J Sci Technol 9(36)
2. Rehman S, Mahir HS, Zaidi K, Hasan AK (2015) Study of ignition characteristics of microemulsion of coconut oil under off diesel engine conditions. Eng Sci Technol Int J 18:318–324
3. Rehman S, Alam F, Adil M (2019) Ignition and combustion characteristics of impinging diesel and biodiesel blended sprays under diesel engine-like operating conditions. In: ICMET

4. Elkelawy M, Bastawissi HA, Esmaeil KK, Radwan AM, Panchal H, Sadasivuni KK, Sadasivuni KK, Ponnamma D, Walvekar R (2019) Experimental studies on the biodiesel production parameters optimization of sunflower and soybean oil mixture and DI engine combustion, performance, and emission analysis fueled with diesel/biodiesel blends. Fuel 255
5. Rehman S, Alam SS (2020) Soot formation characteristic of impinging diesel and biodiesel blended sprays at diesel engine-like conditions. Proc ICOTF 27:481–491
6. Schumacher LG, Borgelt SC, Fosseen D, Goetz W, Hires WG (1996) Heavy-duty engine exhaust emission tests using methyl ester soybean oil/diesel fuel blends. Bioresour Technol 57
7. Sidharth, Naveen K (2019) Performance and emission studies of ternary fuel blends of diesel, biodiesel and octanol. Energy Sources, Part A: Recover, Util, Environ Eff
8. Aydin H, Bayindir H, İlkiliç C (2010) Emissions from an engine fueled with biodiesel-kerosene blends. Energy Sources, Part A: Recover, Util, Environ Eff 33:130–137
9. Devaraj R, Dinesh BM, Sivakumar DB, Durai C (2020) Performance, emission, and combustion analysis on diesel engine fueled with blends of neem biodiesel/diesel/ additives. Energy Sources, Part A: Recover, Util, Environ Eff
10. Abed AK, El Morsi KA, Sayed MM, El Shaib AA, Gad SM (2018) Effect of waste cooking-oil biodiesel on performance and exhaust emissions of a diesel engine. Egypt J Pet 27:985989
11. Hariram V, Mohan Kumar G (2020) Performance and emission characteristic analysis of Cucurbita pepo L. and Tectona grandis seed oil biodiesel blends in CI engine with additive. Energy Sources, Part A: Recover, Util, Environ Eff
12. Vinothkannan V, Pushparaj T (2012) Combustion and heat release analysis of soybean biodiesel in direct injection compression ignition engine. i-Manager's J Futur Eng Technol; Nagercoil 8
13. Zhu L, Cheung SC, Zhang GW, Huang Z (2011) Combustion, performance and emission characteristics of a DI diesel engine fueled with ethanol-biodiesel blends. Fuel 90:1743–1750
14. Annisa B, Eriko M, Jiro S (2019) Performance and emission characteristics of biodiesel waste cooking oil water emulsions under varying engine load condition. Energy Sources, Part A: Recover, Util, Environ Eff
15. Yehliu K, Boehman AL, Armas O (2010) Emissions from different alternative diesel fuels operating with single and split fuel injection. Fuel 89
16. Pasupathy RS, Mohan RT (2019) Optimisation and effective utilisation of esterified rice bran oil in a turbocharged VCR engine by analysing its operating characteristics. Energy Sources, Part A: Recover, Util, Environ Eff
17. Kumar SM, Jaikumar M (2014) A comprehensive study on performance, emission and combustion behavior of a compression ignition engine fuelled with WCO (waste cooking oil) emulsion as fuel. J Energy Inst 1–9
18. Rehman S, Nafis MS, Waheed MA, Khan AA (2022) Experimental investigation on combustion and performance characteristics of a direct-injection diesel engine. 1–17
19. Chaurasiya PK, Singh SK, Dwivedi R, Choudri RV (2019) Combustion and emission characteristics of diesel fuel blended with raw jatropha, soybean and waste cooking oils. Heliyon 5(5)
20. Shelke PS, Sakhare NM, Lahane S (2016) Investigation of combustion characteristics of a cotton seed biodiesel fueled diesel engine. Procedia Technol 25:1049–1055
21. Gautam R, Kumar S (2020) Performance and combustion analysis of diesel and tallow biodiesel in CI engine. Energy Rep 6:2785–2793
22. Mehmet RS, Selman A, Adem Y, Seyfi S (2019) Evaluation of comparative combustion, performance, and emission of soybean based alternative biodiesel fuel blends in a CI engine. Renew Energy 148:1065–1073
23. Kannan GR, Balasubramanian KR, Sivapirakasam SP, Anand R (2011) Studies on biodiesel production and its effect on DI diesel engine performance, emission and combustion characteristics. Int J Ambient Energy 32
24. Karnwal A, Kumar N, Hasan MM, Siddiquee AN, Khan ZA (2010) Performance evaluation of a medium capacity diesel engine on Thumba biodiesel and diesel blends. J Biofuels 1:172–181
25. Pal A, Kachhwaha SS, Maji S, Babu MKG (2010) Thumba (Citrulluscolocyntis) seed oil: a sustainable source of renewable energy for biodiesel production. Renew Energy 69:108–113

26. Qi DH, Geng LM, Chen H, Bian YZH, Liu J, Ren XCH (2009) Combustion and performance evaluation of a diesel engine fueled with biodiesel produced from soybean crude oil. Renew Energy 34(12):2706–2713
27. Hosamani BR, Katti VV (2018) Experimental analysis of combustion characteristics of CI DI VCR engine using mixture of two biodiesel blend with diesel. J Eng Sci Technol
28. Uyumaz A, Solmaz H, Yılmaz E, Yamı H, Polat S (2014) Experimental examination of the effects of military aviation fuel JP-8 and biodiesel fuel blends on the engine performance, exhaust emissions and combustion in a direct injection engine. Fuel Process Technol 158–165
29. Buyukkaya E (2010) Effects of biodiesel on a DI diesel engine performance, emission and combustion characteristics. Fuel 89:3099–3105
30. Muralidharan K, Vasudevan D, Sheeba NK (2011) Performance, emission and combustion characteristics of biodiesel fuelled variable compression ratio engine. Energy 36:5385–5393
31. Wei L, Cheung SC, Ning Z (2017) Influence of Waste cooking oil biodiesel on combustion, unregulated gaseous emissions and particulate emissions of a direct-injection diesel engine. Energy 127:175185
32. Tutak W, Jamrozik A (2014) Generator gas as a fuel to power a diesel engine. Therm Sci 18:205–216
33. Murugesan A, Umarani C, Subramanian R, Nedunchezhian N (2009) Bio-diesel as an alternative fuel for diesel engines—a review. Renew Sustain Energy Rev 13(3):653–662
34. Enweremadu CC, Rutto LH (2010) Combustion, emission and engine performance characteristics of used cooking oil biodiesel—a review. Renew Sustain Energy Rev 14:2863–2873
35. Jorge C, Murari MR, Wilson W (2018) Performance and emissions of a diesel engine fueled by biodiesel-diesel blends with recycled expanded polystyrene and fuel stabilizing additive. Energy 149:204–212
36. Tse H, Leung CW, Cheung CS (2015) Investigation on the combustion characteristics and particulate emissions from a diesel engine fuelled with diesel-biodiesel-ethanol blends. Energy 83:343–350
37. Dowell GT, Feng B, Alvensleben NV, Brown RJ (2015) Combustion analysis of microalgae methyl ester in a common rail direct injection diesel engine. Fuel 143:351–360

Experimental Investigation on Combustion and Performance Characteristics of DI Diesel Engine Using Waste Cooking Oil Biodiesel Blends

Anas Anwar Khan, Sanaur Rehman, Muhammad Aazam Waheed, S. S. Alam, and Mohd Sufiyan Nafis

Abstract The hefty prices of raw materials are the greatest hurdle to biodiesel's adoption as an alternate fuel. While waste cooking oil offers a cost-effective solution, it poses challenges in disposal and the potential for contamination. These factors lead waste cooking oil to a promising alternative. Waste cooking oil biodiesel blends have therefore been developed and employed to evaluate their impact on engine combustion and performance parameters. In the present experimental study, a catalytic transesterification process was used to produce biodiesel (waste cooking oil). Biodiesel was blended with diesel in three proportions (W10–10% waste cooking oil blend 90% diesel, W20–20% waste cooking oil blend 80% diesel, and W30–30% waste cooking oil blend 70% diesel). In-cylinder pressure, rate of pressure rise, net heat release, cumulative heat release, mean gas temperature, ignition delay, brake specific fuel consumption (BSFC), and brake thermal efficiency (BTE) were among the combustion and performance parameters that were analysed on a diesel engine at five engine loads (20%, 40%, 60%, 80%, and 100%). The percentage of biodiesel in the blend was increased, which improved the values of combustion and performance parameters. The optimal blend in terms of combustion and performance parameters was found to be W30. In comparison to diesel, W30 had a 13% shorter ignition delay, 22% lower net heat release, 10% slower rate of pressure rise, 2% higher BTE, and 1% lower BSFC. Therefore, biodiesel blends with a high concentration may prove to be a viable substitute for diesel fuel. 034.

Keywords Biodiesel · Waste cooking oil · Diesel engine · Combustion · Performance

A. A. Khan (✉) · S. Rehman · M. A. Waheed · S. S. Alam · M. S. Nafis
Department of Mechanical Engineering, Aligarh Muslim University, Aligarh, India
e-mail: anas.khan81269@gmail.com

S. Rehman
e-mail: sanaur.rehman@zhcet.ac.in

M. A. Siddiqui et al. (eds.), *Advances in Heat Transfer and Fluid Dynamics*, Lecture Notes in Mechanical Engineering, https://doi.org/10.1007/978-981-99-7213-5_25

Nomenclature

ATDC	After Top Dead Centre
BSFC	Brake Specific Fuel Consumption
BTE	Brake Thermal Efficiency
CA	Crank Angle
CI	Compression Ignition
CR	Compression Ratio
DI	Direct Injection
EOC	End Of Combustion
ICE	Internal Combustion Engines
NHR	Net Heat Release
TDC	Top Dead Centre
ROPR	Rate of Pressure Rise
SOC	Start Of Combustion
SOI	Start Of Injection
WCO	Waste Cooking Oil

1 Introduction

Since the turn of the twentieth century, the industrial revolution, demographic expansion, population growth, and a multiplied rise in energy demand have all contributed to a sharp increase in the use of Energy worldwide. The development of ICEs and ongoing improvements in engine technology have also increased our dependency on petroleum sources. In every area of the world, even contemporary transportation and its dependence on conventional energy sources like fossil fuels exacerbate the depletion of those resources, as stated by Gupta et al. [1]. This enormous demand for Energy derived from the combustion of fossil fuels has led to the widespread exploitation of fuel supply. The consequence of continuous use of fossil fuels in response to the increase in global energy demand has led scientists and researchers to search for innovative and efficient alternative fuels that are both renewable and non-polluting, with the potential to mitigate global warming and pollution [2, 3]. Several studies indicate that biodiesel is the most effective and viable alternative Fuel for diesel engines, as it is non-toxic, biodegradable, and environmentally friendly. In addition, it is composed of fatty acids and may be generated from vegetable oils, animal fats, and waste cooking oil through the catalytic transesterification process. Moreover, it is devoid of sulphur and minerals, and its high oxygen content results in low engine emissions [4, 5].

Numerous earlier studies have demonstrated that biodiesel gives the same or minimally enhanced thermal efficiency when compared to diesel, as well as lower or equivalent BSFC values for larger loads. Siddarth and Naveen [6] examined the performance and emissions of ternary fuel mixes of diesel, biodiesel, and octanol and found that the blend has 10% octanol, 10% biodiesel, and 80% diesel had better BTE and BSEC than diesel. In addition, the mixtures produced fewer emissions (CO, unburned hydrocarbon, NO, and smoke opacity) and improved combustion. Furthermore, Devaraj et al. [7] examined the impact of two oxygenated additives, namely DMC (Dimethyl-carbonate) and Pentanol (n-P), on the ignition patterns of biodiesel/diesel blends in a diesel engine. Adding DMC and Pentanol to base gasoline raised peak pressure by 2.3 and 3.1 bar, decreased BSFC by 0.4 and 0.5 g/kW while increasing BTE by 0.3 and 0.6 percent. Moreover, the addition of DMC and Pentanol to NBD50D50 resulted in an increased heat release rate of 3.9 and 2.1 J/CA. Abed at el. [8] investigated the effect of WCO on the performance of a single-cylinder diesel engine and created three mixtures with concentrations of 10, 20, and 30%. The thermal efficiency of waste cooking oil biodiesel blends was found to be lower than that of diesel fuel; however, the BSFC was higher. In addition, as the mixture's concentration increases, thermal efficiency also increases, and BSFC decreases simultaneously. Di at el. [9] studied waste cooking oil biodiesel blends combined with diesel fuel at an engine speed of 1800 rpm for 5 engine loads of 28, 70, 130, 190, and 230 N meters. The usage of biodiesel boosted BSFC and BTE while reducing the temperature of the exhaust gases. Annisa at el. [10] studies performance and emissions of WCO water emulsions, Four loads were applied to a four-cylinder turbocharged Yanmar diesel engine throughout the testing. The experiment results demonstrated that the BTE of BD5W10 at low load was up to 8.92 percent greater than diesel. At medium load, the BSFC of BD5W10 increased by 11.66 percent compared to diesel. BD5W10 has higher CO emissions than diesel and B10. Compared to diesel under heavy load, BDSW10 CO emissions were reduced by 12.8%. Pasupathy and Mohan [11] investigated the effects on performance, combustion, and emissions of a turbocharged engine by changing compression ratios for the blends and comparing them with the standard diesel. Among compression ratios tested (16:1, 17:1, and 18:1), 18:1 yielded the best performance. Increasing the compression ratio without a turbocharger increases NOx and CO2 emissions while reducing CO and HC emissions. Without a turbocharger, B10 has lower Nox and HC emissions than B20 and diesel but somewhat greater CO emissions. The turbocharger reduces emissions overall. With greater compression ratios, the rate of pressure rises, and heat release is accelerated considerably. Kumar and Jaikumar [12] examined the efficiency and combustion characteristics of a single-cylinder WCO-powered diesel engine. According to the findings, the WCO mixture enhanced the in-cylinder pressure and the delay period. Compared to pure diesel, pure WCO demonstrated lower peak pressure under all operating circumstances. As engine power rose, the BTE of every investigated Fuel increased (BMEP). Clean diesel has a lower BTE content than WCO. At a maximum BMEP of 4.4 bar, the maximum BTE was determined to be 26%, whereas it was 30% for clean diesel. Although multiple research studies on performance characteristics and some combustion parameters for biodiesel blends of waste cooking oil have been undertaken in the past, the literature lacks a comprehensive combustion and performance assessment of these blends. Consequently, with a constant CR of 17.5 and an engine speed of 1500 rpm, this article presents the impact of load fluctuation and mix modification on a variety of combustion parameters, including cylinder pressure, Net Heat Release, Rate of Pressure Rise, Cumulative Heat Release Mean Gas Temperature and Ignition Delay period. Under differing engine loads, variations in these properties have been noted. Finally, performance parameters such as BTE and BSFC are calculated and compared with diesel.

2 Experimental Setup

Figure 1 depicts the block diagram of the experimental arrangement [13]. Table 1 provides the technical characteristics of the engine test bed and the combustion parameters, respectively. In this study, a twin-cylinder, four-stroke, water-cooled, constant-speed DI diesel engine linked to a hydraulic dynamometer was used. The apparatus consists of a diesel engine, an airflow assembly, a pressure transducer, an orifice meter, manometers, a K-type thermocouple, a hydraulic dynamometer, a CA sensor, a piezo-powered unit, and a computer. Using an orifice meter, the rate of air flowing into the engine cylinder through the airbox was determined, and the inclined manometer was used to determine the density of the upstream air. During combustion, a pressure transducer was employed to keep track of the pressure within the cylinder. A temperature sensor was used to gauge the engine's air inflow temperature, and a K-type thermocouple was employed to gauge the cylinder gas temperature. A crank encoder was used to measure the crank angle in degrees, and data was gathered using a data-collecting card. The experimental setup utilized a hydraulic dynamometer to alter the engine load in conjunction with a transparent glass tube fuel meter that was directly linked to the fuel tank, enabling the accurate determination of the mass flow rate of Fuel entering the engine cylinder. All combustion parameters were measured for 1° CA for a mean cycle of 05. Table 2 shows the accuracy of the several types of equipment employed.

2.1 Experimental Methodology

In this experiment, waste cooking oil biodiesel blends were employed as diesel engine fuel. Biodiesel was created in the Combustion and Pollution Control Laboratory (Department of Mechanical Engineering, Zakir Husain College of Engineering and Technology, Aligarh Muslim University) using the conventional catalytic transesterification process. W10 contains 10% WCO and 90% diesel, W20 contains 20% WCO and 80% diesel, and W30 has 30% WCO 30% WCO and 70 percent diesel. The engine was initially run without any load for 10 min before any measurements could be taken so that it could run in a steady state. The following loads were applied: 20%, 40%, 60%, 80%, and 100% at a steady engine speed of 1500 rpm. At a given load on the engine, a pressure sensor in one of the engine cylinders transmitted signals to the computer, enabling the creation of graphs and tables illustrating various combustion

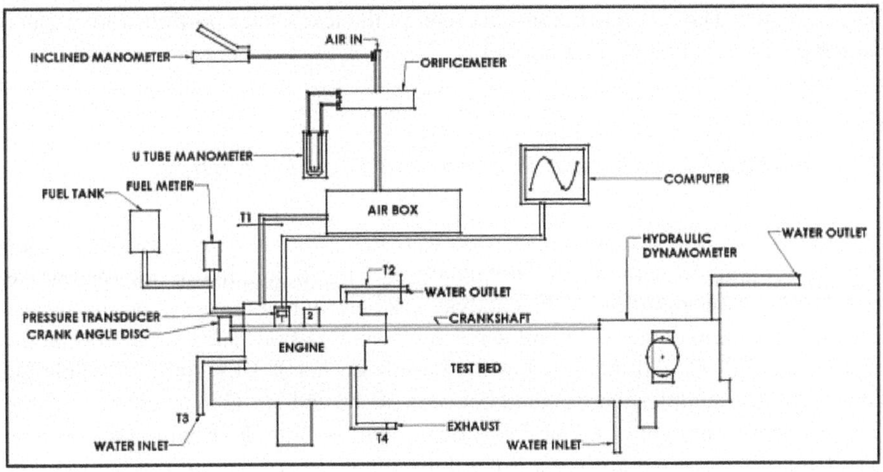

Fig. 1 Block diagram of a DI diesel engine test bed (taken from Ref. [13])

Table 1 Engine test bed details and combustion parameters [13]

Engine type	Kirloskar, TV2, 2-Cylinder, 4-Stroke, water-cooled, Diesel engine
Power	10 kW @ 1500 rpm
Cylinder bore	87.50 mm
Stroke length	110.00 mm
Compression ratio	17.50
Swept volume	661.45 cc
Injection timing	27° bTDC
Orifice diameter	25 mm
Fuel pipe diameter	34.50 mm
Ambient temperature	27 °C
Fuel injection pressure	200 bar
Specific gas constant	1.00 kJ/kg.K
Air density	1.17 kg/m^3
Adiabatic index	1.41
Polytrophic index	1.13
Number of cycles	05

Table 2 Accuracy of instruments used [13]

Instrument	Accuracy
Pressure sensor (Model S111A22)	± 0.1 mV/psi
K-type thermocouple	± 2.2%
Hydraulic dynamometer	± 0.1 kg
CA sensor	1 degree (resolution)

characteristics. The combustion graphs seen in the text's later portion were created using a program called IC Engine Soft.

2.2 Biodiesel Production Process and Fuel Properties

The biodiesel fuels used in this experiment were made via catalytic transesterification of waste cooking oil with ethanol and catalysed with sodium hydroxide (NaOH) as the base. The concentration of the catalyst was determined to be 1% by the weight of the oil. The ratio of alcohol to oil was 1:6 by weight of oil. The temperature of the reaction was 60 °C, and the speed of the stirrer was set at 1000 rpm. To accelerate the process, the samples were heated between 50 °C and 70 °C in an electric heater. The determined amount of ethanol was mixed with sodium hydroxide (NaOH) and stirred until the entire catalyst reacted with the ethanol while the oil was heated. The alcohol–catalyst combination was then added to the oil, which was heated to 60 °C and stirred continuously for one hour. This resulted in the separation of the mixture into two distinct layers: golden liquid ester on top and light brown denser glycerine on the bottom. This immiscible solution was then poured through a transparent funnel to separate glycerine and ester. Finally, the waste cooking oil esters were thoroughly washed many times with room-temperature distilled water until all FFA was removed. Ultimately, this biodiesel was heated for 15 min at 100 °C to remove any leftover water.

Prepared biodiesel characteristics, such as kinematic Viscosity, dynamic Viscosity, and density, were evaluated in the Combustion and Pollution Control Laboratory using a kinematic viscosity bath and hydrometer. The kinematic Viscosity was tested at 40 °C. The apparatus was filled to the appropriate level with 25L of distilled water. Then, a 25 ml sample of Fuel was collected in a clean, dry, and calibrated viscometer. The test fuel was suspended in the viscosity bath for 30 min until it achieved the required temperature of 40 °C. Later, suction was applied to the viscometer's thicker arm, test fuel was made to flow from mark (C) to mark (G), and the flow time was recorded.

Likewise, a hydrometer with a range of 800–850 kg/m^3 was selected to calculate the density. A 500 ml sample of test fuel was poured into a beaker, and the hydrometer was immersed in it for two minutes before settling. Later, the density measurements were noted. Table 3 lists the properties of all test fuels.

Table 3 Properties of all test fuels

Parameters/test fuels	Diesel	W10	W20	W30
Density (kg/m^3)[a]	820	830	839	850
Kinematic viscosity (cSt)[a]	3.65	5.37	6.93	9.77
Dynamic viscosity (mPa-s)[a]	2.993	4.45	5.81	8.30
Calorific value (kJ) [14]	42,200	41,800	41,204	40,954
Cetane number [14]	48	49.5	51	52.5
Flashpoint (°C) [14]	74	84.2	94.4	104.6

A measured property

3 Results and Discussion

3.1 *Combustion Characteristics*

3.1.1 Cylinder Pressure

The premixed burning phase in a diesel engine determines the peak cylinder pressure or pressure generated at the beginning of combustion, which is directly related to the amount of Fuel burned during this phase. Consequently, cylinder pressure indicates how efficiently the air/fuel combination burns in the combustion chamber. Figure 2 shows the variation in cylinder pressure with load for diesel and biodiesel blends of waste cooking oil (W10, W20, and W30). Figure 2a illustrates how the cylinder pressure of a diesel engine rises as the engine load increases from 20 to 100%. Among all the fuels tested, Diesel has a maximum cylinder pressure of 63.6 (bar), which is approximately 12° aTDC (after Top dead centre). Pressure rises with the load as a result of increased fuel flow to the engine [15].

Figure 2b–d illustrates how the cylinder pressure changes with load under the blend comparison and rises as a result of a greater energy release. Diesel has a peak cylinder pressure of 63.6 bar, which is higher than the cylinder pressures of all WCO mixtures (63.13 bar for W10, 62.31 bar for W20, and 62.52 bar for W30). Despite the fact that the W10 blend had pressure values comparable to those of diesel, the other two tested fuels had in-cylinder pressure curves that were identical, suggesting that they had similar combustion processes in the engine, specifically premixed combustion followed by diffusion combustion. WCO blending has improved combustion at all engine loads, with peak pressures for blends approaching 10° aTDC. Furthermore, faster ignition of the blends after injection into the combustion chamber and larger pressure peaks have been achieved by WCO blends because of higher cetane numbers [16]. Ignition delay is the time difference between SOI and SOC, and it comprises physical and chemical delay phases. Physical delay refers to the time it takes for Fuel to evaporate and atomize, while chemical delay specifies the time it takes for Fuel to ignite and a flame to form [17].

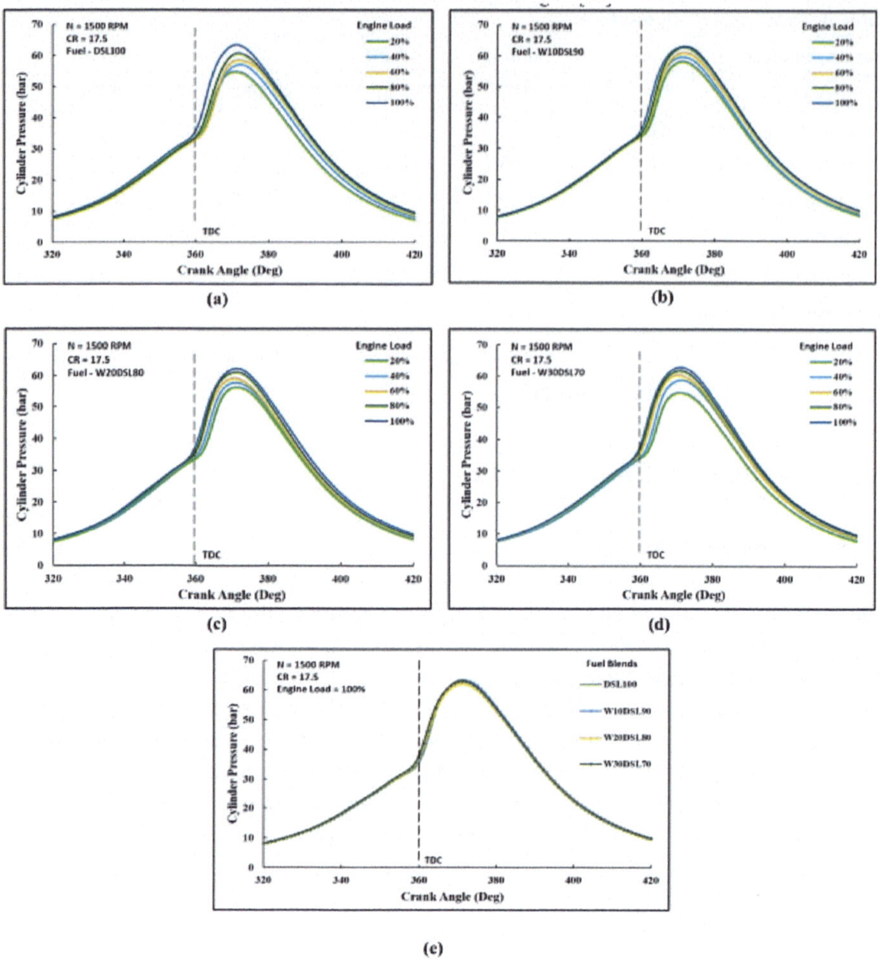

Fig. 2 Variation of cylinder pressure versus crank angle for diesel and waste cooking oil blends

Figure 3 demonstrates that diesel's ignition delay period for increasing load is larger than that of most biodiesel mixes. Diesel fuel has a longer ignition delay period than biodiesel mixes due to its lower cetane number. Since biodiesel is not extracted from crude oil, the high density and Viscosity of biodiesel fuel result in poor atomization, slower mixing, more penetration, and a lowered cone angle. That is why biodiesel has lower volatility and a shorter ignition delay than diesel fuel. This is the result of a complicated, rapid pre-flame chemical process that occurs at high temperatures, which produces lighter compounds and may ignite more rapidly, resulting in a shorter ignition delay.

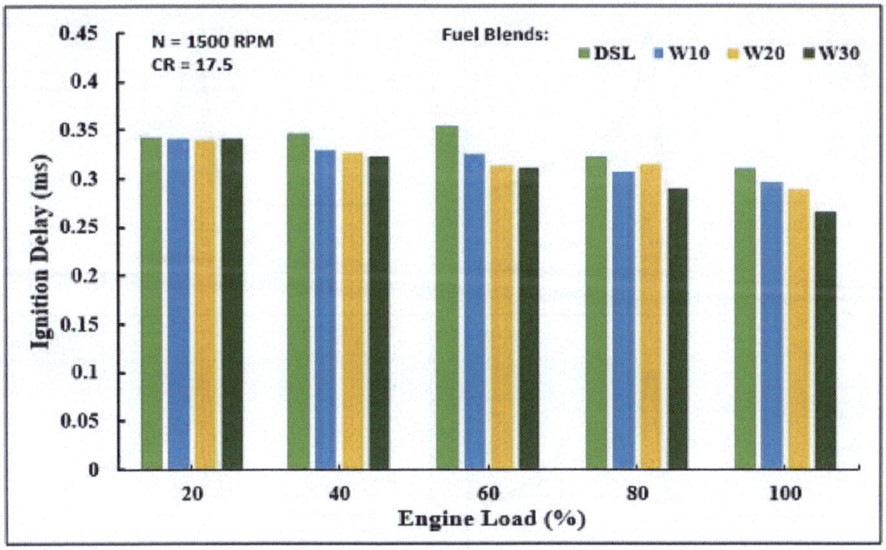

Fig. 3 Variation of ignition delay period for various fuel blends with varying loads

3.1.2 Net Heat Release

NHR is caused by the fast combustion of air–fuel mixture accumulated during the ignition delay interval in the combustion chamber [18]. Figure 4 illustrates the variation in net heat release vs. crank angle for diesel and waste cooking oil biodiesel blends under varied engine loads. For all mixes with an initially rising load, it is likely to see an increase in NHR. A greater NHR value is the result of a greater amount of heat produced by the Energy transformation of the Fuel during combustion. Thus, increasing engine loads result in improved fuel combustion due to the release of more heat energy as cylinder pressures increase. Longer ignition delay periods cause more air–fuel mixture to be collected, leading to the rapid ignition of all the charge present and, thus, a greater NHR at engine loads of 80 to 100 percent. The rate of increase in the NHR is found to remain constant, which is a result of typically constant delay times with even greater loads, as demonstrated [19].

Figure 4b–d illustrate a negative heat release at the start of the compression stage. The negative heat is due to the endothermic chemical and physical reactions that happen during the ignition delay period. In addition, the concentration of Fuel encourages vaporization, which is the phenomenon of heat absorption and subsequent negative heat release. Diesel and biodiesel combinations experience combustion stages that are similar. At the conclusion of the delay phase, the air/fuel mixture ignites rapidly during the premixed combustion period, and heat release becomes positive. Furthermore, the quantity and velocity of air/fuel mixing may be controlled at this stage to manage the burn rate. For a 100% engine load, Fig. 4e illustrates the fluctuation of NHR vs. crank angle for diesel, and WCO blends at 100 percent engine

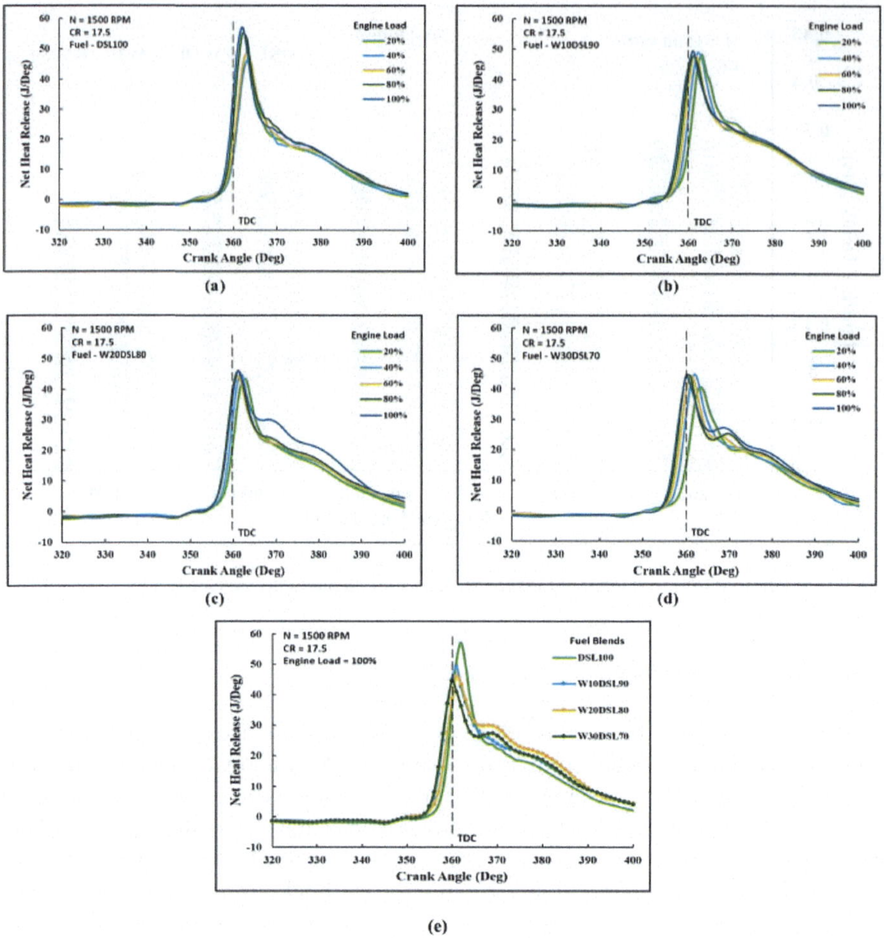

Fig. 4 Variation of net heat release versus crank angle for diesel and waste cooking oil blends

load. Peak values of heat release at full engine load for diesel, W10, W20, and W30 blends are 57.12 J/deg and 362 CA; 49.37 J/deg and 361 CA; 46.12 J/deg and 361 CA; and 44.33 J/deg and 360 CA, respectively. Diesel has a significantly higher peak NHR than its mixes because of its stronger volatility and superior ability to mix with air. Increased accumulation of fuel mixture during the longer delay periods is also a significant cause for the bigger NHR peak, and its most distant position for diesel is 362 CA. These observations were also referenced in [20, 21]. WCO blends provide an appropriate progression of increasing and decreasing NHR. Due to the shortest delay period, as noted in Fig. 3, and the highest cetane number of the W30 blend, the NHR value measured for it is the lowest and is located at TDC, at 360 CA. In addition, the NHR of W20 and W10 mixes increases as the delay duration increases [1, 22]. In terms of heat release and combustion acceleration, biodiesel blends also beat diesel fuel. This is mostly due to biodiesel's higher oxygen concentration, which allows it to burn more swiftly and completely than diesel fuel and causes a greater proportion of biodiesel to ignite more slowly than diesel fuel [1, 23].

3.1.3 Rate of Pressure Rise

The rate of pressure rise during engine operation indicates the roughness of the combustion, the peak cylinder pressure, the operation's smoothness, and the produced power. ROPR is determined by the rate of combustion of the collected air/fuel mixture in the engine cylinder during the premixed combustion phase. Figure 5 illustrates how the rate of pressure rise changes with the crank angle for diesel and waste cooking oil blends operating under various loads. Diesel and all blend records reveal that ROPR increases non-linearly with load and tends to stabilize between 80 and 100 percent engine load. The combustion of a substantial volume of injected Fuel gathered during the ignition delay interval is indicated by growing ROPR measurements as the load increases. Due to the quick and fast combustion that happens after the first flame kernel, longer delay durations result in a faster rate of pressure rise. Consequently, for diesel and all WCO blends, the charges accumulated during the initially longer delay periods burn, resulting in a high heat release, an increase in cylinder pressure, and subsequently, an increase in ROPR.

Figure 5a–d illustrate the influence of combining waste cooking oil with diesel on ROPR for varying engine loads. When the load is raised for all fuel mixes, the ROPR peak switches to TDC as compared to diesel. This is likewise attributable to the earlier onset of combustion, before TDC, for the mixes in comparison to diesel. Moreover, the higher the cetane number of the mixtures, the quicker the air/fuel mixture burns, which advances the ROPR peak towards TDC [24]. According to Fig. 5e, the ROPR value and position for diesel, W10, W20, and W30 under full load circumstances are 4.75 $dP/d\theta$ at 361, 4.63 $dP/d\theta$ at 360, 4.21 $dP/d\theta$ at 360, and 4.25 $dP/d\theta$ at 360, respectively. Full-load diesel ROPR is the highest reported value due to the excellent mixing of air and fuel during the time of premixed combustion. In addition, diesel's premixed combustion NHR is larger, which may account for its quicker ROPR [25]. As the blending fraction of a WCO blend grows, its impact on ROPR and location changes proportionally. Lastly, as the blending % grows, the magnitude of ROPR for the blends decreases, which may be ascribed to the lowered energy content, poor flowability, and decreased volatility of biodiesel in the blends.

3.1.4 Cumulative Heat Release

The entire heat produced from the chemical energy of fuel after combustion is referred to as cumulative heat release. CHR depicts information regarding the engine's combustion progress. Figure 6 shows the variance in CHR versus crank angle for diesel, and WCO blends at various engine loads. CHR rises with increasing engine load, as seen in the graph for all blends, and follows a modest nonlinear pattern.

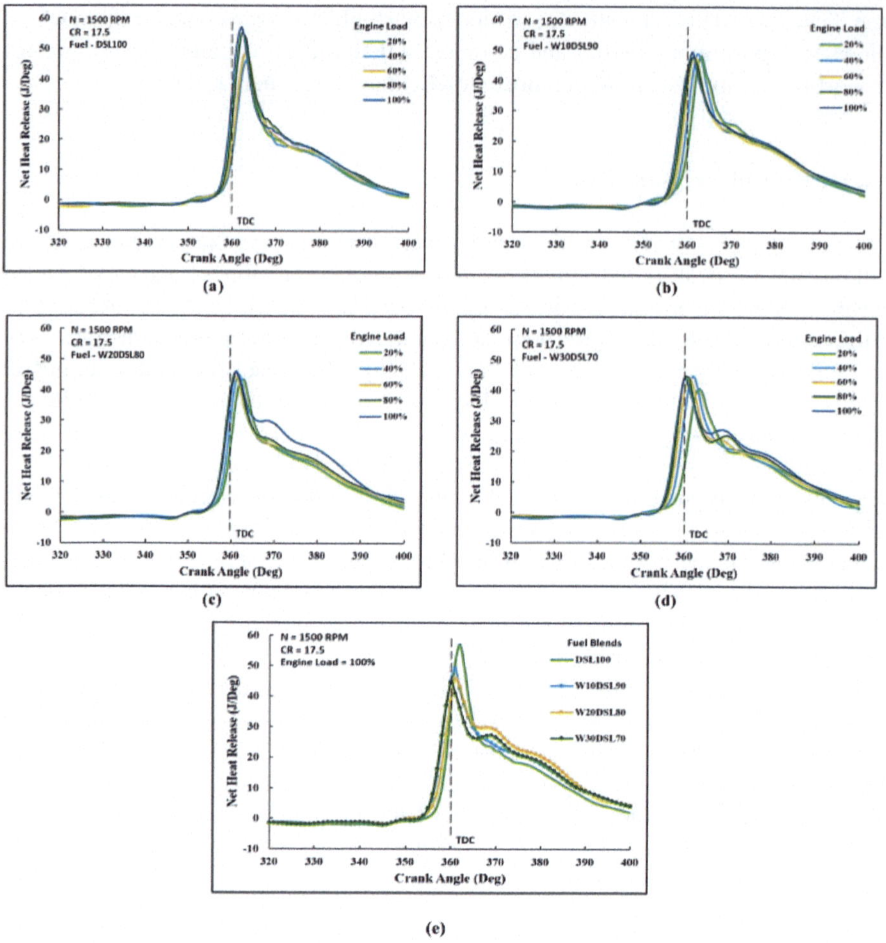

Fig. 5 Variation of the rate of pressure rise versus crank angle for diesel and waste cooking oil blends

Figure 6a–d illustrate the influence of combining waste cooking oil with diesel on CHR for varying engine loads. From Fig. 6e, the maximum value of CHR for diesel, W10, W20, and W30 is 0.95 kJ, 0.99 kJ, 1 kJ, and 0.97 kJ, respectively. When compared to WCO, CHR for diesel fuel is lower during the early stages of combustion. This might be attributed to the fact that diesel takes longer to start igniting than biodiesel mixes owing to the greater delay time. WCO biodiesel also enables early combustion because of its greater cetane number and intrinsic oxygen. Also, even after delayed combustion of diesel compared to WCO blends, the CHR of diesel approaches the W10, W20, and W30 values as the combustion progresses due to the better ignition and mixing property, greater energy content, and altogether a better combustion property of diesel fuel.

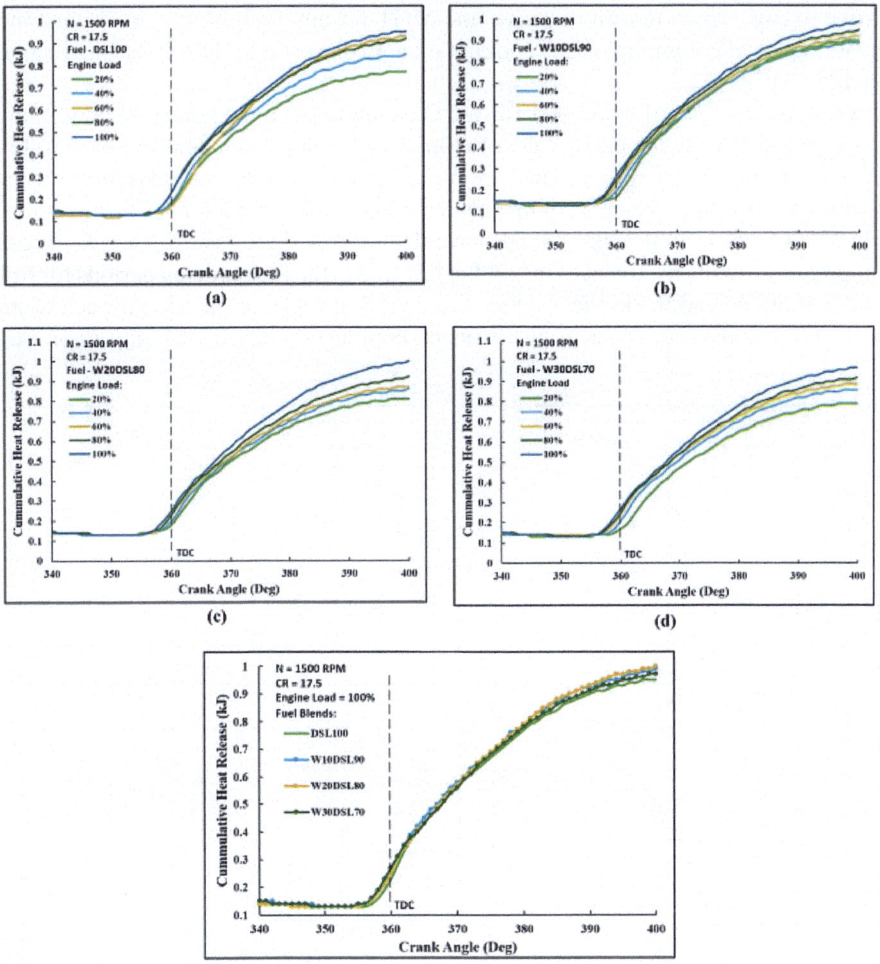

Fig. 6 Variation of cumulative heat release versus crank angle for diesel and waste cooking oil blends(e)

3.1.5 Mean Gas Temperature

The temperature of the gas generated in the engine cylinder following the power stroke is referred to as the mean gas temperature. MGT is a tool for determining combustion efficiency. Figure 7 shows the effect of WCO blends on MGT under various load conditions. The MGT value increases with increased engine load for every operating condition, as shown by the figures for all blends. On increasing the engine load, as the power demand of the engine is increased, more Fuel is injected into the cylinder to compensate, resulting in increased heat release at higher engine loads. This increase in heat release causes the charge temperature during the exhaust stroke to rise, which in turn increases the MGT for any fuel. Figure 7a–d illustrate the influence of combining waste cooking oil with diesel on MGT for varying engine loads.

From Fig. 7e, variation in MGT for WCO blends under full engine load conditions is presented. During the early stage of combustion, the gas temperature of diesel is recorded to be lowest as compared to WCO blends. It is due to delay in the start combustion for diesel because of low cetane number along with longer delay period. All WCO blends, on the other hand, have demonstrated the fastest increase in gas temperature, owing to advances in combustion due to the lowest delay periods for full load conditions and the highest cetane number. The MGT of diesel approaches the W10, W20, and W30 values in the diffusion combustion phase, owing to improved

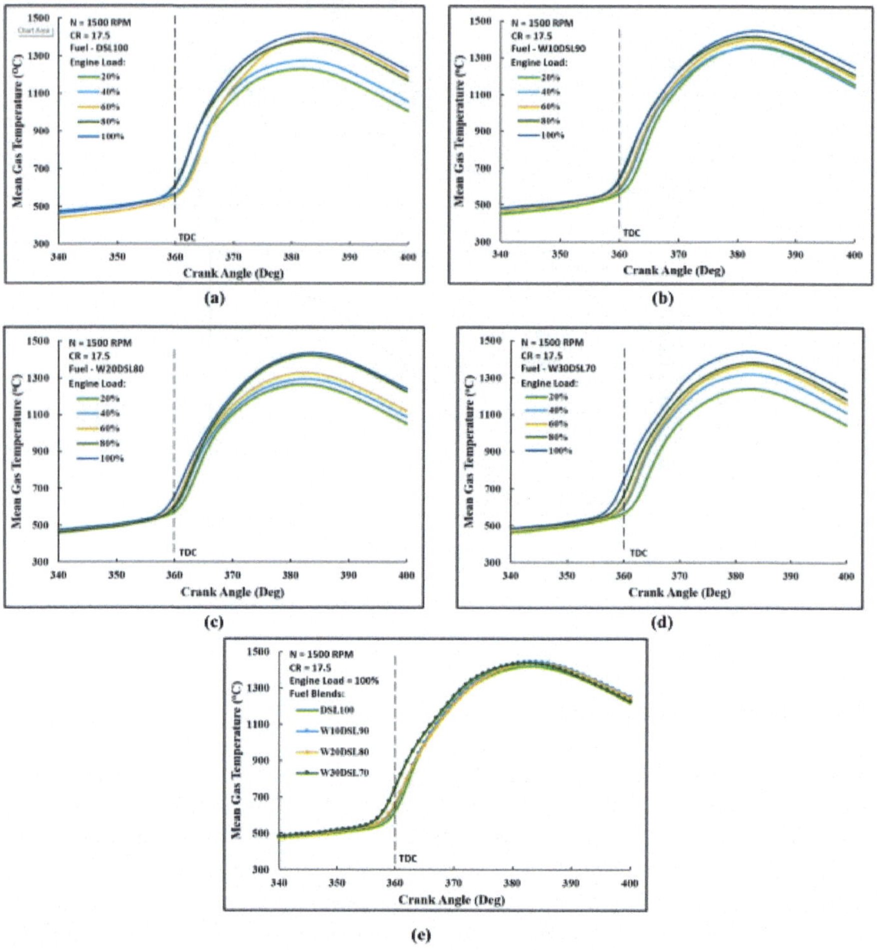

Fig. 7 Variation of mean gas temperature versus crank angle for diesel and waste cooking oil blends

mixing of diesel fuel with the air and higher heat content, all of which contribute to efficient combustion. The MGT values recorded for diesel, W10, W20, and W30 for full engine load conditions are 1421 °C, 1448 °C, 1437 °C, and 1440 °C, respectively.

3.2 *Performance Parameters*

3.2.1 Brake Thermal Efficiency

Brake thermal efficiency is a statistic that measures the efficiency with which an engine transforms heat from the Fuel into mechanical Energy. It refers to the conversion of fuel energy into usable work that is produced at the crankshaft. Figure 8 illustrates the variance in experimental BTE for diesel and various biodiesel blends at W10, W20, and W30 concentrations under increasing engine loads. The BTE bar graph demonstrates that as loads have grown, the BTE has risen for every blend in every operating condition. At the greatest loads, improved spray characteristics, air/fuel mixing, appropriate cylinder pressure, temperature, heat release, and reduced delay periods have led to successful combustion. The BTE for diesel is 8.82% at 20% load and 25.4% at 100% load. For WCO fuels, the best efficiency is attained at 100% engine load when considering the blend changes. However, compared to pure diesel, the efficiency of WCO blends is somewhat lower at lower loads, except for the W30 blend, which has demonstrated about the same efficiency as diesel at lower loads while surpassing it at higher loads. This improvement in BTE for the W30 blends as engine load increases is primarily due to more combustion reactions taking place as combustion pressure and temperature rise. In contrast, the drop in BTE found for other mixes is a result of incomplete combustion caused by a high air-to-fuel ratio. BTE recorded for an 80% engine load indicates values that are nearly equal for all mixes. Similar results are also studied by [1, 16, 25].

3.2.2 Brake Specific Fuel Consumption

The Brake Specific Fuel Consumption (BSFC) is the quantity of fuel required to produce one unit of power per hour by an internal combustion engine. It is a crucial performance parameter that assists in the evaluation of an engine's efficiency and performance. Figure 9 illustrates the fluctuation in BSFC values for diesel and WCO blends under various loads. The results indicate that when load increases, the BSFC value of all blends decreases dramatically, which may be related to the proportionate increase in brake power with load compared to the increase in fuel consumption [26]. When the load is elevated, the cylinder pressure and temperature rise, resulting in an increase in combustion temperature and improved combustion at peak loads. Consequently, when loads increase, the quantity of Fuel required to deliver a unit of brake power decreases, resulting in a lower BSFC. It has been noted that, for lighter loads, WCO mixes provide a greater fuel flow rate to the engine than diesel, also stated

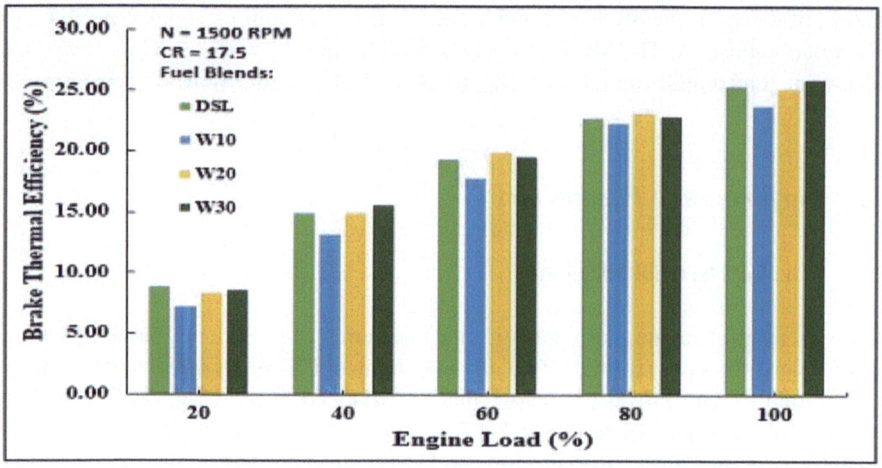

Fig. 8 Brake thermal efficiency for various fuel blends with varying loads

by [27]. When considering 100 percent load, it is noted that the BSFC of all mixes has decreased to the minimum value and is relatively comparable. Nevertheless, across all the blends and engine loads, the W30 blend has demonstrated the closest BSFC value to that of diesel, which is 0.348 kg/kWh (0.337 kg/kWh for diesel). Therefore, higher load values for biodiesels are quite favourable since the BSFC is very close to that of diesel, and there is a rise in diffusion combustion and thermal efficiency [16].

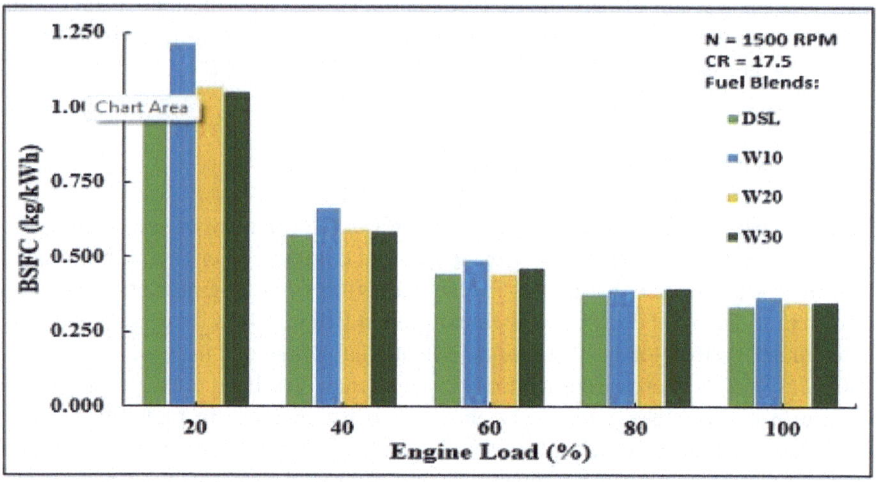

Fig. 9 Brake specific fuel consumption for various fuel blends with varying loads

4 Conclusions

In the current experimental research, different combustion and performance measures were compared between the WCO blends and diesel fuel. At full load, combustion characteristics such as cylinder pressure CHR and MGT were measured to be the greatest for all test fuels. The variation in cylinder pressure for WCO mixes was minor and near the diesel value. In contrast, ROPR and NHR exhibited an initial rise with increasing load up to 60 percent, followed by stagnation at 80 percent and 100 percent load. Also, combustion was seen to advance for higher blending ratio due to their lower cetane number as by the early increase of mean gas temperature in the MGT graph. Performance metrics such as BTE and BSFC were enhanced at full load. At full load conditions, the BSFC is found to be the lowest, whereas the BTE peaked at full load (25%). Ignition delay for all biodiesel blends was lower than that of diesel. In contrast, parameters such as NHR and ROPR decreased in magnitude as the blending fraction of WCO increased in comparison to diesel due to less intense premixed combustion as a result of shorter delay periods. According to the evaluated combustion and performance criteria, W30 proved to be the optimal blend. At maximum engine load, the cylinder pressure for the W30 blend was 62.52 bar compared to 63.6 bar for diesel. At maximum engine load, an NHR of 44.33 J/deg was found, compared to diesel's 57.12 J/deg.

Likewise, the lowest ROPR of 4.25 dP/dθ for W30 as compared to diesel's 4.75 dP/dθ was recorded. In addition, W30 had a 13% reduced delay period than diesel at maximum load. A little improvement in performance characteristics was also recorded for W30, with a 2% rise in BTE and a 1% decrease in BSFC compared to diesel at maximum load.

References

1. Gupta R, Bhalla J, Mourya S (2016) Performance analysis of a diesel engine using the soybean oil based biodiesel. Indian J Sci Technol 9(36)
2. Rehman S, Mahir HS, Zaidi K, Hasan AK (2015) Study of ignition characteristics of microemulsion of coconut oil under off diesel engine conditions. Eng Sci Technol Int J 18:318–324
3. Rehman S, Alam F, Adil M (2019) Ignition and combustion characteristics of impinging diesel and biodiesel blended sprays under diesel engine-like operating conditions. In: ICMET
4. Elkelawy M, Bastawissi HA, Esmaeil KK, Radwan AM, Panchal H, Sadasivuni KK, Sadasivuni KK, Ponnamma D, Walvekar R (2019) Experimental studies on the biodiesel production parameters optimization of sunflower and soybean oil mixture and DI engine combustion, performance, and emission analysis fueled with diesel/biodiesel blends. Fuel 255
5. Tse H, Leung CW, Cheung CS (2015) Investigation on the combustion characteristics and particulate emissions from a diesel engine fuelled with diesel-biodiesel-ethanol blends. Energy 83:343–350
6. Sidharth, Kumar N (2019) Performance and emission studies of ternary fuel blends of diesel, biodiesel and octanol. Energy Sources Part A Recovery Util Environ Effects 42(18)

7. Devaraj R, Dinesh BM, Sivakumar DB, Durai C (2020) Performance, emission, and combustion analysis on diesel engine fueled with blends of neem biodiesel/diesel/ additives. Energy Sources Part A Recovery Util Environ Effects
8. Abed AK, El Morsi KA, Sayed MM, El Shaib AA, Gad SM (2018) Effect of waste cooking-oil biodiesel on performance and exhaust emissions of a diesel engine. Egypt J Petrol 27:985–989
9. Di Y, Cheung SC, Huang Z (2009) Experimental investigation on regulated and unregulated emissions of a diesel engine fueled with ultra-low sulfur diesel fuel blended with biodiesel from waste cooking oil. Sci Total Environ 407:835–846
10. Annisa B, Eriko M, Jiro S (2019) Performance and emission characteristics of biodiesel waste cooking oil water emulsions under varying engine load condition. Energy Sources Part A Recovery Util Environ Effects
11. Pasupathy RS, Mohan RT (2019) Optimisation and effective utilisation of esterified rice bran oil in a turbocharged VCR engine by analysing its operating characteristics. Energy Sources Part A Recovery Util Environ Effects
12. Kumar SM, Jaikumar M (2014) A comprehensive study on performance, emission and combustion behavior of a compression ignition engine fuelled with WCO (waste cooking oil) emulsion as fuel. J Energy Instit:1–9
13. Rehman S, Nafees MS, Waheed MA, Khan AA (2022) Experimental investigation on combustion and performance characteristics of a direct-injection diesel engine. Int J Energy Clean Environ 23:1–17
14. Chaurasiya PK, Singh SK, Dwivedi R, Choudri RV (2019) Combustion and emission characteristics of diesel fuel blended with raw Jatropha, soybean and waste cooking oils. Heliyon 5(5)
15. Shelke PS, Sakhare NM, Lahane S (2016) Investigation of combustion characteristics of a cottonseed biodiesel fueled diesel engine. Proc Technol 25:1049–1055
16. Mehmet RS, Selman A, Adem Y, Seyfi S (2019) Evaluation of comparative combustion, performance, and emission of soybean based alternative biodiesel fuel blends in a CI engine. Renew Energy 148:1065–1073
17. Kannan GR, Balasubramanian KR, Sivapirakasam SP, Anand R (2011) Studies on biodiesel production and its effect on DI diesel engine performance, emission and combustion characteristics. Int J Ambient Energy 32
18. Hosamani BR, Katti VV (2018) Experimental analysis of combustion characteristics of CI DI VCR engine using mixture of two biodiesel blend with diesel. J Eng Sci Technol
19. Uyumaz A, Solmaz H, Yılmaz E, Yamı H, Polat S (2014) Experimental examination of the effects of military aviation fuel JP-8 and biodiesel fuel blends on the engine performance, exhaust emissions and combustion in a direct injection engine. Fuel Process Technol:158–165
20. Buyukkaya E (2010) Effects of biodiesel on a DI diesel engine performance, emission and combustion characteristics. Fuel 89:399–315
21. Muralidharan K, Vasudevan D, Sheeba NK (2011) Performance, emission and combustion characteristics of biodiesel fuelled variable compression ratio engine. Energy 36:5385–5393
22. Gautam R, Kumar S (2020) Performance and combustion analysis of diesel and tallow biodiesel in CI engine. Energy Rep 6:2785–2793
23. Wei L, Cheung SC, Ning Z (2017) Influence of waste cooking oil biodiesel on combustion, unregulated gaseous emissions and particulate emissions of a direct-injection diesel engine. Energy 127:175185
24. Murugesan A, Umarani C, Subramanian R, Nedunchezhian N (2009) Bio-diesel as an alternative fuel for diesel engines—a review. Renew Sustain Energy Rev 13(3):653–662
25. Enweremadu CC, Rutto LH (2010) Combustion, emission and engine performance characteristics of used cooking oil biodiesel—a review. Renew Sustain Energy Rev 14:2863–2873
26. Dowell GT, Feng B, Alvensleben NV, Brown RJ (2015) Combustion analysis of microalgae methyl ester in a common rail direct injection diesel engine. Fuel 143:351–360
27. Işık MZ, Aydın H (2016) Analysis of ethanol RCCI application with safflower biodiesel four blends in a high load diesel power generator. Fuel 184

Solar and Renewable Energy

Internet of Things Connected Hybrid Solar Cooker

Pratyush Kukreti, S. S. Bhandari, and Govind Verma

Abstract The potential of cooking during off sunshine hours is investigated using an energy storage unit in the solar cooker, such as a thermal energy storage material or thermal heating of the absorber plate of solar cooker by photovoltaic (PV) technology. To determine the viability of a solar cooker during the off-sunshine hours, performance experiments on a hybrid solar cooker utilizing phase change materials (PCMs) and PV technology have been carried out. In order to improve user experience, the hybrid solar cooker has also been upgraded with an internet of things-based monitoring interface. The Bureau of Indian Standards are also used to test the thermal characteristics of solar cookers. The duration of cooking in a solar cooker using various PCMs in sunshine and off-sunshine hours being is the main emphasis of the present experimental study.

Keywords Phase change materials (PCM) · PV technology · Finned cooking vessel · Napthalene · Internet of Things

1 Introduction

The importance of energy to human welfare and economic growth cannot be overstated. The primary concern is having access to a sufficient energy supply, particularly in developing nations that lack access to proper infrastructure for harvesting an adequate amount of energy. Due to rising energy consumption in various parts of the world, it is anticipated that by 2050, the demand for primary energy would be increased up to 1.5–3 times [14]. Due to the rise in energy consumption, more natural

P. Kukreti (✉) · S. S. Bhandari
MED, G.B. Pant University of Agriculture and Technology, Pantnagar 263145, Uttarakhand, India
e-mail: 55323_pratyushkukreti@gbpuat-tech.ac.in

S. S. Bhandari
e-mail: susheelsinghbhandari.me@gbpuat-tech.ac.in

G. Verma
ITD, G.B. Pant University of Agriculture and Technology, Pantnagar 263145, Uttarakhand, India

M. A. Siddiqui et al. (eds.), *Advances in Heat Transfer and Fluid Dynamics*, Lecture Notes in Mechanical Engineering, https://doi.org/10.1007/978-981-99-7213-5_26

resources are being exploited. Coal, petrol, and fossil fuels are the primary sources of energy. The dependency on fossil fuels has led to an increasing percentage of CO_2 emission in the upper atmosphere and their extraction process also causes various types of pollution to our natural environment and disrupts the ecological system. Therefore, utilizing a new and clean form of energy is required for industrial and household purposes.

A solar cooker utilizes solar energy as free fuel for operation. Owing to its simplicity, a box-type solar cooker (BTSC) is widely used but its use is limited due to the intermittent nature of solar energy. This drawback can be resolved by the addition of thermal energy storage (TES) unit with the cooker. Domanski et al. [3] studied the effect of two PCM, Magnesium nitrate hexa-hydrate (MNHH) and Stearic acid, in a BTSC. Two concentric vessels were utilized and PCM was filled in the annular spacing between the vessels. The overall thermal efficiency of solar cooker is about 82% (with MNHH as PCM). This study also confirms that MNHH can be used for indoor cooking. Chen et al. [2] theoretically investigated PCM as a TES medium for BTSC. A comparative two dimensional enthalpy based simulation model with calculations for melt fraction was proposed on five different PCM and results suggested that the complete melting of the PCM material is strongly affected by thermal conductivity of the heat exchanger material and the effective thermal conductivity of the PCM. Geddam et al. [6] studied the influence of the geometry of the cooking vessel (by providing fins) and the effectiveness of Paraffin wax as a TES medium in the solar cooker. The study concluded that the food cooked can be kept at elevated temperatures for 3–4 h by using PCM during offsunshine hours. Kanimozhi et al. [10] studied the effect of Commercial grade coconut as a PCM on a BTSC. Copper tubes, filled with PCM, were placed on the absorber plate. The cooking load was kept above these tubes resulting in a significant reduction in cooking time.

From a customer's perspective cooking time is one of the most important aspects and some studies show that the design of cooking vessels has an appreciable impact on cooking time. Gaur et al. [4] reported that if the plain lid of the cooking pots is modified to a concave shape then an increase in stagnation temperature by 3–7 °C can be observed. The utensil having a concave shaped lid shows a 10–13% reduction in achieving the same temperature in comparison to the plain lid. Rao and Subramanyam [12] investigated the influence of elevating the vessel over lugs in the absorber plate and concluded that due to the flow of hot air between the vessel and the floor of the cooker it significantly reduces the cooking time because the effective heat transfer area to the vessel is increased. Rao and Subramanyam [13] have also performed experimental studies on the cooking vessel with an annular cavity that was placed on the lugs. This permits hot air to freely move through the gap between the vessel's bottom surface and absorber plate, as well as through the cylindrical cavity in the vessel. The setup showed a remarkable reduction in cooking time. Harmim et al. [7] conducted a comparative experimental evaluation to determine the impact of the geometry of cooking vessels in a BTSC. The first vessel cylindrical in geometry (traditional vessel), and the second vessel, similar in shape and volume to the first but provided with fins on its external lateral surface (modified vessel). The effective heat transfer area was increased due to the addition of fins, thus reducing the cooking time considerably. It was observed that the water in the modified vessel was always at a temperature higher than the water in the traditional vessel. The time it took to boil the water (99 °C) was 91 min for the modified container and 103 min for the traditional container.

The large scales dissemination of solar cookers due to the intermittent nature of solar energy can also be resolved by integrating the thermal heating of the absorber plate of the solar cooker by photovoltaic technology (PV). The solar cooker using PV technology is termed as the hybrid solar cooker and much research are conducted on the performance of hybrid solar cookers. Joshi and Jani [8] constructed a photo-thermo solar cooker by attaching a PV cell-powered DC heater to a stainless-steel casserole utilizing screws. The hybrid solar cooker is connected to 5 panels of 15 W each and a dc heater and is called a Small Scale Hybrid Box type solar cooker (SSHB). The mean efficiency of SSBH was found to be 30.1%. It was found that water reached its boiling point with 50 min of heating. Later this cooker was further improved by replacing the dish type dc heater with 3 rod-type heaters each of 25 W and efficiency increased to 38%. Gawali and Papade [5] employed a hybrid solar cooker consisting of a TES tank, installed at the parabolic mirror's focal point, containing PCM for storing solar energy as sensible heat, an auxiliary DC heater, operated via PV panel and battery. Different types of oils used for analysis were Olive oil, Soyabean oil, and Thermic fluid (Hytherm-500). This setup was capable enough to prepare meals on cloudy days and in the absence of sunlight. Talbi et al. [15] studied the heating of a thermal resister powered by a photovoltaic panel. The resistance of the heating resistor under observation was 11.5 Ω, supporting a power of 800 W, and reached a temperature of 600 °C. The study showed that as the solar irradiation increased the temperature of the heating unit also increased with a steep slope. For a load of 1 kg of water, the electric and thermal power were found to be 164.5 W and 118.71 W, respectively, and the thermal efficiency was 72%. Watkins et al. [16] have developed an insulated solar electric cooker made up of PV panel Ni–Cr wire with an electrical resistor 814 Ω/m, fire glass, straw, and perlite as the insulating material. The highest recorded temperature of the absorber plate was 502 °C which corresponded to an input of 100 W for 5 gallon drum insulated with fibre glass.

Furthermore, solar cooking is a slow process. The food contents are inside the container for a long time and the only thing the user can do is wait for a specified time before the food is cooked. Therefore, with a hybrid solar cooker coupled with Internet of Things based monitoring interface, the user can get information remotely about the time remaining for complete cooking. Kagale and Jangamshetti [9] presented an IoT based monitoring system for box type solar cooker. The temperature readings were taken via PT100 sensor and sent to an android application through Arduino Uno. Various food products (rice, dal, ground nut, etc.) were cooked in the cooker and data were recorded and then fed into the computer program to sense the time remaining for the contents to be fully cooked and thus making the device more user-friendly. Another observation was the reduction in cooking time by using this interface since the user need not physically check if the contents were cooked by opening the cooker or food container thus reducing the loss of heat. Barik [1] used raspberry pi and HTU 211D sensor device to monitor the temperature and humidity and data were uploaded over the cloud that will be reviewed by cell phone. This system allowed user-friendly monitoring of agricultural parameters. This setup also controls a DC fan, motor, and water level. Based on the water level, the farmer can remotely turn on–off the motor to supply water or DC fan to control temperature and humidity. This system is beneficial in areas of water scarcity or can be used where the crop is extremely sensitive to environmental conditions.

In this study experiments on smart hybrid solar cookers using PCMs and PV technology coupled with an Internet of Things (IoT) interface were carried.

2 Materials and Method

In the present study, the experiments were performed on a box-type solar cooker (BTSC) and hybrid solar cooker, with PCM incorporated in the annular cavity of two concentric cylindrical vessels and the inner vessel filled with water. The temperature–time history of water is recorded and analyzed during sunshine and off-shine hours. Nine experiments divided into 4 categories were conducted to quantify the influence of various physical conditions on the energy storage behavior of PCMs. Figure 1 represents the schematic chart of the tests conducted.

The equipment used in this experimental study are as follows: A pair of BTSC, Pyranometer, K-type Thermocouple, Data logger, PCM (Naphthalene, Acetanilide, and Magnesium nitrate hexa-hydrate), Cooking vessel/container (finned and un-finned), Photovoltaic solar panels (100 W each), Inverter of 1050 VA, Battery of 90 AH, Heating Plate cast iron, Multimeter, IoT Module.

2.1 *Hybrid Solar Cooker*

In this study, experiments were performed on the hybrid solar cooker (Fig. 2). The vessels containing PCM and cooking load were directly placed over a resistance heating plate which was powered by a battery charged with the help of PV panels. Thus, the integrated impact of energy storage with PCM and plate heating by solar panels is investigated, assessing the viability of cooking during off-sunshine hours.

The approximate cost of the smart hybrid solar cooker was ₹35,000 or $432.

2.2 *Internet of Things*

This module helps the user to connect the device, hybrid solar cooker in this case, with the internet. Once the device is connected to the internet the user can remotely observe its functions and can take necessary actions. In the present study, the IoT interface is made possible by using Raspberry pi 4B+. The temperature sensors (DS18B20) were

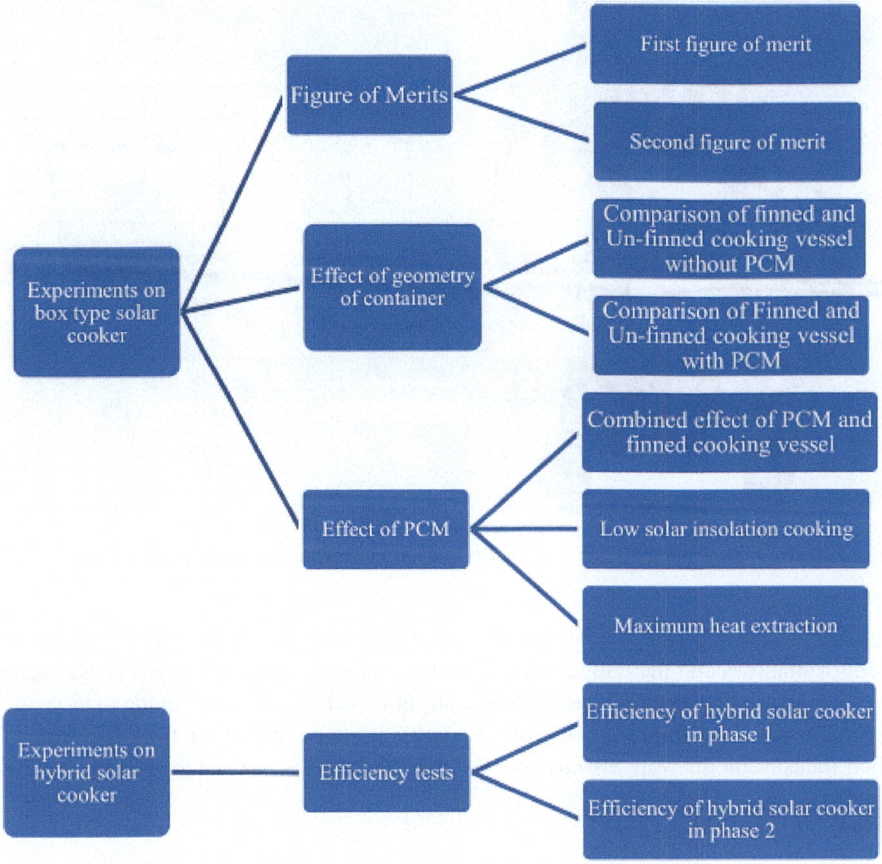

Fig. 1 A schematic chart of the tests conducted

connected with raspberry pi. A computer program was written in python programming language. The raspberry pi was connected to wifi and commands were given by logging in using the same wifi.

2.3 *Experimental Procedure for Box-Type Solar Cooker*

A pair of identical BTSC was used both containing a pair of concentric cooking vessels containing PCM in the annular cavity and water in the inner vessel. The inner container was made up of Copper with fins on its lateral surface. The outer container was a Stainless Steel container. 0.5 kg of PCM and an equal quantity of water were used. The solar cookers with their contents were kept in sunlight in the morning (at around 10 A.M.) and the temperature of water was recorded until it

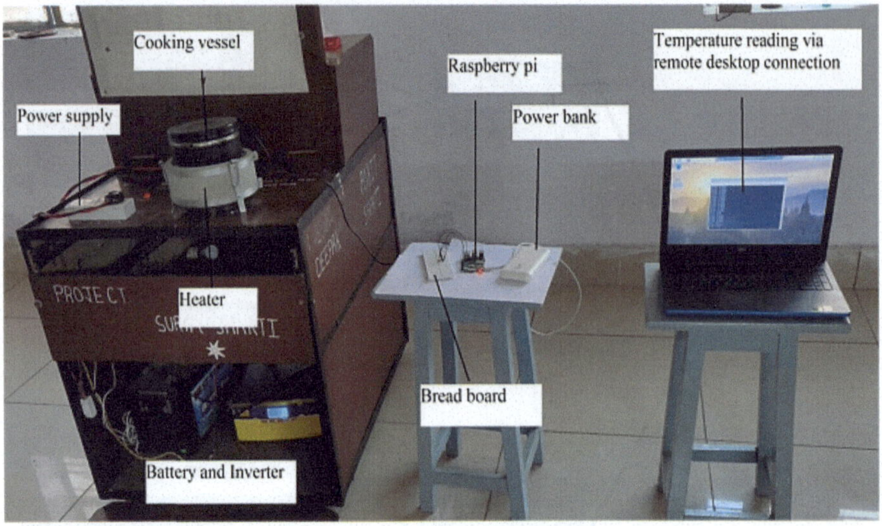

Fig. 2 Hybrid solar cooking setup using IoT based monitoring interface during off-sunshine hours

reaches its boiling point. After attaining the boiling point, the solar cookers were kept in shadows and the temperature of water was allowed to fall until a steady state was achieved. For different experiments, slight modifications were made to measure the required effects. Solar insolation and ambient temperature with the exact time of their measurements were also recorded. Each observation was made at an interval of 10 min.

2.4 *Experimental Procedure in Smart Hybrid Solar Cooker*

Hybrid solar cooker is connected to Internet of Things (IoT) to make it a smart hybrid solar cooker and then its efficiency was measured. The cooking load (water) was heated on the resistance heater. The product of mass, specific heat capacity and rise in temperature of water gives the amount of heat stored in water. The power supplied to the system is equal to the product of voltage and current supplied by the battery.

Therefore, thermal efficiency is:

$$\text{Efficiency } \eta_0 = \frac{P_0}{P_i} \tag{1}$$

$$P_i = V * I \tag{2}$$

$$P_0 = \frac{m_w C_p \Delta T}{\Delta t} \tag{3}$$

The system allows the charging of the battery by the solar panel.

The subsequent procedure is followed for the experimentation on the hybrid solar cooker:

- To test the efficiency of the system, 0.5 kg of PCM with an equal amount of water at ambient temperature is heated to the boiling temperature on the hybrid solar cooker and the power output (P_o) was calculated by Eq. (3).
- Synchronously, current and voltage drawn by the heating element are recorded and input power (P_i) to the system is determined by Eq. (2).
- The efficiency of the system is measured by using Eq. (1).
- Similar experiments were conducted using three different PCMs. The heat stored in PCMs can be used to maintain the water at elevated temperatures for several hours.

3 Results and Discussion

3.1 *First Figure of Merit*

Mullick et al. [11] defined the approach for calculating F1. Three tests were performed on each solar cooker, and the average value for each cooker is 0.1112 and 0.1125. The findings indicate that the solar cooker belongs to the type B class of BTSC.

3.2 *Second Figure of Merit*

The F2 or load test is used to determine the heat transmission factor among the contents of the solar cooker. The test is carried out to establish the impact of both load as well as the number of pots on F2. The graphs in Fig. 3a, b show the outcomes of the preceding experiment.

3.3 *Effect of Fins on Cooking Time*

The traditional and modified (finned) cooking vessels took 130 min and 95 min, respectively, to reach the boiling point (Fig. 4). Because the use of fins enhances the effective heat transfer area, a temperature of 93 °C is reached 35 min faster in the modified vessel than in the regular cooking vessel with a weight of 0.5 kg.

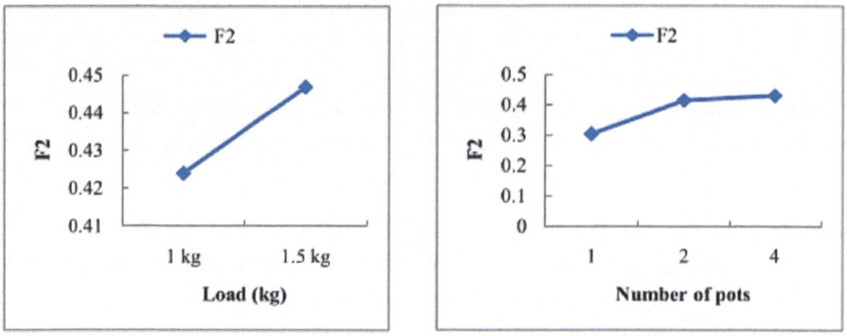

Fig. 3 **a** Effect of load on F2. **b** Effect of number of pots on F2. F_2 equals 0.42937 for 4 pots and 1 kg weight

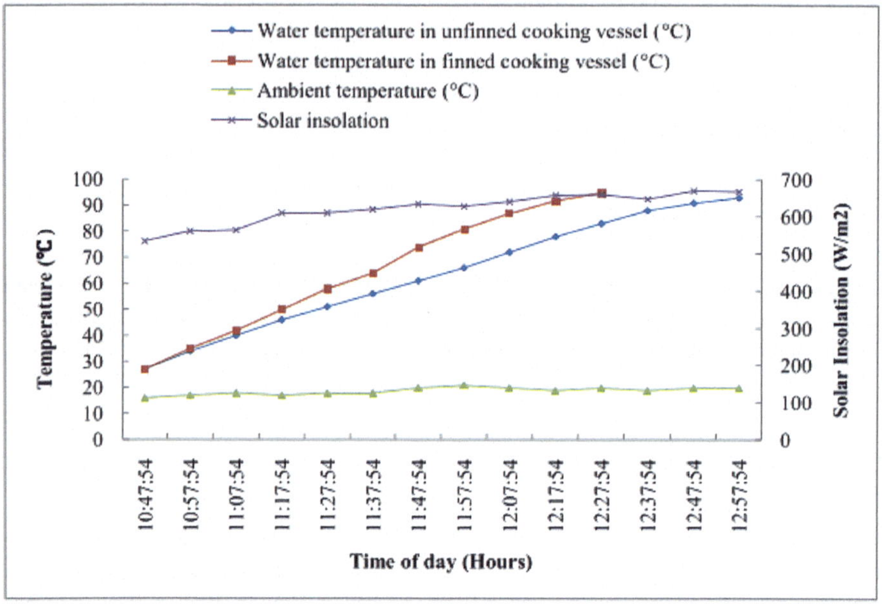

Fig. 4 Variation in water temperature in finned and non-finned vessel for 0.5 kg

3.4 Effect of PCM on Temperature of Water in Finned Vessel

When the solar cookers are kept in shadow after boiling of water, the temperature of the vessels starts to decrease and the heat energy flows from charged PCM to the cooking load thus keeping it warm for several hours. All three PCMs were tested and the results indicate that after 4 h in the off-sunshine condition, the steady state temperature of the water with PCM was 8–15 °C, 12–17 °C and 11–16 °C

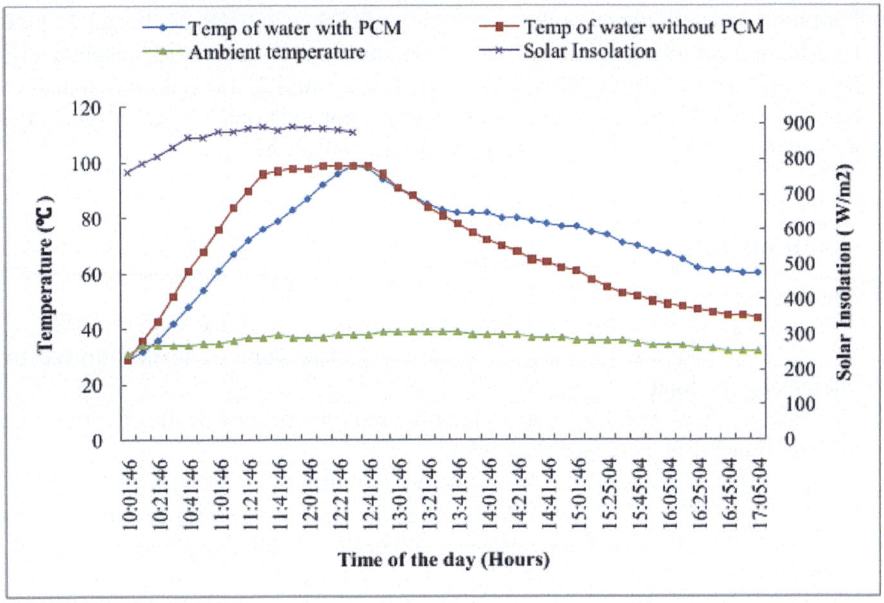

Fig. 5 Variation of water temperature with Naphthalene and without Naphthalene in finned cooking vessel with time of the day (hours) for solar cooker

higher than the steady state temperature of water without PCM for Magnesium nitrate hexa-hydrate, Naphthalene, and Acetanilide respectively (Fig. 5).

3.5 Experiments on Photovoltaic Heating System

The battery was charged during the day by PV panels and the tests were conducted during the evening. The hybrid solar cooker was connected to Raspberry pi for remote observation of data.

- It takes around 40 min for the heating system to boil the water with a mean thermal efficiency of 38.35%. One day of charging by the solar panel yields a battery backup of approximately 90 min. For phase 1, the mean thermal efficiency of the smart hybrid solar cooker employing PCM is 28.05%, 27.38, and 31.39% for Magnesium nitrate hexahydrate, Naphthalene, and Acetanilide. In phase 1 the water and PCM were first boiled in the solar cooker and then allowed to cool. Once the steady state was reached the vessels with their contents were heated on the hybrid cooker until water is boiled and efficiency is calculated, whereas in phase 2 water and PCM were at ambient conditions and directly heated on the hybrid cooker until the water reaches its boiling point.

- For phase 2 the efficiency of the system using PCM is 26.05%, 26.38 and 31.26% for Magnesium nitrate hexa-hydrate, Naphthalene, and Acetanilide respectively.
- By comparing the efficiency results from phases 1 and 2, it is possible to deduce that the mean efficiency of the smart hybrid cooker is unaffected by the initial temperature of the cooking contents, i.e. water and PCM.

4 Conclusions

1. The first figure of merit shows the cookers are of grade B according to BIS.
2. As the pots increases in number, so does F2. The same trend is followed by increasing the load.
3. A cooking vessel with fins reduces the time required to cook as the effective area of heat transfer is increased due to fins.
4. After 4 h in the shade, the steady state temperature of water with PCM was 8–15 °C, 12–17 °C, and 11–16 °C higher than the steady state temperature of water without PCM for Magnesium nitrate hexahydrate, Naphthalene, and Acetanilide, respectively.
5. The un-finned cooking vessel with PCM shows a latency in reaching boiling temperature of 15 min, 20 min, and 10 min, than a finned cooking vessel with PCM for Acetanilide, Naphthalene, and Magnesium nitrate hexa-hydrate respectively.
6. The smart hybrid solar cooker takes around 40 min to boil the water with a mean thermal efficiency of 38.35%. One day of charging by the solar panel yields a battery backup of approximately 90 min. The mean thermal efficiency of the smart hybrid solar cooker using PCM is 27.38, 28.05% and 31.39% for Naphthalene, Magnesium nitrate hexa-hydrate and Acetanilide respectively.

Acknowledgements The author thanks G. B. Pant University of Agriculture and Technology, Pantnagar for funding the required research facilities.

References

1. Barik L (2019) IOT based temperature and humidity controlling using Arduino and Raspberry pi. Int J Adv Comput Sci Appl 10(9):494–502
2. Chen CR, Sharma A, Tyagi SK, Buddhi D (2008) Numerical heat transfer studies of PCMs used in a box-type solar cooker. Renew Energy 33(5):1121–1129
3. Domanski R, El-Sebaii AA, Jaworski M (1995) Cooking during off-sunshine hours using PCMs as storage media. Energy 20(7):607–616
4. Gaur A, Singh OP, Singh SK, Pandey GN (1999) Performance study of solar cooker with modified utensil. Renew Energy 18(1):121–129
5. Gawali SR, Papade CV (2015) Hybrid solar cooker. Int J Eng Res 4(6):763–766

6. Geddam S, Dinesh GK, Sivasankar T (2015) Determination of thermal performance of a box type solar cooker. Sol Energy 113:324–331
7. Harmim A, Boukar M, Amar M (2008) Experimental study of a double exposure solar cooker with finned cooking vessel. Sol Energy 82(4):287–289
8. Joshi SB, Jani AR (2015) Design, development and testing of a small scale hybrid solar cooker. Sol Energy 122:148–155
9. Kagale P, Jangamshetti SH (2020) Development and testing of IoT based monitoring interface for solar box cooker. In: Bangalore humanitarian technology conference, Vijaypur, 8–10 Oct 2020. IEEE, pp 1–4
10. Kanimozhi B, Sanandharya K, Anand S, Kumar S (2015) Experimental study on solar cooker using phase change materials. Appl Mech Mater 766:463–467
11. Mullick SC, Kandpal TC, Saxena AK (1987) Thermal test procedure for box-type solar cookers. Sol Energy 39(4):353–360
12. Rao AN, Subramanyam S (2003) Solar cookers—part I: cooking vessel on lugs. Sol Energy 75(3):181–185
13. Rao AN, Subramanyam S (2005) Solar cookers—part-II—cooking vessel with central annular cavity. Sol Energy 78(1):19–22
14. Shahsavari A, Akbari M (2018) Potential of solar energy in developing countries for reducing energy-related emissions. Renew Sustain Energy Rev 90:275–291
15. Talbi S, Kassmi K, Deblecker O, Bachiri N (2019) Thermal heating by photovoltaic solar energy. Mater Today: Proc 13:1125–1133
16. Watkins T, Arroyo P, Perry R, Wang R, Arriaga O, Fleming M, O'Day C, Stone I, Sekerak J, Mast D, Hayes N (2017) Insulated solar electric cooking–tomorrow's healthy affordable stoves. Dev Eng 2(3):47–52

Performance Evaluation of Solar Box Cooker with Phase Change Materials

Md. Riyaz Anwer, Naiem Akhtar, and Md. Reyaz Arif

Abstract Solar cooker uses solar radiation to cook the food. These are good alternative for cooking as it uses solar radiation which is free and does not cause any environmental pollution. Normally, the box type cooker can cook the food in day time only. On the other hand, the solar box cooker that utilises phase change material (PCM) is able to prepare food both during the day and in the evening. In this study, simulations of box solar cookers with three distinct PCMs have been conducted to compare the performance of the cooker with different PCMs. The 3D model of the box solar cooker has been prepared, grid independence test has been done and the simulations have been performed using ANSYS FLUENT package. In the simulation the maximum temperature achieved by the absorber plate during the charging mode and time to achieve this maximum temperature have been evaluated. Using the maximum temperature achieved the other thermal performance parameters have been calculated. Through the simulation it is found that maximum temperature of the absorber plate is achieved with PCM Erythritol. The drop in temperature of the absorber plate during the discharging mode is more with PCM Quinone. Also, during the discharging mode, the energy provided by Acetanilide and Quinone is less than the energy required for evening cooking, so Acetanilide and Quinone do not support evening cooking. However, the analysis shows that the evening cooking is possible by using Erythritol as the phase change material. All the performance parameters are high with Erythritol as compared to Acetanilide and Quinone. Thus, the performance of the cooker is better with Erythritol.

Keywords Solar box cooker · Phase change material · Simulation · Thermal performance · ANSYS FLUENT

Md. R. Anwer (✉) · N. Akhtar · Md. R. Arif
Department of Mechanical Engineering, Zakir Husain College of Engineering and Technology, Aligarh Muslim University, Aligarh, India
e-mail: mdriyazanwer@gmail.com

N. Akhtar
e-mail: nakhtar.me@amu.ac.in

Md. R. Arif
e-mail: reyazarif@zhcet.ac.in

M. A. Siddiqui et al. (eds.), *Advances in Heat Transfer and Fluid Dynamics*, Lecture Notes in Mechanical Engineering, https://doi.org/10.1007/978-981-99-7213-5_27

1 Introduction

Solar energy is one of the most significant renewable energy sources. It is available in a limitless quantity worldwide. Due to the decline of nonrenewable energy sources, the significance of solar energy increases. Solar energy is a pollution free environment friendly energy source. Lighting, cooking and heating are the major energy requirements of domestic consumers. A solar cooker is a device that uses the energy of the sun to cook food. Sunlight warms the cooking pot which is used for cooking food. Solar cookers are classified into two types; Box solar cooker and Concentrating type solar cooker. The box solar cooker is made up of an insulated box, metallic cooking pots inside the box, an absorber tray, a double glass top on the cooking tray, and a reflecting mirror mounted on the box's edge. Incoming solar rays strike the glass cover, pass through it, and reach the cooking tray. The heat is absorbed by cooking tray and transfers this heat to cooking pots to facilitate cooking. The reflector mirror reflects the solar radiation falling on it into solar cooker. In most cases, up to four cooking pots are placed within the box type solar cooker. Depending upon the availability of solar radiation the solar box cooker takes 2–3 h to cook vegetables, rice, dals, etc. [1]. The solar box cooker can typically only make food during the day. Food cannot be prepared by it in the evening or at night. If we somehow store solar energy during day time and use that energy later, then the cooking in the evening is possible. The three techniques of solar energy storage are thermochemical, sensible heat, and latent heat [2]. Latent heat storage is a more efficient way of storing thermal energy for the solar box cooker. The materials that are capable of storing latent heat are referred to as phase change materials (PCMs). Different PCM materials have been identified which can be used in solar box cooker. Material for storing solar energy is generally placed below the cooking tray in the solar cooker. The amount of PCM used depends on the energy required to cook food.

The study of PCM was started in the 1940s, but the usage of PCM in the solar cooker was started during 1990s. A simple solar box cooker with focusing plane mirrors was constructed by Ramadan et al. [3]. They employed locally available sand as an energy storage medium around the cooking pot. The cooking time attained with using sand as an energy storage medium was six hours per day (3 h of indoor cooking and 3 h of outdoor cooking). Buddhi and Sahoo [4] had developed an energy-storing solar box cooker. They used steric acid as energy storage material. The PCM was kept below the absorber plate which has a depth of 8 cm. The cooking capacity of the cooker is designed for 0.75 kg. Evening cooking is quite slow as the energy stored by steric acid is low. In order to cook evening meals, Sharma et al. [5] have designed a solar cooker that stores energy during the day. They have used acetamide as PCM. The energy storage unit is in cylindrical shape which surrounds the cooking pot. They have used 2 kg of PCM and the result shows that second meal can be cooked if we put the meal in the cooker before 3:30 pm. Vigneswaran et al. [6] have developed a simple solar box cooker with an energy storage unit. The energy storage unit of PCM was made in concentric cylinder shape. Cylinder interspace was filled with oxalic acid dihydrate. It has a melting point of 101 °C. The amount of PCM was calculated according to the energy required for cooking 0.5 kg rice. They have used single and double reflector and compared the performance for both reflectors. The overall utilization efficiency for the double reflector (25.74%) is more as compared to the single reflector (15.74%). Sharma and Rai [7] have performed the experiment on the two models (model-1 and model-2) of the box solar cooker. They used Magnesium nitrate hexahydrate as PCM. The PCM was also placed below the absorber plate. Both the maximum temperature of the cooking pot as well as the overall utilization efficiency of the model-2 was found to be higher which has finned attached at base. Sharma et al. [8] had designed an evacuated tube collector that stores energy. Two hollow, concentric cylinders were used to create the PCM storage component. These hollow cylinders measured 30.0 cm in diameter on the inside and 44.0 cm on the outside. The height of the cylinder is 42.0 cm and 45 kg of Erythritol was used to fill the space between cylinders. This cooker can cook two foods one in daytime and one in evening. However, it requires huge quantity of PCM. Ali and Akhtar [9] have designed and analysed the performance of an energy-storing box cooker. They have used Erythritol having a melting point of 118 °C for storing solar energy. They have also modified the design. Generally, the opening is provided on the top side for keeping the cooking pot, but they have fixed the top glass cover. A rectangular window is provided on the back side for opening and for keeping the cooking pot. By fixing the top glass cover the energy losses from the top decreases.

From the literature it has been discovered that solar cookers operate better when PCM has a greater melting point. All the performance analysis of the box type solar cooker has been done experimentally and simulation work is not reported in the literature. Additionally, there is no information on how changing the PCM quantity may affect the cooker's performance. Therefore, the main objective of the present work is to simulate the performance of the box type solar cooker by selecting different PCMs and also to study the effect of varying the amount of PCM on the performance of cooker.

The performance simulation has been carried out using the ANSYS-FLUENT academic package. First, the 3D model of the box type solar cooker was made. Then that model was imported into ANSYS FLUENT and simulation was performed after mesh validation. The performance of the cooker with three distinct PCMs, Erythritol, Acetanilide and Quinone, were compared. The effect of changing the PCM quantity from 5 to 4 kg on the performance of the solar cooker is also studied. The performance of the solar cooker with these three PCMs on the basis of maximum temperature achieved by the absorber plate, time taken to reach maximum temperature, charging time of PCM, PCM storage efficiency and temperature of the absorber plate during the discharging mode has been compared.

2 Description of Proposed Solar Box Cooker

Generally, the solar cooker available in the market is used to cook food only during day time. It is not able to cook food in the evening since it lacks an energy storage unit. These cookers have opening in the top for keeping the cooking pot so it causes lots of heat losses from the top. The proposed PCM solar cooker has aperture area of 0.5 m^2. It has a double glass cover of length 71.5 cm and width 70 cm. The cooking tray as well as the PCM tray is made of GI sheet. The front side height of the absorber plate is 10 cm and the back side height is 22 cm. The height of the PCM tray was 9 cm and it was placed below the absorber plate. The PCM tray was filled with PCM. The outer box (casing) is composed of wood. The heights of the front and back sides that face the sun are 27 and 41 cm. The inner side of the wooden box is filled with insulation at the bottom and sides. The shape of the cooking pots is cylindrical with a flat base having a diameter of 18.0 cm and a height of 7.5 cm. A rectangular window of width 26.0 cm and height 13.0 cm is provided behind the cooker for placing or removing the cooking pots. A reflector mirror is attached to the top cover. The space between the wooden cover and reflector mirror is filed with insulation to minimize the top heat loss. The absorber plate measures 59 cm in length, 10 cm in front height, 22 cm in back height, 65 cm in front width, and 57 cm in back width. The PCM tray has a length of 64 cm. Considering the dimensions of the above PCM box solar cooker, a 3D model has been developed as shown in Fig. 1.

The solar cooker receives solar energy that reaches the absorber plate while it is in the charging mode. Additionally, the radiation that is reflected by the reflector mirror is received by the absorber plate. As a result of the radiation being absorbed on the absorber plate, the temperature within the cooker ultimately rises. At the time of no-load condition some part of energy received on the absorber plate is used to increase the temperature of the cooker and some energy is transferred to PCM. During loading conditions, a portion of the energy is also used to raise the load's temperature. During the hours when solar radiation is unavailable and the lid of the cooker is closed. Throughout this time, the cooker can prepare food using energy stored in the PCM during the charging time. During evening cooking, some

Fig. 1 3D model of box cooker

of the energy stored in the PCM is used to cook the meal, while some is lost to the environment.

3 Modelling of Solar Box Cooker

The solar box cooker equipped with PCM storage unit is shown in Fig. 1. The energy storage material was placed in the PCM tray, which is placed below absorber plate. During day time the solar energy is absorbed by the absorber plate. Some portion of the absorbed energy is transferred to PCM, and that energy is stored by PCM. In the late evening, PCM uses its stored energy to cook the food.

3.1 Energy Balance for PCM Cooker

The performance of the box type solar cooker with energy storage is defined in terms of charging- discharging time, temperature attained by load, time to attain maximum temperature, storage efficiency, and PCM utilization efficiency. The charging time is the amount of time necessary to completely melt PCM. Discharging time is the amount of time it takes for PCM to solidify completely.

Charging time of PCM:

$$\tau_{\mathrm{ch}} = \tau_{\mathrm{sh}} + \tau_m \tag{1}$$

Discharging time of PCM:

$$\tau_{\mathrm{dis}} = \tau_{\mathrm{sc}} + \tau_s \tag{2}$$

where

τ_{ch} Total charging time of PCM
τ_{sh} Sensible heating time of PCM with initial temperature equal to ambient
τ_m Melting time of PCM
τ_{dis} Total discharging time of PCM
τ_{sc} Time of sensible cooling of PCM
τ_s Solidification time of PCM

The energy stored in the PCM is evaluated by

$$Q_{\mathrm{stored}} = \int_{T_{\mathrm{pi}}}^{T_{\mathrm{pm}}} mC_{\mathrm{ps}}dT + mL + \int_{\mathrm{pm}}^{T_{\mathrm{pf}}} mC_{\mathrm{pl}}dT \tag{3}$$

Energy supplied to the cooker:

$$Q_{\text{in}} = I_{\text{av}} A_p \eta_o \tau_{\text{ch}} \tag{4}$$

where,

m	Mass of PCM
C_{ps}	Specific heat capacity of PCM in solid state
L	Latent heat capacity of PCM
C_{pi}	Specific heat of molten PCM
T_{pi}	Initial PCM temperature
T_{pm}	Melting temperature of PCM
T_{pf}	Maximum temperature of PCM during charging
A_P	Solar cooker aperture area
η_o	Optical efficiency of double glass cover
I_{av}	Average intensity of solar radiation

The storage efficiency of solar cooker is given by:

$$\eta_{\text{stored}} = \frac{Q_{\text{stored}}}{Q_{\text{in}}} = \frac{\int_{T_{\text{Pi}}}^{T_{\text{Pm}}} mC_{\text{ps}}dT + mL + \int_{T_{\text{Pm}}}^{T_{\text{pf}}} mC_{\text{pl}}dT}{I_{\text{av}} A_p \eta_o \tau_{\text{ch}}} \tag{5}$$

The PCM utilization efficiency is given by:

$$\eta_{\text{pu}} = \frac{Q_u}{Q_{\text{storage}}} = \frac{m_f C_f (T_{\text{maxf}} - T_{\text{fi}})}{\int_{T_{\text{Pi}}}^{T_{\text{Pm}}} mC_{\text{ps}}dT + mL + \int_{T_{\text{Pm}}}^{T_{\text{pf}}} mC_{\text{pl}}dT} \tag{6}$$

m_f	Mass of cooking food
C_f	Specific heat capacity of cooking fluid
T_{maxf}	Maximum cooking fluid temperature
T_{fi}	Initial temperature of cooking fluid

4 Numerical Aspects

After importing geometry and naming the component name the next step is to generate the mesh. High quality unconstructed mesh was generated using ANSYS Meshing. First, very fine meshing of the solar cooker was generated. After that element size of the mesh was increased. As very fine mesh will give more accurate result but it will take more computing time so mesh validation or grid size independency test was carried out for obtaining a mesh size which will give an optimum result in an optimum computing time. Figure 2 depicts the mesh of a solar box cooker with an element size of 18 mm.

Fig. 2 Mesh of solar cooker with element size 18 mm

Table 1 Grid independency test

Grid size (mm)	No of elements	Temperature of absorber plate (°C)
10	227,470	129.07
14	128,509	128.68
18	69,162	128.32

4.1 Grid Independence Test

Grid independency test is conducted to obtain a suitable mesh size for the simulation. For grid independency test the absorber plate temperature at different grid sizes is calculated. Table 1 shows the result of the grid independence test.

From Table 1 it is concluded that the mesh with element number 128,509 (grid size 14 mm) is optimum for simulation of the cooker.

4.2 Boundary Conditions

The assigning of the material to each component of the solar cooker is done by the inserting material with their properties. In the input solar loading on the cooker has been applied by inserting the latitude and longitude of the location, date and month and starting time of the solar loading. For the present case the location is Aligarh with latitude and longitude 27.8974° N and 78.0880° E, respectively. The selected date is 10th June with starting time 9:30 am. The data available for the solar radiation on 10th June at Aligarh has been considered which varies from 723.74 W/m^2 at 9:30 a.m. to 835.5 W/m^2 at 1:30 p.m. The overall heat loss factor for the charging mode is calculated for different temperatures using Akhtar and Mullick correlation [9] and then an average heat loss factor was inserted. The same boundary conditions were applied for each PCM material. In the output absorber plate temperature with time and load temperature with time are obtained.

4.3 Validation of Numerical Model

The selected model is validated by Abid et al. [9] with the experimental results obtained on the same model with Erythritol as the PCM material having quantity of 6 kg. The maximum temperature obtained with 6 kg of Erythritol with simulation on 11 march is 116.1 °C while through the experimental result the maximum temperature of the absorber plate is 115.7 °C. The variation of temperature of the absorber plate by experiment and simulation is shown in Fig. 3.

5 Results and Discussion

The boundary conditions, as discussed above were applied and the simulation was initialized. After initialization calculation was run with a suitable time step. The initial temperature for the absorber plate as well as the load was taken as 27 °C. The change in the absorber plate temperature and load temperature with time for the measured

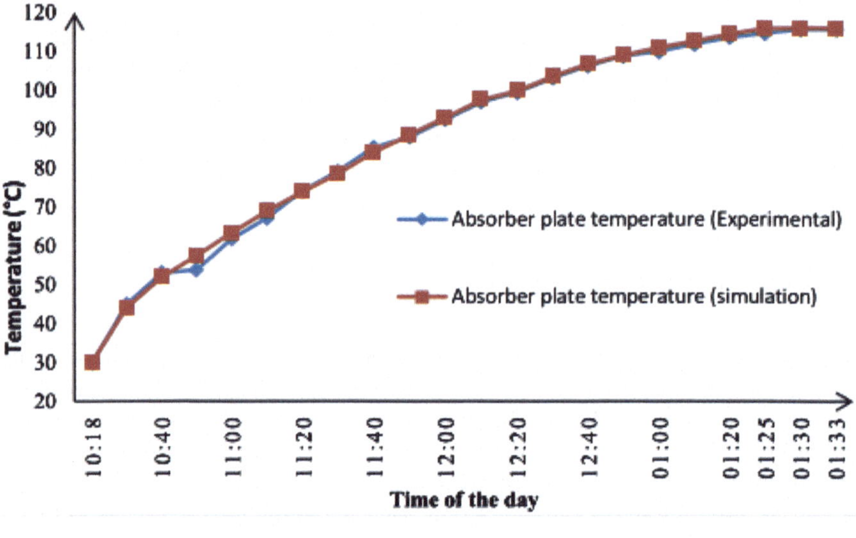

Fig. 3 Absorber plate temperature with the time of day with experiment and simulation

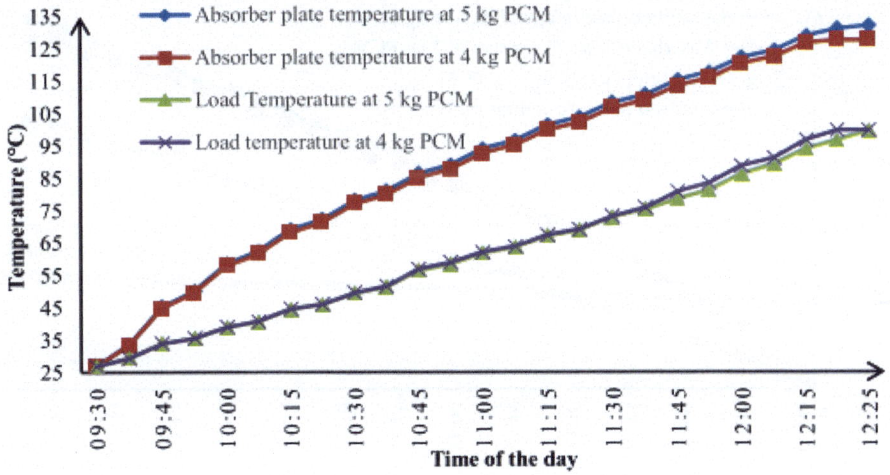

Fig. 4 Absorber plate temperature and load temperature with time of the day for Erythritol

values of solar radiation available on 10th June 2022 at Aligarh, considering 4 and 5 kg of Erythritol as PCM is shown in Fig. 4.

Figure 4 shows that the temperature of the absorber plate and the temperature of the load both rise with the passing of the day and reach their highest values around noon. For 5 kg of Erythritol in 2 h and 55 min, the absorber plate can reach a maximum temperature of 132 °C, and for 4 kg in 2 h and 50 min, it can reach a maximum temperature of 127.51 °C.

The change in the temperature of the absorber plate and load with time of the day for the measured values of solar radiation available on 10th June 2022 at Aligarh, considering 4 and 5 kg of Acetanilide as PCM is shown in Fig. 5.

The maximum temperature achieved by the absorber plate is 126.6 °C with 5 kg of Acetanilide in 3-h 20 min and 121.9 °C with 4 kg of Acetanilide in 3-h 19 min as shown in Fig. 5.

The change in absorber plate temperature and load temperature with time of the day for the measured values of solar radiation available on 10th June 2022 at Aligarh, considering 4 and 5 kg of Quinone as PCM is shown in Fig. 6.

Figure 6 shows that the maximum temperature achieved by the absorber plate is 122.91 °C in 3-h 19 min with 5 kg of Quinone and 119.5 °C in 3-h 15 min for 4 kg of Quinone.

Figure 7 illustrates how the temperature of the absorber plate varies during the day for each of the three PCMs. The greatest temperature of the absorber plate is achieved in the case of Erythritol rather than Acetanilide for the same boundary condition and for the same quantity of PCM that is employed, whereas the minimum temperature is achieved with Quinone for the same boundary condition. The absorber plate may reach a maximum temperature of 132 °C with Erythritol, 126.6 °C with Acetanilide, and 122.9 °C with Quinone.

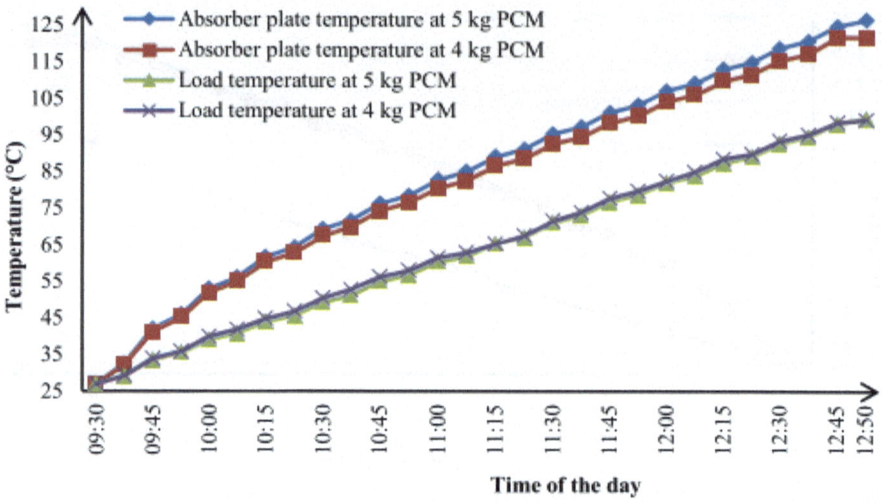

Fig. 5 Absorber plate temperature and load temperature with time of the day for Acetanilide

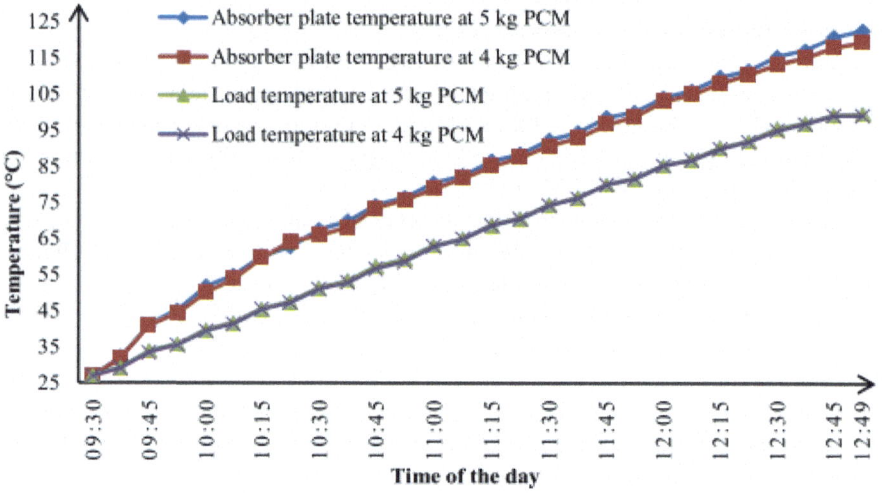

Fig. 6 Absorber plate temperature and load temperature with time of day for Quinone

During evening cooking, the energy stored by the PCM material is the only energy source for cooking. Some amount of energy stored by PCM is used by load for cooking and some is lost to the environment; as a result the temperature of absorber plate continuously decreases. The variation of absorber plate temperature during the discharging mode is shown in Fig. 8.

During the hours when the sun is not shining, the only source of energy available to cook the meal is the energy that has been stored in the PCM. A part of stored energy

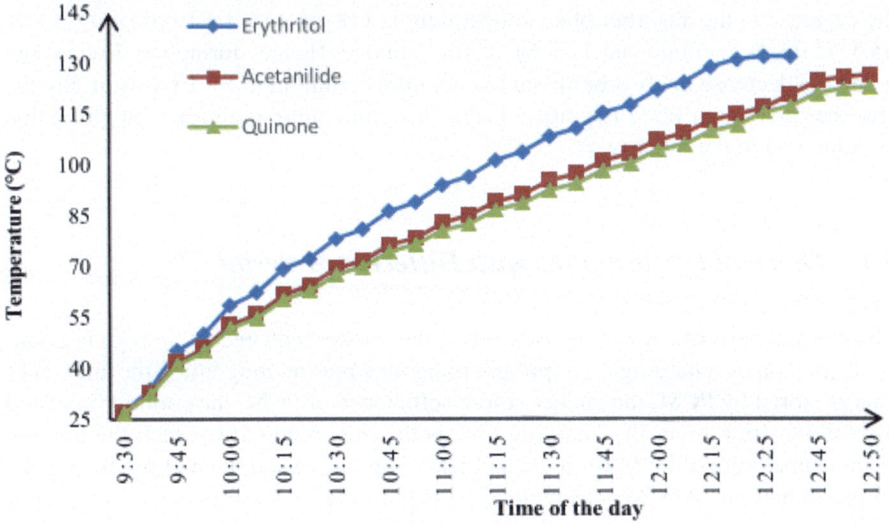

Fig. 7 Absorber plate temperature with time of day for all three PCM of quantity 5 kg

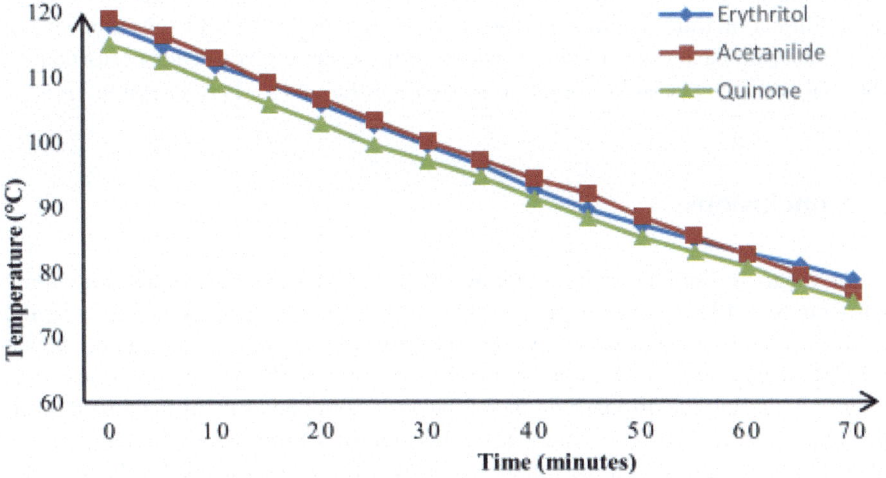

Fig. 8 Absorber plate temperature in discharging mode for all three PCM of quantity 5 kg

is used to cook food and the residual energy is lost to the environment. Hence the temperature of the absorber plate decreases. It has been found that during this period of 70 min duration with 5 kg of PCM the absorber plate temperature decreases from 118 to 78.82 °C for Erythritol, 118.9 to 76.9 °C for Acetanilide and 115 to 75.4 °C for Quinone. With the decrease in the amount of PCM the temperature of the absorber plate during the discharging mode further decreases as with 4 kg of PCM where the decrease in the absorber plate temperature is 118–66.1 °C for Erythritol, 118.9–65.5 °C for Acetanilide and 115–64 °C for Quinone. Hence, during the discharging mode the decrease in absorber plate temperature is minimum for Erythritol. So, the absorber plate with PCM Erythritol keeps the temperature high for a long duration as compared to the other two.

5.1 Thermal Performance with Different Material

The thermal performance of the Box type solar cooker with energy storage is evaluated based on the charging time (pre-charging time plus melting time), the amount of energy stored by PCM, the energy storage efficiency of PCM, the cooling time, and the solidification time. Pre-charging time is the time required to reach the melting point temperature of PCM from the ambient temperature and the melting time is the complete melting of PCM at its melting point temperature. Energy storage efficiency of PCM is the ratio of energy stored by PCM to the total incident radiation falling on the Box type solar cooker. Cooling time is the amount of time required to cool the PCM from its highest temperature to its melting point temperature. The amount of time needed for the PCM to completely solidify at its melting point is known as the solidification time. Thermal performance of each PCM has been investigated and compared with each other to obtain which of these three gives better performance. Thermal performance of box solar cooker with different PCMs is shown in Table 2.

6 Conclusions

According to the findings of the simulation, the absorber Plate can reach its maximum temperature of 132 °C when treated with PCM Erythritol of quantity 5 kg for a period of 2 h and 55 min. Energy stored by 5 kg of Erythritol is significantly greater than that of PCM Acetanilide and Quinone. Total charging time for PCM Quinone is less than other two. During the discharging mode the absorber plate of PCM Erythritol drops to 78.82 °C from 118 °C within a time duration of 70 min while the temperature drops to lower values with the other two PCMs. The energy needed to heat a 4 kg load with Acetanilide and Quinone is significantly more than the energy stored by them. As a result, Acetanilide and Quinone cannot be used to prepare meals in the evening. The performance of a solar cooker with 5 kg PCM is much superior to that of a solar cooker with 4 kg PCM. Overall, the performance of a cooker containing the PCM Erythritol is superior to that of a cooker containing the PCM Acetanilide and Quinone.

Table 2 Thermal performance of solar cooker with different PCMs

	Erythritol (5 kg)	Acetanilide (5 kg)	Quinone (5 kg)	Erythritol (4 kg)	Acetanilide (4 kg)	Quinone (4 kg)
Maximum temperature of absorber plate (°C)	132	126.6	122.9	127.51	121.9	119.5
Time taken to reach maximum temperature	2-h 55 min	3-h 20 min	3-h 19 min	2-h 50 min	3-h 19 min	3-h 15 min
Pre-charging time (min)	61	92	63	48	84	50
Melting time	6-h 29 min	4-h 28 min	2-h 53 min	5-h 11 min	3-h 35 min	2-h 18 min
Charging time	7-h 30 min	6-h	3-h 56 min	5-h 59 min	4-h 59 min	3-h 08 min
Energy absorbed by PCM (kJ)	2499.4	1971.7	1612.55	1999.56	1577.36	1290.04
PCM storage efficiency in percentage	38.65	37.6	47.11	38.18	36.2	46.8
Cooling time (min)	47	23	24	18	8	12
Solidification time	5-h 13 min	3-h 24 min	2-h 39 min	4-h 10 min	2-h 44 min	2-h 07 min

References

1. Arif MR, Khan MA, Azhar M, Meraj M, Akhtar N (2021) Thermal performance comparison and augmentation of two identical box type cookers operating in tropical climatic conditions. Therm Sci Eng Appl ASME 13: 061004-1–10
2. Sharma A, Chen CR, Murty VVS, Shukla A (2009) Solar cooker with latent heat storage systems: a review. Renew Sustain Energy Rev 13:1599–1605
3. Ramadan MRI, Aboul-Enein S, El-Sebaii AA (1988) A model of an improved low cost-indoor solar-cooker in Tanta. Solar Wind Technol 5:387–393
4. Buddhi D, Sahoo LK (1997) Solar cooker with latent heat storage design and experimental testing. Energy Convers Manag 38:493–498
5. Sharma SD, Buddhi D, Sawhney RL, Atul Sharma (2000) Design, development and performance evaluation of a latent heat storage unit for evening cooking in a solar cooker. Energy Convers Manag 41:1497–1508
6. Vigneswaran VS, Kumaresan G, Sudhakar P, Santosh R (2017) Performance evaluation of solar box cooker associated with latent heat energy storage system for cooking. Earth Environ Sci Conf 67:135–141
7. Sharma S, Rai AK (2016) Study of box type solar cooker with energy storage medium. Int J Adv Res Eng Technol 7:37–42

8. Sharma SD, Iwata T, Kitano H, Sagara K (2005) Thermal performance of solar cooker based on an evacuated tube solar collector with a PCM storage unit. Sol Energy 78:416–426
9. Ali A, Akhtar N (2021) Design and performance evaluation of box type solar cooker with energy storage. Springer Nature Singapore Pte Ltd., pp 181–194
10. Akhtar N, Mullick SC (2007) Computation of glass cover temperatures and top heat loss coefficient of solar collector with double glazing. Energy - Int J 32:1067–1074

Experimental Investigation of Photovoltaic Panel Cooling by Uniformly Flowing Water on the Top Surface for Efficiency Improvement

Vikas, Abhishek Pandey, and Ankit Yadav

Abstract The study done in this article furnish the results of an experimental investigation on the cooling of solar photovoltaic (PV) panel with water flowing uniformly over the top surface. An experimental setup was designed, fabricated, and put to testing in real outdoor environment in the geographical region of Chandigarh, India. To overcome non-uniformities in the flow, various trials were made and a cuboidal vessel was designed. About a 50% decrease in temperature was observed. An improvement of about 17% in efficiency was observed by cooling the PV panel, it increased from 11.83% to 13.83%.

1 Introduction

In physics, the quantitative aspect of energy is that it must be conveyed to any matter to perform work on, or to heat, that matter. Indeed, the human race is entirely dependent on energy to stay alive and function. Energy is being derived from various both conventional as well as non-conventional energy resources such as crude oil, coal, solar, hydel, geothermal, etc. The contemporary world is entirely dependent on the capacity of humans to transform energy from one form to another usable form. The countries in the world are categorized as prosperous based on their high energy consumption rate per capita. The biggest challenge that humans are facing is the judicious and efficient use of energy.

1.1 The Energy Scenario

The global human population is increasing by around 1.1% annually. It is estimated to reach up to 8.4 billion by 2030 and 9.6 billion by 2050 in reference to the present

Vikas (✉) · A. Pandey · A. Yadav
Punjab Engineering College, Chandigarh, India
e-mail: vikka612@gmail.com

M. A. Siddiqui et al. (eds.), *Advances in Heat Transfer and Fluid Dynamics*, Lecture Notes in Mechanical Engineering, https://doi.org/10.1007/978-981-99-7213-5_28

census of 7 billion people living on earth today [1]. Every one of these individuals will require energy, which is shooting up the worldwide energy demand to the next level high. What's more, the energy utilization per capita is connected to the expectation for the everyday comforts of a nation. The global primary energy consumption had increased by about 2.9% in 2018 which was the maximum growth ever seen since 2010 [2]. In addition to that the carbon emissions from this much use of energy grew by 2% which is also the fastest rise in the past few years. In many nations, the expectation for daily life comforts is quickly increasing like in China and India, where roughly 2.5 billion individuals are living. Both the expanding population and expanding expectations for everyday comforts will build the energy request. What's more, it is normal that the world energy utilization is relied upon to increment by 33% in the period 2010–2030 [3].

To take care of the energy demand there is a weight on the accessible normal assets, which incorporates non-renewable energy sources like crude oil, coal, and gas. Fossil fuels are nothing just solar energy stored in chemical form for million years. The issue lies in the way that people drain these non-renewable energy sources quicker than they are recharged through the photosynthetic procedure in nature [4]. Besides this, in the process of harnessing energy from fossil fuels, these emit greenhouse gases into the environment. These greenhouse gases are solely responsible for the extremities due to climate change all around the globe. The CO_2 concentration in the Earth's environment is drastically increasing. As the pollution level is rising at an alarming rate, the world cannot rely on conventional techniques for harnessing energy from non-renewable sources of energy. Therefore, opting for efficient green technologies, RES (Renewable energy sources) are some of the potential solutions. Harnessing of energy from RES (solar, wind, hydro, etc.) can be done by various processes [5]. However, in any case, the sun is the vitality hotspot for every one of the procedures on the surface of our planet. Indeed, even breeze is a consequence of temperature distinction in the air actuated by sun-oriented illumination, waves are created by the breeze, mists, and rain are at first shaped by the dissipation of water because of daylight. As the sun is the parent source of energy, we have to move back again to a time when we begin to use the sun to fulfill our energy requirements.

1.2 Solar Energy and Solar Photovoltaics

Solar energy, being the biggest source of energy has immense potential to meet the desired energy requirement. To express this lucidly, the solar energy irradiating the earth's surface per hour is equal to the measure of annual energy requirements of the earth. Sun is a radiating black body with an average surface temperature of around 5800K, which prompts $1366\,W/m^2$ of energy dispersal over the atmosphere [6]. This much energy can be utilized in various ways. Solar energy can be utilized as heat or can be converted to other useful forms of energy. The solar light can directly be converted into electrical energy by using PV panels. Photo represents light and voltaic represents electricity, this is one of the inclusive techniques for solar energy harnessing, which converts sunlight into usable electrical energy. There are mainly four types of PV panels in the market namely amorphous, mono-crystalline, poly-crystalline, and organic cells. The majority of the solar cells in the market are silicon-based. In this research, the purview is on a typical silicon PV panel consisting of numerous solar cells integrated into groups that are arranged in one plane that constitute a solar PV panel or solar PV module [7, 8]. On account of PV cells, a few scientists have broken down the module conduct and they found out that only 15–20% of the sun's incident illumination can be harnessed in electrical form while the remaining is lost as heat dissipation [9]. In this manner, to improve the electrical efficiency of the PV panel we have to investigate the factors influencing its performance.

1.3 Factors Affecting the Solar PV Performance

Various such factors are classified and sub-classified into operational, economic, and other miscellaneous factors [10]. Considering all the factors, nothing can be done much on the factors, which are mostly governed by nature such as irradiation, soiling, dust accumulation, etc. [11]. Given all the above-stated factors, the panel temperature affects the electrical efficiency of the PV panel the most and it diminishes at a rate of 0.4%–0.65% with every degree rise in the panel temperature [12, 13]. Some researchers observed that PV panel temperature can attain values as high as 353.15 K if installed in hot arid regions [14]. However, to the benefit, the panel temperature can be regulated to a possible extent and utilized the very same to maximize the performance of the PV panel.

1.4 Effect of PV Panel Temperature on Performance

Majorly the rise in panel temperature affects the open-circuit voltage (V_{oc}), which diminishes as directly with the rise in panel temperature. The cell voltage diminishes by about 2.2 mV for every 1 °C rise in the panel temperature and therefore the efficiency descends by around 0.5%–1% for crystalline PV panels [15]. By providing proper cooling to the panel, the electrical efficiency can be enhanced, and it also reduces the rate of deterioration or PV cells with time, bringing about augmentation of the longevity of PV panels [16]. The waste heat extracted by the cooling water may be utilized as in many thermal applications [17]. Figure 1 illustrates some methods being used for cooling PV panels to minimize the consequences of the increased temperature. There are various cooling methods available but the water being assuredly available, the more propitious one is to use a technology where water plays a dominant role in cooling the PV panels [18]. Sensible cooling by air and water is usually applied for the thermal management of various systems such as engines, electronic components, etc. Some of the advantages of this type of cooling are low operating

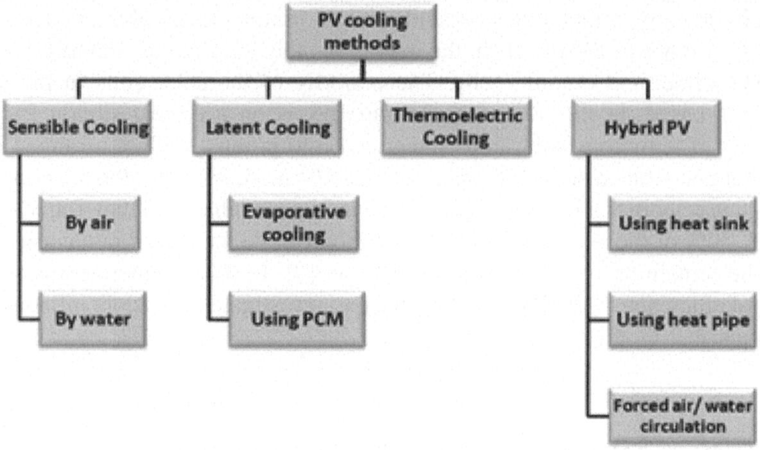

Fig. 1 PV panel cooling methods

costs and low material requirements. A study to improve the performance of PV modules using forced air cooling technique has been presented based on the comparison, it was shown that using the air cooling technique resulted in 7.2% and 6% higher electric output. In a study represented by Kasaeian et al. [19], four fans were used to flow cooling air inside the channels designed for PV cooling.

1.5 Various Top Surface Cooling Frameworks for PV Panels

Some active cooling frameworks used for the PV panel are shown in Fig. 2. These frameworks cannot shower and cover the total surface zone of the PV panel, which implies that some fraction of it is only cooled. There are flow non-uniformity phenomena that were distinguished amid the tuning of the water-spreading framework [20].

Many investigations had been done on the surface cooling of PV panels using water. Abdolzadeh and Ameri [21] in Iran investigated a cooling framework for cooling PV panels using water spray on its top surface. The effect of spray cooling on electrical efficiency was observed and compared with the conventional non-cooled PV panel. The effect of the water pumping rate was also observed. The results concluded that higher pumping of water was required during the peak sunshine hours since the temperature rise was highest. Hosseini et al. [22] used a thin-film water cooling method for reducing the PV surface temperature and reflection losses. Water was made to flow using a slit of a pipe which was supplied with water with the help of a recirculation pump and a finned water cooler combined in series. A temperature gradient of 1.5 °C was observed across the thickness of the PV panel. A significant reduction in surface temperature and improvements in electrical efficiency

Fig. 2 Details of stream non-uniformity phenomena identified in the water spreading framework

was obtained. Nizetic et al. [23] investigated the water spray cooling technique on both the surfaces of the PV panel. The electrical output raised to a maximum of 16.3% as an effect of cooling of PV panel up to 24 °C. Bahaidarah et al. [24] highlighted the need for water flow uniformity in the cooling of PV panels. The results showed that the consistent water flow cooling increased electrical efficiency and lowered the panel temperature better than the non-uniform water cooling. A lot of other similar studies on Photovoltaic Thermal (PV/T) hybrid systems had been performed by many researchers [18]. These systems are used for simultaneous water heating and PV panel cooling. In such systems, the power generated by the PV panel is used to drive a pump for the circulation of water in the system. PV/T water disinfection systems are also used for thermal or photo-disinfection of water against microbes [25, 26].

This article is aiming to fill the observed gap in the previous research done in the cooling methods for removing the flow non-uniformities. It represents an experimental study on a PV panel cooling arrangement using water with uniform flow over the top surface. The main objective of this investigation is to propose an innovative cooling idea with these advantages: (1) It will not consume any external power to pump water like in the case of spray nozzles, as in the proposed arrangement gravity-assisted flow can be possible. (2) Water is readily available so it is quite a cheap cooling fluid. (3) It also serves another purpose of cleaning the top surface of the PV panel and reducing the reflection losses. (4) This type of arrangement can easily be incorporated in naturally flowing waterfalls, dam spillways, rivers, etc. All this adds up to introduce an effective method to lower the surface temperature of the PV panel and thus enhance its performance. The main aim of this study is to reduce the flow non-uniformities of an active cooling system over PV panels. Moreover, the effect of reduction of flow non-uniformity on efficiency is also investigated. The experiments were conducted in real environmental conditions in Chandigarh, India on summer days. The surface temperature, irradiance, open-circuit voltage, and efficiency were recorded every 15 minutes during the observations which continued for about 3 h.

2 Experimental Setup

An experimental exercise had been directed with the point of examining a PV panel cooling system and the sole aim of the exercise was to decrease the stream non-uniformity. With the assistance of a solar module analyzer, the efficiency of the PV board was estimated and it was additionally contrasted with the non-cooled edition of it.

The panel was inclined to 30° for maximum output, with the help of wooden support for maximum output [27, 28]. The remedies for reducing the flow non-uniformity were tested inside the lab and the experimental setup was then exposed to the real outdoor environment as shown in Fig. 3.

In the real outdoor environment, the experiments were carried out on six sunny days of summer in Chandigarh (30.7333°N, 76.7794°E), India. A pyranometer was connected with Lab Quest 2 (a computer sensor interface to collect data) to record the solar irradiance.

(a) Experimental Setup **(b) Uniform flow over PV panel**

Fig. 3 Experimental trials to remove flow discontinuities: **a** Trial 1 with a series of nozzles **b** Trial 2 with slitted pipe section **c** Trial 3 Flat spillways with cuboidal catchment vessel

2.1 Experimentation Process

To overcome the non-uniformity in the flow of water various trials were conducted to find the best possible solution.

Trial 1: Initially, water nozzles, commercially available were used along with the breadth of the PV panel. Nozzles were placed at a distance of 5 cm apart over the complete span (66.6 cm) as shown in Fig. 4a. The nozzles were placed on a pipe at a distance of 5 cm away from the panel. While conducting the experiment, it was found that there was flow non-uniformity, water was not able to spread over the whole surface of the panel. A phenomenon of water film breaking was visible as found by the previous researchers.

Trial 2: To solve this problem, a pipe slitted along the length was used. The shape of the pipe was cylindrical and it was maintained throughout the breadth of the panel as shown in Fig. 4b. Again, the problem was not solved. Water was not coming from the edge evenly, only from a certain region. There may be various reasons: (1) the flow rate of the pipe wasn't ample and (2) the edge of the pipe was not finished properly, had it been finished properly still to implement in real life is impractical.

Trial 3: After contemplating the problem and with the help of various suggestions, the concept used in the dam's spillway was incorporated. An open flat spillway with cubical catchment vessel placed with catchment vessel dimension 67 cm x 8 cm x 5 cm, hoping that as soon as the vessel is over-flooded the water would flow uniformly over the panel as shown in Fig. 4c. The problem was still there and water was flowing mainly in the center part owing to the fact that the vessel was buckling; the level of water was not maintained giving rise to the non-uniformity phenomena. To avoid buckling of the vessel, the size was reduced to (67 cm x 6 cm x 4 cm), besides the center was made a little strong. The source of water in all the above cases was a reservoir placed above the PV panel which was supplied to the vessel under the action of gravity.

After the selection of appropriate cooling arrangements, the setup was taken to outdoors for the collection of performance data in the real environment. The uniformity of the water stream was achieved as shown in Fig. 3. To record the data experiments were performed for 6 sunny days, the first 3 days on a non-cooled setup,

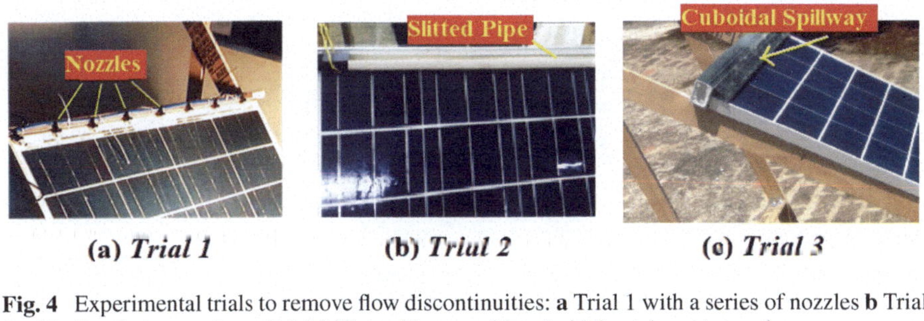

(a) ***Trial 1*** **(b)** ***Trial 2*** **(c)** ***Trial 3***

Fig. 4 Experimental trials to remove flow discontinuities: **a** Trial 1 with a series of nozzles **b** Trial 2 with slitted pipe section **c** Trial 3 Flat spillways with cuboidal catchment vessel and the next three days on a setup with a cooling arrangement. The stipulated schedule for experimenting was 10:45 am–1:45 pm during the peak sunny hours of the day when the maximum temperature of the PV panel was observed. It had been taken care of that the flow uniformity was maintained throughout the process. With the help of the pyranometer, the irradiance level was measured and the input was given at a frequency of 30 minutes to the PV module analyzer. The data of both the setups were collected and efficiency was compared.

3 Uncertainty Analysis

During the experimental analysis, there might be a few errors encountered that need to be identified to ensure the correctness of the results discussed above. So, uncertainty analysis was also carried out for the most important parameter which was the temperature measured on the surface of the PV panel. The total error or uncertainty was calculated as

$$E_t = \sqrt{(E_p)^2 + (E_b)^2} \tag{1}$$

E_b is the biased or systematic error which was specified by the manufacturer and is equal to 1 °C and E_p is the precision or stochastic error which is calculated from the given data in Table 1. by taking five different values of temperature repeatedly measured by the same IR thermometer.

$$E_p = \frac{\sigma}{\sqrt{n}} t_{u,\frac{\phi}{2}} \tag{2}$$

where E_p is the precision error, σ is a standard deviation for n number of measurements taken, $t_{u,\frac{\phi}{2}}$ is the t-distribution corresponding to the degree of freedom $v = n - 1 = 5 - 1 = 4$, at a confidence level of 95% which give $\frac{\phi}{2} = 0.025$, σ of the data is 0.1385. From the t-distribution table, $t_{4,0.025} = 2.776$. Then precision errors are

$$E_p = \frac{0.1385}{\sqrt{5}} 2.776 = 0.172°C \tag{3}$$

Then the total error is calculated as:

$$E_t = \sqrt{1^2 + 0.172^2} = 1.014°C \tag{4}$$

Table 1 Temperature sample for uncertainty analysis

Measurement no.	1	2	3	4	5
Temperature (°C)	85.6	85.72	85.68	86.01	85.76

Since the values of temperature in this study were very large as compared to the total error, it concludes that the results obtained in this study were reliable and correct.

4 Results and Discussion

As the prospect of the experiment was to achieve a condition of uniformity of water stream over the panel, the condition being achieved further provides an opportunity to compare the performance of the cooled and non-cooled setup. Hence, the data was collected for six days, three for non-cooled and three for cooled setup. The first 3 days of experiments were without cooling and the last 3 days were with using the cooling arrangement on the PV panel. The following parameters were measured and compared for the two cases:

4.1 Surface Temperature of PV Panel

The panel surface temperature was measured with the help of an IR thermometer. Figure 5a shows the Irradiance for three hours of peak irradiance. Figure 5b shows the variation of the surface temperature of the panel for the two cases. The first 3 days were for the non-cooled (reference) case and the next 3 days were for the cooled case of the PV panel. The 3 days and 3 hours average temperature of the non-cooling case was recorded at 63.8 °C and that of the cooling case was recorded at 32.5 °C. So, a 31.3 °C (about 50%) decrease in temperature was observed as a result of the cooling arrangement employed in this research.

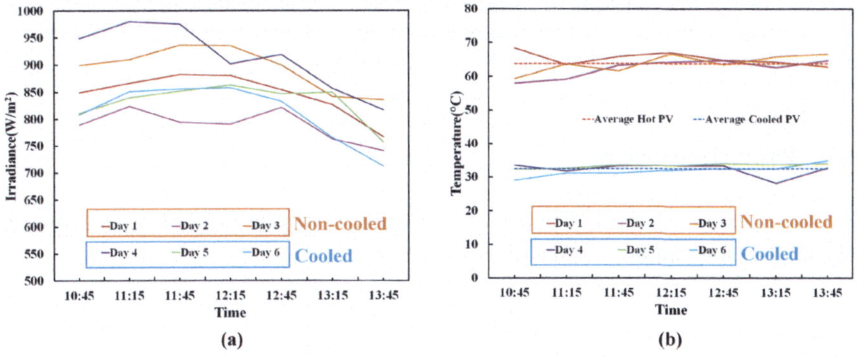

Fig. 5 **a** Solar irradiance measured during the 6 days of experimentation; **b** Surface temperature of the PV panel during reference case and cooled case for peak radiation hours

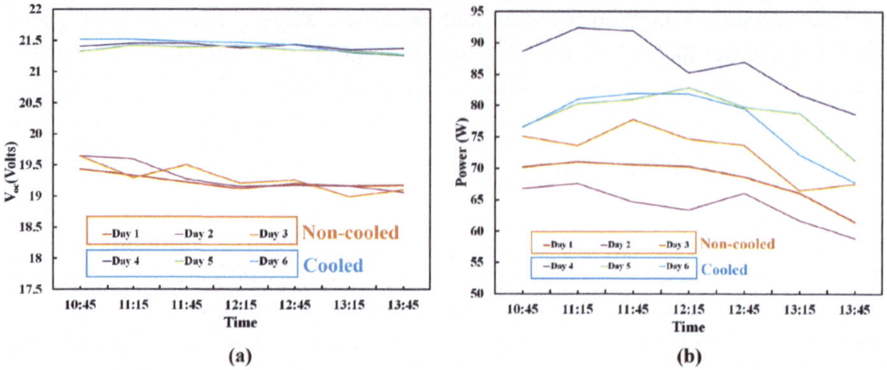

Fig. 6 **a** Open-circuit voltage variation during the peak radiation hours; **b** Power output variation during peak radiation hours

4.2 *Electrical Improvements*

Clear information is given in the previous section about the decrease in the surface temperature of the PV panel. There must be some improvement in the electrical performance of the PV panel.

4.2.1 Improvement in Open-Circuit Voltage V_{oc}

The electrical performance is quantified in terms of V_{oc}. Figure 6a shows the variation of Voc with time and compares the non-cooling case with the cooling case. The value of V_{oc} increased with a decrease in temperature. An increase of about 2.12 Volts was recorded due to the enhancements imparted by the cooling arrangement. This is about a 0.35% increase in V_{oc} /°C decrease in temperature. The V_{oc} in the cooled case was more uniform as compared to that of the non-cooled case.

4.2.2 Power Improvements

The power output of the PV panel was recorded with the help of a solar module analyzer. Also, it can be calculated by this expression.

$$P_{out} = I_{sc} V_{oc} FF \tag{5}$$

$$P_{ideal} = I_{sc} V_{oc} \tag{6}$$

where I_{sc} denotes the short circuit current, V_{oc} denotes the open-circuit voltage and FF is the fill factor. FF is an important parameter to specify the overall performance of a PV panel. It is the ratio of the maximum power output given by the panel to the theoretical power output of the panel. Its value varies between 0 and 1. It was also measured with the help of a solar module analyzer and was 0.71 for non-cooled and 0.75 for cooled setups. Figure 6b shows the variation of the power output of the PV panel for both cases. As we can see the blue lines representing the power output of the cooled panel are above the non-cooled. An 18.1 % increase in average power output was observed as a result of the cooling effect.

4.2.3 Efficiency Improvements

The improvement in V_{oc} improves the efficiency of the PV panel [29]. Which was calculated by the solar module analyzer by using this empirical formula:

$$\eta = \frac{P_{out}}{P_{in}} = \frac{I_{sc} V_{oc} FF}{G A_s} \tag{7}$$

where η is the efficiency, I_{sc} is the short circuit current, V_{oc} is the open-circuit voltage, FF is the fill factor, and G is the solar irradiance and A_s is the surface area of the PV panel. All these quantities were measured and calculated with the help of a solar module analyzer. Figure 7a shows the variation of the efficiency of the PV panel with time for the two cases. It is seen that the efficiency of the cooled setup was improved over the non-cooled setup. An improvement of about 17% in efficiency was observed by cooling the PV panel, it increased from 11.83% to 13.83%.

$$Improvement\ in\ efficiency = \frac{\eta_{cooled} - \eta_{non-cooled}}{\eta_{non-cooled}} \times 100 \tag{8}$$

Table 2 compares the improvements in the efficiency of this work with other studies on the cooling of PV panels using water. Although the design used in this study was

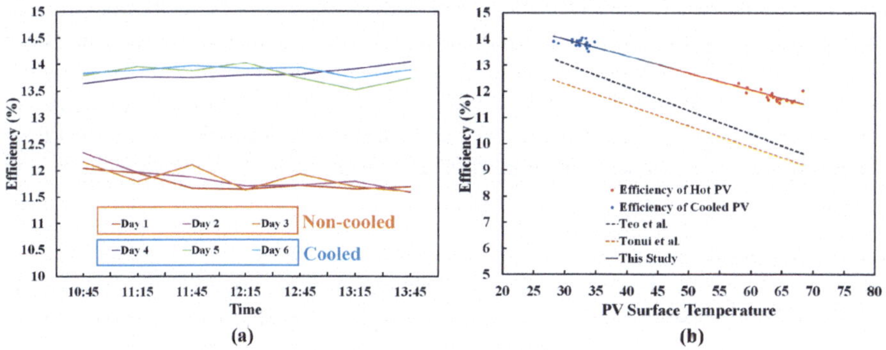

Fig. 7 **a** Efficiency variation during the peak radiation hours; **b** Variation of electrical efficiency with PV surface temperature and validation of this study

Table 2 Comparison of results with the previous studies

Reference	% Improvements
Papadopoulos et al. [31]	7.9
Lucas et al. [32]	10
Haidar et al. [33]	12–14.77
This study	17

simple with cheap material, it helped in the successful removal of flow discontinuity and resulted in effective cooling of the panel. Figure 7b shows the variation of electrical efficiency with the surface temperature of the PV panel and compares the results with previous investigations by Teo et al. [8] and Tonui et al. [30]. A linear correlation of electrical efficiency with surface temperature was observed in the investigation which is in agreement with the previous studies. A mathematical correlation between the electrical efficiency and surface temperature was also established as

$$\eta_{pv} = 15.923 - 0.0641 T_{pv} \tag{9}$$

where η_{pv} is the electrical efficiency of the PV panel. T_{pv} is the surface temperature of the PV panel. The trend lines show that the efficiency of the PV panel decreases with an increase in surface temperature.

5 Conclusions

The electrical performance of a PV panel decreases with an increase in its surface temperature. Out of the different proposed cooling methods, the water cooling system is a cheap and effective method. In this paper, a technique to prevent flow discontinuities in water-based cooling systems is discussed. The results of the experimental investigation enunciate that by maintaining the flow uniformity of an active cooling system over the panel, there was about 50% reduction in the surface temperature of the PV panel, as a result of which there was about 0.35% increase in V_{oc}/°C decrease in temperature and an improvement of about 17% in efficiency was observed by the used cooling technique in this study. It was shown clearly in this study that integrating a water-based cooling system with PV panels and maintaining the continuous flow of cooling water the availability, reliability, and cleanliness of solar power plants can be improved.

Acknowledgements The authors gratefully acknowledge the Department of Mechanical Engineering and the Department of Electrical Engineering in Punjab Engineering College (Deemed to be University), Chandigarh, India.

References

1. Town I, Region A (2018) Population forecasting for design of water supply system, pp 54–65
2. BP (2018) BP statistical review. Stat Rev World Energy:1–56
3. Indicators WD (2017) World development indicators 2017
4. Hasanuzzaman M, Rahim NA, Hosenuzzaman M, Saidur R, Mahbubul IM, Rashid MM (2012) Energy savings in the combustion based process heating in industrial sector. Renew Sustain Energy Rev 16(7):4527–4536
5. Erdiwansyah M, Mamat R, Sani MSM, Sudhakar K (2019) Renewable energy in Southeast Asia: policies and recommendations. Sci Total Environ 670:1095–1102
6. Zeshao C, Songping MO, Peng HU (2008) Recent progress in thermodynamics of radiation–exergy of radiation , effective temperature of photon and entropy constant of photon 51(8):1096–1109
7. Hu J, Chen W, Yang D, Zhao B, Song H, Ge B (2016) Energy performance of ETFE cushion roof integrated photovoltaic/thermal system on hot and cold days. Appl Energy 173:40–51
8. Teo HG, Lee PS, Hawlader MNA (2012) An active cooling system for photovoltaic modules. Appl Energy 90(1):309–315
9. Cen J, du Feu R, Diveky ME, McGill C, Andraos O, Janssen W (2018) Experimental study on a direct water heating PV-T technology. Sol Energy 176(July):604–614
10. Fouad MM, Shihata LA, Morgan EI (2017) An integrated review of factors in fluencing the performance of photovoltaic panels. Renew Sustain Energy Rev 80(May):1499–1511
11. Jaaz AH, Sopian K, Gaaz TS (2018) Study of the electrical and thermal performances of photovoltaic thermal collector-compound parabolic concentrated. Results Phys 9:500–510
12. Reddy SR, Ebadian MA, Lin CX (2015) A review of PV-T systems: thermal management and efficiency with single phase cooling. Int J Heat Mass Transf 91:861–871
13. Haidar ZA, Orfi J, Kaneesamkandi Z (2018) Experimental investigation of evaporative cooling for enhancing photovoltaic panels efficiency. Results Phys 11(October):690–697
14. Skoplaki E, Palyvos JA (2009) On the temperature dependence of photovoltaic module electrical performance: a review of efficiency/power correlations. Sol Energy 83(5):614–624
15. Meral ME, Dinc F (2011) A review of the factors affecting operation and efficiency of photovoltaic based electricity generation systems 15:2176–2184
16. El-Shobokshy MS, Hussein FM (1993) Effect of dust with different physical properties on the performance of photovoltaic cells. Sol Energy 51(6):505–511
17. Qiu L, Ouyang Y, Feng Y, Zhang X (2019) Review on micro/nano phase change materials for solar thermal applications. Renew Energy 140:513–538
18. Osma-Pinto G, Ordóñez-Plata G (2019) Measuring the effect of forced irrigation on the front surface of PV panels for warm tropical conditions. Energy Rep 5:501–514
19. Kasaeian A, Khanjari Y, Golzari S, Mahian O, Wongwises S (2017) Effects of forced convection on the performance of a photovoltaic thermal system: An experimental study. Exp Thermal Fluid Sci 85:13–21
20. Schiro F, Benato A, Stoppato A, Destro N (2017) Improving photovoltaics efficiency by water cooling: Modelling and experimental approach. Energy
21. Abdolzadeh M, Ameri M (2009) Improving the effectiveness of a photovoltaic water pumping system by spraying water over the front of photovoltaic cells. Renew Energy 34(1):91–96
22. Hosseini R, Hosseini N, Khorasanizadeh H (2011) An experimental study of combining a photovoltaic system with a heating system. In: Proceedings of the world renewable energy congress—Sweden, 8–13 May 2011, Linköping, vol 57, pp 2993–3000
23. Nižetić S, Čoko D, Yadav A, Grubišić-Čabo F (2016) Water spray cooling technique applied on a photovoltaic panel: the performance response. Energy Convers Manage 108:287–296
24. Bahaidarah HM, Baloch AA, Gandhidasan P (2016) Uniform cooling of photovoltaic panels: a review. Renew Sustain Energy Rev 57:1520–1544
25. Qin L, Wang Y, Vivar M, Huang Q, Zhu L, Fuentes M, Wang Z (2015) Comparison of photovoltaic and photocatalytic performance of non-concentrating and V-trough SOLWAT (solar water purification and renewable electricity generation) systems for water purification. Energy 85:251–260

26. Wang Y, Jin Y, Huang Q, Zhu L, Vivar M, Qin L (2016) Photovoltaic and disinfection performance study of a hybrid photovoltaic-solar water disinfection system. Energy 106:757–764
27. Ullah A, Imran H, Maqsood Z, Butt NZ (2019) Investigation of optimal tilt angles and effects of soiling on PV energy production in Pakistan. Renew Energy 139:830–843
28. Yadav S, Panda SK, Tripathy M (2018) Performance of building integrated photovoltaic thermal system with PV module installed at optimum tilt angle and influenced by shadow. Renew Energy 127:11–23
29. Shahatha Salim M, Mohammed Najim J, Mohammed Salih S (2013) Practical evaluation of solar irradiance effect on PV performance. Energy Sci Technol 6(62):36–40
30. Tonui JK, Tripanagnostopoulos Y (2007) Air-cooled PV/T solar collectors with low cost performance improvements. Sol Energy 81(4):498–511
31. Papadopoulos AM, Giama E, Niz S (2017) Comprehensive analysis and general economic-environmental evaluation of cooling techniques for photovoltaic panels, part I : passive cooling techniques etic 149:334–354
32. Lucas M, Aguilar FJ, Ruiz J, Cutillas CG, Kaiser AS, Vicente PG (2017) Photovoltaic evaporative chimney as a new alternative to enhance solar cooling. Renew Energy 111:26–37
33. Haidar ZA, Orfi J, Oztop HF, Kaneesamkandi Z (2016) Cooling of solar PV panels using evaporative cooling. J Therm Eng 2(5):928–933

Performance Study of Basin Type Solar Still with Different Concentrations of Water-Al_2O_3 Nanofluids

Farooque Azam and Naiem Akhtar

Abstract In this paper, a comparison of the performance of single basin single slope solar still with water and with water based nanofluid of Aluminium Oxide (Al_2O_3) has been made. For this, two experimental solar stills of the same basin area have been fabricated and tested for their identical thermal performance. These solar stills have been used to compare the performance with water in one solar still and water-based nanofluid in the other solar still by performing experiments at the same location and time. It is found that the solar still with water based Aluminium Oxide (Al_2O_3) nanofluid having concentrations 0.05%, 0.1%,0.15% and 0.2% enhances the output of the solar still as compared to the water by 18.54%, 22.26%, 24.37% and 20.38%, respectively, thus giving an optimum value of enhanced output at 0.15% concentration.

Keywords Solar still · Nanofluid · Aluminium oxide

Nomenclature

Ø	Latitude (Degree)
V	Volume of water (ml)
ρ	Density of nanoparticles (gm/ml)

F. Azam (✉) · N. Akhtar
Department of Mechanical Engineering, ZHCET Aligarh Muslim University, Aligarh, India
e-mail: farooqueazam3615@gmail.com

N. Akhtar
e-mail: nakhtar.me@amu.ac.in

M. A. Siddiqui et al. (eds.), *Advances in Heat Transfer and Fluid Dynamics*, Lecture Notes in Mechanical Engineering, https://doi.org/10.1007/978-981-99-7213-5_29

1 Introduction

One of the main challenge's humans will face in the near future is the availability of drinking water, which is becoming less and less available steadily while the demand is increasing. Although water covers more than two third of the earth, only about 0.014% of the world's water may be used directly for industrial and human needs [1]. Lack of access to proper drinking water is one of the significant issues for people living in dry, rural places around the world. In regions with abundant solar energy, solar stills can help with some of these issues. Solar stills are easy to fabricate, inexpensive, have low maintenance costs and offer the significant benefit of being pollution-free. Although solar stills have poor productivity, these are an environment-friendly and sustainable way to produce pure water. Extensive research is still being done on solar still with the goal of increasing their production. The performance of solar stills is being improved through several theoretical and experimental studies [2]. Recently, nanofluids have gained attention due to their remarkable thermophysical properties and significant enhancement of heat transfer coefficient [3, 4]. Sain and Kumawat [5] conducted experiments with Al_2O_3 nanoparticles in a solar still. They [5] blended the nanoparticles with black paint coated on the basin of the single slope solar still of area $1m^2$and the overall results concluded that the mixed Al_2O_3 nanoparticles in the paint improve the productivity. Elango et al. [6] compared solar still performance with water and water based nanofluid of aluminium oxide (Al_2O_3), zinc oxide (ZnO) and tin oxide (SnO_2) of different concentrations in a single basin single slope solar still and conclude that productivity enhancement is more in aluminium oxide (Al_2O_3) as compared to other nanofluids.

In this present work, the effects of different concentrations of Al_2O_3 nanofluids on the productivity of solar still has been studied experimentally. Experiments have been performed at a concentration of 0.05, 0.10, 0.15, and 0.20% by volume, and the output is compared with identical solar stills placed side by side under the same environmental circumstances. The effect of the concentration of Al_2O_3 nanofluids with the same water depth is analysed in this experimental work.

2 Experimental Setup

Two identical-sized solar stills with proper distillate collection channels, feed water supply, and drainage lines are designed and constructed. Black paint is coated on the basin surface, having an area $1m^2$. The height of the front and rear walls is 8 cm and 38 cm, respectively (Fig. 1). Both stills are made with a galvanised iron sheet that is 1 mm thick and has insulation (glass wool) that is 1 inch thick on all four sides and the bottom. The insulation is covered with a toughened glass cover that is 4 mm thick and tilted at 28° (for Aligarh, $Ø = 27.89°$) [7]. The arrangement of inlet and outlet drain valves simplifies flow across the basin. Silicon sealant was employed as a sealing material between the glass cover and the still's body to check the leakage

Fig. 1 Actual view of experimental setup

of vapour. It was also ensured that there is no leakage of water from the solar stills by pouring water in the basin. The setup is well-equipped with temperature measurement instruments at various positions. A k-type thermocouple with a digital display is mounted with a setup to measure and record basin temperature, water temperature, vapour temperature, glass interior temperature, and glass outside temperature at regular intervals of 5 min. Ambient temperature, wet bulb temperature, and relative humidity are also measured. Kipp and Zonen Pyranometer (CMP-11) with data logger is used to measure and records solar radiation data. The hourly distillate from both stills was estimated using a beaker with a capacity of (0–1000) ml. Both stills are kept on an obstruction-free roof, facing south.

3 Preparation of Nanofluids

The nanoparticles of Aluminium Oxide (Al_2O_3)have been selected because of their good thermal conductivity and low cost [6, 8]. Aluminium Oxide (Al_2O_3) is less expensive when compared to other nanoparticles used to produce nanofluids. Even though some nanoparticles have great thermal conductivity, their exorbitant cost prevents their usage. As a result, the chosen nanoparticles are more affordable and have better thermal conductivity [9]. Average primary particle size of the Aluminium Oxide nanoparticles is 13 nm, specific surface area is 100 ± 15 m^2/g and density is approximately 50 g/l.

The preparation of nanofluids is a critical step in using nanoparticles to boost the thermal conductivity of fluids. Since nanoparticles are hydrophobic, prone to conglomeration, and settle quickly, a surfactant such as sodium dodecyl benzene sulphonate (SDBS) is used to retain the suspension stable and even. The preparation of nanofluids is a critical step in using nanoparticles to increase the thermal conductivity of fluids [10]. In the present work, water based nanofluids ofAl_2O_3(Al_2O_3-water) with four different concentrations are prepared. The known weight of dry nanoparticles of Al_2O_3, is mixed into distilled water. The weight of the nanoparticles for the desired volume concentrations (0.05, 0.1, 0.15 and 0.2%), calculated using Eq. (1).The prepared samples were stirred for 20 min with a magnetic stirrer and ultrasonicated for 1 h with an ultrasonicator. Mass of the nanoparticle is calculated by the following expressions.

$$\text{nanoparticles(gm)} = \frac{\%\text{Conc}}{100} \times V \times \rho \tag{1}$$

4 Experimental Procedure

Tests were carried out in both the solar stills on the 4th, 5th, 6th, and 7th of July 2022, with one of the stills containing water and the other containing water nanofluid. Before conducting the present experimental work, a series of experiments was conducted on both the solar stillss to verify their identity on thermal performance so that a good comparison could be possible. After verification, the present experimental work was started using 10 L nanofluids in one solar still and 10 L distilled water in another solar still. Various instruments used to measure different parameters are listed in Table 1. The tests were carried out at Aligarh Muslim University's (AMU) Solar Energy Laboratory, Mechanical Engineering Department (27.89°N, 78.08°E).

Table 1 Details of the instruments used in the experimental setup

Instrument	Accuracy	Range	% error
Kipp and zonen pyranometer (CMP-11)	±1 W/m^2	0–4000 W/m^2	0.25
Anemometer (QS-FS01)	±0.1 m/s	0.1–30 m/s	5.00
K-type thermocouples	±0.1 °C	0–1260 °C	0.20
Borosil measuring jar	±5 mL	500 mL	5.00

5 Results and Discussion

Solar radiation, ambient temperature, wind speed, basin water temperature, and the glass temperature of stills were all measured with different concentrations of nanofluids as shown in Figs. 2, 3, 4 and 5. It can be seen from these figures that the temperature profiles and solar radiation intensity exhibit the same behaviour. Its value is at its highest around the noon and is at its lowest in the morning and afternoon. Additionally, the temperature profile follows the same pattern as that of the solar radiation. The performance of the solar still is described in terms of distillate output per hour for each of the four cases (four different concentration of Al_2O_3) as shown in Fig. 6a–d. Table 2 shows the enhancement in distillation output due to the use of nanofluids during day as well as night hours.

It is concluded from experimental results thatAl_2O_3 nanoparticles used with water as the base fluid increase solar still productivity. Additionally, it served as a heat-storage medium, released energy into the water, and improved production at night. It

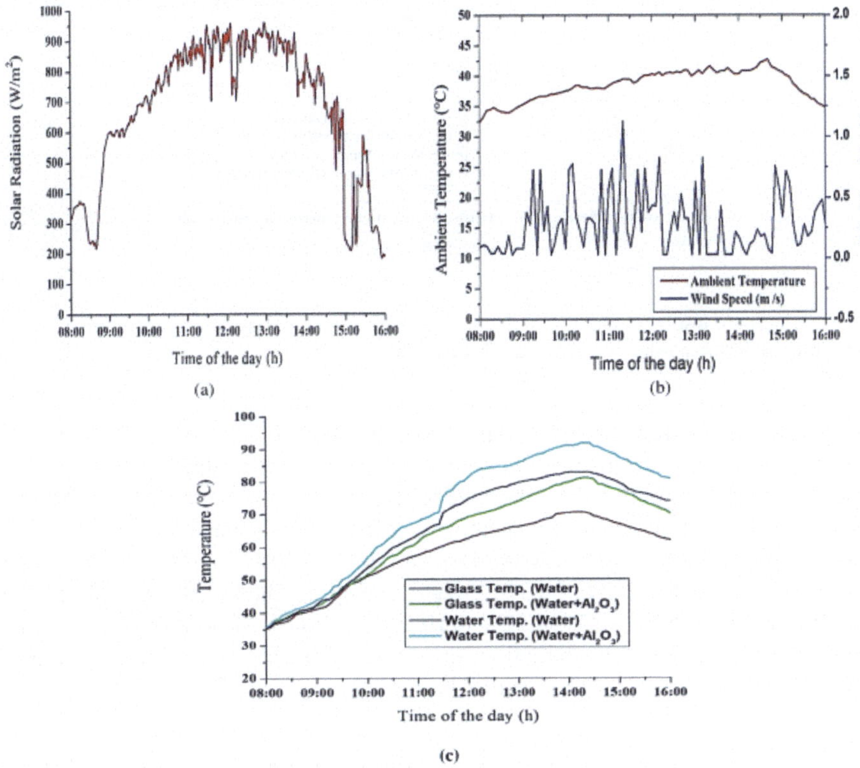

Fig. 2 Variation of (a)solar radiation (b) ambient temperature and wind speed (c) temperatures of solar still for the 0.05% concentration of Al_2O_3on 4th July 2022

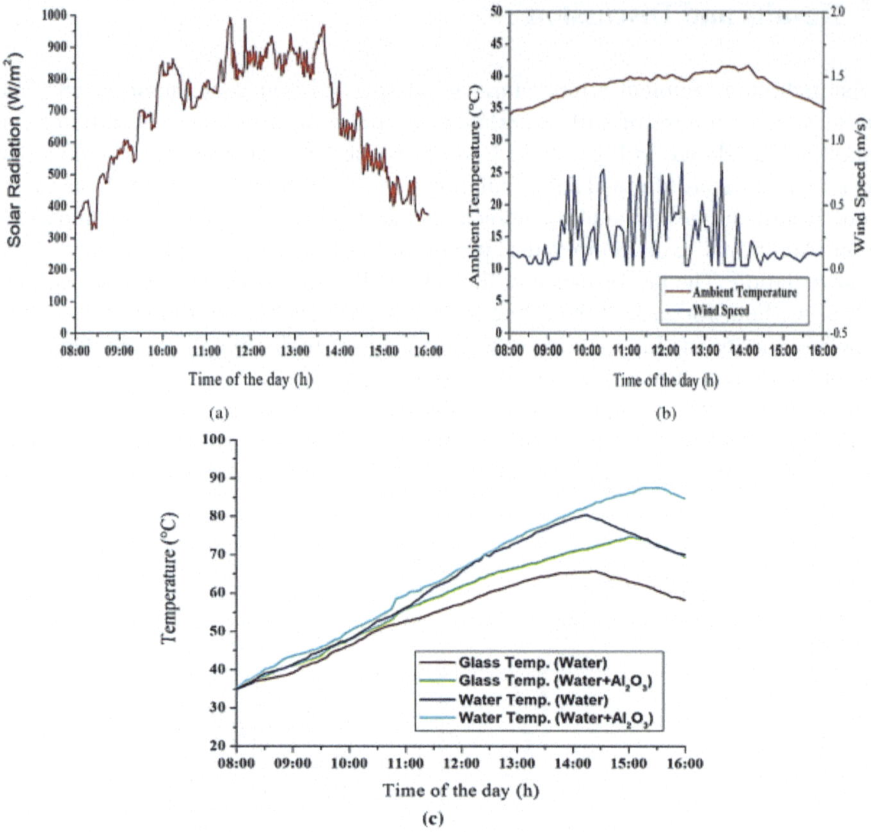

Fig. 3 Variation of **a** solar radiation **b** ambient temperature and wind speed **c** temperatures of solar still for the 0.1% concentration of Al_2O_3on 5th July 2022

has been discovered that, up to a certain point, the concentration of Al_2O_3 nanoparticles enhances the yield of distilled water before it starts to decline as shown in Table 2 The total output of the solar still (distilled water) is increased by 18.54%, 22.26%, 24.37%, and 20.38%, respectively, when using water-based Aluminium Oxide (Al_2O_3) nanofluid at concentrations of 0.05, 0.1, 0.15, and 0.2%. The same enhancement is 18.77, 22.76, 25.57, 23.79% during day hours (08:00 AM–04: 00 PM) and 18.24, 21.63, 22.77, 15.52% during night hours 04:00 PM–07: 00 AM). Result also shows that solar radiation has strong influence on distillation output, more the solar flux more will be the distillation.

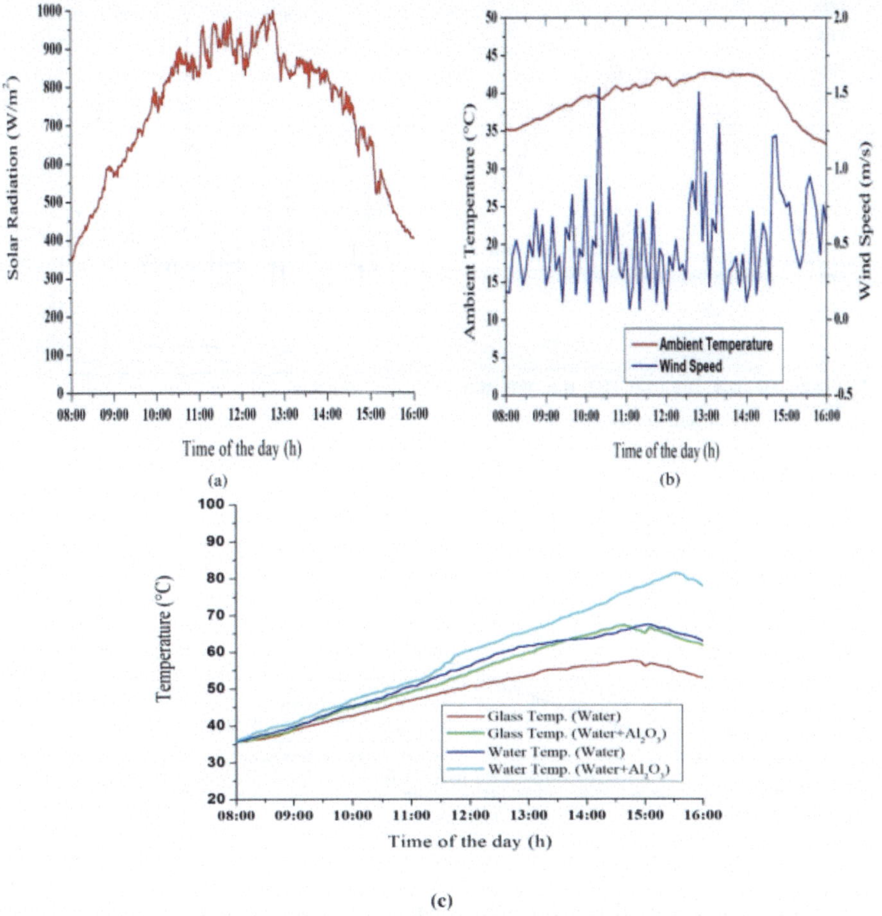

Fig. 4 Variation of **a** solar radiation **b** ambient temperature and wind speed **c** temperatures of solar still for the 0.15% concentration of Al_2O_3on 6th July 2022

6 Testing of Water Quality

The pH and total dissolved solids (TDS) of the generated water as determined by the the solar still were used to assess its quality. Samples of the distillate and feed water were both examined at the Environmental Engineering Laboratory, Department of Civil Engineering, A.M.U., Aligarh (India). Water distilled from different concentrations of nanofluids, 0.05%,0.1%,0.15% and 0.2% of $Al_2O_3$are samples 1,2,3 and 4 respectively, and their properties are shown in Table 3.

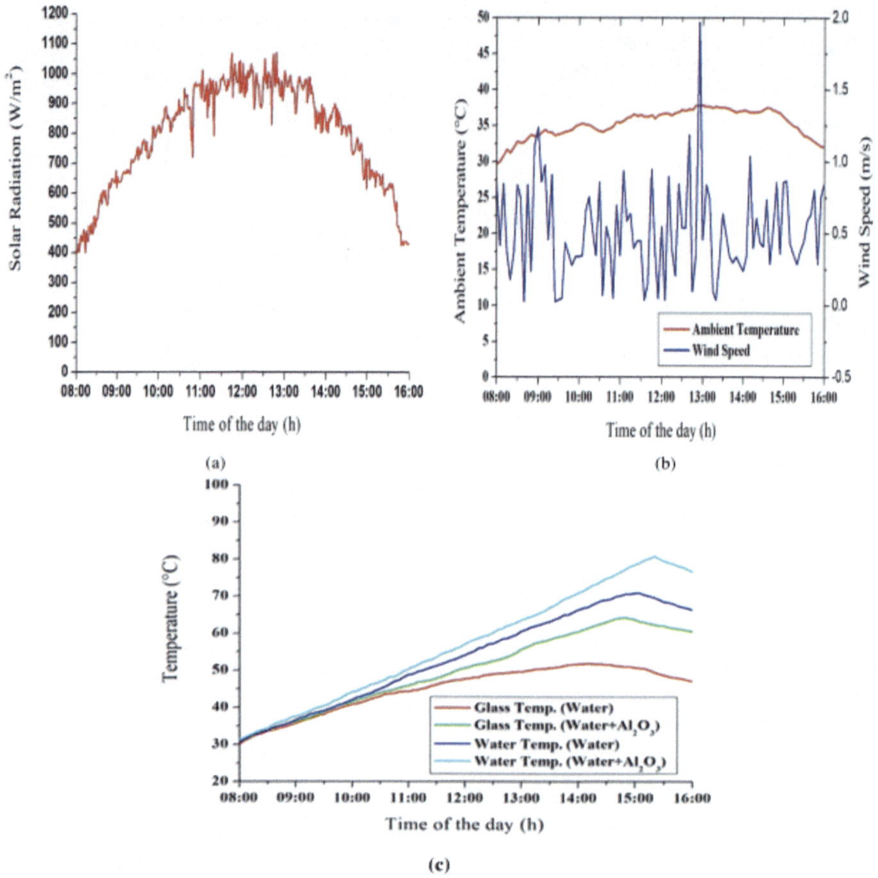

Fig. 5 Variation of **a** solar radiation **b** ambient temperature and wind speed **c** temperatures of solar still for the 0.2% concentration of Al_2O_3on 7th July 2022

7 Conclusions

- Al_2O_3 nanoparticles with water as a base fluid enhance the yielding of solar still. It also worked as a heat storage material, releasing energy into water and increasing output during the night.
- Temperature of basin fluid and glass cover are observed to be more with nanofluids as compared to the conventional one.
- Solar still with water based Aluminium Oxide (Al_2O_3) nanofluid having concentrations of 0.05%, 0.1%, 0.15% and 0.2% increase the total distilled water output of the solar still (distilled water) by 18.54%, 22.26%, 24.37% and 20.38%, respectively whereas same enhancement is 18.77%, 22.76%, 25.57%, 23.79% during

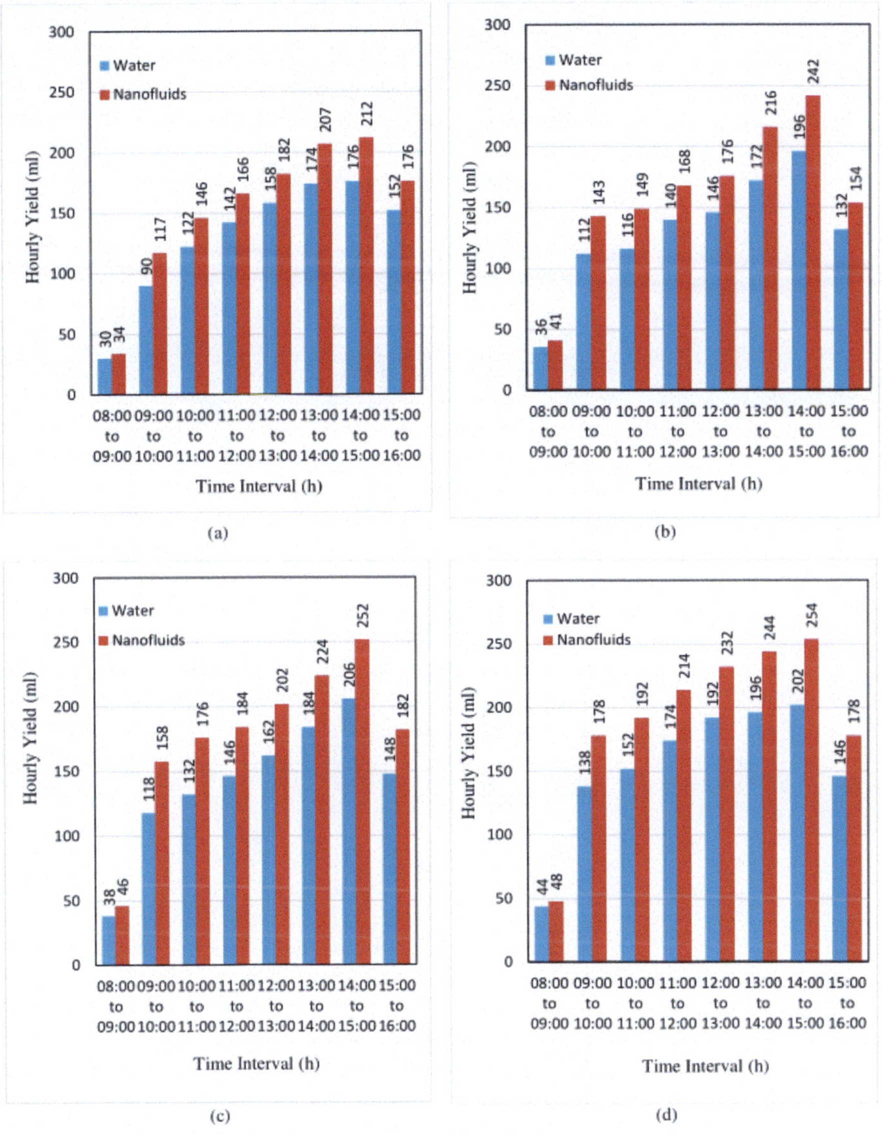

Fig. 6 Comparison in variation of distillate output with time for nanofluids with concentration **a** 0.05% **b** 0.1% **c** 0.15% **d** 0.2%

day hours (08:00 AM–04: 00 PM) and 18.24%, 21.63%, 22.77%, 15.52% during night hours 04:00 PM–07: 00 AM).

- It is found that yielding of distilled water increases with the concentration of Al_2O_3 nanoparticles upto a certain limit and then decreases.

Table 2. Yielding (water distilled) during day and night hours with different concentrations of nanofluids

S. no	Conc. of Al_2O_3 (%)	Yielding (ml) in day hours (08:00 AM to 04: 00 PM)			Yielding (ml) in night hours (04:00 PM to 07: 00 AM)			Total yielding (ml) in day hours (08:00 AM to 07: 00 AM)		
		Water	Water + Al_2O_3	% increment	Water	Water + Al_2O_3	% increment	Water	Water + Al_2O_3	% increment
1	0.05	1044	1240	18.77	822	972	18.24	1866	2212	18.54
2	0.10	1050	1289	22.76	832	1012	21.63	1882	2301	22.26
3	0.15	1134	1424	25.57	852	1046	22.77	1986	2470	24.37
4	0.20	1244	1540	23.79	876	1012	15.52	2120	2552	20.38

Table 3 Quality of distilled water

Property	Sample 1	Sample 2	Sample 3	Sample 4
pH	7.36	7.15	7.91	7.13
TDS (mg/L)	10.18	5.25	7.75	7.27

References

1. Kabeel AE (2009) Performance of solar still with a concave wick evaporation surface. Energy 34:1504–1509
2. Badran O (2010) Theoretical analysis of solar distillation using active solar still. Int J Therm Environ Eng 3:113–120
3. Khan SA, Siddiqui MA, Khan ZA, Asjad M, Husain S (2022) Numerical investigation and implementation of the Taguchi based entropy-ROV method for optimization of the operating and geometrical parameters during natural convection of hybrid nanofluid in annuli. Int J Therm Sci 172:107317
4. Yu W, France DM, Routbort JL, Choi SUS (2011) Review and comparison of nanofluid thermal conductivity and heat transfer enhancements. 29: 432–460https://doi.org/10.1080/014576307 0185085129,432-460
5. Kumar MS, Kumawat G (2015) Performance enhancement of single slope solar still using nano-particles mixed black paint. Adv Nanosci Technol An Int J 1: 55–65
6. Elango T, Kannan A, Kalidasa Murugavel K (2015) Performance study on single basin single slope solar still with different water nanofluids. Desalination 360: 45–51
7. Singh HN, Tiwari GN (2004) Monthly performance of passive and active solar stills for different Indian climatic conditions. Desalination 168:145–150
8. Haddad Z, Abid C, Oztop HF, Mataoui A (2014) A review on how the researchers prepare their nanofluids. Int J Therm Sci 76:168–189
9. Elango T, Kalidasa Murugavel K (2015) The effect of the water depth on the productivity for single and double basin double slope glass solar stills. Desalination 359: 82–91
10. Sharma KV, Sundar LS, Sarma PK (2009) Estimation of heat transfer coefficient and friction factor in the transition flow with low volume concentration ofAl_2O_3 nanofluid flowing in a circular tube and with twisted tape insert. Int Commun Heat Mass Transf 36:503–507

Recent Advances in Applications of Solar Dish Stirling Engine Technology

Vaibhav Singh and Anil Kumar

Abstract In the face of dangers posed by Climate Change in the World today, a shift to renewable sources is the need of the hour. A healthy mix of different energy sources of both renewable and non-renewable nature is the way to move forward. A Solar Stirling Engine has one of the highest thermal efficiency among Solar Thermals. Its applications can play a vital role in contributing to this energy mix of fuel sources. In this paper, recent advancements in the applications of the Solar Dish Stirling Engine System are reviewed. These include Solar Stirling Electric Power Generation, Off Grid Electrification, Combined Heat and Power, Hybridisation and Storage, Water pumping, Water distillation and desalination. It was found that researchers are assessing several new combinations of energy systems, especially in the case of Combined Heat and Power Systems for Residential and Commercial Buildings. Several Studies have also focused on using Solar Stirling Engines in conjunction with other technologies for Water Pumping in the agriculture sector, Water Distillation and Desalination, which is vital for the shortages of clean water.

1 Introduction

The World today faces many dangers. Ever-increasing carbon content in the earth's atmosphere changes the mean temperature, leading to catastrophic consequences, which prominently includes climate change. In real life, the effects are subtle to be noticed immediately, in a few months or years, but they are noticed when the accumulation becomes large enough, and by that time, it becomes too late to avert damages. Subtle changes in rainfall patterns, a change in time and severity of seasons, the untimely appearance of flora and fauna in our surroundings, and events like

V. Singh · A. Kumar (✉)
Department of Mechanical Engineering, Delhi Technological University, Delhi 110042, India
e-mail: anilkumar76@dtu.ac.in

A. Kumar
Centre for Energy and Environment, Delhi Technological University, Delhi 110042, India

M. A. Siddiqui et al. (eds.), *Advances in Heat Transfer and Fluid Dynamics*, Lecture Notes in Mechanical Engineering, https://doi.org/10.1007/978-981-99-7213-5_30

increase in flash floods, increase in droughts, and new waves of diseases are a few of the objective World evidence of climate change [1].

The human need for energy has been increasing rapidly and at an alarming rate. Most of our energy comes from oil, coal, gas, and other hydroelectric and renewable energy sources. Oil, coal and natural Gas are carbon fuels buried inside the earth's crust but burning them releases all that carbon into the atmosphere, disrupting the balance of gases in the atmosphere. Hence we can say that the human need for energy is directly responsible for climate change. The way to tackle this is to reduce our energy requirements, then shift the energy sources from fossil fuel to renewable energy. This shift to renewable energy should be well planned as manufacturing the energy converters for renewable to usable energy also causes emissions. Therefore, the pacing and mix of energy sources are crucial [2].

The Prominent players in renewable energy sources are Wind, Photo Voltaic, and Solar Thermals. Solar thermals include sources like Parabolic Trough Collectors (PTC), Flat plate collectors (FPC), Dish Collectors (DC) & Solar Towers (ST). All these sources can be divided according to the temperature achievable at the collector. FPC can reach up to 100 °C, PTC up to 400–500 °C, DC up to 800–1000 °C and Solar Tower up to 1100 °C [3]. Thus, low-temperature applications like FPC and PTC are suitable for heating requirements, and high-temperature applications like Dish Collector and Solar Towers for power generation. Out of these, Dish collectors can be modular with high-temperature capability, thus reaching higher efficiencies and yet do not require a large field like for Solar Tower.

Stirling Engines are external combustion engines, and they operate on the Stirling cycle which is a closed regenerative thermodynamic cycle consisting of two isothermal and two isochoric processes. They can operate on fluids like Hydrogen, Helium, Air and Nitrogen. Being an external combustion engine, they can operate from various heat sources like Concentrated Solar, Biomass, Geothermal, and fossil fuels [4].

Thus Solar Dish Concentrated System (Fig. 1) combined with Stirling Engine is an attractive option for power generation with multi-fuel and hybridization capability. Solar Dish Stirling System (SDSS) has achieved a maximum efficiency of 32% [5].

This paper presents recent advancements in applications of Solar Dish Stirling Engine System. Those include applications under the Solar Stirling Electric Power Generation, Combined Heat & Power (CHP)/Microgeneration, Hybridisation and Storage, Off-grid Electrification, Water Pumping, Water Distillation and Desalination. Researchers like [5–8] recently reviewed Solar Dish Stirling Engine System applications.

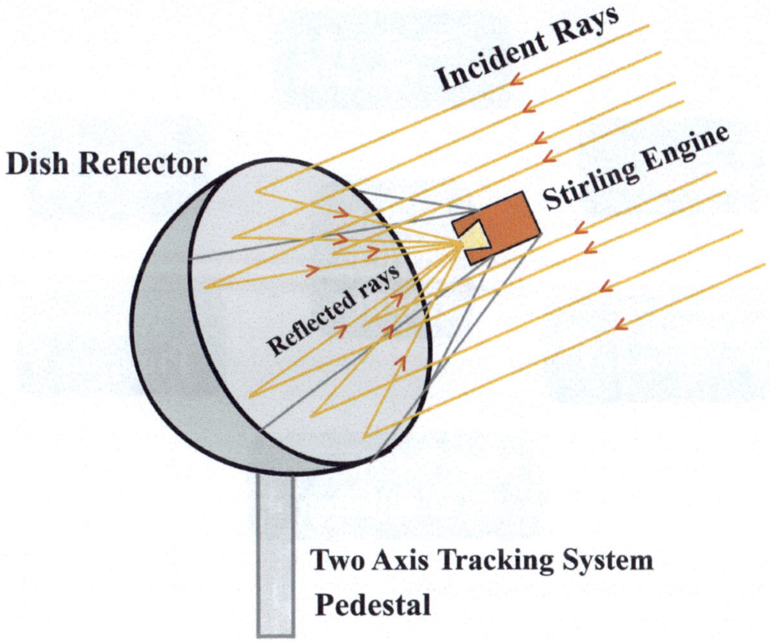

Fig. 1 Schematic diagram of a solar dish stirling system

2 Applications of Solar Dish Stirling Engine

Figure 2 illustrates the primary uses for solar Stirling engine systems. In this section, recent works about these applications are examined.

2.1 Solar Stirling Electric Power Generation

Li et al. [9] created a dynamic model for a solar power plant that allows for temperature variation in the Stirling engine receiver/absorber. Additionally, the capability of the fixed-speed dish-Stirling system to provide frequency control was investigated by varying the operating temperature of the receiver.

Mendoza Castellanos et al. [10] did the experimental analysis and numerical validation of the solar Stirling system with a parabolic dish connected to the electrical grid. The effectiveness of the TRINUM system and the previously published mathematical model showed a variation of 2–12%. At the Federal University of Itajub, Brazil, with solar radiation of 725 w/m^2, the system produced 1 kW of electrical power with a total efficiency of 17.6%.

Lashari et al. [11] used the system advisor model (SAM) to do a power forecast and techno-economic analysis for a 25 kW standalone solar parabolic dish system.

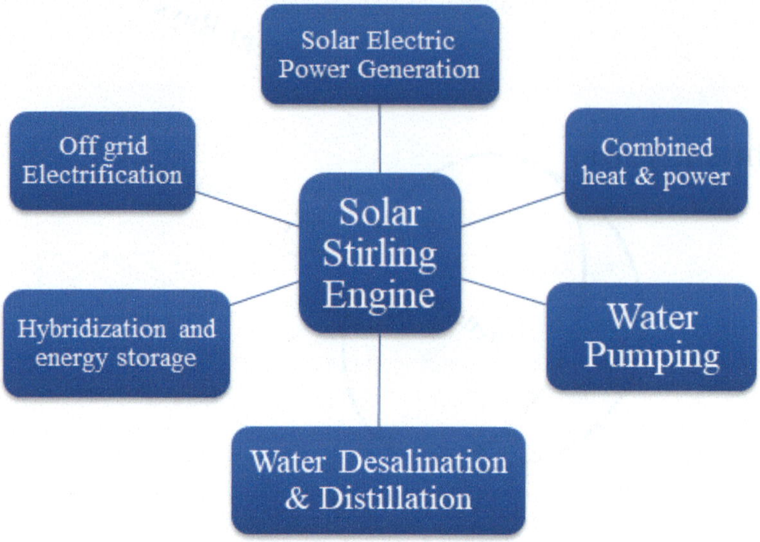

Fig. 2 Applications of solar dish stirling engine

The observed DNI is 1719.15 kWh/m^2/year. 38.6 MWh of power can be produced annually by the system, with a net overall efficiency of 23.39%. Because of high insolation, the month of January produced the most energy, 3.38 MWh.

Vahidi Bidhendi and Abbassi [12] investigated how the Parabolic Stirling Dish System (PSDS) system performed under various operational and meteorological circumstances. It is determined that a 30-degree drop in the temperature of the cold sink will result in a realistic 5% gain in power. With 24% efficiency, the system will yield its most in desert climates. The production of electric power was 50% lower in humid climates.

Zayed et al. [13] used multi-objective particle swarm optimization (MOPSO) to develop, analyze and optimize a thermodynamically balanced dish Stirling system. A sensitivity study was also performed to determine the effects of the rim angle, dish concentrator diameter, and concentrator mirror soiling coefficient. According to sensitivity results, Solar Dish Stirling System with concentrator diameters within 2.5–15 m can provide final ideal output powers between 1.4 and 33.34 kW, with negligible variations in overall efficiency, from 29.80 to 30.20% at the acquired optimal solutions for the optimized dish concentrator designs. The concentrator's mirror soiling factor, which has a significant impact on the optimal electric power and overall efficiency of SDSS, is also highlighted by the results. Achieving a maximum power of 23.46 kW with an ideal total efficiency of 30.15% is doable with the proposed MOPSO technique. Energy flow in a solar dish Stirling engine is depicted in (Fig. 3).

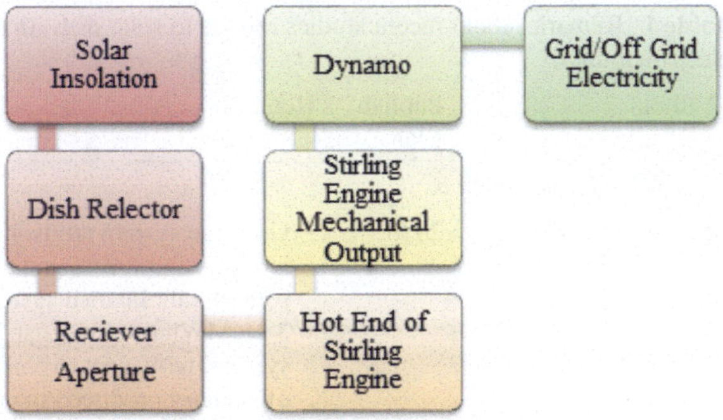

Fig. 3 Energy flow in a solar dish stirling engine power plant

2.2 *Off Grid Electrification*

Kadri and Hadj Abdallah [14] assessed the efficiency of a single Stirling solar dish system placed for rural electrification. Simulation studies and thermodynamic modeling were conducted for the Stirling heat engine that runs on solar electricity. To investigate the viability of the hybrid arrangement, a model for a standalone dish-Stirling with the PMSG system was created in MATLAB software. A comparison study of the Euro dish system was conducted to validate the model's predictions. A variable speed system was shown to be a good option for feeding an unmanageable load under various climatic circumstances during the cold, moderate, and hot seasons.

In order to electrify rural settlements, Bataineh and Taamneh [15] studied the effectiveness of a standalone PSDS system employing SAM. Results indicated that the Levelized energy cost might be reduced while yearly energy production might increase by 19%. The net overall system efficiency was close to 21%, and under ideal conditions, it may even be increased by 3%.

For a 30 kW small-scale power plant, Ezeanya et al. [16] established a performance forecasting model in the System Advisor Model (SAM), for the electrification of Crowley, Louisiana. It is possible to reduce the cost of electricity by around 70%, by running the power plant at the appropriate combination of solar multiple and thermal storage.

The effectiveness of a solar-powered Parabolic Dish Stirling engine for rural electrification in Jordan was assessed [17]. The simulation was carried out in MATLAB software and validated against experimental data with a very excellent match. According to modeling results, overall system efficiency is predicted to be around 30% in summer and 22% in winter, with 230 being the ideal concentration ratio. Under the varied weather conditions of summer, spring, and fall, storage capacity equal to daily energy consumption was determined to be sufficient to handle the uncontrollable load. Remarks about some studies related to solar dish stirling electric generation and off-grid electrification are given in Table 1.

Table 1 Remarks about recent studies related to solar dish stirling electric generation and off-grid electrification

Author	Publish year	Remarks
Solar electric generation		
Lashari et al. [11]	2020	• Used the system advisor model to do techno-economic analysis and forecasted power for a 25 kW standalone solar parabolic dish system • 38.6 MWh of power can be produced annually by the system, with a net overall efficiency of 23.39%. The month of January produced the most energy, 3.38 MWh
Vahidi Bidhendi and Abbassi [12]	2020	• Determined that a 30-degree drop in the temperature of the cold sink will result in a realistic 5% gain in power. With a 24% efficiency, the system will yield its most in desert climates • The production of electric power was 50% lower in humid climates
Zayed et al. [13]	2020	• Developed, analyzed, and optimized thermodynamically balanced dish Stirling system using multi-objective particle swarm optimization (MOPSO) • SDSS with concentrator diameters in the range of 2.5–15 m can provide final ideal output powers between 1.4 and 33.34 kW with negligible variations in overall efficiency, from 29.80 to 30.20% at the acquired optimal solutions for the optimized dish concentrator designs • The concentrator's mirror soiling factor significantly impacts the optimal electric power and overall efficiency of SDSS
Off-grid electrification		
Bataine [15]	2022	• The simulation was carried out in MATLAB software and validated against experimental data with a very good match • The overall system efficiency is predicted to be around 30% in summer and 22% in winter, with 230 being ideal concentration ratio

2.3 Combined Heat and Power

In Forl, Italy, Bianchini et al. [18] conducted an experimental evaluation for the future use of an outdoor 4.5 m diameter dish system for home hot water. They measured the flow rate and temperatures of cooling fluids, receiver temperature, weather data, and power output. The solar dish was discovered to produce heat at a rate of roughly 5440 kWh in 1326 h. According to reports, the planned solar dish would be utilized in the winter to pre-heat domestic hot water and provide space heating. At the same time, in the summer, it would be suitable for hybridization with an absorption chiller to provide space cooling and produce hot water.

A unique cogeneration system based on Stirling solar dish technology was created by Gu et al. [19]. The recommended system has a control strategy that enables switching between heat and power supply utilizing a movable dual receiver. Compared to the normal design, this novel configuration is with an Solar Dish Stirling Cogeneration System (SDSCS) demonstrated superior system running proportion in three different locations in Tianjin, Yinchuan, and Lhasa. A low LCOE power source is also possible with this cogeneration system.

A CFD study of the performance of a solar dish Stirling engine system with micro-cogeneration was conducted by Papurello, Bertino, and Santarelli [20]. COMSOL software was employed. The results showed that a 2.4 m diameter concentrator could generate a value of electrical energy equivalent to 0.99 kWe, starting from 800 W/ m^2 of mean irradiance.

2.3.1 Building Related

A hybrid system that can deliver energy, cooling, and heating simultaneously was proposed by Açıkkalp, Kandemir, and Ahmadi [21]. The system's major components are an absorption refrigeration system, a chemical heat pump, and a solar Stirling engine. The results show that the system performs most effectively at high temperatures for both the collector surface and the working fluids of the Stirling engine. The findings indicate that the suggested system, with a maximum power of 9.463 kW, can improve the overall energy efficiency by more than 13%.

In order to meet the needs of a hotel building in East Azerbaijan, Iran, for cold, pure water and electricity, Jabari et al. [22] created a novel hybrid design of the PSDS system for cooling, potable water, and power Zero Energy Bilding (ZEB). The Stirling Engine and evaporator heat exchanger were coupled to a closed-air open-water Humidification Dehumidification (HDH) desalination process via the suggested configuration.

A Dish Stirling photothermal concentrator was created, built, and virtually tested by Merabet et al. [23] to meet the electrical requirements of a small residential building, in a coastal Algerian city outside the solar belt. A 10.52 m diameter Dish Stirling system was created, utilizing hydrogen as the working fluid and a concentration factor of 1000. It was found to be 23 kW in July during solar noon, which equates to roughly 153% of the needs of the families being met.

In order to satisfy the electric needs of sustainable multi-family buildings with the potential for clean hydrogen production, Allouhi et al. [24] proposed and optimized a Hybrid Renewable Energy System (HRES) based on microgrid power. The proposed hybrid system has a battery bank, an electrolyzer, a hydrogen storage tank, and Solar Dish Stirling (SDS) technology integrated with a wind turbine (WT) for power generation. Results show that the hybrid system design is site dependent. The model's performance was assessed for two cities in Morocco. For Dakhla and Ouarzazate, the optimal HRES design yielded LCOH values of 21.4 and 23.6 euros per kilogram, respectively. This study opens up possibilities for the SDS's use in carbon–neutral buildings and indicates the possibility of its deployment in future HRESs.

Guarino et al. [25] proposed developing a cogeneration plant that would produce heat and electricity by combining a field of dish-Stirling collectors with seasonal geothermal storage and water-to-water pumps. The facility was designed to generate electricity and provide thermal energy for a building's heating system in Palermo. Based on modeling results, it is possible to use solar energy to meet 97% of the building's yearly thermal loads with the optimal plant configuration. Other uses are permitted for the remaining 64% of the electrical energy generated by the electric engines.

2.3.2 Refrigeration

Al Keyyam et al. [26] looked into the efficiency of a novel integrated solar thermal water harvesting system that uses a concentrated photovoltaic thermal unit (CPV/T) to harness solar energy and generate electricity. An alpha-type Stirling engine and a single-effect LiBr/H_2O absorption cooling cycle are powered by rejected heat. The power output of the Stirling engine and the CPV/T is used to drive a vapour compression refrigeration cycle, which uses the cooling capacity of the cooling cycles to chill and dehumidify ambient air and generate drinkable water. The highest water production rate, around 30 L/h, was found to occur in hot, humid areas with high solar radiation, high ambient temperature, and high relative humidity. The suggested method is appropriate for small-scale applications with water demands under 180 L daily. The amount of electrical energy used by this system to create one liter of water is between 225 and 315 Wh/L in general. Remarks about some studies related to combined heat and power are given in Table 2.

2.4 Hybridisation and Storage

In the field of research, the solar dish-Stirling engine is increasingly being combined with thermal energy storage or hybridized with other renewable energy sources. Additionally, hybridization makes it possible for the system to operate more continuously. These methods also lessen the reliance on solar energy's supply. This section discusses recent research in hybridization and storage combined with dish-Stirling systems.

Andraka [27] examined the impact on the Levelized cost of energy after determining whether it was feasible to add thermal storage to a dish-Stirling system. The engine and heat storage units were placed at the back of the parabolic dish to enhance the structure's design. The researchers found that a dish-Stirling storage system can add value and significantly lower the system's life cycle cost of ownership. For an ideal value of a solar multiple of 1.25 and a storage capacity of 6 h, a storage system integrated in a Dish-Stirling can provide value and reduce the Levelized cost of electricity.

Table 2 Remarks about combined heat and power

Author	Publish year	Remarks
Combined heat and power		
Gu et al. [19]	2021	• It has a control strategy that enables switching between heat and power supply utilizing a movable dual receiver • A low LCOE power source is also possible with this type of cogeneration system
Papurello et al. [20]	2021	• A CFD study of the performance of a solar dish Stirling engine system with micro-cogeneration using COMSOL software • A concentrator of 2.4 m diameter could generate a value of electrical energy equivalent to 0.99 kWe, starting with 800 W/m^2 of mean irradiance
Building related		
Merabet et al. [23]	2022	• A 10.52 m-diameter Dish Stirling system was created, utilizing hydrogen as the working fluid and a concentration factor of 1000 • It was generating 23 kW in July during solar noon, which equates to roughly 153% of the needs of the families
Allouhi et al. [24]	2022	• The proposed hybrid system has a battery bank, an electrolyzer, a hydrogen storage tank, and Solar Dish Stirling technology integrated with a wind turbine for power generation • For two cities of Morocco, Dakhla and Ouarzazate, the optimal HRES design yielded LCOH values of 21.4 and 23.6 euros per kilogram, respectively
Guarino et al. [25]	2022	• A cogeneration plant that would produce both heat and electricity by combining a field of dish-stirling collectors with seasonal geothermal storage and water-to-water pumps • Using solar energy to meet 97% of the building's yearly thermal loads with the optimal plant configuration is possible. Other uses are permitted for the remaining 64% of the electrical energy generated by the electric engines
Refrigeration		
Al Keyyam et al. [26]	2021	• The highest water production rate, around 30 L/h, was found to occur in hot, humid areas with high solar radiation, ambient temperature, and relative humidity • The amount of electrical energy used by this system to create one liter of water is between 225 and 315 Wh/L in general

To prevent thermal overloading and boost the system's conversion efficiency, Mohammadnia et al. [28] devised a novel hybrid system that combines a Solar Dish Stirling System with a thermoelectric generator (TEG), which serves as an energy harvester. According to reports, the Stirling engine was shielded from unfavorable temperature increases over the critical temperature, thanks to the temperature regulation of the cavity using the TEG. This was especially true during the evening and morning hours when solar irradiation was low. At maximum sun irradiation, the projected TEG-SDSS produced 14.1 kW. (solar noon). Furthermore, the temperature management strategy increased the system's overall performance by 20–30%.

A system based on a solar Stirling engine and a diesel engine was proposed by Jabari et al. [29] The installation of this system aimed to meet peak energy demand throughout the summer while assuring clean, reliable and independent power generation. The outcomes of a microgrid reduced the daily cost by 15.4 dollars and the amount of electricity used from the main power supply.

Mehrpooya et al. [30] assessed the effectiveness of a combination system made up of a Stirling engine, a parabolic dish concentrator, and a thermoelectric drive under varied conditions. Stirling Engine exhaust is transmitted to TEG hot end, which generates electricity. The system's performance was assessed for several scenarios and the cities of Moscow, Kiev, Tehran, Beijing, and Geneva. Additionally, it was discovered that on June 14, PDC in Moscow produced an average daily useable power of 373.97 W/m^2, roughly 11.1%, 1.55%, 33.3%, and 14.23% greater than Tehran, Beijing, Geneva, and Kiev, respectively. Additionally, Tehran, Beijing, and Moscow systems have greater justifications for the PDC numbers than the systems in the other two cities (Geneva and Kiev).

Li et al. [31] introduced and demonstrated a brand-new energy system that combines an electrolyzer, a photovoltaic field, an alkaline fuel cell (AFC), a Stirling engine, and an absorption chiller. Additionally, they considered a second pumped hydro compressor air system subsystem and assessed them under various working situations. According to the results, a 1 kW (el) AFC produces 1.27 kW and 2.48 kW of power and cooling load from the Stirling engine and absorption chiller, respectively, with a system efficiency of 63.8%.

A unique hybrid system consisting of a horizontal axis wind turbine and concentrated parabolic solar dish Stirling engine was developed, designed, and technologically and economically evaluated by Shboul et al. [32]. The horizontal axis wind turbine works with a battery bank to provide backup power when the primary power source is unavailable. The solar dish Stirling engine is the primary source of electrical power generation. The efficiency of the Stirling engine is 37% at the optimal design point, with a net output power of 1500 kWe. The levelized cost is between $0.13 and $0.15 per kWh, and the hourly cost is approximately $4, making it very competitive with other integrated renewable energy technologies.

2.5 Water Pumping

A Stirling converter model that can pump water under various operating conditions was proposed by Tavakolpour-Saleh and Jokar [33]. The proposed sophisticated mathematical approach was determined to be inappropriate for the real-time implementation of the model-based controller when they implemented an Artificial Neural Network control system. Thus, a less sophisticated neural network model was proposed; experimental findings supported the viability of the new model.

An innovative Stirling engine design by SunPulse [34] uses air as the operating fluid. A Stirling engine and an integrated solar thermal rotary water pump were created. The integrated solar thermal technology has the potential use for water pumping in remote or off-grid locations. The market for independent mechanical applications, such as milling, grinding, and compressing, was shown to exist.

Bekele and Ancha [35] examined a small-scale irrigation system for Ethiopia powered by a solar dish Stirling engine. According to the results, the thermoelectric unit produces 5.2 W of electrical power at a maximum efficiency of 2.78% at a heat source temperature of 413.8 K. At a thermal efficiency of 18.61%, the Stirling engine-driven pump delivers a daily cumulative flow rate of 173,594.95 L. This promotes the nation's food security and enables small farmers to earn more money.

2.6 *Water Distillation and Desalination*

A hybrid energy system was created by Lai et al. [36] that combines an SDSS with an energy storage system based on salinity gradients. To pressurize the seawater into reverse osmosis (RO), the SDSS's generated electricity was used. The ideal operating conditions were chosen to achieve the greatest possible energy efficiency. 9.23% is the highest possible energy conversion efficiency that has been attained.

Al-Nimr and Al-Ammari [37] suggests a cutting-edge solution that combines TEC and Solar Stirling Engine (SE) modules. The saline water has been heated using recycled waste heat from the TEC module's hot and SE's cold sides. Additionally, the condensation and water desalination rate was increased using the TEC modules' cold side. As a result, the desalination rate rises from 2.93 kg/day to 40.96 kg/day, and the efficiency rises from 22.84 to 64.44%.

Soliman et al. [38] investigated an innovative wastewater treatment method for Libya's Al-Marj area's industrial sector. The primary source of power for reverse osmosis operation will be a solar Stirling engine. Instead of batteries or diesel generators, a hydraulic power system will be used as an energy storage and recovery system. The results show that the overall water price was in the range of 0.65 $/m^3, and the particular power consumption was not greater than 4 kWh/m^3. The freshwater produced is estimated to supply 55,000 residents.

A complete parametric study based on finite-time thermodynamics is conducted to assess the system's water productivity and energy and exergy efficiency. Geng et al. [39] introduced a RO desalination system powered by a solar dish-Stirling (DS) engine. The maximum water productivity, energy/exergy efficiency, and average absorber temperature were shown to be linearly correlated with the source side's ideal working fluid temperature. The water productivity and energy/exergy efficiency first rise with rising sink side temperature before falling. Remarks about some studies related to hybridisation and storage, water pumping, water distillation and desalination are given in Table 3.

Table 3 Remarks about hybridisation and storage, water pumping, water distillation and desalination

Author	Publish year	Remarks
Hybridisation and storage		
Mehrpooya et al. [30]	2021	• A hybrid system comprises a Stirling engine, a parabolic dish concentrator, and a thermoelectric drive under varied conditions. Stirling Engine exhaust is transmitted to TEG hot end, which generates electricity • PDC in Moscow produced an average daily useable power of 373.97 W/m^2, roughly 11.1% greater than Tehran, 1.55% greater than Beijing, 33.3% greater than Geneva, and 14.23% greater than Kiev
Li et al. [31]	2021	• A hybrid system that combines an electrolyzer, a photovoltaic field, an alkaline fuel cell, a Stirling engine, and an absorption chiller • 1 kW(el) AFC produces 1.27 and 2.48 kW of power and cooling load from the Stirling engine and absorption chiller, respectively, with a system efficiency of 63.8%
Shboul et al. [32]	2021	• Hybrid system consisting of a horizontal axis wind turbine and concentrated parabolic solar dish Stirling engine. In the event that the main source of power is unavailable, a battery bank and a horizontal axis wind turbine work together to supply backup power • The efficiency of the Stirling engine is 37% at the optimal design point, with a net output power of 1500 kWe. The Levelized cost of energy is found to be between \$0.13 and \$0.15 per kWh
Water pumping		
Bekele and Ancha [35]	2022	• Small-scale irrigation system for Ethiopia powered by a solar dish stirling engine • The thermoelectric unit produces 5.2 W of electrical power at a maximum efficiency of 2.78% at a heat source temperature of 413.8 K. At a thermal efficiency of 18.61%, the Stirling engine-driven pump delivers a daily cumulative flow rate of 173,594.95 L
Water distillation and desalination		
Geng et al. [39]	2021	• A RO desalination system powered by a solar dish-Stirling engine • The maximum water productivity, energy/exergy efficiency, and average absorber temperature were shown to be linearly correlated with the source side's ideal working fluid temperature

3 Commercial Solar Stirling Engine Power Plants

Currently, there are no active utility commercial solar Stirling engine power plants operating in the World. 1.5 MW Maricopa Solar was the Largest. A list of Commercial Solar Stirling Engine Power Plants is given in Table 4.

Table 4 List of commercial solar dish stirling power plants with details

S. no	1	2	3	4	5	6	7
Plant	SAIC/STM	SBP	SES	WGA off grid	Kerman Pilot parabolic dish-Stirling system	Maricopa solar project CSP project	Tooele army depot CSP project
Location	Salt river project near phoenix, Arizona, USA	Plataforma solar de Almeria (PSA), Spain	Boeing, Huntington beach, California, USA	Sandia's national solar thermal test facility, albuquerque, USA	Kerman City, Iran	Peoria, Arizona, United States	Tooele, Tooele county, Utah, United States
Year of installation	1997	2001	2002	2000	2017	2010	2012
Avg daily solar radiation (kWh/m^2)	6.57	5.2	7.1	6.48	5.48	–	–
Aperture diameter (cm)	38	15	20	14	12	–	–
Projected area (m^2)	113.5	56.7	87.7	41.2	–	–	–
Focal length (m)	12	4.5	7.45	5.45	–	–	–
Type of engine	STM 4–120 double acting kinematic	SOLO 161 kinematic	Kockums/SES 4–95 kinematic	SOLO 161 kinematic	Free piston stirling engine	–	–
Working fluid	Hydrogen	Helium	Hydrogen	Hydrogen	Helium	–	Helium
Displacement (cc)	480	160	380	160	–	–	–
No of cylinders	4	2	4	2	–	–	–
Operational speed (RPM)	2200	1500	1800	1800–1890	–	–	–

(continued)

Table 4 (continued)

S. no	1	2	3	4	5	6	7
Control method	Variable stroke	Variable pressure	Variable pressure	Variable pressure	–	–	–
Peak output (kW)	22.9	8.5	25.3	8	0.63	1.5 MW plant output	1.5 MW plant output
Peak efficiency (%)	20	19	29.4	22.5	17		
Misc. remarks		Maximum pressure is 150 bars and working temp 650 °C				60* 25 kW dish receivers	429 * 3.5 kW dish receivers

Sources [5, 40]

4 Conclusion

Substantial progress has been made in recent years to improve the performance and application of solar-powered Stirling engines.

Hybrid systems, which include Dual Receivers, one for heating oil and the other for producing electricity, are better at meeting the Heat and Power loads of a building. While the system, which includes seasonal geothermal storage, improves sustainability by utilizing more renewable sources. But hybrid systems in which many subsystems are involved might improve the overall efficiency but may increase complexity and drive up the setup and maintenance costs. The systems with Solar Stirling Engine mechanical output directly coupled with Pumps are simple, more efficient and thus better for field use as an Agricultural pump and Reverse Osmosis pump, unlike the systems that utilize electric pumps.

References

1. COP 22 | UNFCCC. https://unfccc.int/event/cop-22
2. Hiremath RB, Kumar B, Balachandra P, Ravindranath NH, Raghunandan BN (2009) Decentralised renewable energy: scope, relevance and applications in the Indian context. Energy Sustain Dev 13
3. Sukhatme SPP, Nayak JK, Naik JK (2014) Solar energy principles of thermal collection and storage. Mc Graw Hill
4. Walker G (1973) The Stirling engine. Sci Am 229
5. Singh UR, Kumar A (2018) Review on solar Stirling engine: development and performance. Thermal Sci Eng Progress 8
6. Zayed ME et al (2021) A comprehensive review on Dish/Stirling concentrated solar power systems: design, optical and geometrical analyses, thermal performance assessment, and applications. J Clean Prod 283
7. Malik MZ et al (2022) A review on design parameters and specifications of parabolic solar dish Stirling systems and their applications. Energy Rep 8:4128–4154
8. Allouhi H, Allouhi A, Buker MS, Zafar S, Jamil A (2022) Recent advances, challenges, and prospects in solar dish collectors: designs, applications, and optimization frameworks. Sol Energy Mater Sol Cells 241:111743
9. Li Y, Choi SS, Vilathgamuwa DM (2018) Primary frequency control scheme for a fixed-speed dish-Stirling solar-thermal power plant. IEEE Trans Power Syst 33
10. Mendoza Castellanos LS, et al (2019) Experimental analysis and numerical validation of the solar Dish/Stirling system connected to the electric grid. Renew Energy 135
11. Lashari AA, et al (2021) The performance prediction and techno-economic analyses of a standalone parabolic solar dish/Stirling system, for Jamshoro, Pakistan. Clean Eng Technol 2
12. Vahidi Bidhendi M, Abbassi Y (2020) Exploring dynamic operation of a solar dish-Stirling engine: validation and implementation of a novel TRNSYS type. Sustain Energy Technol Assessments 40
13. Zayed ME, Zhao J, Elsheikh AH, Li W, Elaziz MA (2020) Optimal design parameters and performance optimization of thermodynamically balanced dish/Stirling concentrated solar power system using multi-objective particle swarm optimization. Appl Therm Eng 178
14. Kadri Y, Hadj Abdallah H (2016) Performance evaluation of a standalone solar dish Stirling system for power generation suitable for off-grid rural electrification. Energy Convers Manage 129

15. Bataineh K, Taamneh Y (2017) Performance analysis of standalone solar dish Stirling system for electricity generation. Int J Heat Technol 35
16. Ezeanya EK, Massiha GH, Simon WE, Raush JR, Chambers TL (2018) System advisor model (SAM) simulation modelling of a concentrating solar thermal power plant with comparison to actual performance data. Cogent Eng 5
17. Bataineh K (2022) Performance evaluation of a standalone solar dish Stirling system for off–grid electrification. Energy Sources, Part A Recover Util Environ Eff 44
18. Bianchini A, Guzzini A, Pellegrini M, Saccani C (2019) Performance assessment of a solar parabolic dish for domestic use based on experimental measurements. Renew Energy 133
19. Gu L, et al (2021) Research on a new type of solar dish Stirling cogeneration system based on adjustable receiver and control strategy. Energy Convers Manage 249
20. Papurello D, Bertino D, Santarelli M (2021) CFD performance analysis of a dish-Stirling system for microgeneration. Processes 9
21. Açıkkalp E, Kandemir SY, Ahmadi MH (2019) Solar driven Stirling engine - chemical heat pump - absorption refrigerator hybrid system as environmental friendly energy system. J Environ Manage 232
22. Jabari F, Nazari-heris M, Mohammadi-ivatloo B, Asadi S, Abapour M (2020) A solar dish Stirling engine combined humidification-dehumidification desalination cycle for cleaner production of cool, pure water, and power in hot and humid regions. Sustain Energy Technol Assessments 37
23. Merabet N, Chouichi L, Kerboua K (2021) Numerical design and simulation of a thermodynamic solar solution for a pilot residential building at the edge of the sun-belt region. Environ Dev Sustain. https://doi.org/10.1007/s10668-021-01956-2
24. Allouhi H, Allouhi A, Almohammadi KM, Hamrani A, Jamil A (2022) Hybrid renewable energy system for sustainable residential buildings based on Solar Dish Stirling and wind Turbine with hydrogen production. Energy Convers Manag 270:116261
25. Guarino S, Buscemi A, Messineo A, Brano V (2022) Lo. Energy and environmental assessment of a hybrid dish-Stirling concentrating solar power plant. Sustain 14: 6098
26. Al Keyyam I, Al-Nimr M, Khashan S, Keewan A (2021) A new solar atmospheric water harvesting integrated system using CPV/T—Stirling engine—Absorption cooling cycle and vapor compression refrigeration cycle. Int J Energy Res 45
27. Andraka CE (2014) Dish Stirling advanced latent storage feasibility. Energy Procedia 49
28. Mohammadnia A, Ziapour BM, Sedaghati F, Rosendahl L, Rezania A (2020) Utilizing thermoelectric generator as cavity temperature controller for temperature management in dish-Stirling engine. Appl Therm Eng 165
29. Jabari F, Nazari-Heris M, Mohammadi-Ivatloo B, Asadi S, Abapour M, Toward energy-efficient microgrids under summer peak electrical demand integrating solar dish Stirling heat engine and diesel unit. J Energy Manage Technol 4
30. Mehrpooya M, Ghadimi N, Marefati M, Ghorbanian SA (2021) Numerical investigation of a new combined energy system includes parabolic dish solar collector, Stirling engine and thermoelectric device. Int J Energy Res 45
31. Li D, Guo J, Zhang J, Zhan L, Alizadeh M (2021) Numerical assessment of a hybrid energy generation process and energy storage system based on alkaline fuel cell, solar energy and Stirling engine. J Energy Storage 39
32. Shboul B, et al (2021) Design and techno-economic assessment of a new hybrid system of a solar dish Stirling engine instegrated with a horizontal axis wind turbine for microgrid power generation. Energy Convers Manage 245
33. Tavakolpour-Saleh AR, Jokar H (2016) Neural network-based control of an intelligent solar Stirling pump. Energy 94
34. Mitra Ardron - Natural Innovation - Innovation: SunPulse Water - solar thermal water pump. http://www.naturalinnovation.org/?id=33&alias=sunpulsewater
35. Bekele EA, Ancha VR (2022) Transient performance prediction of solar dish concentrator integrated with Stirling and TEG for small scale irrigation system: a case of Ethiopia. Heliyon 8:e10629

36. Lai X, Yu M, Long R, Liu Z, Liu W (2019) Clean and stable utilization of solar energy by integrating dish solar Stirling engine and salinity gradient technology. Energy 182
37. Al-Nimr MA, Al-Ammari WA (2020) A novel hybrid and interactive solar system consists of Stirling engine/vacuum evaporator/thermoelectric cooler for electricity generation and water distillation. Renew Energy 153
38. Soliman AM, et al (2020) A new system design of using solar dish-hydro combined with reverse osmosis for sewage water treatment: case study Al-Marj, Libya. Desalin Water Treat 193
39. Geng D, Cui J, Fan L (2021) Performance investigation of a reverse osmosis desalination system powered by solar dish-Stirling engine. Energy Rep 7
40. National Renewable Energy Laboratory (NREL) Home Page | NREL. https://www.nrel.gov/

Analysis of Evacuated Tube Consisting Parabolic Concentrator with Al_2O_3-Water Nanofluid

Muntashir Mohammed Razzak, Afzal Husain, Nasser Al-Azri, and Nabeel Al-Rawahi

Abstract Efficiency enhancement has always been a challenge for scientists and researchers. Although the realization of proposed concepts is necessary for development, it consists of several challenges from fabrication to testing the concept. The present study carried out experiments on parabolic trough collectors with the bare tube and evacuated tube receivers. Further, the performance has been tested for water and Al_2O_3-water as nanofluid. The parabolic trough and experimental setup were fabricated to carry out experiments outdoors for the daily solar flux and varying flow rates of the working fluid. The nanofluid with an evacuated receiver tube continuously shows better performance over the range of flow rates investigated in this study.

Keywords Parabolic trough collector · Al_2O_3-water nanofluid · Thermal efficiency · Renewable energy · Evacuated tube receiver

1 Introduction

The last few decades have seen an intense focus on the production and usage of non-renewable forms of energy. The world has started realizing the adverse effects of excessive usage of fossil fuels and non-renewable resources. More attention is focused on producing electrical energy from clean methods and processes. Solar energy is a form of clean energy and abundantly available free of cost like other forms of energy such as wind energy and hydropower and has been looked at as an ideal substitute for fossil fuels. Over the decades, the focus of scientists and engineers

M. M. Razzak · A. Husain (✉) · N. Al-Azri · N. Al-Rawahi
Mechanical and Industrial Engineering Department, Sultan Qaboos University, Muscat, Oman
e-mail: afzal19@squ.edu.om

N. Al-Azri
e-mail: nalazri@squ.edu.om

N. Al-Rawahi
e-mail: alrawahi@squ.edu.om

M. A. Siddiqui et al. (eds.), *Advances in Heat Transfer and Fluid Dynamics*, Lecture Notes in Mechanical Engineering, https://doi.org/10.1007/978-981-99-7213-5_31

has shifted to making renewable energy conversion techniques more affordable and efficient, in order to make the transition from conventional, harmful techniques to non-conventional renewable methods.

Concentrating Solar Collector (CSC) is a promising alternative technology that allows the harnessing of solar energy and consequently converting it into useful forms. Concentrating solar collectors with the use of high-temperature heat transfer fluids such as thermal oils, molten metals, and salts have been tested for their promising performance. Further, enhancement of the performance was highlighted by several researchers with the application of nanofluids. However, nanofluids are to be tested as potential working fluids for such applications. The CSC has reflectors that reflect incoming solar radiation into a specific line, which is where the receiver tube is located. There are several variants of reflector configurations and designs. The receiver is a component available in a variety of configurations. It absorbs the solar energy reflected onto it, heating a flowing fluid within it often called Heat Transfer Fluid (HTF). The HTF can be of many uses in industry including electrical power generation and process steam. For effective heat absorption and transport, an HTF should have a high boiling point and low melting point temperature along with low vapor pressure at high temperatures. Further HTF should have high thermal stability, heat capacity, and thermal conductivity [1, 2]. Nanofluid, which is one of the unconventional fluids, has been proposed as next-generation HTFs [3]. Several studies have been carried out and it is suggested that the nanofluids provide improved thermal characteristics such as thermal conductivity, specific heat [4, 5], and enhanced radiation properties such as absorption and emission [6–9]. Therefore, for an effective solar collector system, a synergetic combination of the concentrator, receiver, and HTF is necessary.

In modern CSCs, mirrors are usually programmed to track the sun. The receiver carries energy transfer fluid often called heat transfer fluid (HTF) which receives solar radiation and increases the temperature of the HTF. The HTF is then used to run the turbine directly or in association with a steam generator-based thermal cycle to produce electric power. The parabolic trough is one of the widely used collectors in the present day. Over the last few decades, several studies have been carried out to realize the potential of this technology. Dudley et al. [10] carried out an experiment to test the efficiency and heat losses from a parabolic trough solar collector (surface absorption method) using cremet and black chrome as selective receiver coating for three-receiver configurations, i.e., glass tube envelope surrounding steel pipe, with vacuum, with air, and the one without an envelope. Vacuum in tube annulus performed best, then air in tube annulus, and the worst performer was the no envelope case. The cremet performed better than black chrome as the receiver coating due to its better radiative properties.

Ahmed et al. [11] conducted an experimental and numerical study on a PV/T collector with water and ethylene-based nanofluid with Al2O3 and Cu. The experiment was done at three different locations and the result showed that the water-based nanofluid performs better than the one with ethylene. Arthur and Karim [12] carried out two-dimensional CFD modeling of high heat flux solar collectors using molten salts such as NaNO3-KNO3 andLi2CO3-K2CO3 based nanofluid as carrier fluids. The author analyzed the different losses that occurred and estimated the solar collector efficiency. He observed an increase in solar collector efficiency with an increase in solar concentration and nanoparticle volume fraction and a decrease with decreasing receiver length. Keeffe et al. [13] mathematically analyzed several variants of direct absorption solar collectors with the nanofluid and as the working fluid using a continuum mechanics approach to resolve heat transfer and flow field. The authors noted an increase in performance with the reflecting base panel as compared to the nonreflecting one and an increase in nanoparticle volume concentration increases the collector performance.

Parabolic trough collectors can be used to heat water in cold areas efficiently. Thermal efficiency is better when the heat transfer fluid has a high temperature as per the experimental study [14]. Fluids at high temperatures tend to have a better Reynolds number, which contributes to increased efficiency. The thermal efficiency is improved with larger fluid flow rates, but high wind velocity lowers the efficiency. A thermal efficiency of 52% was obtained by the authors for a flow rate of 0.6 m^3/h using thermal oil as the HTF.

Parabolic trough solar collectors outperform other types of commonly used solar collectors, such as flat plate collectors [15]. Experimental studies were conducted to investigate the performance of a flat plate solar collector and a parabolic trough solar collector at the same time, location, and operating conditions. The results indicated that the PTC provides higher temperature outputs for the heat transfer fluid and higher thermal efficiencies compared to the flat plate collector. The experiments were conducted in the winter. The efficiencies of the PTC and the flat plate collector were 6% and 12% respectively. The authors also noted that the PTC can produce higher thermal efficiencies while occupying lesser space than the flat plate collector.

A case study on the performance of parabolic trough collectors found that the reflectivity of the reflecting mirrors is severely diminished by up to 25% in dusty and other adverse environmental conditions [16]. Moreover, scatter and absorption during the reflection leads to decreased efficiency of the mirror and cause laser-induced damage. The authors recommended the use of Highly Reflective Coating (HRC) to reduce reflection losses. The effect of dust and salt on the reflective mirror mixing with the condensed water in the early morning hours creates cohesive layers when the water evaporates [17]. Hence, regular, and thorough cleaning of the mirrors are necessary.

Low-temperature PTCs are suitable for use in residential areas. It is important that buildings be integrated with PTCs in the design phase of the building construction. Exergy analyses are required when applying solar energy systems in residential areas. Thermal efficiency of 55% is possible using Al_2O_3/ Water based PTC for homes [18].

Nanofluids are becoming increasingly popular because of their good thermal conductivity, absorptivity, and heat transfer coefficient, which boosts thermal performance. However, increasing the mass fraction may not always result in better performance. Mass fractions greater than 3% may cause corrosion and erosion of the solar collector. The effect of nanoparticle size on performance is unclear and more research is needed in that area [19].

In experiments using Al_2O_3 and Fe_2O_3 nanofluids on PTCs with mass concentrations of 0.20, 0.25, and 0.3%, it was observed that the highest mass concentration produced the best temperature difference [20]. The temperature difference declined at higher flow rates, but thermal efficiency was increased. Nanofluids were reported to have better optical and heat transfer properties. Al_2O_3 nanofluids gave efficiencies 13% higher than water and Fe_2O_3 nanofluids produced an efficiency that was 11% higher than water. Additional studies have also reported that if the volumetric flow rate is reduced, the temperature difference may increase but the thermal efficiency is not improved significantly [21]. A review of the geometry of nanoparticles in nanofluids revealed that for most experimental studies on PTCs, the thermal efficiency is higher when the nanoparticles are smaller in terms of diameter. The larger surface area to volume ratio may help the smaller nanoparticles to conduct more heat [22].

The receiver tube is a critical component of the parabolic trough collector. The receiver tube is generally dark in color to help it better absorb solar radiation. The receiver tube may be made of various types of metal such as stainless steel, copper, and brass. Many contemporary receiver tubes have glass covers/ envelopes to create a vacuum annular space, so the heat is only transmitted from the internal tube via radiation and not convection. The effects of different types of receiver tubes have been studied by [23]. The study compared the use of receiver tubes in 2 configurations—a bare copper tube and a copper tube that was vacuum evacuated with borosilicate glass. The tests employed were conducted on a PTC and used the nanofluid made of multi-wall carbon nanotube (MWCNT) dispersed in ethylene glycol as the HTF. The thermal efficiency of the evacuated tube was around 15% more than the bare receiver tube, consistently producing higher outlet temperatures.

This research study presents an experimental approach to realize the concept of nanofluid-enhanced concentrating solar collection with or without an evacuated receiver tube. Experiments were conducted to measure the energy collection and efficiency of the system.

2 Experimental Setup

The experimental setup for the testing of the performance of the concentrating parabolic solar collector consists of a parabolic mirror-finished stainless-steel parabolic reflector of 0.5 mm thickness. There are two types of receiver tubes. A dark, powder-coated 304-grade bare stainless-steel tube, and an evacuated receiver tube which is made of a darkened 304-grade stainless steel tube and are enclosed in Borosilicate Glass, providing a vacuum environment or a greenhouse effect. The inner diameters of the stainless-steel tubes are 40 mm, with a thickness of 2 mm. For the evacuated receiver tube, the inner diameter of the glass cover is 70 mm, and the thickness of the cover is 2 mm. The length of each of the tubes is 2 m. The solar collector is supported by a mild steel structure. The solar collector uses a manual tracking system.

A 320-L galvanized steel tank is utilized to store the water and the nanofluids. Housed within the tank is a submersible pump of 5-m head capacity. PVC and flexible vinyl tubes are used to connect the pump to the receiver tube. A Ball type valve and a flexible hose valve are used to control the flow rates. The storage tank is fixed below the receiver tubes centerline. The heat transfer fluid is forced into the tube via the submersible pump. At the outlet of the receiver tube, a heat exchanger is placed to collect the nanofluid to cool it down and recirculate the fluid back into the tank, which can then be reused in the system. The system, therefore, is a semi-closed loop system.

3 Results Analysis and Discussion

The temperature at the inlet section is 39.9 °C and at the outlet section, it reaches 43.3 °C, which is a difference of 3.4 °C. This is displayed in Fig. 1. At the periphery, the bottom-most point of the tube shows a temperature of 42.1 °C, where all the solar flux is concentrated. The temperature at the topmost section is 39.8 °C, so there is a difference of 2.3 °C. The temperatures are similar at the left- and right-hand sections of the periphery, at 40.8 °C and 41 °C respectively. Figure 2 shows how the temperature varies from T2 to T1 and to T4.

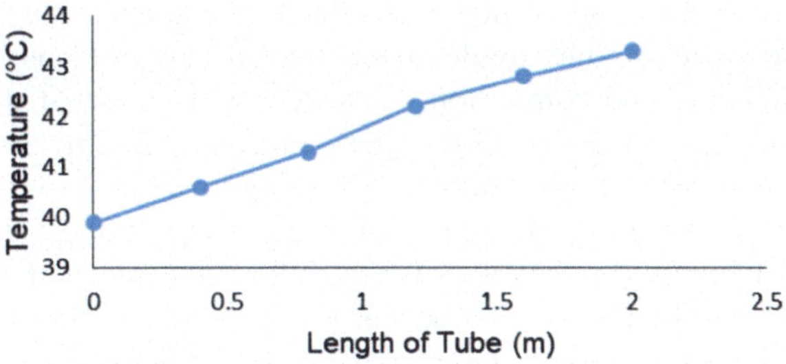

Fig. 1 Temperature variation for stainless-steel tube along the length

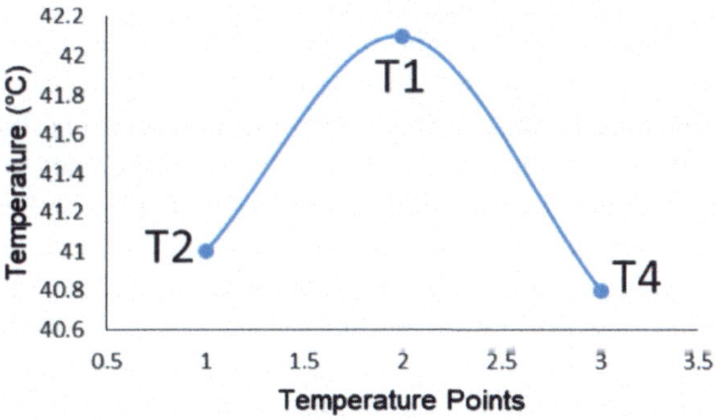

Fig. 2 Temperature variation for stainless-steel tube on the periphery

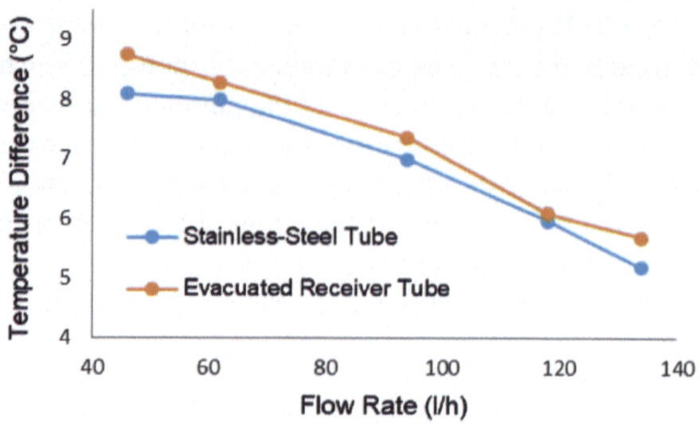

Fig. 3 Variation of temperature difference with respect to flow rate for water

Figure 3 shows the effect of flow rate on the temperature difference when using water as the HTF. This is an approximately inverse relationship. As the flow rate is increased, the temperature difference between the inlet and outlet sections diminishes. This is because, at lower flow rates, the fluid gets more time to pass through the solar flux zones of the receiver tube, whereas at higher flow rates, the time spent is lesser. Furthermore, as per the equation of heat gain, there is an inverse relationship between temperature difference and the mass flow rate, where, if the mass flow rate is raised, the temperature difference must be conversely lower.

The trends are similar for both the stainless-steel tube and the evacuated receiver tube, except that the evacuated receiver tube had a slightly better overall temperature across the range of flow rates, due to the greenhouse effects that the glass envelope provided. For the stainless-steel tube, heat is lost from the tube surface through both radiation and convection. However, as the evacuated receiver tube has a vacuum annular region due to the glass envelope, the internal tube only loses heat to the glass cover via radiation, hence losing lower heat overall compared to the stainless-steel tube. From the glass cover, the heat is transferred to the atmosphere through radiation and convection. For the initial flow rate of 46 l/h, the temperature difference for the stainless-steel tube was 8.1 °C, whereas, for the evacuated receiver tube, the difference between the inlet and outlet temperature was 8.8 °C. At the highest flow rate of 138 l/h, the temperature difference was 5.2 °C for the stainless-steel tube and 5.7 °C for the evacuated receiver tube.

Figure 4 presents the thermal efficiency of the collector against the flow rate. The efficiency is increased at higher flow rates, with the evacuated receiver tube producing better efficiency than the stainless-steel tube. At the lowest flow rate of 46 l/h, the stainless-steel tube exhibited an efficiency of 12.6%, whereas the evacuated receiver tube had an efficiency of 14.4%. For the highest flow rate of 138 l/h, the peak efficiency for the stainless-steel tube was 27% and the efficiency for the evacuated receiver tube was 30%. This is because the thermal efficiency is directly proportional to the mass flow rate according to the equation for thermal efficiency. High mass flow rates also have higher Reynolds numbers, improving heat transfer.

Figure 5 shows the useful heat gain against the volumetric flow rate for water. It can be observed that as the flow rate increases, so does the useful heat gain. However,

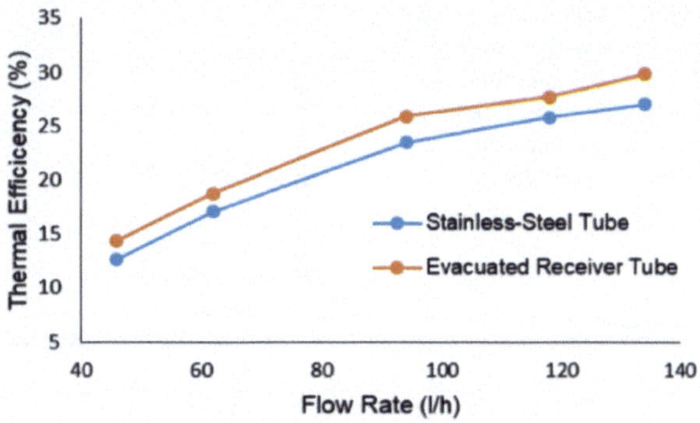

Fig. 4 Variation of thermal efficiency with respect to flow rate for water

the curve appears to be flattening at the higher flow rates, indicating that larger flow rates cannot trap more heat as the temperature difference diminishes. The evacuated receiver tube had slightly better heat gains than the stainless-steel tube. At 46 l/h, the useful heat gain for the stainless-steel tube was 432W and for the evacuated receiver tube the heat gain was 468W. Conversely, at the highest flow rate of 138 l/h, the useful heat gain for the stainless-steel tube was 833W, and the heat gain for the evacuated receiver tube was 914W.

Figure 6 illustrates the thermal efficiency of the stainless-steel tube with respect to flow rates ranging from 46 to 138 l/h. The plot is for the nanofluids and water. As the flow rate increases, so does the thermal efficiency. This phenomenon can be explained due to the higher mass flow of water, which is directly proportional to the useful heat gain. Furthermore, high flow rates have better Reynolds numbers. It can be observed that the thermal efficiencies are higher when higher concentrations of nanoparticles are used, highest at 0.4% and lowest at 0.1%. This is explained due to the higher thermal conductivity of the nanofluids. The nanofluids also have better heat transfer coefficients than water. The highest thermal efficiency was 37.7% at a flow rate of 138 l/h for the 0.4% mass concentration of nanoparticles. The lowest nanofluid thermal efficiency was 14.5% for the 0.1% mass concentration of the nanoparticles, flowing at 46 l/h. Water produced slightly lower thermal efficiency compared to the 0.1% mass fraction nanofluid.

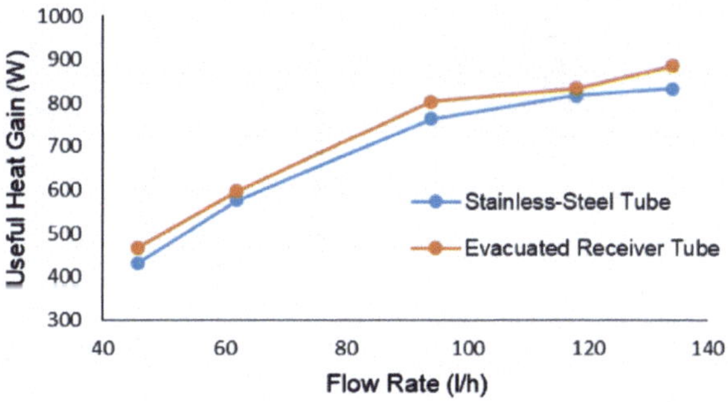

Fig. 5 Variation of useful heat gain with respect to flow rate for water

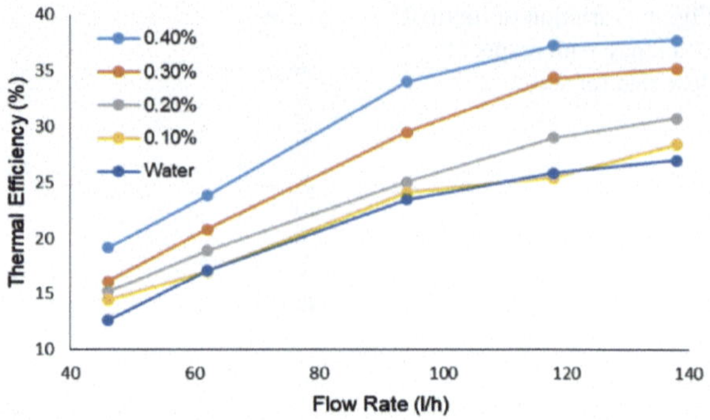

Fig. 6 Variation of thermal efficiency with respect to flow rate for nanofluids of concentrations 0.4, 0.3, 0.2, 0.1%, and water using stainless-steel tube

Figure 7 illustrates the thermal efficiency of the evacuated receiver tube with regard to the flow rates varying from 46 to 138 l/h. The trends are similar to the results with the stainless-steel tube. Higher flow rates produced higher thermal efficiencies. Mass flow rates are proportional to the useful heat gain and high flow rates have higher Reynolds numbers. Nanofluids of higher mass fractions of nanoparticles produced better thermal efficiencies due to their better thermal conductivity and heat transfer coefficient. The highest thermal efficiency was 39.1% for the flow rate of 138 l/h for the 0.4% mass fraction of nanoparticles. The lowest thermal efficiency for the nanofluids was 16.2% for the 0.1% mass fraction of the nanoparticles, flowing at 46 l/h. 0.1% mass fraction nanofluid produced nearly identical results to water.

With regards to thermal efficiency, the evacuated receiver tube once again outperformed the stainless-steel tube at all mass concentrations. For instance, the peak thermal efficiency achieved using the evacuated tube was 39.1%, whereas, for the stainless-steel tube, the figure was 37.7%. The results of the experiments prove that the evacuated receiver tube gives better thermal efficiency than the bare stainless-steel tube, irrespective of the HTF used. It is also evident from the results that the nanofluids outperformed water in terms of giving better thermal efficiencies and temperature differences. As the mass concentration of nanoparticles in the nanofluid increased, the thermal efficiencies also increased.

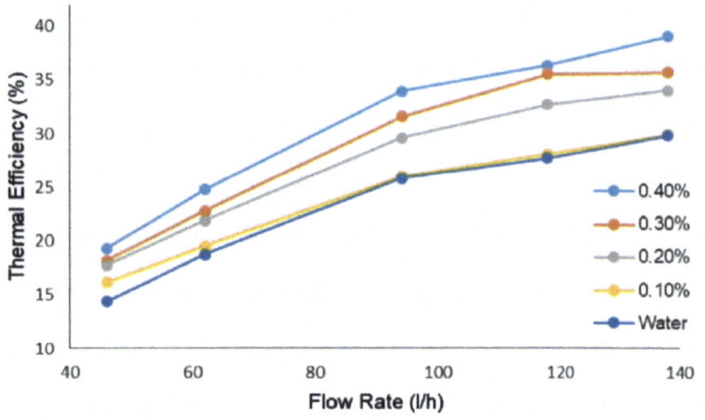

Fig. 7 Variation of thermal efficiency with respect to flow rate for nanofluids of concentrations 0.4, 0.3, 0.2, 0.1%, and water using evacuated receiver tube

4 Conclusion

This research included the fabrication and testing of the nanofluid-based concentrating parabolic trough solar collector. Experiments were done using two receiver tubes, which were the bare stainless-steel receiver tubes and the evacuated receiver tubes. The experiments were carried out using water and aluminum oxide–water nanofluid (Al_2O_3-water) as the heat transfer fluids (HTF).

The stainless-steel receiver tube was compared with the evacuated receiver tube. Experiments conducted with water as a working fluid produced peak thermal efficiencies of 27% and 30% when using the stainless-steel receiver tube and the evacuated receiver tube, respectively. However, with the 0.4% nanofluid mass fraction, the stainless-steel receiver tube and the evacuated receiver tube produced peak thermal efficiencies of 37.7% and 39.1%, respectively.

With water as the HTF, the peak temperature difference for the evacuated receiver tube was 8.8 °C compared to 8.1 °C for the stainless-steel receiver tube. These temperature differences are for a flow rate of 46 l/h.

With the same operating conditions as described in the point above, the peak useful heat gain for the evacuated receiver tube was 914W, whereas it was 833W for the stainless-steel receiver tube.

The improvement for the evacuated receiver tube over the stainless-steel receiver tube is a result of the greenhouse effect provided by the vacuum annular space, allowing the evacuated receiver tube to collect more thermal energy. On using the nanofluids, the thermal efficiency increased as compared to using water as the heat transfer fluid, due to the superior thermo-physical properties of the nanofluids, especially its thermal conductivity.

Acknowledgements The authors acknowledge the support from Sultan Qaboos University, Oman, (Grant No. (IG/ENG/MEID/21/01), and the Ministry of Higher Education, Research and Innovation (RC/URG-ENG/MIED/19/04) for funding this research.

References

1. Cordaro JG, Rubin NC, Bradshaw RW (2011) Multicomponent molten salt mixtures based on nitrate/nitrite anions. J Sol Energy Eng 133:011014. https://doi.org/10.1115/1.4003418
2. Pacio J, Wetzel T (2013) Assessment of liquid metal technology status and research paths for their use as efficient heat transfer fluids in solar central receiver systems. Sol Energy 93:11–22. https://doi.org/10.1016/J.SOLENER.2013.03.025
3. Choi SUS, Li S, Eastman JA (1999) Measuring thermal conductivity of fluids containing oxide nanoparticles. J Heat Transfer 121:280–289. https://doi.org/10.1115/1.2825978
4. Xuan Y, Li Q (2000) Heat transfer enhancement of nanofluids. Int J heat fluid flow 21:58–64. https://doi.org/10.1016/S0142-727X(99)00067-3
5. Choi S (1998) Nanofluid technology: current status and future research. Energy Technol Div Argonne 26

6. Tyagi H, Phelan P, Prasher R (2009) Predicted efficiency of a low-temperature nanofluid-based direct absorption solar collector. J Sol Energy Eng 131:041004. https://doi.org/10.1115/1.3197562
7. Otanicar TP, Phelan PE, Prasher RS, Rosengarten G, Taylor RA (2010) Nanofluid-based direct absorption solar collector. J Renew Sustain Energy 2. https://doi.org/10.1063/1.3429737
8. Taylor RA, Phelan PE, Otanicar TP, Adrian R, Prasher R (2011) Nanofluid optical property characterization: towards efficient direct absorption solar collectors. Nanoscale Res Lett 6:1–11. https://doi.org/10.1186/1556-276X-6-225
9. Puliti G, Paolucci S, Sen M (2011) Nanofluids and their properties. Appl Mech Rev 64:1–23. https://doi.org/10.1115/1.4005492
10. Dudley VE, Kolb GJ, Mahoney RA, Mancini TR, Matthews CW, Sloan M, Kearney D (1994) Test results: SEGS LS-2 solar collector. http://www.nrel.gov/csp/troughnet/pdfs/segs_ls2_solar_collector.pdf%5Cn, http://adsabs.harvard.edu/abs/1994STIN...9611437D
11. Ahmed A, Baig H, Sundaram S, Mallick TK (2019) Review article use of nanofluids in solar PV/Thermal Systems. 2019
12. Arthur O, Karim MA (2016) An investigation into the thermophysical and rheological properties of nanofluids for solar thermal applications. Renew Sustain Energy Rev 55:739–755. https://doi.org/10.1016/J.RSER.2015.10.065
13. O'Keeffe GJ, Mitchell SL, Myers TG, Cregan V (2018) Time-dependent modelling of nanofluid-based direct absorption parabolic trough solar collectors. Sol Energy 174:73–82. https://doi.org/10.1016/j.solener.2018.08.073
14. Zou B, Dong J, Yao Y, Jiang Y (2016) An experimental investigation on a small-sized parabolic trough solar collector for water heating in cold areas. Appl Energy 163:396–407. https://doi.org/10.1016/j.apenergy.2015.10.186
15. Jamadi F, Arabpour M, Abdolzadeh M (2017) Performance comparison of parabolic and flat plate solar collectors utilizing in the heating system of a room-an experimental investigation. Int J Renew Energy Res 7
16. Kumar V, Shrivastava RL, Untawale SP (2015) Fresnel lens: a promising alternative of reflectors in concentrated solar power. Renew Sustain Energy Rev 44:376–390. https://doi.org/10.1016/j.rser.2014.12.006
17. Kazem HA, Chaichan MT (2019) The effect of dust accumulation and cleaning methods on PV panels' outcomes based on an experimental study of six locations in Northern Oman. Sol Energy 187:30–38. https://doi.org/10.1016/j.solener.2019.05.036
18. Ahmadi A, Ehyaei MA, Doustgani A, El Haj Assad M, Hmida A, Jamali DH, Kumar R, Li ZX, Razmjoo A (2021) Recent residential applications of low-temperature solar collector. J Clean Prod 279. https://doi.org/10.1016/j.jclepro.2020.123549
19. Elsheikh AH, Sharshir SW, Mostafa ME, Essa FA, Ahmed Ali MK (2018) Applications of nanofluids in solar energy: a review of recent advances. Renew Sustain Energy Rev 82:3483–3502. https://doi.org/10.1016/j.rser.2017.10.108
20. Rehan MA, Ali M, Sheikh NA, Khalil MS, Chaudhary GQ, ur Rashid T, Shehryar M (2018) Experimental performance analysis of low concentration ratio solar parabolic trough collectors with nanofluids in winter conditions. Renew Energy 118: 742–751. https://doi.org/10.1016/j.renene.2017.11.062
21. Shah SF, Manzoor S, Mazoor A, Iqbal A, Asif M, Jafry AT (2021) Modelling and experimental analysis of a novel evacuated tubes based parabolic trough solar collector for water heating applications. In: 2021 international conference on emerging power technologies, ICEPT 2021. https://doi.org/10.1109/ICEPT51706.2021.9435503
22. Cuce E, Cuce PM, Guclu T, Besir AB (2020) On the use of nanofluids in solar energy applications. J Therm Sci 29:513–534. https://doi.org/10.1007/s11630-020-1269-3
23. Kasaeian A, Daneshazarian R, Pourfayaz F (2017) Comparative study of different nanofluids applied in a trough collector with glass-glass absorber tube. J Mol Liq 234:315–323. https://doi.org/10.1016/j.molliq.2017.03.096

Solar Collector Tilt Angle Optimization for Maximum Solar Irradiation in Lucknow, Uttar Pradesh, India

Abdul Qadeer, Shah Alam, Hasan Zakir Jafri, and Wasim Akram

Abstract This study is oriented about providing the cleanest form of energy for the living species on earth. Solar energy is the cleanest form of energy. Solar plate collector's tilt angle is the main point of interest in this study. This study is about how to find the optimum tilt angle and at the same time the maximum solar energy. This study shows the solar energy falling on the solar plate increased at the optimum tilt angle. For Lucknow (26.87° N, 80.95° E) India, the optimum tilt angle during winter season is 55° (Latitude +28°) and during summer season it is 0° (Latitude − 27°). Solar energy falling on the collector surface during winter is 31.83 MJ/m^2 and during summer season 22.38 MJ/m^2.

Keywords Tilt angle optimization · Solar radiation · Solar collectors · Beam conversion factor · Reflected conversion factor

Nomenclature

H	Daily global radiation incident on a horizontal surface, MJ/m^2-day
Hd	Daily diffuse radiation incident on a horizontal surface, MJ/m^2-day
Hb	Daily beam radiation incident on a horizontal surface, MJ/m^2-day
HT	Daily global radiation on a tilted surface, MJ/m^2-day

A. Qadeer (✉)
Department of Mechanical Engineering, Al-Falah University, Faridabad Haryana-121004, India
e-mail: meetqadeer@gmail.com

S. Alam
University Polytechnic, Jamia Millia Islamia, New Delhi 110025, India
e-mail: salam2@jmi.ac.in

H. Z. Jafri
Director, Jafri Electric Company, Sirsi, Distt Sambal U.P. 244301, India

W. Akram
Department of Mechanical Engineering, Mewat Engineering College, Mewat, Haryana 122107, India

M. A. Siddiqui et al. (eds.), *Advances in Heat Transfer and Fluid Dynamics*, Lecture Notes in Mechanical Engineering, https://doi.org/10.1007/978-981-99-7213-5_32

Rb Ratio of average daily beam radiation incident on an inclined surface to that on a horizontal surface (beam conversion factor).
Rd Ratio of average daily diffuse radiation incident on an inclined surface to that on a horizontal surface (diffused conversion factor).
β Surface slope from the horizontal (degrees)
δ Declination angle (degrees)
ωs Sunrise hour angle (degrees)
ω's Sunrise hour angle for a tilted surface (degrees)
φ Latitude (degrees)
ρ Ground albedo (reflectivity)

1 Introduction

Energy requirement is the key factor for living species survival on earth. Energy is available on earth in both conventional and non-conventional forms. Conventional energy sources are more efficient as compared to non-conventional energy sources but, there is a big hurdle of pollution with conventional energy sources that leads to find some other way to fulfill the need of energy without pollution or with minimum pollution. There is some non-conventional energy sources like: wind energy, solar energy, nuclear energy etc. Solar energy is the cleanest form of energy and mother of all form of energy available on earth. This paper has little approach towards fulfillment of energy requirement by solar energy.

In India, during summer season approximately 90% area receives significant amount of solar radiation of the order of 10.8–23.4 MJ/m^2-day [1]. However, in northern parts of the country this amount reaches up to 27 MJ/m^2-day during summer [2]. Tilt angle of solar plate is the key factor in the performance of the solar system. Remarkable changes can be observed on variation of the tilt angle. To attain the optimum tilt angle, tracking equipment which follow the trajectories of the sun's motion can be installed to get maximum solar radiation on the solar surface [3, 4]. However, according to Vermaak [5] tracking systems are not economical as inclined tracking requires 550% more and 2-axis tracking requires 350% more as compared to static PV panels. Solar radiation data is usually available in the form of global and diffused solar radiation on the horizontal surface at latitude of interest and the problem is how to find the global and diffused form of radiation on the tilted surface. In the northern hemisphere the optimum orientation is due south ($\gamma = 0$) and the optimum tilt angle is dependent upon latitude and day of the year. In winters the optimum tilt is greater (usually latitude $+15^{\circ}$) whereas in summers the optimum tilt is less (usually latitude -15°) [6].

There are many research papers available which offer different recommendations for optimum tilt angle based on the latitude of the station [7, 8]. Practically in the northern hemisphere, solar plates are oriented due south ($\gamma = 0$) with a fixed tilt angle for monthly, seasonaly or annual basis to get maximum solar radiation.

Method of Estimation

As the solar radiation data published on the horizontal surface by IMD Pune (18.52° N, 73.86° E), Maharashtra, India, it is required to find it on the tilted surface. A correlation is developed here to find total solar radiation on the tilted surface.

$$\mathbf{H_T = H_b R_b + H_d R_d + \rho H R_r} \tag{1}$$

H_T is monthly daily solar radiation on the tilted surface whereas H_b, H_d and H are monthly mean daily beam and diffused components of solar radiation respectively. H is the total global solar radiation. R_b, R_d and R_r are known as beam, diffused and reflected conversion factor respectively. ρ is ground albedo, its value is taken as 0.2 for a normal surface.

Beam Conversion Factor

According to Liu and Jordan [9] it can be calculated assuming that there is no atmosphere. In the northern hemisphere, it is sloped toward the equator where the equation of R_b can be written as

$$R_b = \frac{\cos(\Phi - \beta)\cos\delta\sin\omega_s' + \omega_s'\sin(\varphi - \beta)\sin\delta}{\cos\varphi\cos\delta\sin\omega_s + \omega_s\sin\varphi\sin\delta} \tag{2}$$

Here

$$\omega_s' = \min\{\cos^{-1}(-\tan\varphi\tan\delta), \cos^{-1}(-\tan(\varphi - \beta)\tan\delta)\} \tag{3}$$

Φ is latitude, δ is declination angle and ω is the hour angle for the local solar noon.

ω_s is the sunrise or sunset hour angle for the mean day of the month. "min" indicates the smaller value among the two in the bracket. The declination angle can be given as

$$\delta = 23.45\sin\left[\frac{360(284 + n)}{365}\right] \tag{4}$$

Diffused Conversion Factor

Diffused component of solar radiation can be estimated on the basis of two theories. Assuming sky is isotropic or sky is anisotropic. Isotropic condition means solar radiation distributed all over the sky dome evenly whereas anisotropic condition means anisotropy all over the sky dome plus isotropic radiation component. There are many isotropic and anisotropic models that are used to estimate the diffused component of solar radiation. Here we assumed the isotropic condition and use the Liu and Jordan model again for the diffused component of solar radiation. Diffused conversion factor can be written as

$$\boldsymbol{R_d} = \frac{1 + \mathbf{cos}\boldsymbol{\beta}}{2} \tag{5}$$

Reflected Conversion Factor

Assuming isotropic reflection from the ground, the reflected conversion factor can be written as

$$\boldsymbol{R_r} = \boldsymbol{\rho}\frac{1 - \mathbf{cos}\boldsymbol{\beta}}{2} \tag{6}$$

The total radiation on the tilted surface can be estimated by using (1). Since, the Eq. (1) is the function of the tilt angle. This paper is oriented toward the estimation of the optimum tilt angle.

2 Methodology

To estimate the optimum tilt angle, it is advised to partially differentiate the Eq. (1) with respect to tilt angle β and put it equal to zero [12]. Then the obtained tilt angle will be the optimum tilt angle

$$\frac{\partial \mathbf{H_T}}{\partial \boldsymbol{\beta}} = 0 \tag{7}$$

$$\beta_{opt} = \tan^{-1}\left[\frac{H_b(\cos\delta\sin\omega_s'\sin\varphi - \omega_s'\sin\delta\cos\varphi)}{H_b\omega_s'\sin\delta\sin\varphi + \cos\delta\sin\omega_s'\cos\varphi + 1/2(H_d - \rho H)(\omega_s\sin\varphi\sin\delta + \cos\varphi\cos\delta\sin\omega_s)}\right] \tag{8}$$

Put this optimum value of tilt angle in the Eq. (1) and all the values of radiation factors and angles for any Indian station can give the maximum value of solar radiation on the tilted surface. This research paper is to find the optimum tilt angle and maximum solar radiation on the tilted surface for Lucknow, capital city of Uttar Pradesh, India. By, using MSEXCEL graphic software package value of solar radiation is calculated for each month of the year and graphs are plotted.

3 Result and Discussion

Solar energy is the cleanest form of energy available in Earth's atmosphere. The tilt angle of the solar collector is the key factor for the collection solar energy. Tilt angle must be accurate to get maximum amount of solar energy. Tilt angle optimization has been done in this study in order to approach the maximum amount of solar energy and the values are given in Table 1 (Fig. 1).

Table 1 Tilt angle (degree) for different cities of India in each month of the year

Month	New Delhi	Lucknow	Ahmedabad	Bangalore
Jan	57.3197	54.061	52.2915	41.21
Feb	46.9807	46.686	42.1194	32.6
March	31.9795	30.192	27.3238	15.99
April	12.9589	11.329	7.98034	0
May	0	0	0	0
June	0	0	0	0
July	0	0	0	0
Aug	5.27292	2.9557	0.70146	0
Sep	23.6002	19.422	17.2945	7.039
Oct	42.4501	39.31	36.7792	21.64
Nov	54.7698	52.509	49.9173	36.63
Dec	59.5922	58.272	54.5878	42.21

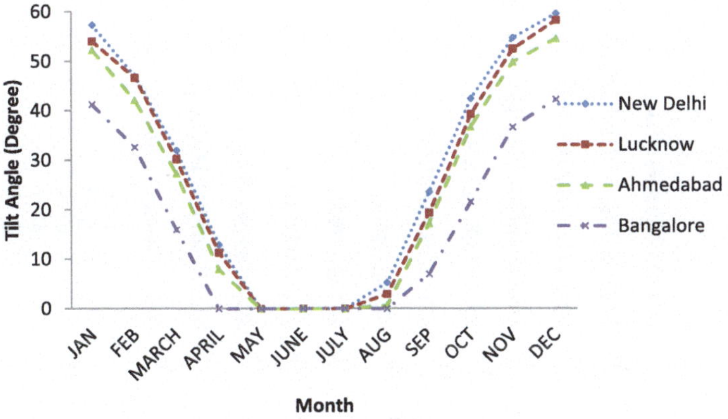

Fig. 1 Values of tilt angle for each month of a year at different stations of India

Research study compares the solar energy falling on flat plat collector for the Liu & Jordan model based on the previous tilt angle vs optimized tilt angle and the result shown in Table 2. H_T is the value of monthly mean daily solar radiation on the tilted surface by the Liu and Jordan model based on the previous tilt angle and H'_T is the same radiation on the optimized tilt angle.

Average value of solar radiation for winter months (November, December and January) is 31.83 MJ/m^2 and for summer (May, June, July) it is 22.38 MJ/m^2 (Fig. 2).

The results show that the tilt angle during winter season is about 55° which is higher (Latitude +28°) and during summer months it is 0° which is lesser (Latitude −27°). These values corroborate well with other studies [10–12].

Table 2 Comparison of monthly mean daily solar radiation (MJ/m^2) falling on tilted surface by Liu and Jordan model at Lucknow

Month	H_T at $\beta = \varphi$	H'_T at $\beta = \beta_{opt}$
Jan	23.38	26.92
Feb	35.62	38.57
March	27.98	28.14
April	28.28	28.56
May	24.40	26.14
June	23.91	24.79
July	15.18	16.21
Aug	16.18	16.80
Sep	19.10	19.16
Oct	23.48	24.41
Nov	26.34	30.03
Dec	31.68	38.54

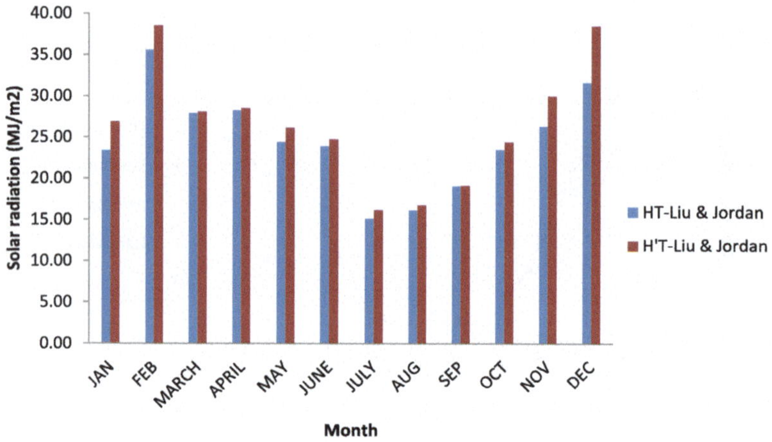

Fig. 2 Comparison of monthly mean daily solar radiation on Liu and Jordan model based on previous tilt angle ($\beta = \phi$) and optimized tilt angle ($\beta = \beta_{opt.}$)

4 Conclusion

This study shows, nearly 1% loss of energy if β_{opt} is adjusted seasonally rather than monthly. If β_{opt} is selected for yearly basis the loss of energy is about 14%. It is concluded that the seasonal optimum tilt angle must be used for power plants and annual optimum tilt angle should be selected for domestic purposes (water heating).

References

1. Jamil B, Siddiqui AT, Akhtar N (2016) Estimation of solar radiation and optimum tilt angles for south-facing surfaces in humid subtropical climatic region of India. Eng Sci Technol, An Int J 19:1826–1835
2. Ramachandra TV, Jain R, Krishnadas G (2011) Hotspots of solar potential in India. Renew Sustain Energy Rev 15:3178–3186
3. Bahrami A (2016) Chiemeka Onyeka Okoye, Ugur Atikol, the effect of latitude on the performance of different solar trackers in Europe and Africa. Appl Energy 177:896–906
4. Okoye CO, Taylan O, Baker DK (2016) Solar energy potentials in strategically located cities in Nigeria: review, resource assessment and PV system design. Renew Sustain Energy Rev 55: 550–566
5. Vermaak HJ (2014) Techno-economic analysis of solar tracking systems in South Africa. Energy Procedia 61:2435–2438
6. Ahmad MJ, Tiwari GN (2009) Optimization of tilt angle for solar collector to receive maximum radiation. The Open Renew Energy J 2:19–24
7. Kern J, Harris L (1975) On the optimum tilt of a solar collector. Sol Energy 17:97
8. lqbal M (1979) Optimum collector slope for residential heating in adverse climates. Solar Energy 22: 77
9. Liu BYH, Jordan RC (1962) Daily insolation on surfaces tilted towards the equator. Trans ASHRAE 53:526–541
10. Elminir HK, Ghitas AE, El-Hussainy F, Hamid R, Beheary MM, Abdel-Moneim KM (2006) Optimum solar flat plate collector slope: case study for Helwan, Egyot. Energy Convers Manage 47: 624–637
11. Huseyin G, Arif H (2007) Determination of optimum tilt angle of solar collectors for building applications. Build Environ 42:779–783
12. Qadeer A, Khan ME, Alam S, Tilt angle optimization by Taylor's series expansion for maximum solar radiation in humid subtropical regions of India. Int J Ambient Energy https://doi.org/10.1080/01430750.2022.2044380

Thermal Performance Analysis of a Double-Pass Solar Air Heater with Discontinuous W-Shaped Ribs

Sudhanshu Dogra and Gaurav Bharadwaj

Abstract Experiments were done on a two-pass solar air heater (SAH) to see the consequence of discontinuous w-shaped ribs on the Thermal characteristics. The absorber plate has discontinuous w-shaped ribs on both sides, with the upper portion painted black to absorb more radiation. Dittus-Boelter and Modified Blasius equations are used to create a rectangular duct and confirm its validity. In this paper, the variation of Thermal and Thermohydraulic/effective efficiency is shown against Reynolds Number. For the analysis, the relative roughness pitch (p/e) ranges from 5 to 20 The relative gap width is taken as 0.5. The angle of attack (α) is taken as 45° and 75°. The Reynolds Number for the experiment was taken from 4900 to 12,000. It has been observed from the experiments that the maximum efficiencies were found out at a relative roughness pitch of 10 with their values a bit higher in the case of 75° than in the case of 45° due to change in its turbulency.

Keywords Reynolds number · Nusselt number · Friction factor

Nomenclature

e	Height of roughness element
p	Pitch
α	Angle of attack
g	Gap
p/e	Relative roughness pitch
g/e	Relative gap width
SAH	Solar air heater

S. Dogra
Lovely Professional University, Phagwara 144411, India
e-mail: sudhanshu.16900@lpu.co.in

G. Bharadwaj (✉)
GLA University, Mathura 281406, India
e-mail: gauravmech2211@gmail.com

M. A. Siddiqui et al. (eds.), *Advances in Heat Transfer and Fluid Dynamics*, Lecture Notes in Mechanical Engineering, https://doi.org/10.1007/978-981-99-7213-5_33

$\eta_{Thermal}$ Thermal efficiency
$\eta_{Thermohydraulic}$ Thermohydraulic efficiency

1 Introduction

Due to increase in the demand of fossil fuels for the purpose of Energy generation, they are depleting at an exponential rate. So, there is a need to turn our focus to some other resources which are capable of producing the similar amount of energy. We can replace fossil fuels with non-conventional energy resources which produce energy without any appreciable change in their amount. There are different types of these resources like Solar energy, Ocean energy, Geothermal energy and so on. These all are the resources which are in abundance and have not been used to their maximum possible extent. Only a small amount of these resources is used till now. Out of various non-conventional energy resources, Solar energy is a type in which exploration is in progress these days at a very rapid pace. The reason being its vast availability and its applications. Before starting the project, SWOT analysis has been done for the solar energy. Starting with its strengths, it is pollution free, less hazardous and it can be utilized in any form of energy quite easily. Its weakness is that it is having a problem of storage and initial investment cost is too high. Coming to the opportunities, is that nowadays due to high energy demand, the Government is providing high subsidies. The last is threats, one of the major threats is climatic condition variations throughout the year, so one may not get the same number of solar radiations throughout the year but still for this there is a provision of solar collectors. So, after doing the SWOT analysis for solar energy, it is concluded that though using solar energy for the energy production has some weakness but the opportunities and its strengths are even more, so that we can rely upon using solar energy for energy production and for other applications too. This paper gives some light to one of the applications of solar energy which is nowadays being used by so many industries for drying purpose. Various researches are going on nowadays in the aspect of using solar energy to heat up air which is then be used for various purposes and the device name is Solar air heater. A typical Solar air heater is a type of heater which uses solar energy falling over it to heat up the air moving in its periphery, generally through a duct and that heated air will then be used for heating up of a room or for drying purposes etc. Generally, there are two types of air heaters. One is the single pass and other is the double pass solar air heater. The basic difference between these two types is that in the single pass air is heated only once while moving through the absorber plate inside the heater. But in double pass, it heats the air two times while moving through the absorber plate. Through researches we get to the conclusion that a double pass Solar air heater is somewhat more efficient than a single pass SAH as it heats the air more due to its two-passes. The diagrammatic representations of both the heaters are shown below in Fig. 1a and b.

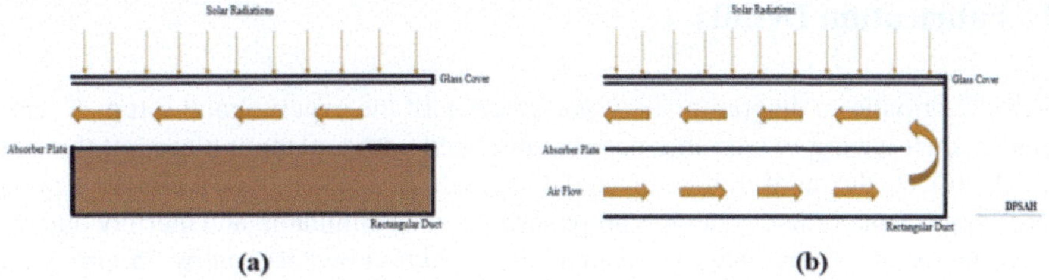

Fig. 1 Single and double pass solar air heater

The use of solar energy in single and double pass SAH has been the subject of extensive research by many researchers. Some of the contributions have been written below. Sodha et al. [1] examined the performance of the single pass and double flow types of SAH for the same mass flow rate and characteristics. He came to the conclusion that double flow type SAH is more effective than single pass under similar mass flow rate conditions. Gupta et al. [2] conducted research on inclined ribs and found that the thermal performance improves with increasing e/D, with a 1.8-fold increase in heat transfer rate and a 2.7-fold increase in friction factor compared to smooth plates. Yeh et al. [3] experimentally determine the collector efficiency of a double flow SAH and came to the conclusion that the collector efficiency always rises following fin placement. Momin et al. [4] employed V-shaped ribs in their research and found that, compared to a smooth plate, the Nusselt Number and friction factor increased by 2.30 and 2.83 times, respectively. Mittal et al. [5] analyzed on the thermohydraulic performance of SAH with wire mesh and concluded that performance gets increased with the wire mesh than the conventional one. Karim et al. [6] investigated on v-corrugated finned air collector and found that v-corrugated type of solar air collector is 10–15% more efficient than single pass and 5–11% more efficient than double pass. Aharwal et al. [7] did their work on inclined discrete ribs and found that at relative gap width of 1.0, p/e of 8 and angle of attack of 60°, heat transfer characteristic gets its maximum value. Arvind et al. [8] did their work on discrete w-shaped ribs in single pass SAH and concluded that at α of 60°, the Nusselt Number and Friction factor becomes 2.16 and 2.75 times respectively than that of the smooth plate. Chabane et al. [9] experimentally studied the heat transfer characteristics of SAH with longitudinal fins below the absorber plate and found that the efficiency is higher with fins than that without fins. Devecioglu et al. [10] investigated on the thermal performance of SAH with porous surface. They concluded that the thermal and thermohydraulic efficiency with a porous absorber surface is higher than the smooth surface by 25–57% and 14–44% respectively. Chii-Dong et al. [11] worked on cross corrugated air collector and found that it creates more turbulence than the conventional one and has more efficiency. Gajendra et al. [12] worked on v-type baffles on double pass solar air heater and found that increasing volume flow rate of air increases thermal performance of SAH.

2 Fabrication Details

Figure 2 provides a diagrammatic representation of the experimental setup. A rectangular duct having two/double pass is fabricated with a plenum placed at the exit which collects the air after it get heated. Other components include a blower, a pressure gauge, a manometer, a temperature gauge, a solar simulator, and thermocouples. The dimensions of a rectangular channel are 2050 mm by 250 mm by 25 mm, with a 50 mm height. At the conclusion of the first pass, there is a gap of 50 mm to allow unrestricted airflow during the subsequent pass. ASHRAE guidelines were followed in the design of the duct. [13].

The total length of the test section is 1600 mm. The absorber plate has discrete w-shaped ribs on both sides, with the upper side painted black to better absorb radiation, which heats the air more. With the aid of a suction blower, air from the atmosphere is drawn into the duct from the bottom. Air then enters the upper side through gap placed at the end of first pass and it then moves through the second pass between the absorber plate and glass sheet. For heating up the absorber plate which is made up of Galvanized iron (GI), a solar simulator is fabricated shown in Fig. 3 which is made up of Halogen bulbs and is kept at some height over the absorber plate so to give equal amount of intensity as given by sun had it been in the outside environment.

For the absorber plate, GI sheet has been taken over which ribs in discontinuous w-shape are placed. The pictorial view of the absorber plate has been shown below in Fig. 4. The plate size where ribs are placed is taken as 1600 mm × 250 mm. To absorb more radiation, the top is painted black. For the experimentation, p/e is taken between the range of 5 and 20, with g/e of 0.5 which is held constant. The experimentation has been done with two angles of attacks (α) of 45°and 75°.

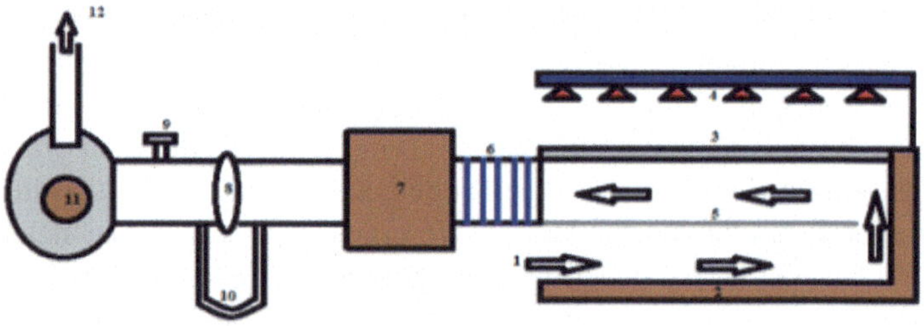

1) Air Inlet 2) Insulation 3) Glass Cover 4) Halogen Bulbs 5) Absorber Plate 6) Flexible pipe 7) Plenum 8) Orifice 9) Control Valve 10) Manometer 11) Blower 12) Air out

Fig. 2 Experimental set-up

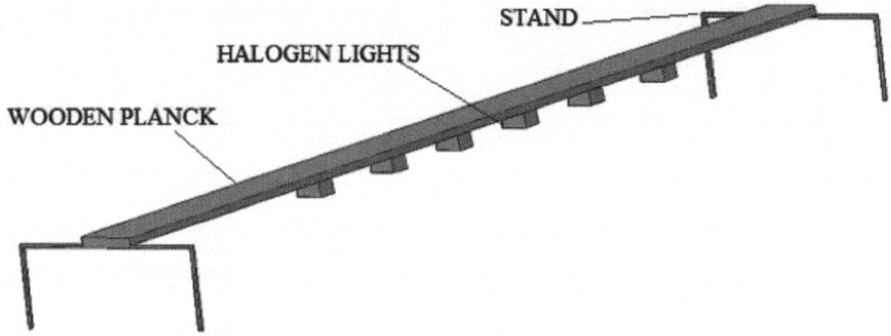

Fig. 3 Schematic of solar simulator used

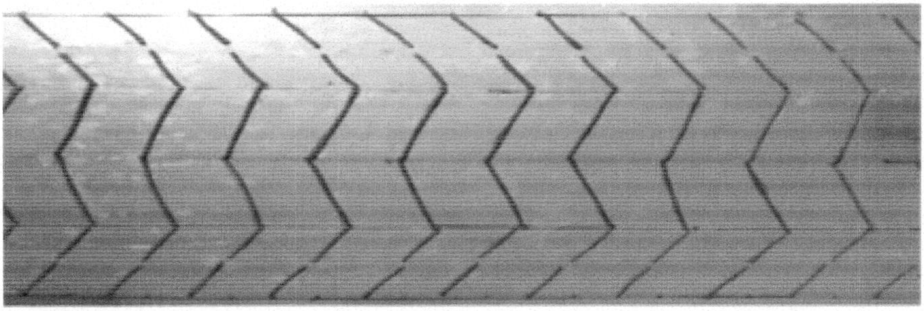

Fig. 4 Pictorial representation of discontinuous w-shaped ribs

3 Experimental Set-Up Validation

Validation of experimental set-up is done before the experiment carried on the ribbed geometry through "Dittus-Boelter equation" [14] and "Modified Blasius equation" [15] for Nusselt Number and Friction factor respectively. The maximum deviation in the predicted and experimental value were found to be 7.1% and 5% for Nusselt Number and Friction factor respectively which is within the limits. Validation has been shown in Fig. 5a and b.

4 Results

4.1 Variation in Thermal Efficiency

In this section, the variation of Thermal efficiency is shown with Reynolds Number with a relative roughness pitch that varies from 5 to 20. Experiments were done for two different angles of attacks of 45° and 75° and is shown below in Figs. 6 and 7

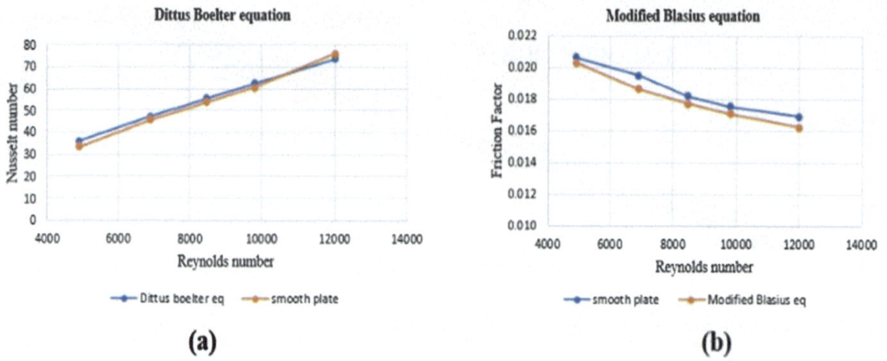

Fig. 5 **a** Comparison among Nusselt number variations **b** Comparison among friction factor variations

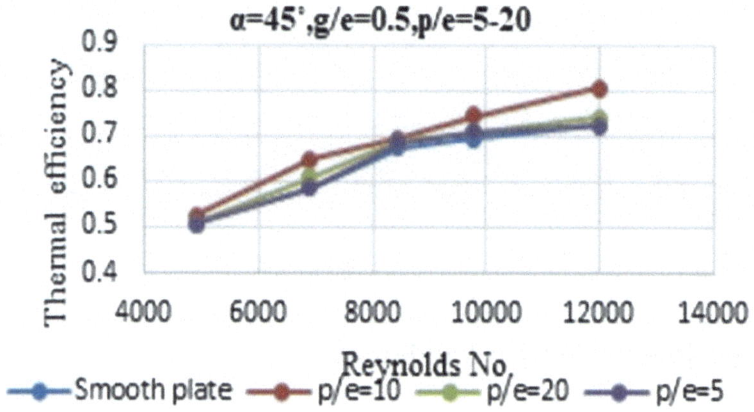

Fig. 6 Graph showing the variation of thermal efficiency with Reynolds no. for a constant angle of attack of 45° with a constant value of g/e of 0.5 and p/e ranges from 5 to 20

respectively. It has been concluded from the below figures that thermal efficiency gets enhanced by placing ribs on the absorber plate and gets its maximum value at p/e of 10 in both the cases due to the flow separation and the formation of reattachment points.

Figure 8 compares the two angles of attack and depicts how thermal efficiency varies with Reynolds number. Additionally, it has been noted that as the angle increases, so does the thermal efficiency. Due to the development of an anti-rotating flow, the angle of attack at 75° is only slightly greater than that at 45°.

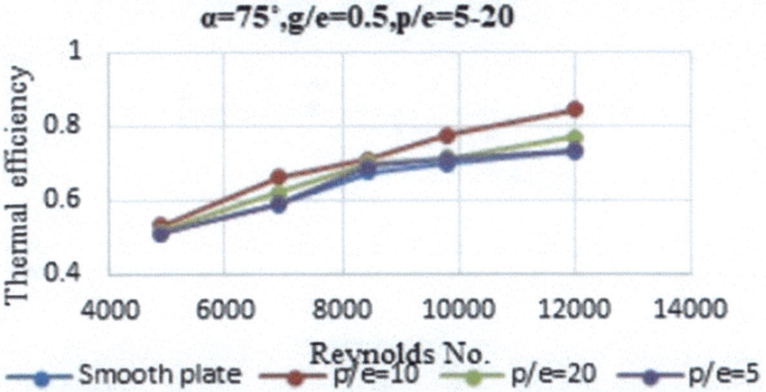

Fig. 7 Graph showing the variation of thermal efficiency with Reynolds no. for a constant angle of attack of 75° with a constant value of g/e of 0.5 and p/e ranges from 5 to 20

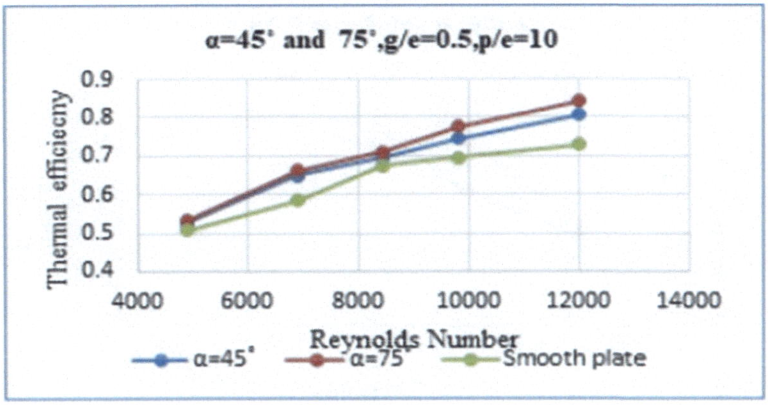

Fig. 8 Graph showing the variation of thermal efficiency with Reynolds no. for a constant angle of attack of 45° and 75° with a constant value of g/e of 0.5 and p/e of 10

4.2 *Thermohydraulic Efficiency*

This section uses Reynolds numbers to illustrate how thermohydraulic efficiency varies. And numbers having p/e that ranges from 5 to 20. Experiments were done for two different angles of attacks of 45° and 75° and are shown below in Figs. 9 and 10 respectively. The conclusions drawn from the data below indicate that adding ribs to the absorber plate increases thermohydraulic efficiency. At p/e of 10, the thermohydraulic efficiency is determined to be at its maximum. Thermohydraulic efficiency increases with Reynolds Number, as shown in Figs. 9 and 10, and then begins to decline after a certain point since the amount of heat collected decreases in comparison to the pump work, which increases significantly.

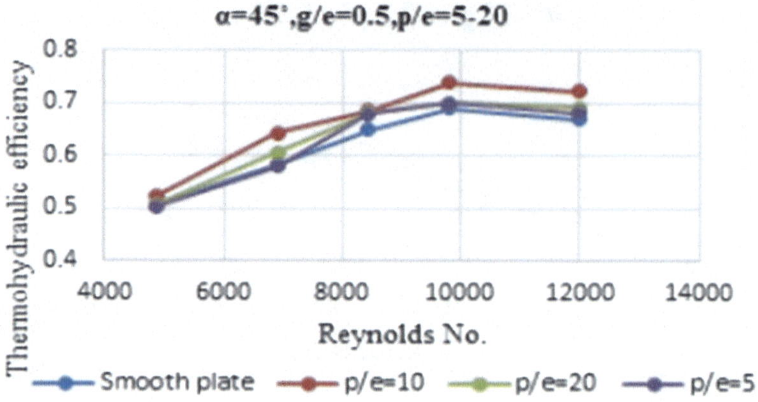

Fig. 9 Graph showing the variation of thermohydraulic efficiency with Reynolds no. for a constant angle of attack of 45° with a constant value of g/e of 0.5 and p/e ranges from 5 to 20

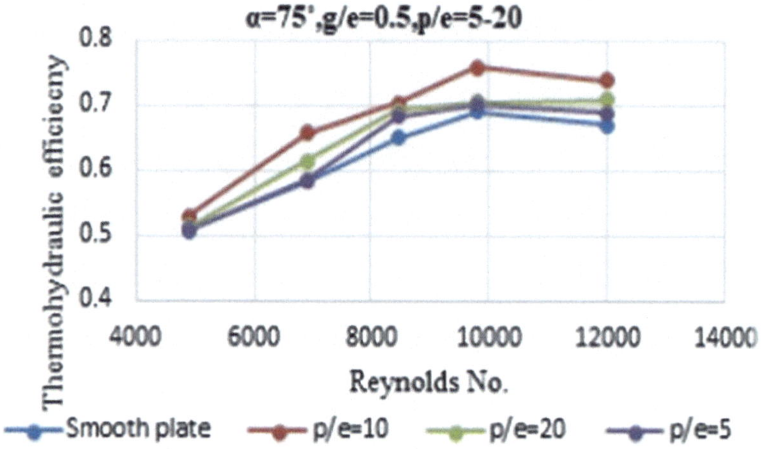

Fig. 10 Graph showing the variation of thermohydraulic efficiency with Reynolds no. for a constant angle of attack of 75° with a constant value of g/e of 0.5 and p/e ranges from 5 to 20

Figure 11 shows the variation of Thermohydraulic efficiency with the Reynolds number and compared the two angles of attacks. It has been noted that as the angle increases, so does the thermohydraulic efficiency, but only up to a point, beyond which it begins to decline. Due to the creation of an anti-rotating flow, α at 75° is only slightly greater than that at 45°.

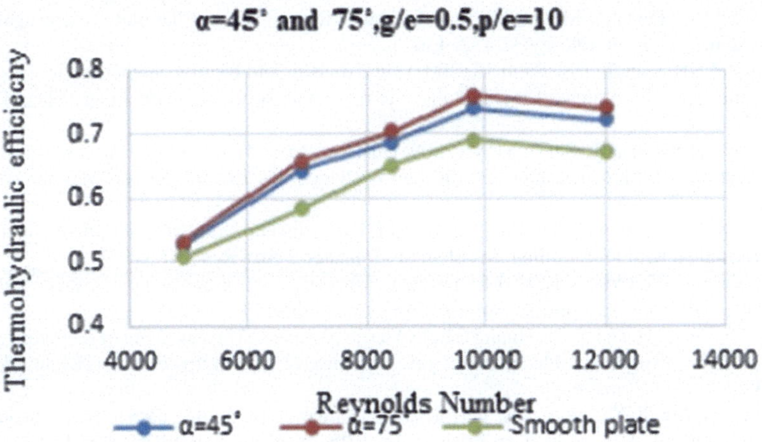

Fig. 11 Graph showing the variation of thermohydraulic efficiency with Reynolds no. for a constant angle of attack of 45° and 75° with a constant value of g/e of 0.5 and p/e of 10

5 Conclusion

The two-pass SAH used in the trials has a rectangular duct with the angle of attack of 45° and 75° and a constant g/e of 0.5. The range for "p/e" is 5 to 20. The figures above show that thermal and thermohydraulic efficiency (up to a certain limit) grow with Reynolds Number and are at their highest at p/e of 10. Additionally, comparisons between 45° and 75° have been made, and it has been discovered that at 75° it is somewhat higher than that of 45°. This is because of the development of anti-rotating flow has a stronger influence at 75° than it does at 45°. Additionally, Thermohydraulic Efficiency improves with Reynolds Number and then begins to decline at a certain point since the amount of collected heat decreases relative to the pump work, which significantly increases.

References

1. Sodha MS, Bansal NK, Singh D (1982) Analysis of a non-porous double-flow solar air heater. Appl Energy 12:251–258
2. Gupta D, Solanki SC, Saini JS (1997) Thermohydraulic performance of solar air heater with roughened absorber plates. Sol Energy 61:33–42
3. Yeh HM, Ho CD, Hou JZ (2002) Collector efficiency of double pass solar air heater. Energy 27:715–727
4. Momin AME, Saini JS, Solanki SC (2002) Heat transfer and friction in solar air heater duct with V-shaped rib roughness on absorber plate. J Heat Mass Transfer 45:3383–3396
5. Mittal MK, Varshney L (2006) Optimal thermohydraulic performance of a wire mesh packed solar air heater. Sol Energy 80:1112–1120

6. Karim MA, Hawladar MNA (2006) Performance investigation of flat plate v- corrugated and finned air collectors. Energy 31:452–470
7. Aharwal KR, Gandhi BK, Saini JS (2009) Heat transfer and friction characteristics of solar air heater ducts having integral inclined discrete ribs on absorber plate. Int J Heat Mass Transfer 52:5970–5977
8. Kumar A, Bhagoria JL, Sarviya RM (2009) Heat transfer and friction correlations for artificially roughened solar air heater duct with discrete W-shaped ribs. Energy Convers Manage 50:2106–2117
9. Chabane F, Moummi N, Benramache S (2014) Experimental study of heat transfer and thermal performance with longitudinal fins of solar air heater. J Adv Res 5:183–192
10. Devecioglu AG, Oruc V (2017) Experimental investigation of thermal performance of a new solar air collector with porous surface. Energy Procedia 113:251–258
11. Ho CD, Chang H, Hsiao CF, Huang CC, Device performance improvement of recycling double-pass cross-corrugated solar air collectors. Energies 11: 338. https://doi.org/10.3390/en1102 0338
12. Kumar G, Singh AK, Verma A, Performance evaluation on double pass solar air heater with broken multiple V-Type Baffles-a review. Int J Res Appl Sci Eng Technol (IJRASET) ISSN: 2321-9653
13. ASHRAE standard 93 7 (1977) Method of testing to determine the thermal performance of solar collector
14. Bhatti MS, Shah RK (1987) Turbulent and transition flow convective heat transfer. In: Kakac S, Shah RK, Aung W (eds) Handbook of single-phase convective heat transfer. Wiley, New York
15. Kays WM, Perkin H, Forced convection internal flow in ducts. In: Rohsenow WM, Hartnett IV (eds) Handbook of heat transfer. McGraw-Hill, New York

SPRINGER NATURE

GPSR Compliance

The European Union's (EU) General Product Safety Regulation (GPSR) is a set of rules that requires consumer products to be safe and our obligations to ensure this.

If you have any concerns about our products, you can contact us on ProductSafety@springernature.com

In case Publisher is established outside the EU, the EU authorized representative is:

Springer Nature Customer Service Center GmbH
Europaplatz 3
69115 Heidelberg, Germany

The manufacturer's authorised representative in the EU is Springer Nature Customer Service Centre GmbH, Europaplatz 3, 69115 Heidelberg, Germany. If you have any concerns regarding our products, please contact ProductSafety@springernature.com

Printed and bound by CPI Group (UK) Ltd, Croydon, CR0 4YY
01/10/2026
02225893-0014